STUDENT SOLUTIONS MANUAL

Richard N. Aufmann
Palomar College

Vernon C. Barker
Palomar College

Joanne S. Lockwood
Plymouth State College

Christine S. Verity

BEGINNING ALGEBRA WITH APPLICATIONS

SIXTH EDITION

Aufmann/Barker/Lockwood

HOUGHTON MIFFLIN COMPANY   BOSTON NEW YORK

Senior Sponsoring Editor: Lynn Cox
Senior Development Editor: Dawn Nuttall
Editorial Associate: Melissa Parkin
Manufacturing Manager: Florence Cadran
Senior Marketing Manager: Ben Rivera

Printed in the U.S.A.

ISBN: 0-618-30607-2

9 – CRS – 07 06 05

Table of Contents

STUDENT SOLUTIONS
MANUAL

Chapter 1: Real Numbers

Prep Test

1. 127.16

2. 49,743

3. 4517

4. 11,396

5. 24

6. 24

7. 4

8. $3 \cdot 7$

9. $\dfrac{4}{10} = \dfrac{2}{5}$

Go Figure

Multiplying by numbers that end in zero will produce a zero at the end of the final answer. Since 10 and 20 are included in the list, they are two numbers that would produce a zero at the end of the number. Also, multiplying 5 and 6 produces 30, and multiplying 2 and 15 also produces 30. So there are 4 zeros at the end of the product of the first 20 natural numbers.

Section 1.1

Concept Review 1.1

1. Sometimes true
   $|0| = 0$ which is not a positive number.

3. Never true
   The absolute value of a number cannot be less than zero.

5. Sometimes true
   The opposite of a positive number is a negative number. The opposite of a negative number is a positive number.

7. Always true

9. Always true

Objective 1.1.1 Exercises

3. $-2 > -5$

5. $-16 < 1$

7. $3 > -7$

9. $0 > -3$

11. $-42 < 27$

13. $21 > -34$

15. $-27 > -39$

17. $-131 < 101$

19. $\{1, 2, 3, 4, 5, 6, 7, 8\}$

21. $\{1, 2, 3, 4, 5, 6, 7, 8\}$

23. $\{-6, -5, -4, -3, -2, -1\}$

25. $-7 < 2$
    $0 < 2$
    $2 = 2$
    $5 > 2$
    The element 5 is greater than 2.

27. $-23 < -8$
    $-18 < -8$
    $-8 = -8$
    $0 > -8$
    The elements -23 and -18 are less than -8.

29. $-35 < -10$
    $-13 < -10$
    $21 > -10$
    $37 > -10$
    The elements 21 and 37 are greater than -10.

31. $-52 < 0$
    $-46 < 0$
    $0 = 0$
    $39 > 0$
    $58 > 0$
    The elements -52, -46, and 0 are less than or equal to 0.

33. $-23 < -17$
    $-17 = -17$
    $0 > -17$
    $4 > -17$
    $29 > -17$
    The elements -17, 0, 4, and 29 are greater than or equal to -17.

35. $1 < 5$
    $2 < 5$
    $3 < 5$
    $4 < 5$
    $5 = 5$
    $6 > 5$
    $7 > 5$
    $8 > 5$
    $9 > 5$
    The elements 5, 6, 7, 8, and 9 are greater than or equal to 5.

37. $-10 < -4$
$-9 < -4$
$-8 < -4$
$-7 < -4$
$-6 < -4$
$-5 < -4$
$-4 = -4$
$-3 > -4$
$-2 > -4$
$-1 > -4$
The elements $-10, -9, -8, -7, -6,$ and -5 are less than -4.

Objective 1.1.2 Exercises

41. -22

43. 31

45. 168

47. -630

49. $-(-18) = 18$

51. $-(49) = -49$

53. $|16| = 16$

55. $|-12| = 12$

57. $-|29| = -29$

59. $-|-14| = -14$

61. $-|0| = 0$

63. $-|34| = -34$

65. a. $-(-8) = 8$
$-(-5) = 5$
$-(-2) = 2$
$-(1) = -1$
$-(3) = -3$

b. $|-8| = 8$
$|-5| = 5$
$|-2| = 2$
$|1| = 1$
$|3| = 3$

67. $|-83| > |58|$

69. $|43| < |-52|$

71. $|-68| > |-42|$

73. $|-45| < |-61|$

Applying Concepts 1.1

75. $-19, -|-8|, |-5|, 6$

77. $-22, -(-3), |-14|, |-25|$

79. a. Strategy
• From the given table, find the wind-chill factor with a temperature of 5°F and a 20-mph wind and the wind-chill factor with a temperature of 10°F and a 15-mph wind.
• Compare the wind-chill factors.

Solution
The wind-chill factor with a temperature of 5°F and a 20-mph wind is –31°F.
The wind-chill factor with a temperature of 10°F and a 15-mph wind is –18°F.
$-31 < -18$
The 5°F temperature with the 20-mph wind feels colder.

b. Strategy
• From the given table, find the wind-chill factor with a temperature of –25°F and a 10-mph wind and the wind-chill factor with a temperature of –15°F and a 20-mph wind.
• Compare the wind-chill factors.

Solution
The wind-chill factor with a temperature of –25°F and a 10-mph wind is –52°F.
The wind-chill factor with a temperature of –15°F and a 20 mph wind is –60°F.
$-52 > -60$
The –15° temperature with the 20-mph wind feels colder

81. -4 and 4

83. -3 and 11

85. negative

87. 0

89. true

Section 1.2

Concept Review 1.2

1. Sometimes true
The sum of two negative integers is less than either integer being added.

3. Always true

5. Always true

7. Never true
$4(-8) = -32$

9. Never true
$8 - 4 \neq 4 - 8$

Objective 1.2.1 Exercises

3. $-3 + (-8) = -11$

5. $-4 + (-5) = -9$

7. $6 + (-9) = -3$

9. $-6 + 7 = 1$

11. $2 + (-3) + (-4) = -1 + (-4)$
$= -5$

13. $-3 + (-12) + (-15) = -15 + (-15)$
$= -30$

15. $-17 + (-3) + 29 = -20 + 29$
$= 9$

17. $-3 + (-8) + 12 = -11 + 12$
$= 1$

19. $13 + (-22) + 4 + (-5) = -9 + 4 + (-5)$
$= -5 + (-5)$
$= -10$

21. $-22 + 10 + 2 + (-18) = -12 + 2 + (-18)$
$= -10 + (-18)$
$= -28$

23. $-126 + (-247) + (-358) + 339$
$= -373 + (-358) + 339$
$= -731 + 339$
$= -392$

Objective 1.2.2 Exercises

27. $16 - 8 = 16 + (-8)$
$= 8$

29. $7 - 14 = 7 + (-14)$
$= -7$

31. $3 - (-4) = 3 + 4$
$= 7$

33. $-4 - (-2) = -4 + 2$
$= -2$

35. $-12 - 16 = -12 + (-16)$
$= -28$

37. $4 - 5 - 12 = 4 + (-5) + (-12)$
$= -1 + (-12)$
$= -13$

39. $-12 - (-3) - (-15) = -12 + 3 + 15$
$= -9 + 15$
$= 6$

41. $13 - 7 - 15 = 13 + (-7) + (-15)$
$= 6 + (-15)$
$= -9$

43. $-30 - (-65) - 29 - 4 = -30 + 65 + (-29) + (-4)$
$= 35 + (-29) + (-4)$
$= 6 + (-4)$
$= 2$

45. $-16 - 47 - 63 - 12 = -16 + (-47) + (-63) + (-12)$
$= -63 + (-63) + (-12)$
$= -126 + (-12)$
$= -138$

47. $-47 - (-67) - 13 - 15 = -47 + 67 + (-13) + (-15)$
$= 20 + (-13) + (-15)$
$= 7 + (-15)$
$= -8$

Objective 1.2.3 Exercises

53. $14 \cdot 3 = 42$

55. $5(-4) = -20$

57. $-8(2) = -16$

59. $(-5)(-5) = 25$

61. $(-7)(0) = 0$

63. $-24 \cdot 3 = -72$

65. $6(-17) = -102$

67. $-4(-35) = 140$

69. $5 \cdot 7(-2) = 35 \cdot (-2)$
$= -70$

71. $(-9)(-9)(2) = 81(2)$
$= 162$

73. $-5(8)(-3) = -40(-3)$
$= 120$

75. $-1(4)(-9) = -4(-9)$
$= 36$

77. $4(-4) \cdot 6(-2) = -16 \cdot 6(-2)$
$= -96(-2)$
$= 192$

79. $-9(4) \cdot 3(1) = -36 \cdot 3(1)$
$= -108(1)$
$= -108$

81. $-6(-5)(12)(0) = 30(12)(0)$
$= 360(0)$
$= 0$

Objective 1.2.4 Exercises

83. $3(-12) = -36$

85. $-5(11) = -55$

87. $12 \div (-6) = -2$

89. $(-72) \div (-9) = 8$

91. $0 \div (-6) = 0$

93. $45 \div (-5) = -9$

95. $-36 \div 4 = -9$

97. $-81 \div (-9) = 9$

99. $72 \div (-3) = -24$

101. $-60 \div 5 = -12$

103. $78 \div (-6) = -13$

105. $-72 \div 4 = -18$

107. $-114 \div (-6) = 19$

109. $-130 \div (-5) = 26$

111. $-132 \div (-12) = 11$

113. $-182 \div 14 = -13$

115. $143 \div 11 = 13$

Objective 1.2.5 Exercises

117. Strategy
To find the temperature, add the rise in temperature ($9°$) to the original temperature ($-6°$).

Solution
$-6 + 9 = 3$
The temperature is 3°C.

119. Strategy
To find the difference, subtract the low temperature ($-4°$C) from the high temperature ($10°$C).

Solution
$$10 - (-4) = 10 + 4$$
$$= 14$$
The difference is 14°C.

121. Strategy
To find the difference, subtract the temperature at which mercury freezes ($-39°$C) from the temperature at which mercury boils ($360°$C).

Solution
$$360 - (-39) = 360 + 39$$
$$= 399$$
The difference is 399°C.

123. Strategy
To find the difference, subtract the elevation of the Caspian Sea (-28 m) from the elevation of Mt. Elbrus (5634 m).

Solution
$$5634 - (-28) = 5634 + 28$$
$$= 5662$$
The difference in elevation is 5662 m.

125. To find the difference, subtract the elevation of the Qattara Depression (-133 m) from the elevation of Mt. Kilimanjaro (5895 m).

Solution
$$5895 - (-133) = 5895 + 133$$
$$= 6028$$
The difference in elevation is 6028 m.

127. Strategy
To find the difference, subtract the elevation of the Dead Sea (-400 m) from the elevation of Mt. Everest (8848 m).

Solution
$$8848 - (-400) = 8848 + 400$$
$$= 9248$$
The difference in elevation is 9248 m.

129. Strategy
To find the average daily high temperature:
• Add the seven temperature readings.
• Divide by 7.

Solution
$$-8 + (-9) + 6 + 7 + (-2) + (-14) + (-1)$$
$$= -17 + 6 + 7 + (-2) + (-14) + (-1)$$
$$= -11 + 7 + (-2) + (-14) + (-1)$$
$$= -4 + (-2) + (-14) + (-1)$$
$$= -6 + (-14) + (-1)$$
$$= -20 + (-1)$$
$$= -21$$
$$-21 \div 7 = -3$$
The average daily high temperature was –3°C.

131. Strategy
To find the drop in temperature, subtract the lower temperature from the higher temperature.

Solution
$$44 - (-56) = 45 + 56$$
$$= 100$$
The temperature dropped 100°F.

133. Strategy
To find the drop in temperature, subtract the lower temperature at 30,000 ft ($-48°$) from the higher temperature at 20,000 ft ($-12°$).

Solution
$$-12 - (-48) = 36$$
The temperature dropped 36°.

135. Strategy
To find the score:
• Multiply the number of correct answers by 7.
• Multiply the number of incorrect answers by -3.
• Multiply the number of questions left blank by -1.
• Add the results.

Solution
$$(17)(7) = 119$$
$$(8)(-3) = -24$$
$$(2)(-1) = -2$$
$$119 + (-24) + (-2) = 93$$
The student's score was 93.

137. Strategy
To calculate the 5-day moving average, determine the average of the stock for days 1 through 5, days 2 through 6, days 3 through 7, and so on.

Solution
Days 1–5
$(-20)+(-20)+(-50)+(-20)+(-10)=-120$
$\dfrac{-120}{5}=-24$
Days 2–5
$(-20)+(-50)+(-20)+(-10)+30=-70$
$\dfrac{-70}{5}=-14$
Days 3–7
$(-50)+(-20)+(-10)+30+(-10)=-60$
$\dfrac{-60}{5}=-12$
Days 4–8
$(-20)+(-10)+30+(-10)+30=20$
$\dfrac{20}{5}=4$
Days 5–9
$(-10)+30+(-10)+30+0=40$
$\dfrac{40}{5}=8$
Days 6–10
$30+(-10)+30+0+50=100$
$\dfrac{100}{5}=20$
The 5-day moving average is –24, –14, –12, 4, 8, and 20.

Applying Concepts 1.2

139. $|13-(-4)|=|17|=17$

141. $|18-21|=|-3|=3$

143. $-4, -9, -14$

145. $-16, 4, -1$

147. 5436

149. a. $|3+4|=|3|+|4|$
$|7|=3+4$
$7=7$
True

  b. $|3-4|=|3|-|4|$
$|-1|=3-4$
$1 \neq -1$
False

  c. $|4+3|=|4|+|3|$
$|7|=4+3$
$7=7$
True

  d. $|4-3|=|4|-|3|$
$|1|=4-3$
$1=1$
True

151. For a, b, and c, if $x=3$ and $y=-2$,

  a. $|x+y| \leq |x|+|y|$
$|3+(-2)| \leq |3|+|-2|$
$|1| \leq 3+2$
$1 \leq 5$
True

  b. $|x+y|=|x|+|y|$
$|3+(-2)|=|3|+|-2|$
$|1|=3+2$
$1 \neq 5$
False

  c. $|x+y| \geq |x|+|y|$
$|3+(-2)| \geq |3|+|-2|$
$|1| \geq 3+2$
$1 \ngeq 5$
False

The answer is a.

153. No. For example, the difference between 10 and –8 is 18, which is greater than both 10 and –8.

Section 1.3

Concept Review 1.3

1. Never true
The rule for multiplying fractions is to multiply the numerators and multiply the denominators.

3. Sometimes true
$\dfrac{1}{3}$ is a rational number and is represented by the repeating decimal $0.\overline{3}$.

5. Always true

7. Never true
To write a decimal as a percent, multiply by 100 and write the percent sign.

9. Always true

6   *Chapter 1: Real Numbers*

Objective 1.3.1 Exercises

1.
$$3\overline{)1.000}$$ with quotient 0.333
$$-\underline{9}$$
$$10$$
$$-\underline{9}$$
$$10$$
$$-\underline{9}$$
$$1$$

$\frac{1}{3} = 0.\overline{3}$

3.
$$4\overline{)1.00}$$ with quotient 0.25
$$-\underline{8}$$
$$20$$
$$-\underline{20}$$
$$0$$

$\frac{1}{4} = 0.25$

5.
$$5\overline{)2.0}$$ with quotient 0.4
$$-\underline{20}$$
$$0$$

$\frac{2}{5} = 0.4$

7.
$$6\overline{)1.000}$$ with quotient 0.166
$$-\underline{6}$$
$$40$$
$$-\underline{36}$$
$$40$$
$$-\underline{36}$$
$$4$$

$\frac{1}{6} = 0.1\overline{6}$

9.
$$8\overline{)1.000}$$ with quotient 0.125
$$-\underline{8}$$
$$20$$
$$-\underline{16}$$
$$40$$
$$-\underline{40}$$
$$0$$

$\frac{1}{8} = 0.125$

11.
$$9\overline{)2.000}$$ with quotient 0.222
$$-\underline{18}$$
$$20$$
$$-\underline{18}$$
$$20$$
$$-\underline{18}$$
$$2$$

$\frac{2}{9} = 0.\overline{2}$

13.
$$11\overline{)5.0000}$$ with quotient 0.4545
$$-\underline{44}$$
$$60$$
$$-\underline{55}$$
$$50$$
$$-\underline{44}$$
$$60$$
$$-\underline{55}$$
$$5$$

$\frac{5}{11} = 0.\overline{45}$

15.
$$12\overline{)7.0000}$$ with quotient 0.5833
$$-\underline{60}$$
$$100$$
$$-\underline{96}$$
$$40$$
$$-\underline{36}$$
$$40$$
$$-\underline{36}$$
$$4$$

$\frac{7}{12} = 0.58\overline{3}$

17.
$$15\overline{)4.000}$$ with quotient 0.266
$$-\underline{30}$$
$$100$$
$$-\underline{90}$$
$$100$$
$$-\underline{90}$$
$$100$$
$$-\underline{90}$$
$$10$$

$\frac{4}{15} = 0.2\overline{6}$

19.
$$\begin{array}{r} 0.4375 \\ 16\overline{)7.0000} \\ -64 \\ \hline 60 \\ -48 \\ \hline 120 \\ -112 \\ \hline 80 \\ -80 \\ \hline 0 \end{array}$$

$$\frac{7}{16} = 0.4375$$

21.
$$\begin{array}{r} 0.24 \\ 25\overline{)6.00} \\ -50 \\ \hline 100 \\ -100 \\ \hline 0 \end{array}$$

$$\frac{6}{25} = 0.24$$

23.
$$\begin{array}{r} 0.225 \\ 40\overline{)9.000} \\ -80 \\ \hline 100 \\ -80 \\ \hline 200 \\ -200 \\ \hline 0 \end{array}$$

$$\frac{9}{40} = 0.225$$

25.
$$\begin{array}{r} 0.68181 \\ 22\overline{)15.0000} \\ -132 \\ \hline 180 \\ -176 \\ \hline 40 \\ -22 \\ \hline 180 \\ -176 \\ \hline 40 \\ -22 \\ \hline 18 \end{array}$$

$$\frac{15}{22} = 0.68\overline{1}$$

27.
$$\begin{array}{r} 0.45833 \\ 24\overline{)11.0000} \\ -96 \\ \hline 140 \\ -120 \\ \hline 200 \\ -192 \\ \hline 80 \\ -72 \\ \hline 80 \\ -72 \\ \hline 8 \end{array}$$

$$\frac{11}{24} = 0.458\overline{3}$$

29.
$$\begin{array}{r} 0.1515 \\ 33\overline{)5.0000} \\ -33 \\ \hline 170 \\ -165 \\ \hline 50 \\ -33 \\ \hline 170 \\ -165 \\ \hline 5 \end{array}$$

$$\frac{5}{33} = 0.\overline{15}$$

31.
$$\begin{array}{r} 0.081081 \\ 37\overline{)3.000000} \\ -296 \\ \hline 40 \\ -37 \\ \hline 30 \\ -0 \\ \hline 300 \\ -296 \\ \hline 40 \\ -37 \\ \hline 3 \end{array}$$

$$\frac{3}{37} = 0.\overline{081}$$

Objective 1.3.2 Exercises

33. $\dfrac{2}{3} + \dfrac{5}{12} = \dfrac{8}{12} + \dfrac{5}{12} = \dfrac{8+5}{12} = \dfrac{13}{12}$

35. $\dfrac{5}{8} - \dfrac{5}{6} = \dfrac{15}{24} - \dfrac{20}{24} = \dfrac{15}{24} + \dfrac{-20}{24}$

$= \dfrac{15+(-20)}{24} = \dfrac{-5}{24} = -\dfrac{5}{24}$

37. $-\dfrac{5}{12} - \dfrac{3}{8} = -\dfrac{10}{24} - \dfrac{9}{24} = \dfrac{-10}{24} + \dfrac{-9}{24}$

$= \dfrac{-10+(-9)}{24} = \dfrac{-19}{24} = -\dfrac{19}{24}$

39. $-\dfrac{6}{13} + \dfrac{17}{26} = -\dfrac{12}{26} + \dfrac{17}{26} = \dfrac{-12}{26} + \dfrac{17}{26}$

$= \dfrac{-12+17}{26} = \dfrac{5}{26}$

41. $-\dfrac{5}{8}-\left(-\dfrac{11}{12}\right)=-\dfrac{5}{8}+\dfrac{11}{12}=-\dfrac{15}{24}+\dfrac{22}{24}$

$\qquad\qquad =\dfrac{-15}{24}+\dfrac{22}{24}=\dfrac{-15+22}{24}=\dfrac{7}{24}$

43. $\dfrac{1}{2}-\dfrac{2}{3}+\dfrac{1}{6}=\dfrac{3}{6}-\dfrac{4}{6}+\dfrac{1}{6}=\dfrac{3}{6}+\dfrac{-4}{6}+\dfrac{1}{6}$

$\qquad\qquad =\dfrac{3+(-4)+1}{6}=\dfrac{0}{6}=0$

45. $-\dfrac{5}{16}+\dfrac{3}{4}-\dfrac{7}{8}=-\dfrac{5}{16}+\dfrac{12}{16}-\dfrac{14}{16}$

$\qquad\qquad =\dfrac{-5}{16}+\dfrac{12}{16}+\dfrac{-14}{16}=\dfrac{-5+12+(-14)}{16}$

$\qquad\qquad =\dfrac{-7}{16}=-\dfrac{7}{16}$

47. $\dfrac{3}{4}-\left(-\dfrac{7}{12}\right)-\dfrac{7}{8}=\dfrac{3}{4}+\dfrac{7}{12}-\dfrac{7}{8}$

$\qquad\qquad =\dfrac{18}{24}+\dfrac{14}{24}-\dfrac{21}{24}=\dfrac{18}{24}+\dfrac{14}{24}+\dfrac{-21}{24}$

$\qquad\qquad =\dfrac{18+14+(-21)}{24}=\dfrac{11}{24}$

49. $\dfrac{2}{3}-\dfrac{1}{2}+\dfrac{5}{6}=\dfrac{4}{6}-\dfrac{3}{6}+\dfrac{5}{6}$

$\qquad\qquad =\dfrac{4}{6}+\dfrac{-3}{6}+\dfrac{5}{6}$

$\qquad\qquad =\dfrac{4+(-3)+5}{6}=\dfrac{6}{6}=1$

51. $\dfrac{5}{8}-\left(-\dfrac{5}{12}\right)+\dfrac{1}{3}=\dfrac{5}{8}+\dfrac{5}{12}+\dfrac{1}{3}$

$\qquad\qquad =\dfrac{15}{24}+\dfrac{10}{24}+\dfrac{8}{24}$

$\qquad\qquad =\dfrac{15+10+8}{24}=\dfrac{33}{24}=\dfrac{11}{8}$

53. $\dfrac{-7}{9}+\dfrac{14}{15}+\dfrac{8}{21}=-\dfrac{245}{315}+\dfrac{294}{315}+\dfrac{120}{315}$

$\qquad\qquad =\dfrac{-245}{315}+\dfrac{294}{315}+\dfrac{120}{315}$

$\qquad\qquad =\dfrac{-245+294+120}{315}$

$\qquad\qquad =\dfrac{169}{315}$

55.
$\quad\begin{array}{r}32.1\\ +\ 6.7\\ \hline 38.8\end{array}$

$-32.1-6.7=-38.8$

57.
$\quad\begin{array}{r}13.092\\ -\ 6.9\\ \hline 6.192\end{array}$

$-13.092+6.9=-6.192$

59.
$\quad\begin{array}{r}5.43\\ +\ 7.925\\ \hline 13.355\end{array}$

$5.43+7.925=13.355$

61.
$\quad\begin{array}{r}8.546\\ -\ 3.87\\ \hline 4.676\end{array}$

$-3.87+8.546=4.676$

63. $2.09-6.72-5.4=-4.63-5.4$

$\qquad\qquad\qquad =-10.03$

65. $-18.39+4.9-23.7=-13.49-23.7$

$\qquad\qquad\qquad =-37.19$

67. $-3.07-(-2.97)-17.4=-3.07+2.97-17.4$

$\qquad\qquad\qquad =-0.10-17.4$

$\qquad\qquad\qquad =-17.5$

69. $317.09-46.902+583.0714$

$\qquad =270.188+583.0714$

$\qquad =853.2594$

Objective 1.3.3 Exercises

71. $\dfrac{1}{2}\left(-\dfrac{3}{4}\right)=-\left(\dfrac{1}{2}\cdot\dfrac{3}{4}\right)$

$\qquad\qquad =-\dfrac{1\cdot3}{2\cdot4}=\dfrac{1\cdot3}{2\cdot2\cdot2}=-\dfrac{3}{8}$

73. $\left(-\dfrac{3}{8}\right)\left(-\dfrac{4}{15}\right)=\left(\dfrac{3}{8}\right)\left(\dfrac{4}{15}\right)=\dfrac{3\cdot4}{8\cdot15}$

$\qquad\qquad =\dfrac{\overset{1}{\cancel{3}}\cdot\overset{1}{\cancel{2}}\cdot\overset{1}{\cancel{2}}}{2\cdot\underset{1}{\cancel{2}}\cdot\underset{1}{\cancel{2}}\cdot\underset{1}{\cancel{3}}\cdot5}=\dfrac{1}{10}$

75. $\left(\dfrac{1}{2}\right)\left(-\dfrac{3}{4}\right)\left(-\dfrac{5}{8}\right)=\dfrac{1}{2}\cdot\dfrac{3}{4}\cdot\dfrac{5}{8}=\dfrac{1\cdot3\cdot5}{2\cdot4\cdot8}$

$\qquad\qquad =\dfrac{3\cdot5}{2\cdot2\cdot2\cdot2\cdot2\cdot2}=\dfrac{15}{64}$

77. $\dfrac{3}{8}\div\dfrac{1}{4}=\dfrac{3}{8}\cdot\dfrac{4}{1}=\dfrac{3\cdot4}{8\cdot1}$

$\qquad\qquad =\dfrac{3\cdot\overset{1}{\cancel{2}}\cdot\overset{1}{\cancel{2}}}{2\cdot\underset{1}{\cancel{2}}\cdot\underset{1}{\cancel{2}}}=\dfrac{3}{2}$

79. $-\dfrac{5}{12}\div\dfrac{15}{32}=-\left(\dfrac{5}{12}\div\dfrac{15}{32}\right)=-\left(\dfrac{5}{12}\cdot\dfrac{32}{15}\right)$

$\qquad\qquad =-\left(\dfrac{5\cdot32}{12\cdot15}\right)=-\left(\dfrac{\overset{1}{\cancel{5}}\cdot\overset{1}{\cancel{2}}\cdot\overset{1}{\cancel{2}}\cdot2\cdot2\cdot2}{2\cdot2\cdot3\cdot3\cdot\underset{1}{\cancel{5}}}\right)=-\dfrac{8}{9}$

81. $-\dfrac{4}{9}\div\left(-\dfrac{2}{3}\right)=\dfrac{4}{9}\div\dfrac{2}{3}=\dfrac{4}{9}\cdot\dfrac{3}{2}$

$\qquad\qquad =\dfrac{4\cdot3}{9\cdot2}=\dfrac{\overset{1}{\cancel{2}}\cdot2\cdot\overset{1}{\cancel{3}}}{\underset{1}{\cancel{3}}\cdot3\cdot\underset{1}{\cancel{2}}}=\dfrac{2}{3}$

83.
$\quad\begin{array}{r}3.47\\ \times\ 1.2\\ \hline 6.94\\ 347\\ \hline 4.164\end{array}$

$(1.2)(3.47)=4.164$

85.
$$\begin{array}{r} 1.89 \\ \times\ 2.3 \\ \hline 567 \\ 378 \\ \hline 4.347 \end{array}$$
$$(-1.89)(-2.3) = 4.347$$

87.
$$\begin{array}{r} 1.06 \\ \times\ 3.8 \\ \hline 848 \\ 318 \\ \hline 4.028 \end{array}$$
$$(1.06)(-3.8) = -4.028$$

89. $(1.2)(-0.5)(3.7) = -0.6(3.7)$
$$= -2.22$$

91. $(-0.8)(3.006)(-5.1) = (-2.4084)(-5.1)$
$$= 12.26448$$

93.
$$\begin{array}{r} 0.747 \\ 1.7\,\overline{)1.2.700} \\ -\,119 \\ \hline 80 \\ -\,68 \\ \hline 120 \\ -\,119 \\ \hline 1 \end{array}$$
$$-1.27 \div (-1.7) \approx 0.75$$

95. $-354.2086 \div 0.1719 \approx -2060.55$

97. $(-3.92) \div (-45.008) \approx 0.09$

Objective 1.3.4 Exercises

101. $75\% = 75\left(\dfrac{1}{100}\right) = \dfrac{75}{100} = \dfrac{3}{4}$
$75\% = 75(0.01) = 0.75$

103. $50\% = 50\left(\dfrac{1}{100}\right) = \dfrac{50}{100} = \dfrac{1}{2}$
$50\% = 50(0.01) = 0.5$

105. $64\% = 64\left(\dfrac{1}{100}\right) = \dfrac{64}{100} = \dfrac{16}{25}$
$64\% = 64(0.01) = 0.64$

107. $175\% = 175\left(\dfrac{1}{100}\right) = \dfrac{175}{100} = 1\dfrac{3}{4}$
$175\% = 175(0.01) = 1.75$

109. $19\% = 19\left(\dfrac{1}{100}\right) = \dfrac{19}{100}$
$19\% = 19(0.01) = 0.19$

111. $5\% = 5\left(\dfrac{1}{100}\right) = \dfrac{5}{100} = \dfrac{1}{20}$
$5\% = 5(0.01) = 0.05$

113. $450\% = 450\left(\dfrac{1}{100}\right) = \dfrac{450}{100} = 4\dfrac{1}{2}$
$450\% = 450(0.01) = 4.5$

115. $8\% = 8\left(\dfrac{1}{100}\right) = \dfrac{8}{100} = \dfrac{2}{25}$
$8\% = 8(0.01) = 0.08$

117. $11\dfrac{1}{9}\% = 11\dfrac{1}{9}\left(\dfrac{1}{100}\right) = \dfrac{100}{9}\left(\dfrac{1}{100}\right) = \dfrac{1}{9}$

119. $31\dfrac{1}{4}\% = 31\dfrac{1}{4}\left(\dfrac{1}{100}\right) = \dfrac{125}{4}\left(\dfrac{1}{100}\right) = \dfrac{5}{16}$

121. $\dfrac{1}{2}\% = \dfrac{1}{2}\left(\dfrac{1}{100}\right) = \dfrac{1}{200}$

123. $6\dfrac{1}{4}\% = 6\dfrac{1}{4}\left(\dfrac{1}{100}\right) = \dfrac{25}{4}\left(\dfrac{1}{100}\right) = \dfrac{1}{16}$

125. $7.3\% = 7.3(0.01) = 0.073$

127. $15.8\% = 15.8(0.01) = 0.158$

129. $9.15\% = 9.15(0.01) = 0.0915$

131. $18.23\% = 18.23(0.01) = 0.1823$

133. $0.15 = 0.15(100\%) = 15\%$

135. $0.05 = 0.05(100\%) = 5\%$

137. $0.175 = 0.175(100\%) = 17.5\%$

139. $1.15 = 1.15(100\%) = 115\%$

141. $0.008 = 0.008(100\%) = 0.8\%$

143. $0.065 = 0.065(100\%) = 6.5\%$

145. $\dfrac{27}{50} = \dfrac{27}{50}(100\%) = \dfrac{2700}{50}\% = 54\%$

147. $\dfrac{1}{3} = \dfrac{1}{3}(100\%) = \dfrac{100}{3}\% \approx 33.3\%$

149. $\dfrac{4}{9} = \dfrac{4}{9}(100\%) = \dfrac{400}{9}\% \approx 44.4\%$

151. $2\dfrac{1}{2} = \dfrac{5}{2}(100\%) = \dfrac{500}{2}\% = 250\%$

153. $\dfrac{3}{8} = \dfrac{3}{8}(100\%) = \dfrac{300}{8}\% = 37\dfrac{1}{2}\%$

155. $\dfrac{5}{14} = \dfrac{5}{14}(100\%) = \dfrac{500}{14}\% = 35\dfrac{5}{7}\%$

157. $1\dfrac{1}{4} = \dfrac{5}{4}(100\%) = \dfrac{500}{4}\% = 125\%$

159. $1\dfrac{5}{9} = \dfrac{14}{9}(100\%) = \dfrac{1400}{9}\% = 155\dfrac{5}{9}\%$

161. $4\% = 4\left(\dfrac{1}{100}\right) = \dfrac{4}{100} = \dfrac{1}{25}$

163. $40\% = 40\left(\dfrac{1}{100}\right) = \dfrac{40}{100}$

$\dfrac{1}{2} = \dfrac{50}{100}$

$\dfrac{40}{100} < \dfrac{50}{100}$

Fewer than one-half found their most recent job in a newspaper.

Applying Concepts 1.3

165. natural number, integer, positive integer, rational number, real number

167. rational number, real number

169. irrational number, real number

171. Strategy
To find the cost, add the cost for the first ounce ($.37) and the cost for the remaining $3\dfrac{1}{2}$ ounces ($.23 times 4).

Solution
$0.37 + 0.23(4) = 0.37 + 0.92$
$\qquad\qquad\qquad = 1.29$
The cost is $1.29.

173. Strategy
To find the difference in deficits, subtract the deficit in 1980 (-73.835) from the deficit in 1985 (-212.334).

Solution
$(-212.334) - (-73.835) = -138.499$
The difference is $138.499 billion.

175. Strategy
To find how many times greater the deficit was, divide the deficit in 1985 (-212.334) by the deficit in 1975 (-53.242).

Solution
$\dfrac{-212.334}{-53.242} = 3.988 \approx 4$

The deficit in 1985 was 4 times greater than the deficit in 1975.

177. Strategy
To find the average, add the values of each year and divide by 6.

Solution
$\begin{array}{r} -163.899 \\ -107.450 \\ -21.940 \\ 69.246 \\ 79.263 \\ +117.305 \\ \hline -27.475 \end{array}$

$\dfrac{-27.475}{6} = -4.579$ billion $= -4,579$ million

The average deficit is $-$4,579 million.

179. The new fraction will be greater than $\dfrac{2}{5}$.

For example, $\dfrac{2+3}{5+3} = \dfrac{5}{8}$, and $\dfrac{5}{8} > \dfrac{2}{5}$.

181. $x + 0.06x = 1x + 0.06x = (1 + 0.06)x = 1.06x$

183. Strategy
To find the price, find a whole number less than 60 that divides evenly into 5459.

Solution
$5459 \div 53 = 103$
The price during the sale was 53¢.

185. Yes, we can find the average of a and b, which would be $c = \dfrac{a+b}{2}$.

Section 1.4

Concept Review 1.4

1. Never true
$(-5)^2 = 25$, $-5^2 = -25$, and $-(5)^2 = -25$

3. Never true
In the expression 3^8, 3 is the base and 8 is the exponent.

5. Always true

7. Never true
The order of Operations Agreement was adopted so that there could be no more than one correct answer to a problem.

9. Never true
$(2^3)^2 = 64$, $2^{(3^2)} = 512$

Objective 1.4.1 Exercises

1. $6^2 = 6 \cdot 6 = 36$

3. $-7^2 = -(7 \cdot 7) = -49$

5. $(-3)^2 = (-3)(-3) = 9$

7. $(-3)^4 = (-3)(-3)(-3)(-3) = 81$

9. $\left(\dfrac{1}{2}\right)^2 = \left(\dfrac{1}{2}\right)\left(\dfrac{1}{2}\right)$

$\qquad = \dfrac{1 \cdot 1}{2 \cdot 2} = \dfrac{1}{4}$

11. $(0.3)^2 = (0.3)(0.3)$
$\qquad\quad = 0.09$

13. $\left(\dfrac{2}{3}\right)^2 \cdot 3^3 = \left(\dfrac{2}{3}\right)\left(\dfrac{2}{3}\right) \cdot 3 \cdot 3 \cdot 3$

$\qquad\qquad = \dfrac{2 \cdot 2 \cdot \overset{1}{\cancel{3}} \cdot \overset{1}{\cancel{3}} \cdot 3}{\underset{1}{\cancel{3}} \cdot \underset{1}{\cancel{3}}} = 12$

15. $(0.3)^3 \cdot 2^3 = (0.3)(0.3)(0.3) \cdot 2 \cdot 2 \cdot 2$
$= (0.027) \cdot 8 = 0.216$

17. $(-3) \cdot 2^2 = (-3) \cdot 2 \cdot 2 = -3 \cdot 4 = -12$

19. $(-2) \cdot (-2)^3 = -2 \cdot (-2)(-2)(-2) = (-2)(-8) = 16$

21. $2^3 \cdot 3^3 \cdot (-4) = 2 \cdot 2 \cdot 2 \cdot 3 \cdot 3 \cdot 3 \cdot (-4) = 8 \cdot 27 \cdot (-4) = 216(-4) = -864$

23. $(-7) \cdot 4^2 \cdot 3^2 = (-7) \cdot 4 \cdot 4 \cdot 3 \cdot 3 = (-7) \cdot 16 \cdot 9 = -112 \cdot 9 = -1008$

25. $\left(\dfrac{2}{3}\right)^2 \cdot \dfrac{1}{4} \cdot 3^3 = \dfrac{2}{3} \cdot \dfrac{2}{3} \cdot \dfrac{1}{4} \cdot 3 \cdot 3 \cdot 3 = \dfrac{2 \cdot 2 \cdot 1 \cdot 3 \cdot 3 \cdot 3}{3 \cdot 3 \cdot 4} = \dfrac{\overset{1}{\cancel{2}} \cdot \overset{1}{\cancel{2}} \cdot \overset{1}{\cancel{3}} \cdot \overset{1}{\cancel{3}} \cdot 3}{\underset{1}{\cancel{3}} \cdot \underset{1}{\cancel{3}} \cdot \underset{1}{\cancel{2}} \cdot \underset{1}{\cancel{2}}} = 3$

27. $8^2 \cdot (-3)^5 \cdot 5 = 8 \cdot 8 \cdot (-3)(-3)(-3)(-3)(-3) \cdot 5 = 64 \cdot (-243) \cdot 5 = -15,552 \cdot 5 = -77,760$

Objective 1.4.2 Exercises

31. $2^2 \cdot 3 - 3 = 4 \cdot 3 - 3$
$= 12 - 3$
$= 9$

33. $16 - 32 \div 2^3 = 16 - 32 \div 8$
$= 16 - 4$
$= 12$

35. $8 - (-3)^2 - (-2) = 8 - 9 - (-2)$
$= 8 - 9 + 2$
$= -1 + 2$
$= 1$

37. $14 - 2^2 - |4 - 7| = 14 - 2^2 - |-3|$
$= 14 - 2^2 - 3$
$= 14 - 4 - 3$
$= 10 - 3$
$= 7$

39. $-2^2 + 4[16 \div (3 - 5)] = -2^2 + 4[16 \div (-2)]$
$= -2^2 + 4[-8]$
$= -4 + 4[-8]$
$= -4 + [-32]$
$= -36$

41. $24 \div \dfrac{3^2}{8 - 5} - (-5) = 24 \div \dfrac{9}{3} - (-5)$
$= 24 \div 3 - (-5)$
$= 8 - (-5)$
$= 8 + 5$
$= 13$

43. $4 \cdot [16 - (7 - 1)] \div 10 = 4 \cdot [16 - 6] \div 10$
$= 4 \cdot [10] \div 10$
$= 40 \div 10$
$= 4$

45. $18 \div \left|9 - 2^3\right| + (-3) = 18 \div |9 - 8| + (-3)$
$= 18 \div |1| + (-3)$
$= 18 \div 1 + (-3)$
$= 18 + (-3)$
$= 15$

47. $4(-8) \div [2(7 - 3)^2] = 4(-8) \div [2(4)^2]$
$= 4(-8) \div [2(16)]$
$= 4(-8) \div [32]$
$= -32 \div 32$
$= -1$

49. $16 - 4 \cdot \dfrac{3^3 - 7}{2^3 + 2} - (-2)^2 = 16 - 4 \cdot \dfrac{27 - 7}{8 + 2} - 4$
$= 16 - 4 \cdot \dfrac{20}{10} - 4$
$= 16 - 4 \cdot 2 - 4$
$= 16 - 8 - 4$
$= 8 - 4$
$= 4$

51. $0.3(1.7 - 4.8) + (1.2)^2 = 0.3(-3.1) + (1.2)^2$
$= 0.3(-3.1) + 1.44$
$= -0.93 + 1.44$
$= 0.51$

53. $(1.65 - 1.05)^2 \div 0.4 + 0.8 = (0.60)^2 \div 0.4 + 0.8$
$= 0.36 \div 0.4 + 0.8$
$= 0.9 + 0.8$
$= 1.7$

55. $\left(\dfrac{3}{4}\right)^2 - \left(\dfrac{1}{2}\right)^3 \div \dfrac{3}{5} = \dfrac{9}{16} - \dfrac{1}{8} \div \dfrac{3}{5}$
$= \dfrac{9}{16} - \dfrac{1}{8} \cdot \dfrac{5}{3}$
$= \dfrac{9}{16} - \dfrac{5}{24}$
$= \dfrac{27}{48} - \dfrac{10}{48}$
$= \dfrac{17}{48}$

Applying Concepts 1.4

57. $(0.9)^3 < 1^5$ because $0.729 < 1$

59. $(-1.1)^2 > (0.9)^2$ because $1.21 > 0.81$

61. $1^3 + 2^3 + 3^3 + 4^3 = 1 + 8 + 27 + 64$
$= 9 + 27 + 64$
$= 36 + 64$
$= 100$

63. $(-2)^2 + (-4)^2 + (-6)^2 + (-8)^2 = 4 + 16 + 36 + 64$
$= 20 + 36 + 64$
$= 56 + 64$
$= 120$

65. $10^8 \div 6,000,000 = 100,000,000 \div 6,000,000$
$= 16\dfrac{2}{3} \approx 17$

It will take the computer approximately 17 s to do the additions.

67. $34^1 = 34$
$34^2 = 1156$
$34^3 = 39,304$
$34^4 = 1,336,336$
$34^5 = 45,435,424$
Even powers of 34 have 6 in the ones digit.
34^{202} has a 6 in the ones digit.

69. $27^1 = 27$
$27^2 = 729$
$27^3 = 19,683$
$27^4 = 531,441$
$27^5 = 14,348,907$
$27^6 = 387,420,489$
$27^{622} = 27^{620} \cdot 27^2$
27^{620} has a 1 in the ones digit.
27^{622} has a 9 in the ones digit.

Chapter Review Exercises

1. $\{1, 2, 3, 4, 5, 6\}$

2. $\dfrac{5}{8} = \dfrac{5}{8}(100\%) = \dfrac{500}{8}\% = 62.5\%$

3. $-|-4| = -4$

4. $16 - (-30) - 42 = 16 + 30 + (-42)$
$= 46 + (-42)$
$= 4$

5. $-561 \div (-33) = 17$

6.
$$\begin{array}{r} 0.777 \\ 9\overline{)7.000} \\ -\underline{6.3} \\ 70 \\ -\underline{63} \\ 70 \\ -\underline{63} \\ 7 \end{array}$$

7.
$$\begin{array}{r} 6.02 \\ \times\ 0.89 \\ \hline 5418 \\ \underline{4816} \\ 5.3578 \end{array}$$
$(6.02)(-0.89) = -5.3578$

8. $\dfrac{-10 + 2}{2 + (-4)} \div 2 + 6 = \dfrac{-8}{-2} \div 2 + 6$
$= 4 \div 2 + 6$
$= 2 + 6$
$= 8$

9. 4

10. $16 - 30 = 16 + (-30)$
$= -14$

11. $0.672 = 0.672(100\%) = 67.2\%$

12. $79\dfrac{1}{2}\% = 79\dfrac{1}{2}\left(\dfrac{1}{100}\right) = \dfrac{159}{2}\left(\dfrac{1}{100}\right) = \dfrac{159}{200}$

13. $-72 \div 8 = -9$

14.
$$\begin{array}{r} 0.85 \\ 20\overline{)17.00} \\ -\underline{160} \\ 100 \\ -\underline{100} \\ 0 \end{array}$$

15. $\dfrac{5}{12} \div \left(-\dfrac{5}{6}\right) = -\left(\dfrac{5}{12} \div \dfrac{5}{6}\right)$
$= -\left(\dfrac{5}{12} \cdot \dfrac{6}{5}\right)$
$= -\left(\dfrac{5 \cdot 6}{12 \cdot 5}\right)$
$= -\dfrac{\overset{1}{\cancel{5}} \cdot \overset{1}{\cancel{2}} \cdot \overset{1}{\cancel{3}}}{\underset{1}{\cancel{2}} \cdot 2 \cdot \underset{1}{\cancel{3}} \cdot \underset{1}{\cancel{5}}} = -\dfrac{1}{2}$

16. $3^2 - 4 + 20 \div 5 = 9 - 4 + 20 \div 5$
$= 9 - 4 + 4$
$= 5 + 4$
$= 9$

17. $-1 < 0$

18. $-22 + 14 + (-8) = -8 + (-8)$
$= -16$

19. $(-5)(-6)(3) = 30(3)$
$= 90$

20.
$$\begin{array}{r} 12.920 \\ -\ 6.039 \\ \hline 6.881 \end{array}$$
$6.039 - 12.92 = -6.881$

21. $-5 \le -3$ true
$-3 \le -3$ true
$0 \le -3$ false
-5 and -3 are less than or equal to -3.

22. $7\% = 7(0.01) = 0.07$

23. $\dfrac{3}{4} \cdot (4)^2 = \dfrac{3}{4} \cdot (4)(4)$
$= \dfrac{3}{4} \cdot 16 = \dfrac{3}{4} \cdot \dfrac{16}{1}$
$= \dfrac{3 \cdot \overset{1}{\cancel{2}} \cdot \overset{1}{\cancel{2}} \cdot 2 \cdot 2}{\underset{1}{\cancel{2}} \cdot \underset{1}{\cancel{2}}} = 12$

24. $-2 > -40$

25. $13 + (-16) = -3$

26. $(-4)(12) = -48$

27. $-\dfrac{2}{5} + \dfrac{7}{15} = -\dfrac{6}{15} + \dfrac{7}{15}$
$= \dfrac{-6}{15} + \dfrac{7}{15} = \dfrac{-6+7}{15}$
$= \dfrac{1}{15}$

28. $(-3^3) \cdot 2^2 = -(3 \cdot 3 \cdot 3) \cdot (2 \cdot 2)$
$= -27 \cdot 4 = -108$

29. $2\dfrac{7}{9} = 2\dfrac{7}{9}(100\%) = \dfrac{25}{9}(100\%)$
$= \dfrac{2500}{9} \approx 277.8\%$

30. $240\% = 240(0.01) = 2.4$

31. 2

32. $7 - 21 = 7 + (-21)$
$= -14$

33. $96 \div (-12) = -8$

34.
$$\begin{array}{r} 0.35 \\ 20\overline{)7.00} \\ -\underline{6.0} \\ 100 \\ -\underline{100} \\ 0 \end{array}$$

35. $-\dfrac{7}{16} \div \dfrac{3}{8} = -\left(\dfrac{7}{16} \div \dfrac{3}{8}\right) = -\left(\dfrac{7}{16} \cdot \dfrac{8}{3}\right) = -\left(\dfrac{7 \cdot 8}{16 \cdot 3}\right)$
$= -\dfrac{7 \cdot \overset{1}{\cancel{2}} \cdot \overset{1}{\cancel{2}} \cdot \overset{1}{\cancel{2}}}{2 \cdot \underset{1}{\cancel{2}} \cdot \underset{1}{\cancel{2}} \cdot \underset{1}{\cancel{2}} \cdot 3} = -\dfrac{7}{6}$

36. $2^3 \div 4 - 2(2 - 7) = 8 \div 4 - 2(2 - 7)$
$= 8 \div 4 - 2(-5)$
$= 2 - (-10)$
$= 2 + 10$
$= 12$

37. $|-3| = 3$

38. $12 - (-10) - 4 = 12 + 10 + (-4)$
$= 22 + (-4)$
$= 18$

39. $2\dfrac{8}{9} = 2\dfrac{8}{9}(100\%) = \dfrac{26}{9}(100\%)$
$= \dfrac{2600}{9}\% = 288\dfrac{8}{9}\%$

40. **a.** $-(-12) = 12$
$-(-8) = 8$
$-(-1) = 1$
$-(7) = -7$

b. $|-12| = 12$
$|-8| = 8$
$|-1| = 1$
$|7| = 7$

41. $(-204) \div (-17) = 12$

42.
$$\begin{array}{r} 0.6363 \\ 11\overline{)7.0000} \\ -\underline{6.6} \\ 40 \\ -\underline{33} \\ 70 \\ -\underline{66} \\ 40 \\ -\underline{33} \\ 7 \end{array}$$
$\dfrac{7}{11} = 0.\overline{63}$

43.
$$\begin{array}{r} 11.53 \\ 0.023.\overline{)0.265.40} \\ -\underline{23} \\ 35 \\ -\underline{23} \\ 124 \\ -\underline{115} \\ 90 \\ -\underline{69} \\ 21 \end{array}$$
$0.2654 \div (-0.023) \approx -11.5$

44. $(7 - 2)^2 - 5 - 3 \cdot 4 = 5^2 - 5 - 3 \cdot 4$
$= 25 - 5 - 3 \cdot 4$
$= 25 - 5 - 12$
$= 20 - 12$
$= 8$

45. $8 > -10$

46. $-12 + 8 + (-4) = -4 + (-4)$
$= -8$

47. $2(-3)(-12) = -6(-12)$
$= 72$

48. $-\dfrac{5}{8}+\dfrac{1}{6}=-\dfrac{15}{24}+\dfrac{4}{24}$

$\qquad =\dfrac{-15}{24}+\dfrac{4}{24}=\dfrac{-15+4}{24}=\dfrac{-11}{24}$

$\qquad =-\dfrac{11}{24}$

49. $\begin{aligned}-24&>-19 &&\text{false}\\-17&>-19 &&\text{true}\\-9&>-19 &&\text{true}\\0&>-19 &&\text{true}\\4&>-19 &&\text{true}\end{aligned}$

The numbers -17, -9, 0, and 4 are greater than -19.

50. $0.002=0.002(100\%)=0.2\%$

51. $-4^2\cdot\left(\dfrac{1}{2}\right)^2=-(4\cdot4)\cdot\left(\dfrac{1}{2}\cdot\dfrac{1}{2}\right)$

$\qquad =-16\cdot\dfrac{1}{4}=-\dfrac{16}{1}\cdot\dfrac{1}{4}=-4$

52. $\begin{aligned}&4.890\\&\underline{-\,1.329}\\&3.561\end{aligned}$

$-1.329+4.89=3.561$

53. $-|17|=-17$

54. $-5-22-(-13)-19-(-6)$
$=5+(-22)+13+(-19)+6$
$=-27+13+(-19)+6$
$=-14+(-19)+6$
$=-33+6$
$=-27$

55. $\left(\dfrac{1}{3}\right)\left(-\dfrac{4}{5}\right)\left(\dfrac{3}{8}\right)=-\left(\dfrac{1}{3}\cdot\dfrac{4}{5}\cdot\dfrac{3}{8}\right)=-\left(\dfrac{1\cdot4\cdot3}{3\cdot5\cdot8}\right)$

$\qquad =-\left(\dfrac{\overset{1}{\cancel{1}}\cdot\overset{1}{\cancel{2}}\cdot\overset{1}{\cancel{3}}}{\underset{1}{\cancel{3}}\cdot5\cdot\underset{1}{\cancel{2}}\cdot\underset{1}{\cancel{2}}\,2}\right)=-\dfrac{1}{10}$

56. $-43<-34$

57. $25\overline{)18.00}$
$\quad\;\; 0.72$
$\quad\underline{-17.5}$
$\qquad 50$
$\quad\;\underline{-\,50}$
$\qquad\;\; 0$

$\dfrac{18}{25}=0.72$

58. $(-2)^3\cdot4^2=(-2)(-2)(-2)\cdot(4\cdot4)$
$\qquad =-8\cdot16=-128$

59. $0.075=0.075(100\%)=7.5\%$

60. $\dfrac{19}{35}=\dfrac{19}{35}(100\%)=\dfrac{1900}{35}\%=54\dfrac{2}{7}\%$

61. $14+(-18)+6+(-20)=-4+6+(-20)$
$\qquad\qquad\qquad\qquad\quad =2+(-20)$
$\qquad\qquad\qquad\qquad\quad =-18$

62. $-4(-8)(12)(0)=32(12)(0)$
$\qquad\qquad\qquad =384(0)$
$\qquad\qquad\qquad =0$

63. $2^3-7+16\div(-3+5)=2^3-7+16\div2$
$\qquad\qquad\qquad\qquad =8-7+16\div2$
$\qquad\qquad\qquad\qquad =8-7+8$
$\qquad\qquad\qquad\qquad =1+8$
$\qquad\qquad\qquad\qquad =9$

64. $\dfrac{3}{4}+\dfrac{1}{2}-\dfrac{3}{8}=\dfrac{6}{8}+\dfrac{4}{8}-\dfrac{3}{8}$

$\qquad =\dfrac{6}{8}+\dfrac{4}{8}+\dfrac{-3}{8}=\dfrac{6+4-3}{8}$

$\qquad =\dfrac{7}{8}$

65. $-128\div(-8)=16$

66. $-57<28$

67. $\left(-\dfrac{1}{3}\right)^3\cdot9^2=\left(-\dfrac{1}{3}\right)\left(-\dfrac{1}{3}\right)\left(-\dfrac{1}{3}\right)\cdot(9\cdot9)$

$\qquad =-\dfrac{1}{27}\cdot81=-\dfrac{1}{27}\cdot\dfrac{81}{1}=-3$

68. $-7+(-3)+(-12)+16=-10+(-12)+16$
$\qquad\qquad\qquad\qquad\quad =-22+16$
$\qquad\qquad\qquad\qquad\quad =-6$

69. $5(-2)(10)(-3)=-10(10)(-3)$
$\qquad\qquad\quad =-100(-3)$
$\qquad\qquad\quad =300$

70. $\{-3,-2,-1)]$

71. Strategy
To find the temperature, add the rise in temperature ($14°$) to the original temperature ($-6°$).

Solution
$-6+14=8$
The temperature was $8°C$.

72. Strategy
To find the average low temperature:
• Add the three temperature readings:
• Divide by 3.

Solution
$\begin{aligned}-8+7+(-5)&=-1+(-5)\\&=-6\end{aligned}$
$-6\div3=-2$

73. Strategy
To find the difference, subtract the low temperature ($-44°$) from the high temperature ($62°$).

Solution
$$62 - (-44) = 64 + 44$$
$$= 106$$

74. Strategy
To find the temperature, add the rise in temperature ($7°$) to the original temperature ($-13°$).

Solution
$$-13 + 7 = -6$$
The temperature was $-6°C$.

75. Strategy
To find the difference, subtract the temperature on Pluto ($-234°$) from the temperature on Venus ($480°$).

Solution
$$480 - (-234) = 480 + 234$$
$$= 714$$
The difference is $714°C$.

Chapter Test

1. $55\% = 55\left(\dfrac{1}{100}\right) = \dfrac{55}{100} = \dfrac{11}{20}$

2. $-8 < -5$ true
$-6 < -5$ true
$-4 < -5$ false
$-2 < -5$ false
-8 and -6 are less than -5.

3. $-9 - (-6) = -9 + 6$
$\qquad\qquad = -3$

4.
$$20\overline{)3.00}^{\,0.15}$$
$$\underline{-20}$$
$$100$$
$$\underline{-100}$$
$$0$$

$\dfrac{3}{20} = 0.15$

5. $\dfrac{3}{4}\left(-\dfrac{2}{21}\right) = -\left(\dfrac{3}{4} \cdot \dfrac{2}{21}\right)$

$\qquad = -\left(\dfrac{3 \cdot 2}{4 \cdot 21}\right) = -\left(\dfrac{\overset{1}{\cancel{3}} \cdot \overset{1}{\cancel{2}}}{\underset{1}{\cancel{2}} \cdot 2 \cdot \underset{1}{\cancel{3}} \cdot 7}\right)$

$\qquad = -\dfrac{1}{14}$

6. $-75 \div 5 = -15$

7. $\left(-\dfrac{2}{3}\right)^3 \cdot 3^2 = \left(-\dfrac{2}{3}\right)\left(-\dfrac{2}{3}\right)\left(-\dfrac{2}{3}\right) \cdot (3 \cdot 3)$

$\qquad = -\dfrac{8}{27} \cdot 9 = -\dfrac{8}{27} \cdot \dfrac{9}{1} = -\dfrac{8}{3}$

8. $-7 + (-3) + 12 = -10 + 12$
$\qquad\qquad\qquad\quad = 2$

9. $\{1, 2, 3, 4, 5, 6\}$

10. $1.59 = 1.59(100\%) = 159\%$

11. $|-29| = 29$

12. $-47 > -68$

13. $-\dfrac{4}{9} - \dfrac{5}{6} = -\dfrac{8}{18} - \dfrac{15}{18} = \dfrac{-8}{18} + \dfrac{-15}{18}$

$\qquad = \dfrac{-8 + (-15)}{18} = \dfrac{-23}{18} = -\dfrac{23}{18}$

14. $-6(-43) = 258$

15. $8 + \dfrac{12 - 4}{3^2 - 1} - 6 = 8 + \dfrac{8}{9 - 1} - 6$

$\qquad\qquad\qquad = 8 + \dfrac{8}{8} - 6$

$\qquad\qquad\qquad = 8 + 1 - 6$

$\qquad\qquad\qquad = 9 - 6$

$\qquad\qquad\qquad = 3$

16. $-\dfrac{5}{8} \div \left(-\dfrac{3}{4}\right) = \dfrac{5}{8} \div \dfrac{3}{4}$

$\qquad = \dfrac{5}{8} \cdot \dfrac{4}{3} = \dfrac{5 \cdot 4}{8 \cdot 3} = \dfrac{5 \cdot \overset{1}{\cancel{2}} \cdot \overset{1}{\cancel{2}}}{\underset{1}{\cancel{2}} \cdot \underset{1}{\cancel{2}} \cdot 2 \cdot 3}$

$\qquad = \dfrac{5}{6}$

17. $\dfrac{3}{13} = \dfrac{3}{13}(100\%) = \dfrac{300}{13}\%$

$\qquad = 23.1\%$

18. $6.2\% = 6.2(0.01) = 0.062$

19. $13 - (-5) - 4 = 13 + 5 + (-4)$
$\qquad\qquad\qquad = 18 + (-4)$
$\qquad\qquad\qquad = 14$

20.
$$30\overline{)13.00}^{\,0.433}$$
$$\underline{-120}$$
$$100$$
$$\underline{-90}$$
$$100$$
$$\underline{-90}$$
$$10$$

$\dfrac{13}{30} = 0.4\overline{3}$

21.
$$\begin{array}{r} 2.7 \\ \times\ 0.9 \\ \hline 2.43 \end{array}$$
$(-0.9)(2.7) = -2.43$

22. $-180 \div (-12) = 15$

23. $2^2 \cdot (-4)^2 \cdot 10 = -(2 \cdot 2) \cdot (-4)(-4) \cdot 10$
$$= 4 \cdot 16 \cdot 10 = 64 \cdot 10$$
$$= 640$$

24. $15 + (-8) + (-19) = 7 + (-19)$
$$= -12$$

25. $-|-34| = -34$

26. $53 > -92$

27. a. $-(-17) = 17$
$$-(-6) = 6$$
$$-(5) = -5$$
$$-(9) = -9$$

   b. $|-17| = 17$
$$|-6| = 6$$
$$|-5| = 5$$
$$|-9| = 9$$

28. $\dfrac{16}{23} = \dfrac{16}{23}(100\%) = \dfrac{1600}{23}\%$
$$= 69\dfrac{13}{23}\%$$

29. $\begin{array}{r} 18.354 \\ -\ \underline{6.97} \\ 11.384 \end{array}$

$-18.354 + 6.97 = -11.384$

30. $-4(8)(-5) = -32(-5)$
$$= 160$$

31. $9(-4) \div [2(8-5)^2] = 9(-4) \div [2(3)^2]$
$$= 9(-4) \div [2(9)]$$
$$= 9(-4) \div (18)$$
$$= -36 \div 18$$
$$= -2$$

32. Strategy
To find the temperature, add the rise in
temperature (12°) to the original temperature
(−8°).

Solution
$-8 + 12 = 4$
The temperature was 4°C.

33. Strategy
To find the average low temperature:
• Add the four temperatures
• Divide by 4.

Solution

$-61 + (-58) + (-49) + (-24) = -192$

$-192 \div 4 = -48$
The average low temperature was −48°F.

Chapter 2: Variable Expressions

Prep Test

1. $-12 - (-15) = -12 + 15 = 3$ [1.2.2]

2. $-36 \div (-9) = \frac{-36}{-9} = 4$ [1.2.4]

3. $-\frac{3}{4} + \frac{5}{6} = -\frac{9}{12} + \frac{10}{12} = \frac{1}{12}$ [1.3.2]

4. $-\frac{4}{9}$ [1.3.3]

5. $\left(-\frac{3}{4}\right) \div \left(-\frac{5}{2}\right) = -\frac{3}{4} \times \left(-\frac{2}{5}\right) = \frac{6}{20} = \frac{3}{10}$ [1.3.3]

6. $-2^4 = -2 \times 2 \times 2 \times 2 = -16$ [1.4.1]

7. $\left(\frac{2}{3}\right)^3 = \frac{2}{3} \times \frac{2}{3} \times \frac{2}{3} = \frac{8}{27}$ [1.4.1]

8. $3 \cdot 4^2 = 3 \cdot 4 \cdot 4 = 48$ [1.4.1]

9. $7 - 2 \cdot 3 = 7 - (2 \cdot 3) = 7 - 6 = 1$ [1.4.2]

10. $5 - 7(3 - 2^2) = 5 - 7(3 - 4)$
$$= 5 - 7(-1)$$
$$= 5 + 7 = 12 \quad [1.4.2]$$

Go Figure

If two fractions are inserted between $\frac{1}{4}$ and $\frac{1}{2}$,

then there are three steps from $\frac{1}{4}$ to $\frac{1}{2}$. First, to

find the two fractions, we find the difference:

$$\frac{1}{2} - \frac{1}{4} = \frac{1}{4}$$

Second, since there are 3 steps between the fractions, divide the difference by 3.

$$\frac{1}{4} \cdot \frac{1}{3} = \frac{1}{12}$$ This is the difference between each

step.

Next, find the value of each fraction

$$\frac{1}{4} + \frac{1}{12} = \frac{3}{12} + \frac{1}{12} = \frac{4}{12}$$
$$\frac{4}{12} + \frac{1}{12} = \frac{5}{12}$$

The four fractions are $\frac{1}{4}, \frac{4}{12}, \frac{5}{12}, \frac{1}{2}$. Each

consecutive fraction has the same difference.
Finally find the sum of these four fractions.

$$\frac{1}{4} + \frac{4}{12} + \frac{5}{12} + \frac{1}{2} = \frac{3 + 4 + 5 + 6}{12} = \frac{18}{12} = \frac{3}{2}$$

Section 2.1

Concept Review 2.1

1. Always true

3. Sometimes true
Any real number could be substituted for x.

5. Always true

Objective 2.1.1 Exercises

1. $2x^2$, $5x$, $\underline{-8}$

3. $\underline{6}$, $-a^4$

5. $7\underline{x^2y}$, $6\underline{xy^2}$

7. coefficient of x^2: 1
coefficient of $-9x$: -9

9. coefficient of n^3: 1
coefficient of $-4n^2$: -4
coefficient of $-n$: -1

13. $a - 2c$
$2 - 2(-4) = 2 - (-8)$
$$= 2 + 8$$
$$= 10$$

15. $2c^2$
$2(-4)^2 = 2(16)$
$$= 32$$

17. $3b - 3c$
$3(3) - 3(-4) = 9 - (-12)$
$$= 9 + 12$$
$$= 21$$

19. $-3c + 4$
$-3(-4) + 4 = 12 + 4$
$$= 16$$

21. $6b \div (-a)$
$6(3) \div (-2) = 18 \div (-2)$
$$= -9$$

23. $-2ab \div c$
$-2(2)(3) \div (-4) = -4(3) \div (-4)$
$$= -12 \div (-4)$$
$$= 3$$

25. $b^2 - c^2$
$3^2 - (-4)^2 = 9 - 16$
$$= -7$$

27. $a^2 + b^2$
$2^2 + 3^2 = 4 + 9$
$$= 13$$

29. $(b - a)^2 + 4c$
$(3 - 2)^2 + 4(-4) = 1^2 + 4(-4)$
$$= 1 + (-16)$$
$$= -15$$

31. $\dfrac{5ab}{6} - 3cb$

$\dfrac{5(2)(3)}{6} - 3(-4)(3) = \dfrac{30}{6} - (-36)$

$\qquad\qquad\qquad = 5 - (-36)$

$\qquad\qquad\qquad = 5 + 36$

$\qquad\qquad\qquad = 41$

33. $\dfrac{b+c}{d}$

$\dfrac{4+(-1)}{3} = \dfrac{3}{3}$

$\qquad\qquad = 1$

35. $\dfrac{2d+b}{-a}$

$\dfrac{2(3)+4}{-(-2)} = \dfrac{6+4}{2}$

$\qquad\qquad = \dfrac{10}{2}$

$\qquad\qquad = 5$

37. $\dfrac{b-d}{c-a}$

$\dfrac{4-3}{-1-(-2)} = \dfrac{-4-3}{-1+2}$

$\qquad\qquad = \dfrac{1}{1}$

$\qquad\qquad = 1$

39. $(b+d)^2 - 4a$

$(4+3)^2 - 4(-2) = 7^2 - 4(-2)$

$\qquad\qquad\qquad = 49 - (-8)$

$\qquad\qquad\qquad = 49 + 8$

$\qquad\qquad\qquad = 57$

41. $(d-a)^2 \div 5$

$[3-(-2)]^2 \div 5 = [3+2]^2 \div 5$

$\qquad\qquad\qquad = 5^2 \div 5$

$\qquad\qquad\qquad = 25 \div 5$

$\qquad\qquad\qquad = 5$

43. $b^2 - 2b + 4$

$4^2 - 2(4) + 4 = 16 - 8 + 4$

$\qquad\qquad\qquad = 8 + 4$

$\qquad\qquad\qquad = 12$

45. $\dfrac{bd}{a} \div c$

$\dfrac{4(3)}{-2} \div (-1) = \dfrac{12}{-2} \div (-1)$

$\qquad\qquad\qquad = -6 \div (-1)$

$\qquad\qquad\qquad = 6$

47. $2(b+c) - 2a$

$2[4+(-1)] - 2(-2) = 2(3) - 2(-2)$

$\qquad\qquad\qquad\qquad = 6 - (-4)$

$\qquad\qquad\qquad\qquad = 6 + 4$

$\qquad\qquad\qquad\qquad = 10$

49. $\dfrac{b-2a}{bc^2 - d}$

$\dfrac{4-2(-2)}{4(-1)^2 - 3} = \dfrac{4-(-4)}{4(1)-3}$

$\qquad\qquad\qquad = \dfrac{4+4}{4(1)-3}$

$\qquad\qquad\qquad = \dfrac{8}{4-3}$

$\qquad\qquad\qquad = \dfrac{8}{1}$

$\qquad\qquad\qquad = 8$

51. $\dfrac{1}{3}d^2 - \dfrac{3}{8}b^2$

$\dfrac{1}{3}(3)^2 - \dfrac{3}{8}(4)^2 = \dfrac{1}{3}(9) - \dfrac{3}{8}(16)$

$\qquad\qquad\qquad = 3 - 6$

$\qquad\qquad\qquad = -3$

53. $\dfrac{-4bc}{2a-b}$

$\dfrac{-4(4)(-1)}{2(-2)-4} = \dfrac{16}{-4-4}$

$\qquad\qquad\qquad = \dfrac{16}{-8}$

$\qquad\qquad\qquad = -2$

55. $a^3 - 3a^2 + a$

$(-2)^3 - 3(-2)^2 + (-2) = -8 - 3(4) + (-2)$

$\qquad\qquad\qquad\qquad\qquad = -8 - 12 + (-2)$

$\qquad\qquad\qquad\qquad\qquad = -20 + (-2)$

$\qquad\qquad\qquad\qquad\qquad = -22$

57. $-\dfrac{3}{4}b + \dfrac{1}{2}(ac+bd)$

$-\dfrac{3}{4}(4) + \dfrac{1}{2}[(-2)(-1)+4(3)] = -\dfrac{3}{4}(4) + \dfrac{1}{2}[2+12]$

$\qquad\qquad\qquad\qquad\qquad\qquad = -\dfrac{3}{4}(4) + \dfrac{1}{2}(14)$

$\qquad\qquad\qquad\qquad\qquad\qquad = -3 + 7$

$\qquad\qquad\qquad\qquad\qquad\qquad = 4$

59. $(b-a)^2 - (d-c)^2$

$[4-(-2)]^2 - [3-(-1)]^2 = [4+2]^2 - [3+1]^2$

$\qquad\qquad\qquad\qquad\qquad = 6^2 - 4^2$

$\qquad\qquad\qquad\qquad\qquad = 36 - 16$

$\qquad\qquad\qquad\qquad\qquad = 20$

61. $4ac + (2a)^2$

$4(-2)(-1) + [2(-2)]^2 = 4(-2)(-1) + (-4)^2$

$\qquad\qquad\qquad\qquad\qquad = 4(-2)(-1) + 16$

$\qquad\qquad\qquad\qquad\qquad = 8 + 16$

$\qquad\qquad\qquad\qquad\qquad = 24$

63. $c^2 - ab$

$(-0.8)^2 - (2.7)(-1.6) = 0.64 - (-4.32)$

$\qquad\qquad\qquad\qquad = 0.64 + 4.32$

$\qquad\qquad\qquad\qquad = 4.96$

65. $\dfrac{b^3}{c} - 4a$

$\dfrac{(-1.6)^3}{-0.8} - 4(2.7) = \dfrac{-4.096}{-0.8} - 4(2.7)$

$= 5.12 - 10.8$

$= -5.68$

67. Strategy

To find the volume, use the formula for the volume of a cylinder.

Solution

$V = \pi r^2 h$

$V = \pi (1.25^2)(5.25)$

$V \approx 25.8$

The volume is approximately 25.8 in^3.

69. Strategy

To find the area, use the formula for the area of a trapezoid.

Solution

$A = \dfrac{1}{2} h(b_1 + b_2)$

$A = \dfrac{1}{2}(6.75)(17.5 + 10.25)$

$A = \dfrac{1}{2}(6.75)(27.75)$

$A \approx 93.7 \text{ cm}^2$

The area is approximately 93.7 cm^2.

71. Strategy

To find the volume, use the formula for the volume of a cylinder.

$r = \dfrac{1}{2} d = \dfrac{1}{2}(7) = 3.5$

Solution

$V = \pi r^2 h$

$V = \pi (3.5)^2 (12.6)$

$V \approx 484.9$

The volume is approximately 484.9 m^3.

73.

$3x^2 - 4x - 5 = 3(-2)^2 - 4(-2) - 5$

$= 3(4) + 8 - 5$

$= 12 + 3$

$= 15 = a$

$3a - 4 = 3(15) - 4$

$= 45 - 4$

$= 41$

75.

$3w^2 - 2v = 3(-1)^2 - 2(3)$

$= 3 - 6$

$= -3 = d$

$d^2 - 4d = (-3)^2 - 4(-3)$

$= 9 + 12$

$= 21$

Applying Concepts 2.1

77. $\left| -4ab \right|$

$\left| -4(-2)(-3) \right| = \left| -24 \right|$

$= 24$

79. $\dfrac{1}{3} a^5 b^6$

$\dfrac{1}{3}\left(\dfrac{2}{3}\right)^5 \left(-\dfrac{3}{2}\right)^6 = \dfrac{1}{3}\left(\dfrac{32}{243}\right)\left(\dfrac{729}{64}\right)$

$= \dfrac{32}{729}\left(\dfrac{729}{64}\right)$

$= \dfrac{1}{2}$

81. $\left| 5ab - 8a^2 b^2 \right|$

$\left| 5 \cdot \dfrac{2}{3}\left(-\dfrac{3}{2}\right) - 8\left(\dfrac{2}{3}\right)^2 \left(-\dfrac{3}{2}\right)^2 \right| = \left| -5 - 8 \right|$

$= \left| -13 \right|$

$= 13$

83. $2^y - y^2 = 2^3 - 3^2$

$= 8 - 9$

$= -1$

85. $z^x = (-2)^2 = 4$

87. $y^{(x^2)} = 3^{2^2}$

$= 3^4$

$= 81$

Section 2.2

Concept Review

1. Never true

The Commutative Property of Addition states that two terms can be added in either order and the sum will be the same.

3. Always true

5. Always true

7. Never true

$3(x + 4) = 3x + 12$ is the correct application of the Distributive Property.

Objective 2.2.1 Exercises

1. 2

3. 5

5. 6

7. -8

9. -4

11. The Inverse Property of Addition

13. The Commutative Property of Addition

15. The Associative Property of Addition

17. The Commutative Property of Multiplication

19. The Associative Property of Multiplication

Objective 2.2.2 Exercises

23. $3x, 5x$

25. $6x + 8x = 14x$

27. $9a - 4a = 5a$

29. $4y - 10y = -6y$

31. $-3b - 7 = -3b - 7$

33. $-12a + 17a = 5a$

35. $5ab - 7ab = -2ab$

37. $-12xy + 17xy = 5xy$

39. $-3ab + 3ab = 0$

41. $-\dfrac{1}{2}x - \dfrac{1}{3}x = -\dfrac{3}{6}x - \dfrac{2}{6}x = -\dfrac{5}{6}x$

43. $\dfrac{3}{8}x^2 - \dfrac{5}{12}x^2 = \dfrac{9}{24}x^2 - \dfrac{10}{24}x^2 = -\dfrac{1}{24}x^2$

45. $3x + 5x + 3x = 11x$

47. $5a - 3a + 5a = 7a$

49. $-5x^2 - 12x^2 + 3x^2 = -14x^2$

51. $7x - 8x + 3y = -x + 3y$

53. $7x - 3y + 10x = 17x - 3y$

55. $3a - 7b - 5a + b = -2a - 6b$

57. $3x - 8y - 10x + 4x = -3x - 8y$

59. $x^2 - 7x - 5x^2 + 5x = -4x^2 - 2x$

Objective 2.2.3 Exercises

61. $4(3x) = 12x$

63. $-3(7a) = -21a$

65. $-2(-3y) = 6y$

67. $(4x)2 = 8x$

69. $3a(-2) = -6a$

71. $(-3b)(-4) = 12b$

73. $-5(3x^2) = -15x^2$

75. $\dfrac{1}{3}(3x^2) = x^2$

77. $\dfrac{1}{8}(8x) = x$

79. $-\dfrac{1}{7}(-7n) = n$

81. $\dfrac{12x}{5}\left(\dfrac{5}{12}\right) = x$

83. $(-10n)\left(-\dfrac{1}{10}\right) = n$

85. $\dfrac{1}{7}(14x) = 2x$

87. $-\dfrac{1}{8}(16x) = -2x$

89. $-\dfrac{5}{8}(24a^2) = -15a^2$

91. $-\dfrac{3}{4}(-8y) = 6y$

93. $(33y)\left(\dfrac{1}{11}\right) = 3y$

95. $(-10x)\left(\dfrac{1}{5}\right) = -2x$

Objective 2.2.4 Exercises

97. $-(x + 2) = -x - 2$

99. $2(4x - 3) = 8x - 6$

101. $-2(a + 7) = -2a - 14$

103. $-3(2y - 8) = -6y + 24$

105. $(5 - 3b)7 = 35 - 21b$

107. $-3(3 - 5x) = -9 + 15x$

109. $3(5x^2 + 2x) = 15x^2 + 6x$

111. $-2(-y + 9) = 2y - 18$

113. $(-3x - 6)5 = -15x - 30$

115. $2(-3x^2 - 14) = -6x^2 - 28$

117. $-3(2y^2 - 7) = -6y^2 + 21$

119. $-(6a^2 - 7b^2) = -6a^2 + 7b^2$

121. $4(x^2 - 3x + 5) = 4x^2 - 12x + 20$

123. $-3(y^2 - 3y - 7) = -3y^2 + 9y + 21$

125. $4(-3a^2 - 5a + 7) = -12a^2 - 20a + 28$

127. $-3(-4x^2 + 3x - 4) = 12x^2 - 9x + 12$

129. $5(2x^2 - 4xy - y^2) = 10x^2 - 20xy - 5y^2$

131. $-(8b^2 - 6b + 9) = -8b^2 + 6b - 9$

Objective 2.2.5 Exercises

133. $6a - (5a + 7) = 6a - 5a - 7 = a - 7$

135. $10 - (11x - 3) = 10 - 11x + 3 = -11x + 13$

137. $8 - (12 + 4y) = 8 - 12 - 4y = -4y - 4$

139. $2(x-4)-4(x+2)=2x-8-4x-8$
$\qquad\qquad\qquad = -2x-16$

141. $6(2y-7)-3(3-2y)=12y-42-9+6y$
$\qquad\qquad\qquad\quad = 18y-51$

143. $2(a+2b)-(a-3b)=2a+4b-a+3b$
$\qquad\qquad\qquad\quad = a+7b$

145. $2[x+2(x+7)]=2[x+2x+14]$
$\qquad\qquad\qquad = 2[3x+14]$
$\qquad\qquad\qquad = 6x+28$

147. $-5[2x+3(5-x)]=-5[2x+15-3x]$
$\qquad\qquad\qquad\quad = -5[-x+15]$
$\qquad\qquad\qquad\quad = 5x-75$

149. $-2[3x-(5x-2)]=-2[3x-5x+2]$
$\qquad\qquad\qquad\quad = -2[-2x+2]$
$\qquad\qquad\qquad\quad = 4x-4$

151. $-7x+3[x-7(3-2x)]=-7x+3[x-21+14x]$
$\qquad\qquad\qquad\qquad = -7x+3[15x-21]$
$\qquad\qquad\qquad\qquad = -7x+45x-63$
$\qquad\qquad\qquad\qquad = 38x-63$

153. $4a-2[2b-(b-2a)]+3b$
$= 4a-2[2b-b+2a]+3b$
$= 4a-2[b+2a]+3b$
$= 4a-2b-4a+3b$
$= b$

155. $5y-2(y-3x)+2(7x-y)$
$= 5y-2y+6x+14x-2y$
$= 20x+y$

Applying Concepts 2.2

157. $\frac{1}{3}(3x+y)-\frac{2}{3}(6x-y)=x+\frac{1}{3}y-4x+\frac{2}{3}y$
$\qquad\qquad\qquad\qquad = -3x+y$

159. 0

161. $-(a-b)$
$-a+b$ is the additive inverse of $a-b$.

163. a. Yes, $\otimes$ is a commutative operation.
$b\otimes a=(b\cdot a)-(b+a)$
$\qquad = (a\cdot b)-(a+b)$
$\qquad = a\otimes b$
because $\cdot$ and $+$ are commutative.

b. No, $\otimes$ is not associative.
For example,
$(7\otimes5)\otimes2=[(7\cdot5)-(7+5)]\otimes2$
$\qquad\qquad = (35-12)\otimes2$
$\qquad\qquad = 23\otimes2$
$\qquad\qquad = (23\cdot2)-(23+2)$
$\qquad\qquad = 45-25$
$\qquad\qquad = 21$

$7\otimes(5\otimes2)=7\otimes[(5\cdot2)-(5+2)]$
$\qquad\qquad = 7\otimes(10-7)$
$\qquad\qquad = 7\otimes3$
$\qquad\qquad = (7\cdot3)-(7+3)$
$\qquad\qquad = 21-10$
$\qquad\qquad = 11$

Section 2.3

Concept Review 2.3

1. Never true
"Five less than n" is translated as $n-5$."

3. Never true
The other number can be expressed as $12-x$.

5. Always true

7. Never true
"Five times the sum of x and y" is written as $5(x+y)$. "The sum of five times x and y" is written as $5x+y$.

Objective 2.3.1 Exercises

1. d less than 19
$19-d$

3. r decreased by 12
$r-12$

5. a multiplied by 28
$28a$

7. 5 times the difference between n and 7
$5(n-7)$

9. y less the product of 3 and y
$y-3y$

11. the product of -6 and b
$-6b$

13. 4 divided by the difference between p and 6
$\dfrac{4}{p-6}$

15. the quotient of 9 less than x and twice x
$\dfrac{x-9}{2x}$

17. 21 <u>less than</u> the <u>product</u> of *s* and –4
$-4s - 21$

19. the <u>ratio</u> of 8 <u>more than</u> *d* to *d*
$\dfrac{d+8}{d}$

21. three-eighths <u>of</u> the <u>sum</u> of *t* and 15
$\dfrac{3}{8}(t+15)$

23. *w* <u>increased by</u> the <u>quotient</u> of 7 and *w*
$w + \dfrac{7}{w}$

25. *d* <u>increased by</u> the <u>difference</u> between 16 <u>times</u> *d* and 3
$d + (16d - 3)$

27. the unknown number: *n*
$\dfrac{n}{19}$

29. the unknown number: *n*
$n + 40$

31. the unknown number: *n*
the difference between the number and ninety:
$n - 90$
$(n - 90)^2$

33. the unknown number: *n*
four-ninths of the number: $\dfrac{4}{9}n$
$\dfrac{4}{9}n + 20$

35. the unknown number: *n*
ten more than the number: $n + 10$
$n(n + 10)$

37. the unknown number: *n*
the product of seven and the number: $7n$
$7n + 14$

39. the unknown number: *n*
the sum of the number and two: $n + 2$
$\dfrac{12}{n+2}$

41. the unknown number: *n*
the sum of the number and one: $n + 1$
$\dfrac{2}{n+1}$

43. the unknown number: *n*
the quotient of the number and fifty: $\dfrac{n}{50}$
$60 - \dfrac{n}{50}$

45. the unknown number: *n*
the square of the number: n^2
three times the number: $3n$
$n^2 + 3n$

47. the unknown number: *n*
three more than the number: $n + 3$
the cube of the number: n^3
$(n + 3) + n^3$

49. the unknown number: *n*
the square of the number: n^2
one-fourth of the number: $\dfrac{1}{4}n$
$n^2 - \dfrac{1}{4}n$

51. the unknown number: *n*
twice the number: $2n$
the quotient of seven and the number: $\dfrac{7}{n}$
$2n - \dfrac{7}{n}$

53. the unknown number: *n*
the cube of the number: n^3
the product of twelve and the number: $12n$
$n^3 - 12n$

Objective 2.3.2 Exercises

55. the unknown number: *n*
the total of the number and ten: $n + 10$
$n + (n + 10) = n + n + 10$
$\qquad\qquad\quad = 2n + 10$

57. the unknown number: *n*
the difference between nine and the number:
$9 - n$
$n - (9 - n) = n - 9 + n$
$\qquad\qquad\quad = 2n - 9$

59. the unknown number: *n*
one-fifth of the number: $\dfrac{1}{5}n$
three-eighths of the number: $\dfrac{3}{8}n$
$\dfrac{1}{5}n - \dfrac{3}{8}n = \dfrac{8}{40}n - \dfrac{15}{40}n$
$\qquad\qquad\quad = -\dfrac{7}{40}n$

61. the unknown number: *n*
the total of the number and nine: $n + 9$
$(n + 9) + 4 = n + 9 + 4$
$\qquad\qquad\quad = n + 13$

63. the unknown number: *n*
three times the number: $3n$
the sum of three times the number and 40:
$3n + 40$
$2(3n + 40) = 6n + 80$

65. the unknown number: n
the product of five and the number: $5n$
$7(5n) = 35n$

67. the unknown number: n
seventeen times the number: $17n$
twice the number: $2n$
$17n + 2n = 19n$

69. the unknown number: n
the product of the number and twelve: $12n$
$n + 12n = 13n$

71. the unknown number: n
the square of the number: n^2
the sum of the square of the number and four:
$n^2 + 4$
$3(n^2 + 4) = 3n^2 + 12$

73. the unknown number: n
sixteen times the number: $16n$
the sum of sixteen times the number and four:
$16n + 4$
$\frac{3}{4}(16n + 4) = 12n + 3$

75. the unknown number: n
the sum of the number and 9: $n + 9$
$16 - (n + 9) = 16 - n - 9$
$\qquad = -n + 7$

77. the unknown number: n
four times the number: $4n$
the quotient of four times the number and two:
$\frac{4n}{2}$
$\frac{4n}{2} + 5 = 2n + 5$

79. the unknown number: n
the total of the number and eight: $n + 8$
$6(n + 8) = 6n + 48$

81. the unknown number: n
the sum of the number and two: $n + 2$
$7 - (n + 2) = 7 - n - 2$
$\qquad = -n + 5$

83. the unknown number: n
six times the number: $6n$
the sum of the number and six times the number:
$n + 6n$
$\frac{1}{3}(n + 6n) = \frac{1}{3}(7n) = \frac{7n}{3}$

85. the unknown number: n
the cube of the number: n^3
eight increased by the cube of the number: $8 + n^3$
twice the cube of the number: $2n^3$
$(8 + n^3) + 2n^3 = 8 + n^3 + 2n^3$
$\qquad = 3n^3 + 8$

87. the unknown number: n
the difference between the number and six: $n - 6$
twelve more than the number: $n + 12$
$(n - 6) + (n + 12) = n - 6 + n + 12$
$\qquad = 2n + 6$

89. the unknown number: n
the difference between the number and twenty:
$n - 20$
the sum of the number and nine: $n + 9$
$(n - 20) + (n + 9) = n - 20 + n + 9$
$\qquad = 2n - 11$

91. the unknown number: n
three less than the number: $n - 3$
the product of three less than a number and ten:
$(n - 3)10$
$14 + (n - 3)10 = 14 + (10n - 30)$
$\qquad = 14 + 10n - 30$
$\qquad = 10n - 16$

Objective 2.3.3 Exercises

93. Let x be one number; x and $18 - x$

95. number of genes in the round worm genome: G
number of genes in human: $G + 11{,}000$

97. total number of Americans: N
number of Americans who want spending on
space exploration of Mars: $\frac{3}{4}N$

99. number of points awarded for a safety: s
number of points awarded for a touchdown: $3s$

101. number of moons Jupiter has: m
number of moons Saturn has: $m + 9$

103. pounds of pecans produced in Texas: p
pounds of pecans produced in Alabama: $\frac{1}{2}p$

105. World population in 1980: p
World population in 2050: $2p$

107. measure of the largest angle: L
measure of the smallest angle: $\frac{1}{2}L - 10$

109. the number of nickels: n
number of dimes: $35 - n$

111. number of hours of overtime worked: h
employee's weekly pay: $720 + 27h$

113. number of minutes of phone calls: m
monthly cost of phone service: $19.95 + 0.08m$

Applying Concepts 2.3

115. length of wire: x
length of side of square: $\frac{1}{4}x$

Chapter Review Exercises

1. $-7y^2 + 6y^2 - (-2y^2) = -7y^2 + 6y^2 + 2y^2$
$\quad\quad = y^2$

2. $(12x)\left(\dfrac{1}{4}\right) = 3x$

3. $\dfrac{2}{3}(-15a) = -10a$

4. $-2(2x - 4) = -4x + 8$

5. $5(2x + 4) - 3(x - 6) = 10x + 20 - 3x + 18$
$\quad\quad\quad\quad\quad = 7x + 38$

6. $a^2 - 3b = 2^2 - 3(-4)$
$\quad\quad\quad = 4 - 3(-4)$
$\quad\quad\quad = 4 - (-12)$
$\quad\quad\quad = 16$

7. 9

8. $-4(-9y) = 36y$

9. $-2(-3y + 9) = 6y - 18$

10. $3[2x - 3(x - 2y)] + 3y = 3[2x - 3x + 6y] + 3y$
$\quad\quad\quad\quad\quad = 3[-x + 6y] + 3y$
$\quad\quad\quad\quad\quad = -3x + 18y + 3y$
$\quad\quad\quad\quad\quad = -3x + 21y$

11. $-4(2x^2 - 3y^2) = -8x^2 + 12y^2$

12. $3x - 5x + 7x = 5x$

13. $b^2 - 3ab = (-2)^2 - 3(3)(-2)$
$\quad\quad\quad\quad = 4 - 3(3)(-2)$
$\quad\quad\quad\quad = 4 - (-18)$
$\quad\quad\quad\quad = 4 + 18$
$\quad\quad\quad\quad = 22$

14. $\dfrac{1}{5}(10x) = 2x$

15. $5(3 - 7b) = 15 - 35b$

16. $2x + 3[4 - (3x - 7)] = 2x + 3[4 - 3x + 7]$
$\quad\quad\quad\quad\quad = 2x + 3[11 - 3x]$
$\quad\quad\quad\quad\quad = 2x + 33 - 9x$
$\quad\quad\quad\quad\quad = -7x + 33$

17. The Commutative Property of Multiplication

18. $3(8 - 2x) = 24 - 6x$

19. $-2x^2 - (-3x^2) + 4x^2 = -2x^2 + 3x^2 + 4x^2$
$\quad\quad\quad\quad\quad\quad = 5x^2$

20. $-3x - 2(2x - 7) = -3x - 4x + 14$
$\quad\quad\quad\quad\quad = -7x + 14$

21. $-3(3y^2 - 3y - 7) = -9y^2 + 9y + 21$

22. $-2[x - 2(x - y)] + 5y = -2[x - 2x + 2y] + 5y$
$\quad\quad\quad\quad\quad\quad = -2[-x + 2y] + 5y$
$\quad\quad\quad\quad\quad\quad = 2x - 4y + 5y$
$\quad\quad\quad\quad\quad\quad = 2x + y$

23. $\dfrac{-2ab}{2b - a} = \dfrac{-2(-4)(6)}{2(6) - (-4)}$
$\quad\quad\quad = \dfrac{48}{12 + 4}$
$\quad\quad\quad = \dfrac{48}{16} = 3$

24. $(-3)(-12y) = 36y$

25. $4(3x - 2) - 7(x + 5) = 12x - 8 - 7x - 35$
$\quad\quad\quad\quad\quad\quad = 5x - 43$

26. $(16x)\left(\dfrac{1}{8}\right) = 2x$

27. $-3(2x^2 - 7y^2) = -6x^2 + 21y^2$

28. $3(a - c) - 2ab = 3[2 - (-4)] - 2(2)(3)$
$\quad\quad\quad\quad\quad = 3(2 + 4) - 2(2)(3)$
$\quad\quad\quad\quad\quad = 3(6) - 2(2)(3)$
$\quad\quad\quad\quad\quad = 18 - 12$
$\quad\quad\quad\quad\quad = 6$

29. $2x - 3(x - 2) = 2x - 3x + 6$
$\quad\quad\quad\quad\quad = -x + 6$

30. $2a - (-3b) - 7a - 5b = 2a + 3b - 7a - 5b$
$\quad\quad\quad\quad\quad\quad = -5a - 2b$

31. $-5(2x^2 - 3x + 6) = -10x^2 + 15x - 30$

32. $3x - 7y - 12x = -9x - 7y$

33. $\dfrac{1}{2}(12a) = 6a$

34. $2x + 3[x - 2(4 - 2x)] = 2x + 3[x - 8 + 4x]$
$\quad\quad\quad\quad\quad\quad = 2x + 3[5x - 8]$
$\quad\quad\quad\quad\quad\quad = 2x + 15x - 24$
$\quad\quad\quad\quad\quad\quad = 17x - 24$

35. $3x + (-12y) - 5x - (-7y) = 3x - 12y - 5x + 7y$
$\quad\quad\quad\quad\quad\quad = -2x - 5y$

36. $\left(-\dfrac{5}{6}\right)(-36b) = 30b$

37. 21

38. $4x^2 + 9x - 6x^2 - 5x = -2x^2 + 4x$

39. $-\dfrac{3}{8}(16x^2) = -6x^2$

40. $-3[2x - (7x - 9)] = -3[2x - 7x + 9]$
$\quad\quad\quad\quad\quad = -3[-5x + 9]$
$\quad\quad\quad\quad\quad = 15x - 27$

41. $-(8a^2 - 3b^2) = -8a^2 + 3b^2$

42. The Multiplication Property of Zero

43. b <u>decreased by</u> the <u>product</u> of 7 and b
$b - 7b$

44. the unknown number: n
the square of the number: n^2
twice the square of the number: $2n^2$
$n + 2n^2$

45. the unknown number: n
the quotient of six and the number: $\dfrac{6}{n}$
$\dfrac{6}{n} - 3$

46. 10 <u>divided by</u> the <u>difference</u> between y and 2
$\dfrac{10}{y - 2}$

47. the unknown number: n
twice the number: $2n$
the quotient of twice the number and sixteen: $\dfrac{2n}{16}$
$8\left(\dfrac{2n}{16}\right) = 8\left(\dfrac{n}{8}\right) = n$

48. the unknown number: n
five times the number: $5n$
the sum of two and five times the number:
$2 + 5n$
$4(2 + 5n) = 8 + 20n$

49. height of the triangle: h
length of the base of the triangle: $h + 15$

50. amount of the mocha Java beans: b
amount of the espresso beans: $20 - b$

Chapter Test

1. $(9y)4 = 36y$

2. $7x + 5y - 3x - 8y = 4x - 3y$

3. $8n - (6 - 2n) = 8n - 6 + 2n$
$\qquad\qquad\quad = 10n - 6$

4. $3ab - (2a)^2$
$3(-2)(-3) - [2(-2)]^2 = 3(-2)(-3) - (-4)^2$
$\qquad\qquad\qquad\quad = 3(-2)(-3) - 16$
$\qquad\qquad\qquad\quad = 18 - 16$
$\qquad\qquad\qquad\quad = 2$

5. The Multiplication Property of One

6. $-4(-x + 10) = 4x - 40$

7. $\dfrac{2}{3}x^2 - \dfrac{7}{12}x^2 = \dfrac{8}{12}x^2 - \dfrac{7}{12}x^2 = \dfrac{1}{12}x^2$

8. $(-10x)\left(-\dfrac{2}{5}\right) = 4x$

9. $(-4y^2 + 8)6 = -24y^2 + 48$

10. 19

11. $\dfrac{-3ab}{2a + b} = \dfrac{-3(-1)(4)}{2(-1) + 4}$
$\qquad\quad = \dfrac{12}{-2 + 4}$
$\qquad\quad = \dfrac{12}{2}$
$\qquad\quad = 6$

12. $5(x + y) - 8(x - y) = 5x + 5y - 8x + 8y$
$\qquad\qquad\qquad\qquad = -3x + 13y$

13. $6b - 9b + 4b = b$

14. $13(6a) = 78a$

15. $3(x^2 - 5x + 4) = 3x^2 - 15x + 12$

16. $4(b - a) + bc = 4(-3 - 2) + (-3)(4)$
$\qquad\qquad\qquad = 4(-5) + (-3)(4)$
$\qquad\qquad\qquad = -20 + (-12)$
$\qquad\qquad\qquad = -32$

17. $6x - 3(y - 7x) + 2(5x - y)$
$= 6x - 3y + 21x + 10x - 2y$
$= 37x - 5y$

18. the <u>quotient</u> of 8 <u>more than</u> n and 17
$\dfrac{n + 8}{17}$

19. the <u>difference</u> between the <u>sum</u> of a and b and the <u>square</u> of b
$(a + b) - b^2$

20. the unknown number: n
the square of the number: n^2
the product of the number and eleven: $11n$
$n^2 + 11n$

21. the unknown number: n
the sum of the number and nine: $n + 9$
$20(n + 9) = 20n + 180$

22. the unknown number: n
the difference between the number and three:
$n - 3$
two more than the number: $n + 2$
$(n - 3) + (n + 2) = n - 3 + n + 2$
$\qquad\qquad\qquad\quad = 2n - 1$

23. the unknown number: n
twice the number: $2n$
the product of one-fourth and twice the number:
$\dfrac{1}{4}(2n)$
$n - \dfrac{1}{4}(2n) = n - \dfrac{1}{2}n = \dfrac{1}{2}n$

24. the distance from Earth to the sun: d
the distance from Neptune to the sun: $30d$

25. the length of one piece of board: L
the length of the other piece: $9 - L$

Cumulative Review Exercises

1. $-4+7+(-10)=3+(-10)=-7$

2. $-16-(-25)-4=-16+25+(-4)$
$$=9+(-4)$$
$$=5$$

3. $(-2)(3)(-4)=-6(-4)=24$

4. $(-60)\div 12=-5$

5. $1\frac{1}{4}=\frac{5}{4}$

$$\begin{array}{r} 1.25 \\ 4\overline{)5.00} \\ \underline{-4} \\ 10 \\ \underline{-8} \\ 20 \\ \underline{-20} \\ 0 \end{array}$$

$$1\frac{1}{4}=1.25$$

6. $60\%=60\left(\frac{1}{100}\right)=\frac{60}{100}=\frac{3}{5}$
$$60\%=60(0.01)=0.60$$

7. $\{-4,-3,-2,-1\}$

8. $\frac{2}{25}=\frac{2}{25}(100\%)=\frac{200}{25}\%=8\%$

9. $\frac{7}{12}-\frac{11}{16}-\left(-\frac{1}{3}\right)$
$$=\frac{7}{12}-\frac{11}{16}+\frac{1}{3}$$
$$=\frac{28}{48}-\frac{33}{48}+\frac{16}{48}$$
$$=\frac{28}{48}+\frac{-33}{48}+\frac{16}{48}$$
$$=\frac{28+(-33)+16}{48}$$
$$=\frac{11}{48}$$

10. $\frac{5}{12}\div\left(\frac{3}{2}\right)=\frac{5}{12}\cdot\frac{2}{3}$
$$=\frac{5\cdot 2}{12\cdot 3}$$
$$=\frac{5\cdot\overset{1}{2}}{\underset{1}{2}\cdot 2\cdot 3\cdot 3}=\frac{5}{18}$$

11. $\left(-\frac{9}{16}\right)\left(\frac{8}{27}\right)\left(-\frac{3}{2}\right)$
$$=\left(\frac{9}{16}\right)\left(\frac{8}{27}\right)\left(\frac{3}{2}\right)$$
$$=\frac{9\cdot 8\cdot 3}{16\cdot 27\cdot 2}$$
$$=\frac{\overset{1}{\cancel{3}}\cdot\overset{1}{\cancel{3}}\cdot\overset{1}{\cancel{2}}\cdot\overset{1}{\cancel{2}}\cdot\overset{1}{\cancel{2}}\cdot 3}{\underset{1}{\cancel{2}}\cdot 2\cdot 2\cdot 2\cdot 3\cdot\underset{1}{\cancel{3}}\cdot\underset{1}{\cancel{3}}\cdot 2}$$
$$=\frac{1}{4}$$

12. $-3^2\cdot\left(-\frac{2}{3}\right)^3=-(3\cdot 3)\cdot\left(-\frac{2}{3}\right)\left(-\frac{2}{3}\right)\left(-\frac{2}{3}\right)$
$$=-9\cdot\left(-\frac{8}{27}\right)$$
$$=\frac{8}{3}$$

13. $-2^5\div(3-5)^2-(-3)=-32\div(3-5)^2-(-3)$
$$=-32\div(-2)^2-(-3)$$
$$=-32\div 4-(-3)$$
$$=-32\div 4+3$$
$$=-8+3$$
$$=-5$$

14. $\left(-\frac{3}{4}\right)^2-\left(\frac{3}{8}-\frac{11}{12}\right)$
$$=\frac{9}{16}-\left(\frac{9}{24}-\frac{22}{24}\right)$$
$$=\frac{9}{16}-\left(-\frac{13}{24}\right)$$
$$=\frac{9}{16}+\frac{13}{24}$$
$$=\frac{27}{48}+\frac{26}{48}$$
$$=\frac{53}{48}$$

15. $a-3b^2=4-3(-2)^2$
$$=4-3(4)$$
$$=4-12$$
$$=-8$$

16. $-2x^2-(-3x^2)+4x^2=-2x^2+3x^2+4x^2$
$$=5x^2$$

17. $8a-12b-9a=-a-12b$

18. $\frac{1}{3}(9a)=3a$

19. $\left(-\frac{5}{8}\right)(-32b)=20b$

20. $5(4-2x)=20-10x$

21. $-3(-2y+7)=6y-21$

22. $-2(3x^2-4y^2)=-6x^2+8y^2$

23. $-4(2y^2 - 5y - 8) = -8y^2 + 20y + 32$

24. $-4x - 3(2x - 5) = -4x - 6x + 15$
$= -10x + 15$

25. $3(4x - 1) - 7(x + 2) = 12x - 3 - 7x - 14$
$= 5x - 17$

26. $3x + 2[x - 4(2 - x)] = 3x + 2[x - 8 + 4x]$
$= 3x + 2[5x - 8]$
$= 3x + 10x - 16$
$= 13x - 16$

27. $3[4x - 2(x - 4y)] + 5y = 3[4x - 2x + 8y] + 5y$
$= 3[2x + 8y] + 5y$
$= 6x + 24y + 5y$
$= 6x + 29y$

28. the unknown number: n
the product of the number and twelve: $12n$
$6 - 12n$

29. the unknown number: n
the difference between the number and seven:
$n - 7$
$5 + (n - 7) = 5 + n - 7 = n - 2$

30. the speed of dial-up connection: s
the speed of DSL connection: $10s$

Chapter 3: Solving Equations and Inequalities

Prep Test

1. -4 [1.2.2]

2. $-\dfrac{3}{4}\left(-\dfrac{4}{3}\right) = \dfrac{12}{12} = 1$ [1.3.3]

3. $-\dfrac{5}{8}(16) = -\dfrac{80}{8} = -10$ [1.3.3]

4. $90\% = 90(0.01) = 0.9$ [1.3.4]

5. $0.75 = 0.75(100\%) = 75\%$ [1.3.4]

6. $3x^2 - 4x - 1$
$$3(-4)^2 - 4(-4) - 1 = 3(16) + 16 - 1$$
$$= 48 + 15 \qquad [2.1.1]$$
$$= 63$$

7. $10x - 5$ [2.2.2]

8. -9 [2.2.2]

9. $6x - 3(6 - x) = 6x - 18 + 3x = 9x - 18$ [2.2.5]

Go Figure

With the donut on a table, slice the donut parallel to the table. Now cut the halved donut into quarters.

Section 3.1

Concept Review 3.1

1. Sometimes true
   The exception is multiplying both sides of an equation by zero.

3. Always true

5. Always true

7. Always true

Objective 3.1.1 Exercises

3.
$$\begin{array}{c|c} 2x = 8 \\ \hline 2(4) & 8 \\ 8 = 8 \end{array}$$
Yes, 4 is a solution.

5.
$$\begin{array}{c|c} 2b - 1 = 3 \\ \hline 2(-1) - 1 & 3 \\ -2 - 1 & 3 \\ -3 \neq 3 \end{array}$$
No, -1 is not a solution.

7.
$$\begin{array}{c|c} 4 - 2m = 3 \\ \hline 4 - 2(1) & 3 \\ 4 - 2 & 3 \\ 2 \neq 3 \end{array}$$
No, 1 is not a solution.

9.
$$\begin{array}{c|c} 2x + 5 = 3x \\ \hline 2(5) + 5 & 3(5) \\ 10 + 5 & 15 \\ 15 = 15 \end{array}$$
Yes, 5 is a solution.

11.
$$\begin{array}{c|c} 4a + 5 = 3a + 5 \\ \hline 4(0) + 5 & 3(0) + 5 \\ 0 + 5 & 0 + 5 \\ 5 = 5 \end{array}$$
Yes, 0 is a solution.

13.
$$\begin{array}{c|c} 4 - 2n = n + 10 \\ \hline 4 - 2(-2) & -2 + 10 \\ 4 - (-4) & 8 \\ 4 + 4 & 8 \\ 8 = 8 \end{array}$$
Yes, -2 is a solution.

15.
$$\begin{array}{c|c} z^2 + 1 = 4 + 3z \\ \hline 3^2 + 1 & 4 + 3(3) \\ 9 + 1 & 4 + 9 \\ 10 \neq 13 \end{array}$$
No, 3 is not a solution.

17.
$$\begin{array}{c|c} y^2 - 1 = 4y + 3 \\ \hline (-1)^2 - 1 & 4(-1) + 3 \\ 1 - 1 & -4 + 3 \\ 0 \neq -1 \end{array}$$
No, -1 is not a solution.

19.
$$\begin{array}{c|c} x(x + 1) = x^2 + 5 \\ \hline 4(4 + 1) & 4^2 + 5 \\ 4(5) & 16 + 5 \\ 20 \neq 21 \end{array}$$
No, 4 is not a solution.

21.
$$\begin{array}{c|c} 8t + 1 = -1 \\ \hline 8\left(-\dfrac{1}{4}\right) + 1 & -1 \\ -2 + 1 & -1 \\ -1 = -1 \end{array}$$
Yes, $-\dfrac{1}{4}$ is a solution.

23.
$$5m + 1 = 10m - 3$$

$5\left(\dfrac{2}{5}\right) + 1$	$10\left(\dfrac{2}{5}\right) - 3$
$2 + 1$	$4 - 3$

$$3 \neq 1$$

No, $\dfrac{2}{5}$ is not a solution.

25.
$$x^2 - 4x = x + 1.89$$

$(2.1)^2 - 4(2.1)$	$2.1 + 1.89$
$4.41 - 8.4$	3.99

$$-3.99 \neq 3.99$$

No, 2.1 is not a solution.

Objective 3.1.2 Exercises

31.
$$x + 5 = 7$$
$$x + 5 - 5 = 7 - 5$$
$$x = 2$$
The solution is 2.

33.
$$b - 4 = 11$$
$$b - 4 + 4 = 11 + 4$$
$$b = 15$$
The solution is 15.

35.
$$2 + a = 8$$
$$2 - 2 + a = 8 - 2$$
$$a = 6$$
The solution is 6.

37.
$$m + 9 = 3$$
$$m + 9 - 9 = 3 - 9$$
$$m = -6$$
The solution is -6.

39.
$$n - 5 = -2$$
$$n - 5 + 5 = -2 + 5$$
$$n = 3$$
The solution is 3.

41.
$$b + 7 = 7$$
$$b + 7 - 7 = 7 - 7$$
$$b = 0$$
The solution is 0.

43.
$$a - 3 = -5$$
$$a - 3 + 3 = -5 + 3$$
$$a = -2$$
The solution is -2.

45.
$$z + 9 = 2$$
$$z + 9 - 9 = 2 - 9$$
$$z = -7$$
The solution is -7.

47.
$$10 + m = 3$$
$$10 - 10 + m = 3 - 10$$
$$m = -7$$
The solution is -7.

49.
$$9 + x = -3$$
$$9 - 9 + x = -3 - 9$$
$$x = -12$$
The solution is -12.

51.
$$b - 5 = -3$$
$$b - 5 + 5 = -3 + 5$$
$$b = 2$$
The solution is 2.

53.
$$2 = x + 7$$
$$2 - 7 = x + 7 - 7$$
$$-5 = x$$
The solution is -5.

55.
$$4 = m - 11$$
$$4 + 11 = m - 11 + 11$$
$$15 = m$$
The solution is 15.

57.
$$12 = 3 + w$$
$$12 - 3 = 3 - 3 + w$$
$$9 = w$$
The solution is 9.

59.
$$4 = -10 + b$$
$$4 + 10 = -10 + 10 + b$$
$$14 = b$$
The solution is 14.

61.
$$m + \frac{2}{3} = -\frac{1}{3}$$
$$m + \frac{2}{3} - \frac{2}{3} = -\frac{1}{3} - \frac{2}{3}$$
$$m = -1$$
The solution is -1.

63.
$$x - \frac{1}{2} = \frac{1}{2}$$
$$x - \frac{1}{2} + \frac{1}{2} = \frac{1}{2} + \frac{1}{2}$$
The solution is 1.

65.
$$\frac{5}{8} + y = \frac{1}{8}$$
$$\frac{5}{8} - \frac{5}{8} + y = \frac{1}{8} - \frac{5}{8}$$
$$y = -\frac{4}{8} = -\frac{1}{2}$$
The solution is $-\dfrac{1}{2}$.

67.
$$m + \frac{1}{2} = -\frac{1}{4}$$
$$m + \frac{1}{2} - \frac{1}{2} = -\frac{1}{4} - \frac{1}{2}$$
$$m = -\frac{1}{4} - \frac{2}{4}$$
$$m = -\frac{3}{4}$$
The solution is $-\dfrac{3}{4}$.

69.
$$x + \frac{2}{3} = \frac{3}{4}$$
$$x + \frac{2}{3} - \frac{2}{3} = \frac{3}{4} - \frac{2}{3}$$
$$x = \frac{9}{12} - \frac{8}{12}$$
$$x = \frac{1}{12}$$
The solution is $\frac{1}{12}$.

71.
$$-\frac{5}{6} = x - \frac{1}{4}$$
$$-\frac{5}{6} + \frac{1}{4} = x - \frac{1}{4} + \frac{1}{4}$$
$$-\frac{10}{12} + \frac{3}{12} = x$$
$$-\frac{7}{12} = x$$
The solution is $-\frac{7}{12}$.

73.
$$-\frac{1}{21} = m + \frac{2}{3}$$
$$-\frac{1}{21} - \frac{2}{3} = m + \frac{2}{3} - \frac{2}{3}$$
$$-\frac{1}{21} - \frac{14}{21} = m$$
$$-\frac{15}{21} = m$$
$$-\frac{5}{7} = m$$
The solution is $-\frac{5}{7}$.

75.
$$\frac{5}{12} = n + \frac{3}{4}$$
$$\frac{5}{12} - \frac{3}{4} = n + \frac{3}{4} - \frac{3}{4}$$
$$\frac{5}{12} - \frac{9}{12} = n$$
$$-\frac{4}{12} = n$$
$$-\frac{1}{3} = n$$
The solution is $-\frac{1}{3}$.

77.
$$w + 2.932 = 4.801$$
$$w + 2.932 - 2.932 = 4.801 - 2.932$$
$$w = 1.869$$
The solution is 1.869.

79.
$$-1.926 + t = -1.042$$
$$-1.926 + 1.926 + t = -1.042 + 1.926$$
$$t = 0.884$$
The solution is 0.884.

Objective 3.1.3 Exercises

85.
$$4y = 28$$
$$\frac{4y}{4} = \frac{28}{4}$$
$$y = 7$$
The solution is 7.

87.
$$2a = -14$$
$$\frac{2a}{2} = \frac{-14}{2}$$
$$a = -7$$
The solution is -7.

89.
$$-5m = 20$$
$$\frac{-5m}{-5} = \frac{20}{-5}$$
$$m = -4$$
The solution is -4.

91.
$$-6n = -30$$
$$\frac{-6n}{-6} = \frac{-30}{-6}$$
$$n = 5$$
The solution is 5.

93.
$$18 = 2t$$
$$\frac{18}{2} = \frac{2t}{2}$$
$$9 = t$$
The solution is 9.

95.
$$-56 = 7x$$
$$\frac{-56}{7} = \frac{7x}{7}$$
$$-8 = x$$
The solution is -8.

97.
$$-5x = 0$$
$$\frac{-5x}{-5} = \frac{0}{-5}$$
$$x = 0$$
The solution is 0.

99.
$$35 = -5x$$
$$\frac{35}{-5} = \frac{-5x}{-5}$$
$$-7 = x$$
The solution is -7.

101.
$$-32 = -4y$$
$$\frac{-32}{-4} = \frac{-4y}{-4}$$
$$8 = y$$
The solution is 8.

103.
$$-12m = -144$$
$$\frac{-12m}{-12} = \frac{-144}{-12}$$
$$m = 12$$
The solution is 12.

105. $\dfrac{x}{4} = 3$

$4\left(\dfrac{1}{4}x\right) = 4(3)$

$x = 12$

The solution is 12.

107. $-\dfrac{b}{3} = 6$

$-3\left(-\dfrac{1}{3}b\right) = -3(6)$

$b = -18$

The solution is -18.

109. $\dfrac{t}{6} = -3$

$6\left(\dfrac{1}{6}t\right) = 6(-3)$

$t = -18$

The solution is -18.

111. $-\dfrac{4}{3}c = -8$

$-\dfrac{3}{4}\left(-\dfrac{4}{3}c\right) = -\dfrac{3}{4}(-8)$

$c = 6$

The solution is 6.

113. $-\dfrac{2}{3}d = 8$

$-\dfrac{3}{2}\left(-\dfrac{2}{3}d\right) = -\dfrac{3}{2}(8)$

$d = -12$

The solution is -12.

115. $\dfrac{2n}{3} = 2$

$\dfrac{3}{2}\left(\dfrac{2}{3}n\right) = \dfrac{3}{2}(2)$

$n = 3$

The solution is 3.

117. $\dfrac{-3z}{8} = 9$

$-\dfrac{8}{3}\left(-\dfrac{3}{8}z\right) = -\dfrac{8}{3}(9)$

$z = -24$

The solution is -24.

119. $-6 = -\dfrac{2}{3}y$

$-\dfrac{3}{2}(-6) = -\dfrac{3}{2}\left(-\dfrac{2}{3}y\right)$

$9 = y$

The solution is 9.

121. $\dfrac{2}{9} = \dfrac{2}{3}y$

$\dfrac{3}{2}\left(\dfrac{2}{9}\right) = \dfrac{3}{2}\left(\dfrac{2}{3}\right)y$

$\dfrac{1}{3} = y$

The solution is $\dfrac{1}{3}$.

123. $-\dfrac{2}{5}m = -\dfrac{6}{7}$

$-\dfrac{5}{2}\left(-\dfrac{2}{5}m\right) = -\dfrac{5}{2}\left(-\dfrac{6}{7}\right)$

$m = \dfrac{15}{7}$

The solution is $\dfrac{15}{7}$.

125. $3n + 2n = 20$

$5n = 20$

$\dfrac{5n}{5} = \dfrac{20}{5}$

$n = 4$

The solution is 4.

127. $10y - 3y = 21$

$7y = 21$

$\dfrac{7y}{7} = \dfrac{21}{7}$

$y = 3$

The solution is 3.

129. $\dfrac{x}{1.4} = 3.2$

$1.4\left(\dfrac{x}{1.4}\right) = 1.4(3.2)$

$x = 4.48$

The solution is 4.48.

131. $3.4a = 7.004$

$\dfrac{3.4a}{3.4} = \dfrac{7.004}{3.4}$

$a = 2.06$

The solution is 2.06.

133. $-3.7x = 7.77$

$\dfrac{-3.7x}{-3.7} = \dfrac{7.77}{-3.7}$

$x = -2.1$

The solution is -2.1.

Objective 3.1.4 Exercises

135. Strategy
To find the how far the train travels, solve the equation $d = rt$ find the distance it travels for the first 3 hours and adding the distance the train travels for the next two hours.

Solution
Using $d = rt$
For the first 3 hours,
$d = 45 \cdot 3 = 135$ mi
For the next two hours,
$d = 55 \cdot 2 = 110$ mi
$135 + 110 = 145$
The distance is 245 mi.

137. Strategy
To find the how fast the dietician travels, solve the equation $d = rt$ find the rate where the distance is 20 mi and the time is 40 min.

Solution
First convert 40 min to hours,
$40 \text{ min} = \dfrac{40}{60} = \dfrac{2}{3}$ hour
Using $d = rt$
$20 = r \cdot \dfrac{2}{3}$
$r = 20 \cdot \dfrac{3}{2} = 30$ mph
The average rate of speed is 30 mph.

139. Strategy
To find the how long to complete the trip, solve the equation $d = rt$ find the time, where the distance is 36 mi and the rate is the 12 mph. Add one hour to the time for the lunch break.

Solution
Using $d = rt$
$36 = 12t$
$t = \dfrac{36}{12} = 3$ h
$3 + 1 = 4$
The trip will take 4 h to complete.

141. Strategy
To find the how long to walk on the moving sidewalk, solve the equation $d = rt$ find the time, where the distance is 250 ft and the rate is the sum of your walking rate (5ft/s) and the sidewalk rate (3ft/s).

Solution
$5 + 3 = 8$ ft/s
Using $d = rt$
$250 = 8t$
$t = \dfrac{250}{8} = 31.25$ s
It will take 31.25 s to walk on the moving sidewalk.

143. Strategy
To find when the joggers will meet, solve the equation $d = rt$ find the time, where the distance is 8 mi and the rate is the sum of the joggers' rates (5 mph and 7 mph).

Solution
$5 + 7 = 12$ mph
Using $d = rt$
$8 = 12t$
$t = \dfrac{8}{12} = \dfrac{2}{3}$ h
Convert hours to minutes,
$\dfrac{2}{3} \text{ h} = \dfrac{2}{3} \cdot 60 = 40$ min
They will meet in 40 minutes.

145. Strategy
Find how long the trip lasts against the current by solving the equation $d = rt$, where the rate is the difference in Petra's canoe speed (10 mph) and the current (2 mph) and the distance is 4 mi.

Solution
$10 - 2 = 8$ mph
Using $d = rt$
$4 = 8t$
$t = \dfrac{4}{8} = 0.5$h
It will take her 0.5 h.

Objective 3.1.5 Exercises

149. $P(50) = 12$
$\dfrac{50P}{50} = \dfrac{12}{50}$
$P = 0.24$
The percent is 24%.

151. $0.18(40) = A$
$7.20 = A$
18% of 40 is 7.2.

153. $(0.12)B = 48$
$\dfrac{0.12B}{0.12} = \dfrac{48}{0.12}$
$B = 400$
The number is 400.

155. $\dfrac{1}{3}(27) = A \qquad \left(33\dfrac{1}{3}\% = \dfrac{1}{3}\right)$
$9 = A$
$33\dfrac{1}{3}\%$ of 27 is 9.

157. $P(12) = 3$
$\dfrac{12P}{12} = \dfrac{3}{12}$
$P = 0.25$
The percent is 25%.

159. $(0.60)B = 3$

$\dfrac{0.60B}{0.60} = \dfrac{3}{0.60}$

$B = 5$

The number is 5.

161. $P(6) = 12$

$\dfrac{6P}{6} = \dfrac{12}{6}$

$P = 2$

12 is 200% of 6.

163. $(0.0525)B = 21$

$\dfrac{0.0525B}{0.0525} = \dfrac{21}{0.0525}$

$B = 400$

The number is 400.

165. $(0.154)(50) = A$

$7.7 = A$

15.4% of 50 is 7.7.

167. $(0.005)B = 1$

$\dfrac{0.005B}{0.005} = \dfrac{1}{0.005}$

$B = 200$

The number is 200.

169. $(0.0075)B = 3$

$\dfrac{0.0075B}{0.0075} = \dfrac{3}{0.0075}$

$B = 400$

The number is 400.

171. $(1.25)(16) = A$

$20 = A$

125% of 16 is 20.

173. $P(20.4) = 16.4$

$\dfrac{20.4P}{20.4} = \dfrac{16.4}{20.4}$

$P \approx 0.8039215$

16.4 is 80% of 20.4.

175. equal to since $40\% \cdot 80 = 80\% \cdot 40$

177. Strategy
To find the number of seats, solve the basic percent equation using $P = 1.11\% = 0.0111$ and $B = 22,500$. The amount is unknown.

Solution

$PB = A$

$0.0111(22,500) = A$

$249.75 = A$

250 seats are reserved for wheelchair accessibility.

179. Strategy
To find the total water used, solve the basic percent equation using $P = 17.8\% = 0.178$ and $A = 13.2$. The base is unknown.

Solution

$PB = A$

$0.178B = 13.2$

$\dfrac{0.178B}{0.178} = \dfrac{13.2}{0.178}$

$B \approx 74$

The total water used 74 gallons a day.

181. Strategy
To find the percent, solve the basic percent equation using $B = 1442$ and $A = 735$. The percent is the unknown.

Solution

$PB = A$

$P(1442) = 735$

$\dfrac{1142P}{1442} = \dfrac{735}{1442}$

$P \approx 0.51$

51% of the vacation bill is paid in cash.

183. There is insufficient information to solve this problem. Either the base or the amount must be given.

185. Strategy
To find the percent, solve the basic percent equation using $B = 30 + 47 + 200 + 1950 = 2227$ and $A = 30 + 47 + 200 = 277$. The percent is unknown.

Solution

$PB = A$

$P(2227) = 277$

$\dfrac{2227P}{2227} = \dfrac{277}{2227}$

$P \approx 0.12$

12% of the accidental deaths were not attributed to motor vehicle accidents.

187. Strategy
To find the total electricity used, solve the basic percent equation using $P = 33\% = 0.33$ and $A = 31.7$. The base is unknown.

Solution

$PB = A$

$0.33B = 31.7$

$\dfrac{0.33B}{0.33} = \dfrac{31.7}{0.33}$

$B \approx 96.1$

The total electricity used for home lighting is 96.1 billion kilowatt hours a year.

189a. Strategy
  To find the percent,
  First find the base, B, by determining the number of cans sold to cover the advertising cost by dividing $2.2 million by $0.08.
  Then solve the basic percent equation using $A = 125$ million. The percent is unknown.

  Solution
  $$B = \frac{2200000}{0.08} = 27,500,000$$
  $PB = A$
  $P(125) = 27.5$ in millions
  $$\frac{125P}{125} = \frac{27.5}{125}$$
  $P = 0.22$
  22% of the people would have to buy one can.

189b. Strategy
  Using the information from part a, divide the solution by 6, since the people are buying 6 cans rather than just one.

  Solution
  $$\frac{22}{6} \approx 3.7\%$$
  3.7% of the people would have to buy a six pack.

191. Strategy
  To find the price, solve the basic percent equation using $P = 111\% = 1.11$ and $A = 1579.99$. The base is the unknown.

  Solution
  $PB = A$
  $1.11B = 1579.99$
  $$\frac{1.11B}{1.11} = \frac{1579.99}{1.11}$$
  $B \approx 1423.41$
  The less expensive model costs $1423.41.

193. Strategy
  To find the simple interest rate, solve the simple interest equation using $I = \$72$, $P = \$1200$, and
  $t = 8$ months $= \dfrac{8}{12}$ years, for r.

  Solution
  $I = Prt$
  $$72 = (1200)r\left(\frac{8}{12}\right)$$
  $72 = 800r$
  $$\frac{72}{800} = \frac{800r}{800}$$
  $0.09 = r$
  The annual simple interest rate is 9%.

195. Strategy
  To find the interest, solve the simple interest equation for each account:
  First, using $P = \$1000$, $r = 7.5\% = 0.075$, and $t = 1$ year, for I.
  Second, using $P = 3000 - 1000 = \$2000$, $r = 8.25\% = 0.0825$, and $t = 1$ year, for I.
  Finally, find the total interest by adding the interest earned in each account.

  Solution
  $I = Prt$
  $I = (1000)(0.075)(1)$
  $I = 75$

  $I = Prt$
  $I = (2000)(0.0825)(1)$
  $I = 165$
  $75 + 165 = \$240$
  Sal earned $240 after one year.

197. Strategy
  To determine how much Makana earned,
  First find the simple interest rate Marlys earned by solving the simple interest equation using $I = \$51$, $P = \$850$, and $t = 1$ year.
  Second, add $1\% = 0.01$ to the interest rate Marlys earned.
  Finally, to find the interest Makana earned, solve the simple interest equation using P = $900, and $t = 1$ year, for I.

  Solution
  $I = Prt$
  $51 = (850)r(1)$
  $51 = 850r$
  $$\frac{51}{850} = \frac{850r}{850}$$
  $0.06 = r$
  $0.06 + 0.01 = 0.07$
  $I = Prt$
  $I = (900)(0.07)(1)$
  $I = 63$
  Makana earned $63.

199. The principal for each investment is the same amount. The time the interest accrued is the same for each account. If one account earns 6% and the other earns 9%, the combined interest earned is between 6% and 9%.
To find simple interest rate on the combined accounts, solve the simple interest equation for each account:
First, using $P = \$1000$, $r = 9\% = 0.09$, and $t = 1$ year, for I.
Second, using $P = \$1000$, $r = 6\% = 0.06$, and $t = 1$ year, for I.
Finally, to find the combined interest rate, add the value of $P = 1000 + 1000 = \$2000$, and total the interest earned in both accounts, using the simple interest equation to find r.

Solution
$I = Prt$
$I = (1000)(0.09)(1)$
$I = 90$
$I = Prt$
$I = (1000)(0.06)(1)$
$I = 60$

$90 + 60 = (2000)r(1)$
$\dfrac{150}{2000} = \dfrac{2000r}{2000}$
$0.075 = r$

The interest rate earned on the combined accounts is between 6% and 9%.

201. Strategy
To find the percent, solve the basic percent equation using $B = 250$ and $A = 5$. The percent is the unknown.

Solution
$PB = A$
$P(250) = 5$
$\dfrac{250P}{250} = \dfrac{5}{250}$
$P = 0.02$

There is a 2% concentration of hydrogen peroxide.

203. Strategy
To find which brand has the greater concentration, solve the basic percent equation for Apple Dan's using $B = 32$ and $A = 8$. The percent is the unknown.
Then solve the basic percent equation for the generic brand using $B = 40$ and $A = 9$. The percent is the unknown. Compare the percent of concentration.

Solution
$PB = A$
$P(32) = 8$
$\dfrac{32P}{32} = \dfrac{8}{32}$
$P = 0.25$ Apple Dan's

$PB = A$
$P(40) = 9$
$\dfrac{40P}{40} = \dfrac{9}{40}$
$P = 0.225$ generic

$25\% > 22.5\%$
Apple Dan's has the greater concentration.

205. Strategy
To find the amount that is not glycerin, solve the basic percent equation, find the percent that is not glycerin using $P = 100 - 75\% = 0.25$ and $B = 50$ g. The amount is unknown.

Solution
$PB = A$
$0.25(50) = A$
$12.5 = A$

There is 12.5 g of cream that is not glycerin.

207. Strategy
To find the percent, solve the basic percent equation using $B = 500 - 100 = 400$ and $A = 50$. The percent is unknown.

Solution
$PB = A$
$P(400) = 50$
$\dfrac{400P}{400} = \dfrac{50}{400}$
$P = 0.125$

The percent concentration is 12.5%.

Applying Concepts 3.1

209. $\dfrac{3y - 8y}{7} = 15$

$-\dfrac{5y}{7} = 15$

$-\dfrac{7}{5}\left(-\dfrac{5}{7}y\right) = -\dfrac{7}{5}(15)$

$y = -21$

The solution is -21.

211.
$$\frac{1}{\frac{1}{x}} + 8 = -19$$
$$1 \div \frac{1}{x} + 8 = -19$$
$$1 \cdot \frac{x}{1} + 8 = -19$$
$$x + 8 = -19$$
$$x + 8 - 8 = -19 - 8$$
$$x = -27$$
The solution is -27.

213.
$$\frac{5}{\frac{7}{a}} - \frac{3}{\frac{7}{a}} = 6$$
$$5 \div \frac{7}{a} - 3 \div \frac{7}{a} = 6$$
$$5 \cdot \frac{a}{7} - 3 \cdot \frac{a}{7} = 6$$
$$\frac{5}{7}a - \frac{3}{7}a = 6$$
$$\frac{2}{7}a = 6$$
$$\frac{7}{2}\left(\frac{2}{7}a\right) = \frac{7}{2}(6)$$
$$a = 21$$
The solution is 21.

215. The sum of the angles x and $3x$ is $88°$.
$$x + 3x = 88$$
$$4x = 88$$
$$x = 22$$
The angle x has a measure of $22°$.

217. Strategy
To find the cost of the dinner, solve the basic percent equation using
$P = (100\% + 15\%)(100\% + 6\%) = (115\%)(106\%) = (1.15)(1.06)$ and $A = 97.52$.
The base is unknown.

Solution
$$PB = A$$
$$(1.15)(1.06)B = 97.52$$
$$1.219B = 97.52$$
$$\frac{1.219B}{1.219} = \frac{97.52}{1.219}$$
$$B = 80$$
The cost of the dinner was $80.

219. If a quantity increases by 100%, then the same quantity is added to itself. The new value will be twice the original value.

221. One equation with a solution of -2 is $5x = -10$.

Section 3.2

Concept Review 3.2

1. Always true

3. Never true
First use the Addition Property of Equations to remove the constant term from the left side of the equation. Then use the Multiplication Property of Equations to multiply both sides of the equation by $\frac{1}{a}$.

5. Never true
Division by zero is undefined.

7. Sometimes true
If $a = c$, the equation will have no variables when cx is subtracted from both sides.

Objective 3.2.1 Exercises

1.
$$3x + 1 = 10$$
$$3x + 1 - 1 = 10 - 1$$
$$3x = 9$$
$$\frac{3x}{3} = \frac{9}{3}$$
$$x = 3$$
The solution is 3.

3.
$$2a - 5 = 7$$
$$2a - 5 + 5 = 7 + 5$$
$$2a = 12$$
$$\frac{2a}{2} = \frac{12}{2}$$
$$a = 6$$
The solution is 6.

5.
$$5 = 4x + 9$$
$$5 - 9 = 4x + 9 - 9$$
$$-4 = 4x$$
$$\frac{-4}{4} = \frac{4x}{4}$$
$$-1 = x$$
The solution is -1.

7.
$$13 = 9 + 4z$$
$$13 - 9 = 9 - 9 + 4z$$
$$4 = 4z$$
$$\frac{4}{4} = \frac{4z}{4}$$
$$1 = z$$
The solution is 1.

9.
$$2 - x = 11$$
$$2 - 2 - x = 11 - 2$$
$$-x = 9$$
$$-1(-x) = -1 \cdot 9$$
$$x = -9$$
The solution is -9.

11.
$$5 - 6x = -13$$
$$5 - 5 - 6x = -13 - 5$$
$$-6x = -18$$
$$\frac{-6x}{-6} = \frac{-18}{-6}$$
$$x = 3$$
The solution is 3.

13.
$$-5d + 3 = -12$$
$$-5d + 3 - 3 = -12 - 3$$
$$-5d = -15$$
$$\frac{-5d}{-5} = \frac{-15}{-5}$$
$$d = 3$$
The solution is 3.

15.
$$-7n - 4 = -25$$
$$-7n - 4 + 4 = -25 + 4$$
$$-7n = -21$$
$$\frac{-7n}{-7} = \frac{-21}{-7}$$
$$n = 3$$
The solution is 3.

17.
$$-13 = -11y + 9$$
$$-13 - 9 = -11y + 9 - 9$$
$$-22 = -11y$$
$$\frac{-22}{-11} = \frac{-11y}{-11}$$
$$2 = y$$
The solution is 2.

19.
$$3 = 11 - 4n$$
$$3 - 11 = 11 - 11 - 4n$$
$$-8 = -4n$$
$$\frac{-8}{-4} = \frac{-4n}{-4}$$
$$2 = n$$
The solution is 2.

21.
$$-8x + 3 = -29$$
$$-8x + 3 - 3 = -29 - 3$$
$$-8x = -32$$
$$\frac{-8x}{-8} = \frac{-32}{-8}$$
$$x = 4$$
The solution is 4.

23.
$$7x - 3 = 3$$
$$7x - 3 + 3 = 3 + 3$$
$$7x = 6$$
$$\frac{7x}{7} = \frac{6}{7}$$
$$x = \frac{6}{7}$$
The solution is $\frac{6}{7}$.

25.
$$6a + 5 = 9$$
$$6a + 5 - 5 = 9 - 5$$
$$6a = 4$$
$$\frac{6a}{6} = \frac{4}{6}$$
$$a = \frac{4}{6} = \frac{2}{3}$$
The solution is $\frac{2}{3}$.

27.
$$11 = 15 + 4n$$
$$11 - 15 = 15 - 15 + 4n$$
$$-4 = 4n$$
$$\frac{-4}{4} = \frac{4n}{4}$$
$$-1 = n$$
The solution is −1.

29.
$$9 - 4x = 6$$
$$9 - 9 - 4x = 6 - 9$$
$$-4x = -3$$
$$\frac{-4x}{-4} = \frac{-3}{-4}$$
$$x = \frac{3}{4}$$
The solution is $\frac{3}{4}$.

31.
$$1 - 3x = 0$$
$$1 - 1 - 3x = 0 - 1$$
$$-3x = -1$$
$$\frac{-3x}{-3} = \frac{-1}{-3}$$
$$x = \frac{1}{3}$$
The solution is $\frac{1}{3}$.

33.
$$8b - 3 = -9$$
$$8b - 3 + 3 = -9 + 3$$
$$8b = -6$$
$$\frac{8b}{8} = \frac{-6}{8}$$
$$b = -\frac{6}{8} = -\frac{3}{4}$$
The solution is $-\frac{3}{4}$.

35.
$$7 - 9a = 4$$
$$7 - 7 - 9a = 4 - 7$$
$$-9a = -3$$
$$\frac{-9a}{-9} = \frac{-3}{-9}$$
$$a = \frac{3}{9} = \frac{1}{3}$$
The solution is $\frac{1}{3}$.

37.
$$10 = -18x + 7$$
$$10 - 7 = -18x + 7 - 7$$
$$3 = -18x$$
$$\frac{3}{-18} = \frac{-18x}{-18}$$
$$-\frac{3}{18} = x$$
$$-\frac{1}{6} = x$$
The solution is $-\frac{1}{6}$.

39.
$$3x - \frac{5}{6} = \frac{13}{6}$$
$$3x - \frac{5}{6} + \frac{5}{6} = \frac{13}{6} + \frac{5}{6}$$
$$3x = \frac{18}{6}$$
$$3x = 3$$
$$\frac{3x}{3} = \frac{3}{3}$$
$$x = 1$$
The solution is 1.

41.
$$9x + \frac{4}{5} = \frac{4}{5}$$
$$9x + \frac{4}{5} - \frac{4}{5} = \frac{4}{5} - \frac{4}{5}$$
$$9x = 0$$
$$\frac{1}{9}(9x) = \frac{1}{9}(0)$$
$$x = 0$$
The solution is 0.

43.
$$7 = 9 - 5a$$
$$7 - 9 = 9 - 9 - 5a$$
$$-2 = -5a$$
$$\frac{-2}{-5} = \frac{-5a}{-5}$$
$$\frac{2}{5} = a$$

The solution is $\frac{2}{5}$.

45.
$$-4x + 3 = 9$$
$$-4x + 3 - 3 = 9 - 3$$
$$-4x = 6$$
$$\frac{-4x}{-4} = \frac{6}{-4}$$
$$x = -\frac{6}{4} = -\frac{3}{2}$$

The solution is $-\frac{3}{2}$.

47.
$$\frac{1}{3}m - 1 = 5$$
$$\frac{1}{3}m - 1 + 1 = 5 + 1$$
$$\frac{1}{3}m = 6$$
$$3\left(\frac{1}{3}m\right) = 3 \cdot 6$$
$$m = 18$$
The solution is 18.

49.
$$\frac{3}{4}n + 7 = 13$$
$$\frac{3}{4}n + 7 - 7 = 13 - 7$$
$$\frac{3}{4}n = 6$$
$$\frac{4}{3}\left(\frac{3}{4}n\right) = \frac{4}{3}(6)$$
$$n = 8$$
The solution is 8.

51.
$$-\frac{3}{8}b + 4 = 10$$
$$-\frac{3}{8}b + 4 - 4 = 10 - 4$$
$$-\frac{3}{8}b = 6$$
$$-\frac{8}{3}\left(-\frac{3}{8}b\right) = -\frac{8}{3}(6)$$
$$b = -16$$
The solution is −16.

53.
$$\frac{y}{5} - 2 = 3$$
$$\frac{y}{5} - 2 + 2 = 3 + 2$$
$$\frac{y}{5} = 5$$
$$5\left(\frac{1}{5}y\right) = 5 \cdot 5$$
$$y = 25$$
The solution is 25.

55.
$$\frac{3c}{7} - 1 = 8$$
$$\frac{3c}{7} - 1 + 1 = 8 + 1$$
$$\frac{3c}{7} = 9$$
$$\frac{7}{3}\left(\frac{3}{7}c\right) = \frac{7}{3}(9)$$
$$c = 21$$
The solution is 21.

57.
$$3 - \frac{4}{5}w = -9$$
$$3 - 3 - \frac{4}{5}w = -9 - 3$$
$$-\frac{4}{5}w = -12$$
$$-\frac{5}{4}\left(-\frac{4}{5}w\right) = -\frac{5}{4}(-12)$$
$$w = 15$$
The solution is 15.

59.
$$17+\frac{5}{8}x=7$$
$$17-17+\frac{5}{8}x=7-17$$
$$\frac{5}{8}x=-10$$
$$\frac{8}{5}\left(\frac{5}{8}x\right)=\frac{8}{5}(-10)$$
$$x=-16$$
The solution is -16.

61.
$$\frac{2}{3}=y-\frac{1}{2}$$
$$\frac{2}{3}+\frac{1}{2}=y-\frac{1}{2}+\frac{1}{2}$$
$$\frac{4}{6}+\frac{3}{6}=y$$
$$\frac{7}{6}=y$$

The solution is $\frac{7}{6}$.

63.
$$\frac{3}{8}=\frac{5}{12}-\frac{1}{3}b$$
$$\frac{3}{8}-\frac{5}{12}=\frac{5}{12}-\frac{5}{12}-\frac{1}{3}b$$
$$\frac{9}{24}-\frac{10}{24}=-\frac{1}{3}b$$
$$-\frac{1}{24}=-\frac{1}{3}b$$
$$-3\left(-\frac{1}{24}\right)=-3\left(-\frac{1}{3}b\right)$$
$$\frac{1}{8}=b$$

The solution is $\frac{1}{8}$.

65.
$$7=\frac{2x}{5}+4$$
$$7-4=\frac{2x}{5}+4-4$$
$$3=\frac{2x}{5}$$
$$\frac{5}{2}(3)=\frac{5}{2}\left(\frac{2}{5}x\right)$$
$$\frac{15}{2}=x$$

The solution is $\frac{15}{2}$.

67.
$$7-\frac{5}{9}y=9$$
$$7-7-\frac{5}{9}y=9-7$$
$$-\frac{5}{9}y=2$$
$$-\frac{9}{5}\left(-\frac{5}{9}y\right)=-\frac{9}{5}(2)$$
$$y=-\frac{18}{5}$$
The solution is $-\frac{18}{5}$.

69.
$$5y+9+2y=23$$
$$7y+9=23$$
$$7y+9-9=23-9$$
$$7y=14$$
$$\frac{7y}{7}=\frac{14}{7}$$
$$y=2$$
The solution is 2.

71.
$$11z-3-7z=9$$
$$4z-3=9$$
$$4z-3+3=9+3$$
$$4z=12$$
$$\frac{4z}{4}=\frac{12}{4}$$
$$z=3$$
The solution is 3.

73.
$$b-8b+1=-6$$
$$-7b+1=-6$$
$$7b+1-1=-6-1$$
$$-7b=-7$$
$$\frac{-7b}{-7}=\frac{-7}{-7}$$
$$b=1$$
The solution is 1.

75.
$$8=4n-6+3n$$
$$8=7n-6$$
$$8+6=7n-6+6$$
$$14=7n$$
$$\frac{14}{7}=\frac{7n}{7}$$
$$2=n$$
The solution is 2.

77.
$$1.2x-3.44=1.3$$
$$1.2x-3.44+3.44=1.3+3.44$$
$$1.2x=4.74$$
$$\frac{1.2x}{1.2}=\frac{4.74}{1.2}$$
$$x-3.95$$
The solution is 3.95.

79.
$$-6.5 = 4.3y - 3.06$$
$$-6.5 + 3.06 = 4.3y - 3.06 + 3.06$$
$$-3.44 = 4.3y$$
$$\frac{-3.44}{4.3} = \frac{4.3y}{4.3}$$
$$-0.8 = y$$
The solution is -0.8.

81.
$$3x + 5 = -4 \qquad 2x - 5 = 2(-3) - 5$$
$$3x + 5 - 5 = -4 - 5 \qquad\qquad = -6 - 5$$
$$3x = -9 \qquad\qquad\qquad = -11$$
$$\frac{3x}{3} = \frac{-9}{3}$$
$$x = -3$$
The answer is -11.

83.
$$2 - 3x = 11 \qquad x^2 + 2x - 3 = (-3)^2 + 2(-3) - 3$$
$$2 - 2 - 3x = 11 - 2 \qquad\qquad = 9 - 6 - 3$$
$$-3x = 9 \qquad\qquad\qquad = 3 - 3$$
$$\frac{-3x}{-3} = \frac{9}{-3} \qquad\qquad\qquad = 0$$
$$x = -3$$
The answer is 0.

85.
$$6y + 2 = y + 17$$
$$6y - y + 2 = y - y + 17$$
$$5y + 2 = 17$$
$$5y + 2 - 2 = 17 - 2$$
$$5y = 15$$
$$\frac{5y}{5} = \frac{15}{5}$$
$$y = 3$$
The solution is 3.

87.
$$11n + 3 = 10n + 11$$
$$11n - 10n + 3 = 10n - 10n + 11$$
$$n + 3 = 11$$
$$n + 3 - 3 = 11 - 3$$
$$n = 8$$
The solution is 8.

89.
$$9a - 10 = 3a + 2$$
$$9a - 3a - 10 = 3a - 3a + 2$$
$$6a - 10 = 2$$
$$6a - 10 + 10 = 2 + 10$$
$$6a = 12$$
$$\frac{6a}{6} = \frac{12}{6}$$
$$a = 2$$
The solution is 2.

91.
$$13b - 1 = 4b - 19$$
$$13b - 4b - 1 = 4b - 4b - 19$$
$$9b - 1 = -19$$
$$9b - 1 + 1 = -19 + 1$$
$$9b = -18$$
$$\frac{9b}{9} = \frac{-18}{9}$$
$$b = -2$$
The solution is -2.

93.
$$7a - 5 = 2a - 20$$
$$7a - 2a - 5 = 2a - 2a - 20$$
$$5a - 5 = -20$$
$$5a - 5 + 5 = -20 + 5$$
$$5a = -15$$
$$\frac{5a}{5} = \frac{-15}{5}$$
$$a = -3$$
The solution is -3.

95.
$$n - 2 = 6 - 3n$$
$$n + 3n - 2 = 6 - 3n + 3n$$
$$4n - 2 = 6$$
$$4n - 2 + 2 = 6 + 2$$
$$4n = 8$$
$$\frac{4n}{4} = \frac{8}{4}$$
$$n = 2$$
The solution is 2.

97.
$$4y - 2 = -16 - 3y$$
$$4y + 3y - 2 = -16 - 3y + 3y$$
$$7y - 2 = -16$$
$$7y - 2 + 2 = -16 + 2$$
$$7y = -14$$
$$\frac{7y}{7} = \frac{-14}{7}$$
$$y = -2$$
The solution is -2.

99.
$$m + 4 = 3m + 8$$
$$m - 3m + 4 = 3m - 3m + 8$$
$$-2m + 4 = 8$$
$$-2m + 4 - 4 = 8 - 4$$
$$-2m = 4$$
$$\frac{-2m}{-2} = \frac{4}{-2}$$
$$m = -2$$
The solution is -2.

101.
$$6d - 2 = 7d + 5$$
$$6d - 7d - 2 = 7d - 7d + 5$$
$$-d - 2 = 5$$
$$-d - 2 + 2 = 5 + 2$$
$$-d = 7$$
$$(-1)(-d) = (-1)(7)$$
$$d = -7$$
The solution is -7.

103.
$$5a + 7 = 2a + 7$$
$$5a - 2a + 7 = 2a - 2a + 7$$
$$3a + 7 = 7$$
$$3a + 7 - 7 = 7 - 7$$
$$3a = 0$$
$$\frac{3a}{3} = \frac{0}{3}$$
$$a = 0$$
The solution is 0.

105.
$$10 - 4n = 16 - n$$
$$10 - 4n + n = 16 - n + n$$
$$10 - 3n = 16$$
$$10 - 10 - 3n = 16 - 10$$
$$-3n = 6$$
$$\frac{-3n}{-3} = \frac{6}{-3}$$
$$n = -2$$
The solution is −2.

107.
$$2b - 10 = 7b$$
$$2b - 2b - 10 = 7b - 2b$$
$$-10 = 5b$$
$$\frac{-10}{5} = \frac{5b}{5}$$
$$-2 = b$$
The solution is −2.

109.
$$9y = 5y + 16$$
$$9y - 5y = 5y - 5y + 16$$
$$4y = 16$$
$$\frac{4y}{4} = \frac{16}{4}$$
$$y = 4$$
The solution is 4.

111.
$$8 - 4x = 18 - 5x$$
$$8 - 4x + 5x = 18 - 5x + 5x$$
$$8 + x = 18$$
$$8 - 8 + x = 18 - 8$$
$$x = 10$$
The solution is 10.

113.
$$5 - 7m = 2 - 6m$$
$$5 - 7m + 6m = 2 - 6m + 6m$$
$$5 - m = 2$$
$$5 - 5 - m = 2 - 5$$
$$-m = -3$$
$$-1(-3) = -1(-3)$$
$$m = 3$$
The solution is 3.

115.
$$6y - 1 = 2y + 2$$
$$6y - 2y - 1 = 2y - 2y + 2$$
$$4y - 1 = 2$$
$$4y - 1 + 1 = 2 + 1$$
$$4y = 3$$
$$\frac{4y}{4} = \frac{3}{4}$$
$$y = \frac{3}{4}$$
The solution is $\frac{3}{4}$.

117.
$$10x - 3 = 3x - 1$$
$$10x - 3x - 3 = 3x - 3x - 1$$
$$7x - 3 = -1$$
$$7x - 3 + 3 = -1 + 3$$
$$7x = 2$$
$$\frac{7x}{7} = \frac{2}{7}$$
$$x = \frac{2}{7}$$
The solution is $\frac{2}{7}$.

119.
$$8a - 2 = 4a - 5$$
$$8a - 4a - 2 = 4a - 4a - 5$$
$$4a - 2 = -5$$
$$4a - 2 + 2 = -5 + 2$$
$$4a = -3$$
$$\frac{4a}{4} = \frac{-3}{4}$$
$$a = -\frac{3}{4}$$
The solution is $-\frac{3}{4}$.

121.
$$\frac{3}{4}x = \frac{1}{12}x + 2$$
$$\frac{3}{4}x - \frac{1}{12}x = \frac{1}{12}x - \frac{1}{12}x + 2$$
$$\frac{9}{12}x - \frac{1}{12}x = 2$$
$$\frac{8}{12}x = 2$$
$$\frac{12}{8}\left(\frac{8}{12}x\right) = \frac{12}{8}(2)$$
$$x = 3$$
The solution is 3.

123.
$$\frac{4}{5}c - 7 = \frac{1}{10}c$$
$$\frac{4}{5}c - \frac{4}{5}c - 7 = \frac{1}{10}c - \frac{4}{5}c$$
$$-7 = \frac{1}{10}c - \frac{8}{10}c$$
$$-7 = -\frac{7}{10}c$$
$$-\frac{10}{7}(-7) = -\frac{10}{7}\left(-\frac{7}{10}c\right)$$
$$10 = c$$
The solution is 10.

125.
$$\frac{2}{3}b = 2 - \frac{5}{6}b$$
$$\frac{2}{3}b + \frac{5}{6}b = 2 - \frac{5}{6}b + \frac{5}{6}b$$
$$\frac{4}{6}b + \frac{5}{6}b = 2$$
$$\frac{9}{6}b = 2$$
$$\frac{6}{9}\left(\frac{9}{6}b\right) = \frac{6}{9}(2)$$
$$b = \frac{4}{3}$$
The solution is $\frac{4}{3}$.

127.
$$4.5x - 5.4 = 2.7x$$
$$4.5x - 4.5x - 5.4 = 2.7x - 4.5x$$
$$-5.4 = -1.8x$$
$$\frac{-5.4}{-1.8} = \frac{-1.8x}{-1.8}$$
$$3 = x$$
The solution is 3.

129.
$$5x = 3x - 8$$
$$5x - 3x = 3x - 3x - 8$$
$$2x = -8$$
$$\frac{2x}{2} = \frac{-8}{2}$$
$$x = -4$$
The answer is –14.

$$4x + 2 = 4(-4) + 2$$
$$= -16 + 2$$
$$= -14$$

131.
$$2 - 6a = 5 - 3a$$
$$2 - 6a + 3a = 5 - 3a + 3a$$
$$2 - 3a = 5$$
$$2 - 2 - 3a = 5 - 2$$
$$-3a = 3$$
$$\frac{-3a}{-3} = \frac{3}{-3}$$
$$a = -1$$
The answer is 7.

$$4a^2 - 2a + 1$$
$$= 4(-1)^2 - 2(-1) + 1$$
$$= 4(1) - 2(-1) + 1$$
$$= 4 + 2 + 1$$
$$= 6 + 1$$
$$= 7$$

133.
$$5x + 2(x + 1) = 23$$
$$5x + 2x + 2 = 23$$
$$7x + 2 = 23$$
$$7x + 2 - 2 = 23 - 2$$
$$7x = 21$$
$$\frac{7x}{7} = \frac{21}{7}$$
$$x = 3$$
The solution is 3.

135.
$$9n - 3(2n - 1) = 15$$
$$9n - 6n + 3 = 15$$
$$3n + 3 = 15$$
$$3n + 3 - 3 = 15 - 3$$
$$3n = 12$$
$$\frac{3n}{3} = \frac{12}{3}$$
$$n = 4$$
The solution is 4.

137.
$$7a - (3a - 4) = 12$$
$$7a - 3a + 4 = 12$$
$$4a + 4 = 12$$
$$4a + 4 - 4 = 12 - 4$$
$$4a = 8$$
$$\frac{4a}{4} = \frac{8}{4}$$
$$a = 2$$
The solution is 2.

139.
$$5(3 - 2y) + 4y = 3$$
$$15 - 10y + 4y = 3$$
$$15 - 6y = 3$$
$$15 - 15 - 6y = 3 - 15$$
$$-6y = -12$$
$$\frac{-6y}{-6} = \frac{-12}{-6}$$
$$y = 2$$
The solution is 2.

141.
$$10x + 1 = 2(3x + 5) - 1$$
$$10x + 1 = 6x + 10 - 1$$
$$10x + 1 = 6x + 9$$
$$10x - 6x + 1 = 6x - 6x + 9$$
$$4x + 1 = 9$$
$$4x + 1 - 1 = 9 - 1$$
$$4x = 8$$
$$\frac{4x}{4} = \frac{8}{4}$$
$$x = 2$$
The solution is 2.

143.
$$4 - 3a = 7 - 2(2a + 5)$$
$$4 - 3a = 7 - 4a - 10$$
$$4 - 3a = -3 - 4a$$
$$4 - 3a + 4a = -3 - 4a + 4a$$
$$4 + a = -3$$
$$4 - 4 + a = -3 - 4$$
$$a = -7$$
The solution is –7.

145.
$$3y - 7 = 5(2y - 3) + 4$$
$$3y - 7 = 10y - 15 + 4$$
$$3y - 7 = 10y - 11$$
$$3y - 10y - 7 = 10y - 10y - 11$$
$$-7y - 7 = -11$$
$$-7y - 7 + 7 = -11 + 7$$
$$-7y = -4$$
$$\frac{-7y}{-7} = \frac{-4}{-7}$$
$$y = \frac{4}{7}$$
The solution is $\frac{4}{7}$.

147.
$$5-(9-6x)=2x-2$$
$$5-9+6x=2x-2$$
$$-4+6x=2x-2$$
$$-4+6x-2x=2x-2x-2$$
$$-4+4x=-2$$
$$-4+4+4x=-2+4$$
$$4x=2$$
$$\frac{4x}{4}=\frac{2}{4}$$
$$x=\frac{1}{2}$$

The solution is $\frac{1}{2}$.

149.
$$3[2-4(y-1)]=3(2y+8)$$
$$3[2-4y+4]=6y+24$$
$$3[6-4y]=6y+24$$
$$18-12y=6y+24$$
$$18-12y-6y=6y-6y+24$$
$$18-18y=24$$
$$18-18-18y=24-18$$
$$-18y=6$$
$$\frac{-18y}{-18}=\frac{6}{-18}$$
$$y=-\frac{1}{3}$$

The solution is $-\frac{1}{3}$.

151.
$$3a+2[2+3(a-1)]=2(3a+4)$$
$$3a+2[2+3a-3]=6a+8$$
$$3a+2[-1+3a]=6a+8$$
$$3a-2+6a=6a+8$$
$$9a-2=6a+8$$
$$9a-6a-2=6a-6a+8$$
$$3a-2=8$$
$$3a-2+2=8+2$$
$$3a=10$$
$$\frac{3a}{3}=\frac{10}{3}$$
$$a=\frac{10}{3}$$

The solution is $\frac{10}{3}$.

153.
$$-2[4-(3b+2)]=5-2(3b+6)$$
$$-2[4-3b-2]=5-6b-12$$
$$-2[2-3b]=-7-6b$$
$$-4+6b=-7-6b$$
$$-4+6b+6b=-7-6b+6b$$
$$-4+12b=-7$$
$$-4+4+12b=-7+4$$
$$12b=-3$$
$$\frac{12b}{12}=\frac{-3}{12}$$
$$b=-\frac{1}{4}$$

The solution is $-\frac{1}{4}$.

155.
$$0.3x-2(1.6x)-8=3(1.9x-4.1)$$
$$0.3x-3.2x-8=5.7x-12.3$$
$$-2.9x-8=5.7x-12.3$$
$$-2.9x-5.7x-8=5.7x-5.7x-12.3$$
$$-8.6x-8=-12.3$$
$$-8.6x-8+8=-12.3+8$$
$$-8.6x=-4.3$$
$$\frac{-8.6x}{-8.6}=\frac{-4.3}{-8.6}$$
$$x=0.5$$

The solution is 0.5.

157.
$$4-3a=7-2(2a+5)$$
$$4-3a=7-4a-10$$
$$4-3a=-3-4a$$
$$4-3a+4a=-3-4a+4a$$
$$4+a=-3$$
$$4-4+a=-3-4$$
$$a=-7$$

$$a^2+7a$$
$$=(-7)^2+(7)(-7)$$
$$=49+(7)(-7)$$
$$=49-49$$
$$=0$$

The answer is 0.

159.
$$2z-5=3(4z+5)$$
$$2z-5=12z+15$$
$$2z-12z-5=12z-12z+15$$
$$-10z-5=15$$
$$-10z-5+5=15+5$$
$$-10z=20$$
$$\frac{-10z}{-10}=\frac{20}{-10}$$
$$z=-2$$

$$\frac{z^2}{z-2}$$
$$=\frac{(-2)^2}{-2-2}$$
$$=\frac{4}{-4}$$
$$=-1$$

The answer is −1.

Objective 3.2.4 Exercises

161. Strategy

To find the distance, replace C by −3 and solve for D.

Solution
$$C=\frac{1}{4}D-45$$
$$-3=\frac{1}{4}D-45$$
$$-3+45=\frac{1}{4}D-45+45$$
$$42=\frac{1}{4}D$$
$$4(42)=4\left(\frac{1}{4}D\right)$$
$$168=D$$

The car will slide 168 ft.

163. Strategy

To find the depth, replace P by 35 and solve for D.

Solution

$$P = \frac{1}{2}D + 15$$
$$35 = \frac{1}{2}D + 15$$
$$35 - 15 = \frac{1}{2}D + 15 - 15$$
$$20 = \frac{1}{2}D$$
$$2(20) = 2\left(\frac{1}{2}D\right)$$
$$40 = D$$

The depth of the diver is 40 ft.

165. Strategy

To find the initial velocity, replace s by 80 and t by 2 and solve for v.

Solution

$$s = 16t^2 + vt$$
$$80 = 16(2)^2 + v(2)$$
$$80 = 16(4) + 2v$$
$$80 = 64 + 2v$$
$$80 - 64 = 64 - 64 + 2v$$
$$16 = 2v$$
$$\frac{16}{2} = \frac{2v}{2}$$
$$8 = v$$

The initial velocity is 8 ft/s.

167. Strategy

To find the length, replace H by 66 and solve for L.

Solution

$$H = 1.2L + 27.8$$
$$66 = 1.2L + 27.8$$
$$66 - 27.8 = 1.2L + 27.8 - 27.8$$
$$38.2 = 1.2L$$
$$\frac{38.2}{1.2} = \frac{1.2L}{1.2}$$
$$31.8 \approx L$$

The approximate length of the humerus is 31.8 in.

169. Strategy

To find the number miles, replace F by 6.25 and solve for m.

Solution

$$F = 1.50 + 0.95(m - 1)$$
$$6.25 = 1.50 + 0.95m - 0.95$$
$$6.25 = 0.55 + 0.95m$$
$$6.25 - 0.55 = 0.55 - 0.55 + 0.95m$$
$$5.70 = 0.95m$$
$$\frac{5.70}{0.95} = \frac{0.95m}{0.95}$$
$$6 = m$$

The passenger was driven 6 mi.

171. Strategy

To find the number of errors, replace S by 35 and W by 390 and solve for e.

Solution

$$S = \frac{W - 5e}{10}$$
$$35 = \frac{390 - 5e}{10}$$
$$10(35) = 10\left(\frac{390 - 5e}{10}\right)$$
$$350 = 390 - 5e$$
$$350 - 390 = 390 - 390 - 5e$$
$$-40 = -5e$$
$$\frac{-40}{-5} = \frac{-5e}{-5}$$
$$8 = e$$

The candidate made 8 errors.

173. Strategy

To find the population, replace P_1 by 48,000, N by 1,100,000, and d by 75, and solve for P_2.

Solution

$$N = \frac{2.51 P_1 P_2}{d^2}$$
$$1,100,000 = \frac{2.51(48,000)P_2}{75^2}$$
$$1,100,000 = \frac{120,480 P_2}{5625}$$
$$\frac{5625}{120,480}(1,100,000) = \frac{5625}{120,480}\left(\frac{120,480 P_2}{5625}\right)$$
$$51,357 \approx P_2$$

The population is approximately 51,000 people.

175. Strategy
To find the break-even point, replace the variables P, C, and F in the cost equation by the given values and solve for x.

Solution
$$Px = Cx + F$$
$$250x = 165x + 29{,}750$$
$$250x - 165x = 165x - 165x + 29{,}750$$
$$85x = 29{,}750$$
$$\frac{85x}{85} = \frac{29{,}750}{85}$$
$$x = 350$$
The break-even point is 350 televisions.

179. Strategy
To determine whether or not the see-saw is balanced, replace the variables F_1, F_2, d, and x in the lever system equation by the given values. Evaluate each side of the equation. If it is a true equation, the see-saw is balanced. If it is not a true equation, the see-saw is not balanced.

Solution
$$F_1 x = F_2(d - x)$$

$60(3.5)$	$50(8 - 3.5)$
210	$50(4.5)$
	$210 \neq 225$

The see-saw is not balanced.

181. Strategy
To find the force when the system balances, replace the variables F_1, x, and d in the lever system equation by the given values and solve for F_2.

Solution
$$F_1 x = F_2(d - x)$$
$$100 \cdot 2 = F_2(10 - 2)$$
$$100 \cdot 2 = F_2 \cdot 8$$
$$200 = 8F_2$$
$$\frac{200}{8} = \frac{8F_2}{8}$$
$$25 = F_2$$
A 25-lb force must be applied to the other end.

183. Strategy
To find the location of the fulcrum when the system balanced, replaces the variables F_1, F_2, and d in the lever system equation by the given values and solve for x.

Solution
$$F_1 x = F_2(d - x)$$
$$128x = 160(18 - x)$$
$$128x = 2880 - 160x$$
$$128x + 160x = 2880 - 160x + 160x$$
$$288x = 2880$$
$$\frac{288x}{288} = \frac{2880}{288}$$
$$x = 10$$
The fulcrum is 10 ft from the 128–lb acrobat.

Applying Concepts 3.2

185. $3(2x - 1) - (6x - 4) = -9$
$$6x - 3 - 6x + 4 = -9$$
$$6x - 6x + 1 = -9$$
$$1 \neq -9$$
No solution

187. $3[4(w + 2) - (w + 1)] = 5(2 + w)$
$$3[4w + 8 - w - 1] = 10 + 5w$$
$$3[3w + 7] = 10 + 5w$$
$$9w + 21 = 10 + 5w$$
$$9w - 5w + 21 = 10 + 5w - 5w$$
$$4w + 21 = 10$$
$$4w + 21 - 21 = 10 - 21$$
$$4w = -11$$
$$\frac{4w}{4} = \frac{-11}{4}$$
$$w = -\frac{11}{4}$$
The solution is $-\dfrac{11}{4}$.

189. To solve, remember

Dividend	=	Quotient	×	Divisor	+	Remainder
$32{,}166$	=	518	×	x	+	50
$32{,}166 - 50$	=	$518x$			+	$50 - 50$
$32{,}116$	=	$518x$				
$\dfrac{32{,}116}{518}$	=	$\dfrac{518x}{518}$				
62	=	x				

191. One possible answer is
$$3x - 6 = 2x - 2$$
$$x = 4$$

Section 3.3

Concept Review 3.3

1. Always true

3. Sometimes true
Multiplying both sides of the inequality by a negative number without reversing the inequality changes the solution. Multiplying both sides by zero also changes the inequality.

5. Never true
The product of a positive number and a negative number is always negative.

7. Always true

9. Sometimes true
This is true only if c is a positive number. If c is negative, then $ac > bc$. If c is zero, then $ac = bc$.

Objective 3.3.1 Exercises

1. $x > 2$

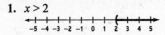

3. $x \leq 0$

5. $\begin{aligned} x+7 &\leq -3 \\ x+7-7 &\leq -3-7 \\ x &\leq -10 \end{aligned}$
a, c

7. $\begin{aligned} x+1 &< 3 \\ x+1-1 &< 3-1 \\ x &< 2 \end{aligned}$

9. $\begin{aligned} x-5 &> -2 \\ x-5+5 &> -2+5 \\ x &> 3 \end{aligned}$

11. $\begin{aligned} n+4 &\geq 7 \\ n+4-4 &\geq 7-4 \\ n &\geq 3 \end{aligned}$

13. $\begin{aligned} x-6 &\leq -10 \\ x-6+6 &\leq -10+6 \\ x &\leq -4 \end{aligned}$

15. $\begin{aligned} 5+x &\geq 4 \\ 5-5+x &\geq 4-5 \\ x &\geq -1 \end{aligned}$

17. $\begin{aligned} y-3 &\geq -12 \\ y-3+3 &\geq -12+3 \\ y &\geq -9 \end{aligned}$

19. $\begin{aligned} 3x-5 &< 2x+7 \\ 3x-2x-5 &< 2x-2x+7 \\ x-5 &< 7 \\ x-5+5 &< 7+5 \\ x &< 12 \end{aligned}$

21. $\begin{aligned} 8x-7 &\geq 7x-2 \\ 8x-7x-7 &\geq 7x-7x-2 \\ x-7 &\geq -2 \\ x-7+7 &\geq -2+7 \\ x &\geq 5 \end{aligned}$

23. $\begin{aligned} 2x+4 &< x-7 \\ 2x-x+4 &< x-x-7 \\ x+4 &< -7 \\ x+4-4 &< -7-4 \\ x &< -11 \end{aligned}$

25. $\begin{aligned} 4x-8 &\leq 2+3x \\ 4x-3x-8 &\leq 2+3x-3x \\ x-8 &\leq 2 \\ x-8+8 &\leq 2+8 \\ x &\leq 10 \end{aligned}$

27. $\begin{aligned} 6x+4 &\geq 5x-2 \\ 6x-5x+4 &\geq 5x-5x-2 \\ x+4 &\geq -2 \\ x+4-4 &\geq -2-4 \\ x &\geq -6 \end{aligned}$

29. $\begin{aligned} 2x-12 &> x-10 \\ 2x-x-12 &> x-x-10 \\ x-12 &> -10 \\ x-12+12 &> -10+12 \\ x &> 2 \end{aligned}$

31. $\begin{aligned} d+\frac{1}{2} &< \frac{1}{3} \\ d+\frac{1}{2}-\frac{1}{2} &< \frac{1}{3}-\frac{1}{2} \\ d &< \frac{2}{6}-\frac{3}{6} \\ d &< -\frac{1}{6} \end{aligned}$

33. $\begin{aligned} x+\frac{5}{8} &\geq -\frac{2}{3} \\ x+\frac{5}{8}-\frac{5}{8} &\geq -\frac{2}{3}-\frac{5}{8} \\ x &\geq -\frac{16}{24}-\frac{15}{24} \\ x &\geq -\frac{31}{24} \end{aligned}$

35.

$$2x - \frac{1}{2} < x + \frac{3}{4}$$

$$2x - x - \frac{1}{2} < x - x + \frac{3}{4}$$

$$x - \frac{1}{2} < \frac{3}{4}$$

$$x - \frac{1}{2} + \frac{1}{2} < \frac{3}{4} + \frac{1}{2}$$

$$x < \frac{3}{4} + \frac{2}{4}$$

$$x < \frac{5}{4}$$

37.

$$3x + \frac{5}{8} > 2x + \frac{5}{6}$$

$$3x - 2x + \frac{5}{8} > 2x - 2x + \frac{5}{6}$$

$$x + \frac{5}{8} > \frac{5}{6}$$

$$x + \frac{5}{8} - \frac{5}{8} > \frac{5}{6} - \frac{5}{8}$$

$$x > \frac{20}{24} - \frac{15}{24}$$

$$x > \frac{5}{24}$$

39.

$$x + 5.8 \le 4.6$$

$$x + 5.8 - 5.8 \le 4.6 - 5.8$$

$$x \le -1.2$$

41.

$$x - 0.23 \le 0.47$$

$$x - 0.23 + 0.23 \le 0.47 + 0.23$$

$$x \le 0.70$$

43.

$$1.2x < 0.2x - 7.3$$

$$1.2x - 0.2x < 0.2x - 0.2x - 7.3$$

$$x < -7.3$$

Objective 3.3.2 Exercises

45.

$$-4x \le 12$$

$$\frac{-4x}{-4} \le \frac{12}{-4}$$

$$x \ge -3$$

Since 0, 3, and –3 are greater than or equal to –3, the solutions are a, b, and c.

47.

$$8x \le -24$$

$$\frac{8x}{8} \le \frac{-24}{8}$$

$$x \le -3$$

49.

$$24x > -48$$

$$\frac{24x}{24} > \frac{-48}{24}$$

$$x > -2$$

51.

$$3x > 0$$

$$\frac{3x}{3} > \frac{0}{3}$$

$$x > 0$$

53.

$$-2n \le -8$$

$$\frac{-2n}{-2} \ge \frac{-8}{-2}$$

$$n \ge 4$$

55.

$$-4x < 8$$

$$\frac{-4x}{-4} > \frac{8}{-4}$$

$$x > -2$$

57.

$$3x < 5$$

$$\frac{3x}{3} < \frac{5}{3}$$

$$x < \frac{5}{3}$$

59.

$$-8x \le -40$$

$$\frac{-8x}{-8} \ge \frac{-40}{-8}$$

$$x \ge 5$$

61.

$$10x > -25$$

$$\frac{10x}{10} > \frac{-25}{10}$$

$$x > -\frac{5}{2}$$

63.

$$-5x \ge \frac{10}{3}$$

$$\frac{-5x}{-5} \le \frac{\frac{10}{3}}{-5}$$

$$x \le -\frac{2}{3}$$

65.

$$\frac{2}{3}x < -12$$

$$\frac{3}{2}\left(\frac{2}{3}x\right) < \frac{3}{2}(-12)$$

$$x < -18$$

67.

$$-\frac{3}{8}x < 6$$

$$-\frac{8}{3}\left(-\frac{3}{8}x\right) > -\frac{8}{3}(6)$$

$$x > -16$$

69.

$$\frac{2}{3}y \ge 4$$

$$\frac{3}{2}\left(\frac{2}{3}y\right) \ge \frac{3}{2}(4)$$

$$y \ge 6$$

71. $-\dfrac{2}{3}x \le 4$

$-\dfrac{3}{2}\left(-\dfrac{2}{3}x\right) \ge -\dfrac{3}{2}(4)$

$x \ge -6$

73. $-\dfrac{2}{11}b \ge -6$

$-\dfrac{11}{2}\left(-\dfrac{2}{11}b\right) \le -\dfrac{11}{2}(-6)$

$b \le 33$

75. $\dfrac{2}{3}n < \dfrac{1}{2}$

$\dfrac{3}{2}\left(\dfrac{2}{3}n\right) < \dfrac{3}{2}\left(\dfrac{1}{2}\right)$

$n < \dfrac{3}{4}$

77. $-\dfrac{2}{3}x \ge \dfrac{4}{7}$

$-\dfrac{3}{2}\left(-\dfrac{2}{3}x\right) \le -\dfrac{3}{2}\left(\dfrac{4}{7}\right)$

$x \le -\dfrac{6}{7}$

79. $-\dfrac{3}{4}y \ge -\dfrac{5}{8}$

$-\dfrac{4}{3}\left(-\dfrac{3}{4}y\right) \le -\dfrac{4}{3}\left(-\dfrac{5}{8}\right)$

$y \le \dfrac{5}{6}$

81. $\dfrac{2}{3}x \le \dfrac{9}{14}$

$\dfrac{3}{2}\left(\dfrac{2}{3}x\right) \le \dfrac{3}{2}\left(\dfrac{9}{14}\right)$

$x \le \dfrac{27}{28}$

83. $-\dfrac{3}{5}y < \dfrac{9}{10}$

$-\dfrac{5}{3}\left(-\dfrac{3}{5}y\right) > -\dfrac{5}{3}\left(\dfrac{9}{10}\right)$

$y > -\dfrac{3}{2}$

85. $-0.27x < 0.135$

$\dfrac{-0.27x}{-0.27} > \dfrac{0.135}{-0.27}$

$x > -0.5$

87. $8.4y \ge -6.72$

$\dfrac{8.4y}{8.4} \ge \dfrac{-6.72}{8.4}$

$y \ge -0.8$

89. $1.5x \le 6.30$

$\dfrac{1.5x}{1.5} \le \dfrac{6.30}{1.5}$

$x \le 4.2$

91. $-3.9x \ge -19.5$

$\dfrac{-3.9x}{-3.9} \le \dfrac{-19.5}{-3.9}$

$x \le 5$

93. $0.07x < -0.378$

$\dfrac{0.07x}{0.07} < \dfrac{-0.378}{0.07}$

$x < -5.4$

Objective 3.3.3 Exercises

97. $4x - 8 < 2x$

$4x - 2x - 8 < 2x - 2x$

$2x - 8 < 0$

$2x - 8 + 8 < 0 + 8$

$2x < 8$

$\dfrac{2x}{2} < \dfrac{8}{2}$

$x < 4$

99. $2x - 8 > 4x$

$2x - 4x - 8 > 4x - 4x$

$-2x - 8 > 0$

$-2x - 8 + 8 > 0 + 8$

$-2x > 8$

$\dfrac{-2x}{-2} < \dfrac{8}{-2}$

$x < -4$

101. $8 - 3x \le 5x$

$8 - 3x - 5x \le 5x - 5x$

$8 - 8x \le 0$

$8 - 8 - 8x \le 0 - 8$

$-8x \le -8$

$\dfrac{-8x}{-8} \ge \dfrac{-8}{-8}$

$x \ge 1$

103. $3x + 2 \ge 5x - 8$

$3x - 5x + 2 \ge 5x - 5x - 8$

$-2x + 2 \ge -8$

$-2x + 2 - 2 \ge -8 - 2$

$-2x \ge -10$

$\dfrac{-2x}{-2} \le \dfrac{-10}{-2}$

$x \le 5$

105. $5x - 2 < 3x - 2$

$5x - 3x - 2 < 3x - 3x - 2$

$2x - 2 < -2$

$2x - 2 + 2 < -2 + 2$

$2x < 0$

$\dfrac{2x}{2} < \dfrac{0}{2}$

$x < 0$

107.
$$0.1(180 + x) > x$$
$$18 + 0.1x > x$$
$$18 + 0.1x - x > x - x$$
$$18 - 0.9x > 0$$
$$18 - 18 - 0.9x > 0 - 18$$
$$-0.9x > -18$$
$$\frac{-0.9x}{-0.9} < \frac{-18}{-0.9}$$
$$x < 20$$

109.
$$0.15x + 55 > 0.10x + 80$$
$$0.15x - 0.10x + 55 > 0.10x - 0.10x + 80$$
$$0.05x + 55 > 80$$
$$0.05x + 55 - 55 > 80 - 55$$
$$0.05x > 25$$
$$\frac{0.05x}{0.05} > \frac{25}{0.05}$$
$$x > 500$$

111.
$$2(3x - 1) > 3x + 4$$
$$6x - 2 > 3x + 4$$
$$6x - 3x - 2 > 3x - 3x + 4$$
$$3x - 2 > 4$$
$$3x - 2 + 2 > 4 + 2$$
$$3x > 6$$
$$\frac{3x}{3} > \frac{6}{3}$$
$$x > 2$$

113.
$$3(2x - 5) \geq 8x - 5$$
$$6x - 15 \geq 8x - 5$$
$$6x - 8x - 15 \geq 8x - 8x - 5$$
$$-2x - 15 \geq -5$$
$$-2x - 15 + 15 \geq -5 + 15$$
$$-2x \geq 10$$
$$\frac{-2x}{-2} \leq \frac{10}{-2}$$
$$x \leq -5$$

115.
$$2(2y - 5) \leq 3(5 - 2y)$$
$$4y - 10 \leq 15 - 6y$$
$$4y + 6y - 10 \leq 15 - 6y + 6y$$
$$10y - 10 \leq 15$$
$$10y - 10 + 10 \leq 15 + 10$$
$$10y \leq 25$$
$$\frac{10y}{10} \leq \frac{25}{10}$$
$$y \leq \frac{5}{2}$$

117.
$$5(2 - x) > 3(2x - 5)$$
$$10 - 5x > 6x - 15$$
$$10 - 5x - 6x > 6x - 6x - 15$$
$$10 - 11x > -15$$
$$10 - 10 - 11x > -15 - 10$$
$$-11x > -25$$
$$\frac{-11x}{-11} < \frac{-25}{-11}$$
$$x < \frac{25}{11}$$

119.
$$5(x - 2) > 9x - 3(2x - 4)$$
$$5x - 10 > 9x - 6x + 12$$
$$5x - 10 > 3x + 12$$
$$5x - 3x - 10 > 3x - 3x + 12$$
$$2x - 10 > 12$$
$$2x - 10 + 10 > 12 + 10$$
$$2x > 22$$
$$\frac{2x}{2} > \frac{22}{2}$$
$$x > 11$$

121.
$$4 - 3(3 - n) \leq 3(2 - 5n)$$
$$4 - 9 + 3n \leq 6 - 15n$$
$$-5 + 3n \leq 6 - 15n$$
$$-5 + 3n + 15n \leq 6 - 15n + 15n$$
$$-5 + 18n \leq 6$$
$$-5 + 5 + 18n \leq 6 + 5$$
$$18n \leq 11$$
$$\frac{18n}{18} \leq \frac{11}{18}$$
$$n \leq \frac{11}{18}$$

123.
$$2x - 3(x - 4) \geq 4 - 2(x - 7)$$
$$2x - 3x + 12 \geq 4 - 2x + 14$$
$$-x + 12 \geq 18 - 2x$$
$$-x + 2x + 12 \geq 18 - 2x + 2x$$
$$x + 12 \geq 18$$
$$x + 12 - 12 \geq 18 - 12$$
$$x \geq 6$$

125.
$$\frac{1}{2}(9x - 10) \leq -\frac{1}{3}(12 - 6x)$$
$$\frac{9}{2}x - 5 \leq -4 + 2x$$
$$\frac{9}{2}x - 2x - 5 \leq -4 + 2x - 2x$$
$$\frac{5}{2}x - 5 \leq -4$$
$$\frac{5}{2}x - 5 + 5 \leq -4 + 5$$
$$\frac{5}{2}x \leq 1$$
$$\frac{2}{5}\left(\frac{5}{2}x\right) \leq \frac{2}{5}(1)$$
$$x \leq \frac{2}{5}$$

127.
$$\frac{2}{3}(9t - 15) + 4 < 6 + \frac{3}{4}(4 - 12t)$$
$$6t - 10 + 4 < 6 + 3 - 9t$$
$$6t - 6 < 9 - 9t$$
$$6t + 9t - 6 < 9 - 9t + 9t$$
$$15t - 6 < 9$$
$$15t - 6 + 6 < 9 + 6$$
$$15t < 15$$
$$\frac{15t}{15} < \frac{15}{15}$$
$$t < 1$$

129.
$$3[4(n-2)-(1-n)] > 5(n-4)$$
$$3[4n-8-1+n] > 5n-20$$
$$3[5n-9] > 5n-20$$
$$15n-27 > 5n-20$$
$$15n-5n-27 > 5n-5n-20$$
$$10n-27 > -20$$
$$10n-27+27 > -20+27$$
$$10n > 7$$
$$\frac{10n}{10} > \frac{7}{10}$$
$$n > \frac{7}{10}$$

131.
$$3x-4 \geq 5$$
$$3x-4+4 \geq 5+4$$
$$\frac{3x}{3} \geq \frac{9}{3}$$
$$x \geq 3$$

The difference between this inequality and $3x - 4 > 5$ is that the first inequality includes 3 in the solution set.

137.

$$5x-12 \leq x+8 \qquad 3x-4 \geq 2+x$$
$$5x-x-12 \leq x-x+8 \quad 3x-x-4 \geq 2+x-x$$
$$4x-12 \leq 8 \qquad 2x-4 \geq 2$$
$$4x-12+12 \leq 8+12 \quad 2x-4+4 \geq 2+4$$
$$4x \leq 20 \qquad 2x \geq 6$$
$$\frac{4x}{4} \leq \frac{20}{4} \qquad \frac{2x}{2} \geq \frac{6}{2}$$
$$x \leq 5 \qquad x \geq 3$$

The integers common to the solution sets of both inequalities: {3, 4, 5}

139.

$$4(x-2) \leq 3x+5 \qquad 7(x-3) \geq 5x-1$$
$$4x-8 \leq 3x+5 \qquad 7x-21 \geq 5x-1$$
$$4x-3x-8 \leq 3x-3x+5 \quad 7x-5x-21 \geq 5x-5x-1$$
$$x-8 \leq 5 \qquad 2x-21 \geq -1$$
$$x-8+8 \leq 5+8 \qquad 2x-21+21 \geq -1+21$$
$$x \leq 13 \qquad 2x \geq 20$$
$$\frac{2x}{2} \geq \frac{20}{2}$$
$$x \geq 10$$

The integers common to the solution sets of both inequalities: {10, 11, 12, 13}

141.

$$\text{—}|\ \ |\ \ (\ |\ \ |\ \ |\ \ |\ \ |\ \ |\ \ |\ \ |\ \ )\ \ |\ \ |\text{—}$$
$$\text{-5 -4 -3 -2 -1 0 1 2 3 4 5}$$

143.

$$\text{—}|\ \ |\ \ |\ \ |\ \ |\ \ |\ \ |\ \ |\ \ (\ |\ \ |\ \ |\text{—}$$
$$\text{-5 -4 -3 -2 -1 0 1 2 3 4 5}$$

Applying Concepts 3.3

133.
$$7-2b \leq 15-5b$$
$$7-2b+5b \leq 15-5b+5b$$
$$7+3b \leq 15$$
$$-7+7+3b \leq -7+15$$
$$3b \leq 8$$
$$\frac{3b}{3} \leq \frac{8}{3}$$
$$b \leq \frac{8}{3}$$
$$b \leq 2\frac{2}{3}$$

{1, 2}

135.
$$2(2c-3) < 5(6-c)$$
$$4c-6 < 30-5c$$
$$4c+5c-6 < 30-5c+5c$$
$$9c-6 < 30$$
$$9c-6+6 < 30+6$$
$$9c < 36$$
$$\frac{9c}{9} < \frac{36}{9}$$
$$c < 4$$

{1, 2, 3}

Chapter Review Exercises

1.
$$\underline{5x-2 = 4x+5}$$
$$5(3)-2 \ \big|\ 4(3)+5$$
$$15-2 \ \big|\ 12+5$$
$$13 \neq 17$$

No, 3 is not a solution.

2.
$$x-4 = 16$$
$$x-4+4 = 16+4$$
$$x = 20$$
The solution is 20.

3. $8x = -56$

$\dfrac{8x}{8} = \dfrac{-56}{8}$

$x = -7$

The solution is -7.

4. $5x - 6 = 29$

$5x - 6 + 6 = 29 + 6$

$5x = 35$

$\dfrac{5x}{5} = \dfrac{35}{5}$

$x = 7$

The solution is 7.

5. $5x + 3 = 10x - 17$

$5x - 10x + 3 = 10x - 10x - 17$

$-5x + 3 = -17$

$-5x + 3 - 3 = -17 - 3$

$-5x = -20$

$\dfrac{-5x}{-5} = \dfrac{-20}{-5}$

$x = 4$

The solution is 4.

6. $3(5x + 2) + 2 = 10x + 5[x - (3x - 1)]$

$15x + 6 + 2 = 10x + 5[x - 3x + 1]$

$15x + 8 = 10x + 5[-2x + 1]$

$15x + 8 = 10x - 10x + 5$

$15x + 8 = 5$

$15x + 8 - 8 = 5 - 8$

$15x = -3$

$\dfrac{15x}{15} = \dfrac{-3}{15}$

$x = -\dfrac{1}{5}$

The solution is $-\dfrac{1}{5}$.

7. $PB = A$

$0.81(500) = A$

$405 = A$

The number is 405.

8. $PB = A$

$0.72B = 18$

$\dfrac{0.72B}{0.72} = \dfrac{18}{0.72}$

$B = 25$

The number is 25.

9. $PB = A$

$P(40) = 27$

$\dfrac{40P}{40} = \dfrac{27}{40}$

$P = 0.675$

The percent is 67.5%.

10.

11. $x - 3 > -1$

$x - 3 + 3 > -1 + 3$

$x > 2$

12. $3x > -12$

$\dfrac{3x}{3} > \dfrac{-12}{3}$

$x > -4$

13. $3x + 4 \geq -8$

$3x + 4 - 4 \geq -8 - 4$

$3x \geq -12$

$\dfrac{3x}{3} \geq \dfrac{-12}{3}$

$x \geq -4$

14. $7x - 2(x + 3) \geq x + 10$

$7x - 2x - 6 \geq x + 10$

$5x - 6 \geq x + 10$

$5x - x - 6 \geq x - x + 10$

$4x - 6 \geq 10$

$4x - 6 + 6 \geq 10 + 6$

$4x \geq 16$

$\dfrac{4x}{4} \geq \dfrac{16}{4}$

$x \geq 4$

15. $\begin{array}{c|c} x^2 + 4x + 1 = 3x + 7 \\ (2)^2 + 4(2) + 1 \ | \ 3(2) + 7 \\ 4 + 8 + 1 \ | \ 6 + 7 \\ 13 = 13 \end{array}$

Yes, 2 is a solution.

16. $4.6 = 2.1 + x$

$4.6 - 2.1 = 2.1 - 2.1 + x$

$2.5 = x$

The solution is 2.5

17. $\dfrac{x}{7} = -7$

$7\left(\dfrac{1}{7}x\right) = 7(-7)$

$x = -49$

The solution is -49.

18. $14 + 6x = 17$

$14 - 14 + 6x = 17 - 14$

$6x = 3$

$\dfrac{6x}{6} = \dfrac{3}{6}$

$x = \dfrac{1}{2}$

The solution is $\dfrac{1}{2}$.

19.
$$12y - 1 = 3y + 2$$
$$12y - 3y - 1 = 3y - 3y + 2$$
$$9y - 1 = 2$$
$$9y - 1 + 1 = 2 + 1$$
$$9y = 3$$
$$\frac{9y}{9} = \frac{3}{9}$$
$$y = \frac{1}{3}$$
The solution is $\frac{1}{3}$.

20.
$$x + 5(3x - 20) = 10(x - 4)$$
$$x + 15x - 100 = 10x - 40$$
$$16x - 100 = 10x - 40$$
$$16x - 10x - 100 = 10x - 10x - 40$$
$$6x - 100 = -40$$
$$6x - 100 + 100 = -40 + 100$$
$$6x = 60$$
$$\frac{6x}{6} = \frac{60}{6}$$
$$x = 10$$
The solution is 10.

21.
$$PB = A$$
$$66\frac{2}{3}\% = \frac{2}{3}$$
$$\frac{2}{3}(24) = A$$
$$16 = A$$
The number is 16.

22.
$$PB = A$$
$$0.48B = 60$$
$$\frac{0.48B}{0.48} = \frac{60}{0.48}$$
$$B = 125$$
The number is 125.

23.
$$PB = A$$
$$P(3) = 0.5$$
$$\frac{3P}{3} = \frac{0.5}{3}$$
$$P = \frac{1}{6} = 16\frac{2}{3}\%$$
0.5 is $16\frac{2}{3}\%$ of 3.

24.
$$2 + x < -2$$
$$2 - 2 + x < -2 - 2$$
$$x < -4$$

-5 -4 -3 -2 -1 0 1 2 3 4 5

25.
$$5x \le -10$$
$$\frac{5x}{5} \le \frac{-10}{5}$$
$$x \le -2$$

-5 -4 -3 -2 -1 0 1 2 3 4 5

26.
$$6x + 3(2x - 1) = -27$$
$$6x + 6x - 3 = -27$$
$$12x - 3 = -27$$
$$12x - 3 + 3 = -27 + 3$$
$$12x = -24$$
$$\frac{12x}{12} = \frac{-24}{12}$$
$$x = -2$$
The solution is –2.

27.
$$a - \frac{1}{6} = \frac{2}{3}$$
$$a - \frac{1}{6} + \frac{1}{6} = \frac{2}{3} + \frac{1}{6}$$
$$a = \frac{4}{6} + \frac{1}{6}$$
$$a = \frac{5}{6}$$
The solution is $\frac{5}{6}$.

28.
$$\frac{3}{5}a = 12$$
$$\frac{5}{3}\left(\frac{3}{5}a\right) = \frac{5}{3}(12)$$
$$a = 20$$
The solution is 20.

29.
$$32 = 9x - 4 - 3x$$
$$32 = 6x - 4$$
$$32 + 4 = 6x - 4 + 4$$
$$36 = 6x$$
$$\frac{36}{6} = \frac{6x}{6}$$
$$6 = x$$
The solution is 6.

30.
$$-4[x + 3(x - 5)] = 3(8x + 20)$$
$$-4[x + 3x - 15] = 24x + 60$$
$$-4[4x - 15] = 24x + 60$$
$$-16x + 60 = 24x + 60$$
$$-16x - 24x + 60 = 24x - 24x + 60$$
$$-40x + 60 = 60$$
$$-40x + 60 - 60 = 60 - 60$$
$$-40x = 0$$
$$\frac{-40x}{-40} = \frac{0}{-40}$$
$$x = 0$$
The solution is 0.

31.
$$4x - 12 < x + 24$$
$$4x - x - 12 < x - x + 24$$
$$3x - 12 < 24$$
$$3x - 12 + 12 < 24 + 12$$
$$3x < 36$$
$$\frac{3x}{3} < \frac{36}{3}$$
$$x < 12$$

32. $\frac{1}{2}\% = 0.5\% = 0.005$

$$PB = A$$
$$0.005(3000) = A$$
$$15 = A$$
The number is 15.

33. $3x + 7 + 4x = 42$
$$7x + 7 = 42$$
$$7x + 7 - 7 = 42 - 7$$
$$7x = 35$$
$$\frac{7x}{7} = \frac{35}{7}$$
$$x = 5$$
The solution is 5.

34. $5x - 6 > 19$
$$5x - 6 + 6 > 19 + 6$$
$$5x > 25$$
$$\frac{5x}{5} > \frac{25}{5}$$
$$x > 5$$

35. $PB = A$
$$P(200) = 8$$
$$\frac{200P}{200} = \frac{8}{200}$$
$$P = 0.04$$
The percent is 4%.

36. $6x - 9 < 4x + 3(x + 3)$
$$6x - 9 < 4x + 3x + 9$$
$$6x - 9 < 7x + 9$$
$$6x - 7x - 9 < 7x - 7x + 9$$
$$-x - 9 < 9$$
$$-x - 9 + 9 < 9 + 9$$
$$-x < 18$$
$$-1(-x) > -1(18)$$
$$x > -18$$

37. $5 - 4(x + 9) > 11(12x - 9)$
$$5 - 4x - 36 > 132x - 99$$
$$-4x - 31 > 132x - 99$$
$$-4x - 132x - 31 > 132x - 132x - 99$$
$$-136x - 31 > -99$$
$$-136x - 31 + 31 > -99 + 31$$
$$-136x > -68$$
$$\frac{-136x}{-136} < \frac{-68}{-136}$$
$$x < \frac{1}{2}$$

38. Strategy
To find the measure of the third angle, replace the variables A and B in the given equation by the given values and solve for C.

Solution
$$A + B + C = 180$$
$$20 + 50 + C = 180$$
$$70 + C = 180$$
$$70 - 70 + C = 180 - 70$$
$$C = 110$$
The measure of the third angle is 110°.

39. Strategy
To find the force when the system balances, replace the variables F_1, x_1, and d in the lever system equation by the given values and solve for F_2.

Solution
$$F_1 x = F_2(d - x)$$
$$120(2) = F_2(12 - 2)$$
$$240 = F_2(10)$$
$$240 = 10F_2$$
$$\frac{240}{10} = \frac{10F_2}{10}$$
$$24 = F_2$$
A 24-lb force must be applied to the other end.

40. Strategy
To find the width, replace the variables P and L in the equation by the given values and solve for W.

Solution
$$P = 2L + 2W$$
$$49 = 2(18.5) + 2W$$
$$49 = 37 + 2W$$
$$49 - 37 = 37 - 37 + 2W$$
$$12 = 2W$$
$$\frac{12}{2} = \frac{2W}{2}$$
$$6 = W$$
The width of the rectangle is 6 ft.

41. Strategy
To find the discount, replace the variables S and R in the equation by the given values and solve for D.

Solution
$$S = R - D$$
$$128 = 159.99 - D$$
$$128 - 159.99 = 159.99 - 159.99 - D$$
$$-31.99 = -D$$
$$-1(31.99) = -1(-D)$$
$$31.99 = D$$
The discount is $31.99.

42. Strategy
To find the total number at risk, solve the basic percent equation, where the amount is 1184 and the percent is 10.7% = 0.107.

Solution
$$PB = A$$
$$0.107B = 1184$$
$$\frac{0.107B}{0.107} = \frac{1184}{0.107}$$
$$B \approx 11,065$$
The total number of plants and animals at risk of extinction is 11,065.

43. Strategy

To find the depth, replace the variable P with the given value and solve for D.

Solution

$$P = 15 + \frac{1}{2}D$$
$$55 = 15 + \frac{1}{2}D$$
$$55 - 15 = 15 - 15 + \frac{1}{2}D$$
$$40 = \frac{1}{2}D$$
$$2(40) = 2\left(\frac{1}{2}D\right)$$
$$80 = D$$

The depth is 80 ft.

44. Strategy

To find the location of the fulcrum, replace the variables F_1, F_2, and d in the lever system equation by the given values, and solve for x.

Solution

$$F_1 x = F_2(d - x)$$
$$25x = 15(8 - x)$$
$$25x = 120 - 15x$$
$$25x + 15x = 120 - 15x + 15x$$
$$40x = 120$$
$$\frac{40x}{40} = \frac{120}{40}$$
$$x = 3$$

The fulcrum is 3 ft from the 25-lb force.

45. Strategy

To find the length, replace the variables P and W in the equation by the given values and solve for L.

Solution

$$P = 2L + 2W$$
$$84 = 2L + 2(18)$$
$$84 = 2L + 36$$
$$84 - 36 = 2L + 36 - 36$$
$$48 = 2L$$
$$\frac{48}{2} = \frac{2L}{2}$$
$$24 = L$$

The length is 24 ft.

46. Strategy

Find how long the trip lasts with the current by solving the equation $d = rt$, where the rate is the sum of the motorboat's speed (15 mph) and the current (5 mph) and the distance is 30 mi.

Solution

$$15 + 5 = 20 \text{ mph}$$
Using $d = rt$
$$30 = 20t$$
$$t = \frac{30}{20} = 1.5 \text{ h}$$

It will take 1.5 h.

47. Strategy

To find the amount invested at 8%, first find the amount of interest earned in the second investment and solve the simple interest equation using $r = 8\% = 0.08$, and $t = 1$ year, for P.

Solution

$$125 - 75 = 50$$
$$I = Prt$$
$$50 = P(0.08)(1)$$
$$\frac{50}{0.08} = \frac{0.08P}{0.08}$$
$$625 = P$$

Cathy must invest $625 at 8%.

48. Strategy

To find the percent, solve the basic percent equation using $B = 150$ and $A = 9$. The percent is unknown.

Solution

$$PB = A$$
$$P(150) = 9$$
$$\frac{150P}{150} = \frac{9}{150}$$
$$P = 0.06$$

The percent concentration is 6%.

Chapter Test

1.
$$\frac{3}{4}x = -9$$
$$\frac{4}{3}\left(\frac{3}{4}x\right) = \frac{4}{3}(-9)$$
$$x = -12$$
The solution is −12.

2.
$$6 - 5x = 5x + 11$$
$$6 - 5x - 5x = 5x - 5x + 11$$
$$6 - 10x = 11$$
$$6 - 6 - 10x = 11 - 6$$
$$-10x = 5$$
$$\frac{-10x}{-10} = \frac{5}{-10}$$
$$x = -\frac{1}{2}$$
The solution is $-\frac{1}{2}$.

3.
$$3x - 5 = -14$$
$$3x - 5 + 5 = -14 + 5$$
$$3x = -9$$
$$\frac{3x}{3} = \frac{-9}{3}$$
$$x = -3$$
The solution is −3.

4.

$x^2 - 3x$	$= 2x - 6$
$(-2)^2 - 3(-2)$	$2(-2) - 6$
$4 - 3(-2)$	$-4 - 6$
$4 + 6$	-10
10	$\neq -10$

No, −2 is not a solution.

5.
$$x + \frac{1}{2} = \frac{5}{8}$$
$$x + \frac{1}{2} - \frac{1}{2} = \frac{5}{8} - \frac{1}{2}$$
$$x = \frac{5}{8} - \frac{4}{8}$$
$$x = \frac{1}{8}$$
The solution is $\frac{1}{8}$.

6.
$$5x - 2(4x - 3) = 6x + 9$$
$$5x - 8x + 6 = 6x + 9$$
$$-3x + 6 = 6x + 9$$
$$-3x - 6x + 6 = 6x - 6x + 9$$
$$-9x + 6 = 9$$
$$-9x + 6 - 6 = 9 - 6$$
$$-9x = 3$$
$$\frac{-9x}{-9} = \frac{3}{-9}$$
$$x = -\frac{1}{3}$$
The solution is $-\frac{1}{3}$.

7.
$$7 - 4x = -13$$
$$7 - 7 - 4x = -13 - 7$$
$$-4x = -20$$
$$\frac{-4x}{-4} = \frac{-20}{-4}$$
$$x = 5$$
The solution is 5.

8.
$$11 - 4x = 2x + 8$$
$$11 - 4x - 2x = 2x - 2x + 8$$
$$11 - 6x = 8$$
$$11 - 11 - 6x = 8 - 11$$
$$-6x = -3$$
$$\frac{-6x}{-6} = \frac{-3}{-6}$$
$$x = \frac{1}{2}$$
The solution is $\frac{1}{2}$.

9.
$$x - 3 = -8$$
$$x - 3 + 3 = -8 + 3$$
$$x = -5$$
The solution is –5.

10.
$$3x - 2 = 5x + 8$$
$$3x - 5x - 2 = 5x - 5x + 8$$
$$-2x - 2 = 8$$
$$-2x - 2 + 2 = 8 + 2$$
$$-2x = 10$$
$$\frac{-2x}{-2} = \frac{10}{-2}$$
$$x = -5$$
The solution is –5.

11.
$$-\frac{3}{8}x = 5$$
$$-\frac{8}{3}\left(-\frac{3}{8}x\right) = -\frac{8}{3}(5)$$
$$x = -\frac{40}{3}$$
The solution is $-\frac{40}{3}$.

12.
$$6x - 3(2 - 3x) = 4(2x - 7)$$
$$6x - 6 + 9x = 8x - 28$$
$$15x - 6 = 8x - 28$$
$$15x - 8x - 6 = 8x - 8x - 28$$
$$7x - 6 = -28$$
$$7x - 6 + 6 = -28 + 6$$
$$7x = -22$$
$$\frac{7x}{7} = \frac{-22}{7}$$
$$x = -\frac{22}{7}$$
The solution is $-\frac{22}{7}$.

13.
$$6 - 2(5x - 8) = 3x - 4$$
$$6 - 10x + 16 = 3x - 4$$
$$22 - 10x = 3x - 4$$
$$22 - 10x - 3x = 3x - 3x - 4$$
$$22 - 13x = -4$$
$$22 - 22 - 13x = -4 - 22$$
$$-13x = -26$$
$$\frac{-13x}{-13} = \frac{-26}{-13}$$
$$x = 2$$
The solution is 2.

14.
$$9 - 3(2x - 5) = 12 + 5x$$
$$9 - 6x + 15 = 12 + 5x$$
$$-6x + 24 = 12 + 5x$$
$$-6x - 5x + 24 = 12 + 5x - 5x$$
$$-11x + 24 = 12$$
$$-11x + 24 - 24 = 12 - 24$$
$$-11x = -12$$
$$\frac{-11x}{-11} = \frac{-12}{-11}$$
$$x = \frac{12}{11}$$
The solution is $\frac{12}{11}$.

15.
$$3(2x - 5) = 8x - 9$$
$$6x - 15 = 8x - 9$$
$$6x - 8x - 15 = 8x - 8x - 9$$
$$-2x - 15 = -9$$
$$-2x - 15 + 15 = -9 + 15$$
$$-2x = 6$$
$$\frac{-2x}{-2} = \frac{6}{-2}$$
$$x = -3$$
The solution is –3.

16.
$$PB = A$$
$$P(16) = 20$$
$$\frac{P(16)}{16} = \frac{20}{16}$$
$$P = 1.25$$
The percent is 125%.

17.
$$PB = A$$
$$0.30B = 12$$
$$\frac{0.30B}{0.30} = \frac{12}{0.30}$$
$$B = 40$$
The number is 40.

18.

19.
$$-2 + x \le -3$$
$$-2 + 2 + x \le -3 + 2$$
$$x \le -1$$

20.
$$\frac{3}{8}x > -\frac{3}{4}$$
$$\frac{8}{3}\left(\frac{3}{8}x\right) > \frac{8}{3}\left(-\frac{3}{4}\right)$$
$$x > -2$$

21.
$$x + \frac{1}{3} > \frac{5}{6}$$
$$x + \frac{1}{3} - \frac{1}{3} > \frac{5}{6} - \frac{1}{3}$$
$$x > \frac{5}{6} - \frac{2}{6}$$
$$x > \frac{3}{6}$$
$$x > \frac{1}{2}$$

22.
$$3(x - 7) \ge 5x - 12$$
$$3x - 21 \ge 5x - 12$$
$$3x - 5x - 21 \ge 5x - 5x - 12$$
$$-2x - 21 \ge -12$$
$$-2x - 21 + 21 \ge -12 + 21$$
$$-2x \ge 9$$
$$\frac{-2x}{-2} \le \frac{9}{-2}$$
$$x \le -\frac{9}{2}$$

23.
$$-\frac{3}{8} \le 6$$
$$-\frac{8}{3}\left(-\frac{3}{8}x\right) \ge -\frac{8}{3}(6)$$
$$x \ge -16$$

24.
$$4x - 2(3 - 5x) \le 6x + 10$$
$$4x - 6 + 10x \le 6x + 10$$
$$14x - 6 \le 6x + 10$$
$$14x - 6x - 6 \le 6x - 6x + 10$$
$$8x - 6 \le 10$$
$$8x - 6 + 6 \le 10 + 6$$
$$8x \le 16$$
$$\frac{8x}{8} \le \frac{16}{8}$$
$$x \le 2$$

25.
$$3(2x - 5) \ge 8x - 9$$
$$6x - 15 \ge 8x - 9$$
$$6x - 8x - 15 \ge 8x - 8x - 9$$
$$-2x - 15 \ge -9$$
$$-2x - 15 + 15 \ge -9 + 15$$
$$-2x \ge 6$$
$$\frac{-2x}{-2} \le \frac{6}{-2}$$
$$x \le -3$$

26.
$$15 - 3(5x - 7) < 2(7 - 2x)$$
$$15 - 15x + 21 < 14 - 4x$$
$$-15x + 36 < 14 - 4x$$
$$-15x + 4x + 36 < 14 - 4x + 4x$$
$$-11x + 36 < 14$$
$$-11x + 36 - 36 < 14 - 36$$
$$-11x < -22$$
$$\frac{-11x}{-11} > \frac{-22}{-11}$$
$$x > 2$$

27.
$$-6x + 16 = -2x$$
$$-6x + 6x + 16 = -2x + 6x$$
$$16 = 4x$$
$$\frac{16}{4} = \frac{4x}{4}$$
$$4 = x$$
The solution is 4.

28.
$$P = 83\frac{1}{3}\% = \frac{5}{6}$$
$$PB = A$$
$$\frac{5}{6}B = 20$$
$$\frac{6}{5} \cdot \frac{5}{6}B = \frac{6}{5} \cdot 20$$
$$B = 24$$
The number is 24.

29.
$$\frac{2}{3}x \ge 2$$
$$\frac{3}{2}\left(\frac{2}{3}x\right) \ge \frac{3}{2}(2)$$
$$x \ge 3$$

30.
$$\begin{array}{c|c} x^2 + 2x + 1 & = (x + 1)^2 \\ \hline 5^2 + 2(5) + 1 & (5 + 1)^2 \\ 25 + 10 + 1 & 6^2 \\ 36 & = 36 \end{array}$$
Yes, 5 is a solution.

31. Strategy

To find the weight, solve the basic percent equation using $B = 180$ and $P = 16\frac{2}{3}\% = \frac{1}{6}$. The amount is unknown.

Solution

$$PB = A$$
$$\frac{1}{6}(180) = A$$
$$30 = A$$

The astronaut would weigh 30 lb on the moon.

32. Strategy

To find the final temperature, replace each of the variables by its given value and solve for T.

Solution

$$m_1(T_1 - T) = m_2(T - T_2)$$
$$100(80 - T) = 50(T - 20)$$
$$8000 - 100T = 50T - 1000$$
$$8000 + 1000 - 100T = 50T - 1000 + 1000$$
$$9000 - 100T = 50T$$
$$9000 - 100T + 100T = 50T + 100T$$
$$9000 = 150T$$
$$\frac{9000}{150} = \frac{150T}{150}$$
$$60 = T$$

The final temperature is 60°C.

33. Strategy

To find the number of calculators produced, replace each of the variables by its given value and solve for N.

Solution

$$T = U \cdot N + F$$
$$5000 = 15N + 2000$$
$$3000 = 15N$$
$$200 = N$$

200 calculators were produced.

34. Strategy

To find when the hikers will meet, solve the equation $d = rt$ find the time, where the distance is 15 mi and the rate is the sum of the hikers' rates (3.5 mph and 4 mph).

Solution

$$3.5 + 4 = 7.5 \text{ mph}$$
Using $d = rt$
$$15 = 7.5t$$
$$t = \frac{15}{7.5} = 2 \text{ h}$$

They will meet in 2 hours.

35. Strategy

To find the amount invested at 5%, first find the amount of interest earned in the first investment and solve the simple interest equation using $r = 5\% = 0.05$, and $t = 1$ year, for P.

Solution

$$I = Prt$$
$$I = 750(0.062)(1)$$
$$I = 46.50$$

$$I = Prt$$
$$46.50 = P(0.05)(1)$$
$$\frac{46.50}{0.05} = \frac{0.05P}{0.05}$$
$$930 = P$$

Victor must invest $930 at 5%.

36. Strategy

To find the percent, solve the basic percent equation using $B = 8$ and $A = 2$. The percent is unknown.

Solution

$$PB = A$$
$$P(8) = 2$$
$$\frac{8P}{8} = \frac{2}{8}$$
$$P = 0.25$$

The percent concentration is 25%.

Cumulative Review Exercises

1. $-6 - (-20) - 8 = -6 + 20 - 8$
$$= 14 - 8$$
$$= 6$$

2. $(-2)(-6)(-4) = 12(-4)$
$$= -48$$

3. $-\frac{5}{6} - \left(-\frac{7}{16}\right) = -\frac{5}{6} + \frac{7}{16}$
$$= -\frac{40}{48} + \frac{21}{48}$$
$$= \frac{-40 + 21}{48}$$
$$= \frac{-19}{48}$$
$$= -\frac{19}{48}$$

4. $-2\frac{1}{3} \div 1\frac{1}{6} = -\frac{7}{3} \div \frac{7}{6}$
$$= -\frac{7}{3} \cdot \frac{6}{7}$$
$$= -\frac{7 \cdot 6}{3 \cdot 7}$$
$$= -\frac{\overset{1}{7} \cdot 2 \cdot \overset{1}{3}}{\underset{1}{3} \cdot \underset{1}{7}}$$
$$= -2$$

5. $-4^2 \cdot \left(-\frac{3}{2}\right)^3 = -16 \cdot \left(-\frac{27}{8}\right)$

$$= 16 \cdot \frac{27}{8}$$

$$= \frac{16 \cdot 27}{8}$$

$$= \frac{\overset{1}{2} \cdot \overset{1}{2} \cdot \overset{1}{2} \cdot 2 \cdot 3 \cdot 3 \cdot 3}{\underset{1}{2} \cdot \underset{1}{2} \cdot \underset{1}{2}}$$

$$= 54$$

6. $25 - 3 \cdot \frac{(5-2)^2}{2^3 + 1} - (-2) = 25 - 3 \cdot \frac{3^2}{2^3 + 1} - (-2)$

$$= 25 - 3 \cdot \frac{9}{8+1} - (-2)$$

$$= 25 - 3 \cdot \frac{9}{9} - (-2)$$

$$= 25 - 3 \cdot 1 - (-2)$$

$$= 25 - 3 - (-2)$$

$$= 25 + (-3) + 2$$

$$= 22 + 2$$

$$= 24$$

7. $3(a - c) - 2ab = 3[2 - (-4)] - 2(2)(3)$

$$= 3(2 + 4) - 2(2)(3)$$

$$= 3(6) - 2(2)(3)$$

$$= 18 - 12$$

$$= 6$$

8. $3x - 8x + (-12x) = -5x + (-12x)$

$$= -17x$$

9. $2a - (-3b) - 7a - 5b = 2a + 3b - 7a - 5b$

$$= 2a - 7a + 3b - 5b$$

$$= (2 - 7)a + (3 - 5)b$$

$$= -5a - 2b$$

10. $16x\left(\frac{1}{8}\right) = \frac{16x}{8} = 2x$

11. $-4(-9y) = 36y$

12. $-2(-x^2 - 3x + 2) = -2(-x^2) - 2(-3x) - 2(2)$

$$= 2x^2 + 6x - 4$$

13. $-2(x - 3) + 2(4 - x) = -2x + 6 + 8 - 2x$

$$= -2x - 2x + 6 + 8$$

$$= -4x + 14$$

14. $-3[2x - 4(x - 3)] + 2 = -3[2x - 4x + 12] + 2$

$$= -3[-2x + 12] + 2$$

$$= 6x - 36 + 2$$

$$= 6x - 34$$

15. $\{-7, -6, -5, -4, -3, -2, -1\}$

16. $\frac{7}{8} = \frac{7}{8}(100\%) = \frac{700}{8}\% = 87\frac{1}{2}\%$

17. $342\% = 342(0.01) = 3.42$

18. $62\frac{1}{2}\% = 62\frac{1}{2}\left(\frac{1}{100}\right) = \frac{125}{2}\left(\frac{1}{100}\right) = \frac{5}{8}$

19.
$$\frac{x^2 + 6x + 9 = x + 3}{(-3)^2 + 6(-3) + 9 \,\Big|\, -3 + 3}$$
$$9 - 18 + 9 \,\Big|\, 0$$
$$0 = 0$$

Yes, -3 is a solution.

20. $x - 4 = -9$

$$x - 4 + 4 = -9 + 4$$

$$x = -5$$

The solution is -5.

21. $\frac{3}{5}x = -15$

$$\frac{5}{3}\left(\frac{3}{5}x\right) = \frac{5}{3}(-15)$$

$$x = -25$$

The solution is -25.

22. $13 - 9x = -14$

$$13 - 13 - 9x = -14 - 13$$

$$-9x = -27$$

$$\frac{-9x}{-9} = \frac{-27}{-9}$$

$$x = 3$$

The solution is 3.

23. $5x - 8 = 12x + 13$

$$5x - 12x - 8 = 12x - 12x + 13$$

$$-7x - 8 = 13$$

$$-7x - 8 + 8 = 13 + 8$$

$$-7x = 21$$

$$\frac{-7x}{-7} = \frac{21}{-7}$$

$$x = -3$$

The solution is -3.

24. $8x - 3(4x - 5) = -2x - 11$

$$8x - 12x + 15 = -2x - 11$$

$$-4x + 15 = -2x - 11$$

$$-4x + 2x + 15 = -2x + 2x - 11$$

$$-2x + 15 = -11$$

$$-2x + 15 - 15 = -11 - 15$$

$$-2x = -26$$

$$\frac{-2x}{-2} = \frac{-26}{-2}$$

$$x = 13$$

The solution is 13.

25. $-\frac{3}{4}x > \frac{2}{3}$

$$-\frac{4}{3}\left(-\frac{3}{4}x\right) < -\frac{4}{3}\left(\frac{2}{3}\right)$$

$$x < -\frac{8}{9}$$

26. $5x - 4 \geq 4x + 8$

$$5x - 4x - 4 \geq 4x - 4x + 8$$

$$x - 4 \geq 8$$

$$x - 4 + 4 \geq 8 + 4$$

$$x \geq 12$$

27.
$$3x + 17 < 5x - 1$$
$$3x - 5x + 17 < 5x - 5x - 1$$
$$-2x + 17 < -1$$
$$-2x + 17 - 17 < -1 - 17$$
$$-2x < -18$$
$$\frac{-2x}{-2} > \frac{-18}{-2}$$
$$x > 9$$

28. the unknown number: n

the quotient of a number and 12: $\frac{n}{12}$

the difference between eight and the quotient of a number and 12

$$8 - \frac{n}{12}$$

29. the unknown number: n
two more than the number: $n + 2$
the sum of a number and two more than the number

$$n + (n + 2) = n + n + 2$$
$$= 2n + 2$$

30. number of five-dollar bills: b
number of ten-dollar bills: $35 - b$

31. length of the longer piece: L
length of the shorter piece: $3 - L$

32. Strategy
To find the percent of the computer programmer's salary deducted for income tax, solve the basic percent equation using $B = \$650$ and $A = \$110.50$.

Solution
$$PB = A$$
$$P(650) = 110.50$$
$$\frac{P(650)}{650} = \frac{110.50}{650}$$
$$P = 0.17$$

The percent is 17%.

33. Strategy
To find the year, replace t by 4 in the equation and solve for y.

Solution
$$t = 17.08 - 0.0067y$$
$$4 = 17.08 - 0.0067y$$
$$4 - 17.08 = 17.08 - 17.08 - 0.0067y$$
$$-13.08 = -0.0067y$$
$$\frac{-13.08}{-0.0067} = \frac{-0.0067y}{-0.0067}$$
$$1952 \approx y$$

The predicted year is 1952.

34. Strategy
To find the final temperature, replace each of the variables by its given value and solve for T.

Solution
$$m_1(T_1 - T) = m_2(T - T_2)$$
$$300(75 - T) = 100(T - 15)$$
$$22,500 - 300T = 100T - 1500$$
$$22,500 + 1500 - 300T = 100T - 1500 + 1500$$
$$24,000 - 300T = 100T$$
$$24,000 - 300T + 300T = 100T + 300T$$
$$24,000 = 400T$$
$$\frac{24,000}{400} = \frac{400T}{400}$$
$$60 = T$$

The final temperature is 60°C.

35. Strategy
To find the force that will balance the lever, replace the variables F_1, x, and d in the lever system equation with the given values, and solve for F_2.

Solution
$$F_1 x = F_2(d - x)$$
$$26(12) = F_2(25 - 12)$$
$$312 = F_2(13)$$
$$\frac{312}{13} = \frac{F_2(13)}{13}$$
$$24 = F_2$$

The system will balance if a force of 24 lb is applied.

Chapter 4: Solving Equations and Inequalities: Applications

Prep Test

1. 0.65R [2.2.2]

2. $0.08x + 0.05(400 - x) = 0.08x + 20 - 0.05x$
   $= 0.03x + 20$

   [2.2.5]

3. $3n + 6$ [2.2.2]

4. $5 - 2x$ [2.3.1]

5. $0.4(100\%) = 40\%$ [1.3.4]

6. $25x + 10(9 - x) = 120$ [3.2.3]
   $25x + 90 - 10x = 120$
   $15x + 90 - 90 = 120 - 90$
   $\dfrac{15x}{15} = \dfrac{30}{15}$
   $x = 2$

7. $36 = 48 - 48r$ [3.2.1]
   $36 - 48 = 48 - 48 - 48r$
   $\dfrac{-12}{-48} = \dfrac{-48r}{-48}$
   $0.25 = r$

8. $4(2x - 5) < 12$ [3.3.3]
   $8x - 20 < 12$
   $8x - 20 + 20 < 12 + 20$
   $\dfrac{8x}{8} < \dfrac{32}{8}$
   $x < 4$

9. the number of nuts: n [2.3.3]
   the number of pretzels: $20 - n$

Go Figure

If $x + y$, xy, and $\dfrac{x}{y}$ all equal the same number,

then they also equal each other. So, $xy = \dfrac{x}{y}$.

Solving for y, divide both sides by x, and multiply both sides by y to get $y^2 = 1$. Then y must be either 1 or -1.
Use $x + y = xy$ check our solution,
If $y = 1$, then $x + 1 = x(1)$
Subtracting x from both sides we have
$1 = 0$, which is never true. So, $y = 1$ is not a valid solution.
If $y = -1$, then
$x - 1 = x(-1)$
$x - 1 = -x$ add x to both sides
$2x = 1$ divide both sides by 2
$x = \dfrac{1}{2}$

So the values are $x = \dfrac{1}{2}$ and $y = -1$.

Section 4.1

Concept Review 4.1

1. Always true

3. Always true

5. Never true
   The length of the other piece is expressed as $(19 - x)$ in.

Objective 4.1.1 Exercises

1. the unknown number: n

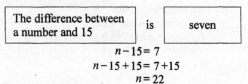

   $n - 15 = 7$
   $n - 15 + 15 = 7 + 15$
   $n = 22$

   The number is 22.

3. the unknown number: n

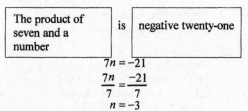

   $7n = -21$
   $\dfrac{7n}{7} = \dfrac{-21}{7}$
   $n = -3$

   The number is -3.

5. the unknown number: n

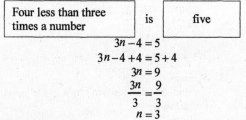

   $3n - 4 = 5$
   $3n - 4 + 4 = 5 + 4$
   $3n = 9$
   $\dfrac{3n}{3} = \dfrac{9}{3}$
   $n = 3$

   The number is 3.

7. the unknown number: n

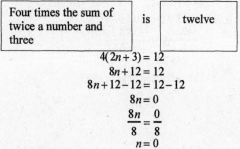

   $4(2n + 3) = 12$
   $8n + 12 = 12$
   $8n + 12 - 12 = 12 - 12$
   $8n = 0$
   $\dfrac{8n}{8} = \dfrac{0}{8}$
   $n = 0$

   The number is 0.

9. the unknown number: n

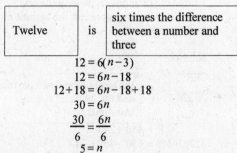

$$12 = 6(n-3)$$
$$12 = 6n - 18$$
$$12 + 18 = 6n - 18 + 18$$
$$30 = 6n$$
$$\frac{30}{6} = \frac{6n}{6}$$
$$5 = n$$

The number is 5.

11. the unknown number: n

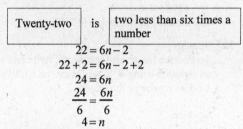

$$22 = 6n - 2$$
$$22 + 2 = 6n - 2 + 2$$
$$24 = 6n$$
$$\frac{24}{6} = \frac{6n}{6}$$
$$4 = n$$

The number is 4.

13. the unknown number: n

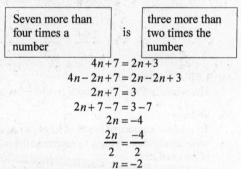

$$4n + 7 = 2n + 3$$
$$4n - 2n + 7 = 2n - 2n + 3$$
$$2n + 7 = 3$$
$$2n + 7 - 7 = 3 - 7$$
$$2n = -4$$
$$\frac{2n}{2} = \frac{-4}{2}$$
$$n = -2$$

The number is –2.

15. the unknown number: n

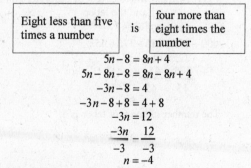

$$5n - 8 = 8n + 4$$
$$5n - 8n - 8 = 8n - 8n + 4$$
$$-3n - 8 = 4$$
$$-3n - 8 + 8 = 4 + 8$$
$$-3n = 12$$
$$\frac{-3n}{-3} = \frac{12}{-3}$$
$$n = -4$$

The number is –4.

17. the unknown number: n

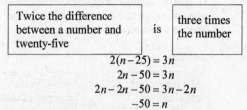

$$2(n - 25) = 3n$$
$$2n - 50 = 3n$$
$$2n - 2n - 50 = 3n - 2n$$
$$-50 = n$$

The number is –50.

19. the smaller number: n
the larger number: $20 - n$

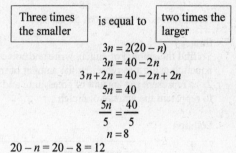

$$3n = 2(20 - n)$$
$$3n = 40 - 2n$$
$$3n + 2n = 40 - 2n + 2n$$
$$5n = 40$$
$$\frac{5n}{5} = \frac{40}{5}$$
$$n = 8$$

$20 - n = 20 - 8 = 12$
The smaller number is 8.
The larger number is 12.

21. the smaller number: n
the larger number: $18 - n$

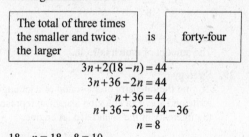

$$3n + 2(18 - n) = 44$$
$$3n + 36 - 2n = 44$$
$$n + 36 = 44$$
$$n + 36 - 36 = 44 - 36$$
$$n = 8$$

$18 - n = 18 - 8 = 10$
The smaller number is 8.
The larger number is 10.

Objective 4.1.2 Exercises

23. Strategy
To find the original value of the car, write and solve an equation using v to represent the original value of the car.

Solution

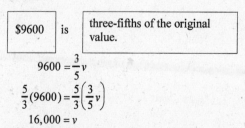

$$9600 = \frac{3}{5}v$$
$$\frac{5}{3}(9600) = \frac{5}{3}\left(\frac{3}{5}v\right)$$
$$16,000 = v$$

The original value of the car was $16,000.

25. Strategy
To find the number of calories, write and solve an equation using c to represent the number of calories in a medium-sized orange.

Solution

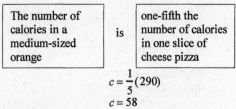

| The number of calories in a medium-sized orange | is | one-fifth the number of calories in one slice of cheese pizza |

$$c = \frac{1}{5}(290)$$
$$c = 58$$
A medium-sized orange has 58 calories.

27. Strategy
To find the amount of mulch, write and solve an equation using x to represent the amount of iron, $2x$ to represent the amount of potassium, and $5x$ to represent the amount of mulch.

Solution

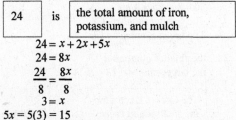

| 24 | is | the total amount of iron, potassium, and mulch |

$$24 = x + 2x + 5x$$
$$24 = 8x$$
$$\frac{24}{8} = \frac{8x}{8}$$
$$3 = x$$
$$5x = 5(3) = 15$$
The amount of mulch is 15 lb.

29. Strategy
To find the intensity of the sound of a jet engine, write and solve an equation using j to represent the intensity of the sound of a jet engine.

Solution

| Intensity of the sound of a jet engine | is | 40 less than twice the intensity of sound of a blender |

$$j = 2(70 + 20) - 40$$
$$j = 180 - 40$$
$$j = 140$$
The intensity of the sound of a jet engine is 140 decibels.

31. Strategy
To find the monthly payment, write and solve an equation using x to represent the amount of the monthly payment.

Solution

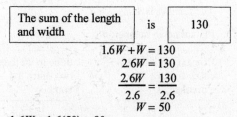

| 3276 | is | the total of the down payment and the 24 equal monthly installments |

$$3276 = 450 + 24x$$
$$3276 - 450 = 450 - 450 + 24x$$
$$2826 = 24x$$
$$\frac{2826}{24} = \frac{24x}{24}$$
$$117.75 = x$$
The monthly payment is \$117.75.

33. Strategy
To find the length and the width, write and solve an equation using W to represent the width and $1.6W$ to represent the length.

Solution

| The sum of the length and width | is | 130 |

$$1.6W + W = 130$$
$$2.6W = 130$$
$$\frac{2.6W}{2.6} = \frac{130}{2.6}$$
$$W = 50$$
$$1.6W = 1.6(50) = 80$$
The width is 50 ft. The length is 80 ft.

35. Strategy
To find the number of hours of labor, write and solve an equation using x to represent the number of hours of labor.

Solution

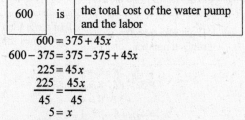

| 600 | is | the total cost of the water pump and the labor |

$$600 = 375 + 45x$$
$$600 - 375 = 375 - 375 + 45x$$
$$225 = 45x$$
$$\frac{225}{45} = \frac{45x}{45}$$
$$5 = x$$
The number of hours of labor is 5.

37. Strategy

To find the number of tickets, write and solve an equation using n to represent the number of tickets purchased.

Solution

343	is	the total cost of the $5.50 fee plus the cost of the tickets

$$343 = 5.50 + 37.5n$$
$$337.5 = 37.5n$$
$$\frac{337.5}{37.5} = \frac{37.5n}{37.5}$$
$$9 = n$$

9 tickets are purchased.

39. Strategy

To find the length of the patio, write and solve an equation using W to represent the width and $3W - 3$ to represent the length.

Solution

The sum of the length and width	is	21

$$(3W - 3) + W = 21$$
$$4W - 3 = 21$$
$$4W - 3 + 3 = 21 + 3$$
$$4W = 24$$
$$\frac{4W}{4} = \frac{24}{4}$$
$$W = 6$$

$$3W - 3 = 3(6) - 3 = 15$$
The length is 15 ft.

41. Strategy

To find the number of hours, write and solve an equation using n to represent the number of hours worked.

Solution

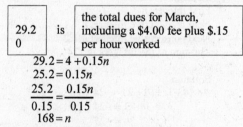

29.20	is	the total dues for March, including a $4.00 fee plus $.15 per hour worked

$$29.2 = 4 + 0.15n$$
$$25.2 = 0.15n$$
$$\frac{25.2}{0.15} = \frac{0.15n}{0.15}$$
$$168 = n$$

The union member worked 168 h.

43. Strategy

To find the number of minutes, write and solve an equation using m to represent the number of minutes the service was used.

Solution

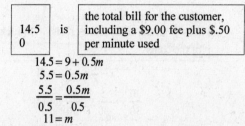

14.50	is	the total bill for the customer, including a $9.00 fee plus $.50 per minute used

$$14.5 = 9 + 0.5m$$
$$5.5 = 0.5m$$
$$\frac{5.5}{0.5} = \frac{0.5m}{0.5}$$
$$11 = m$$

The customer used the hotline for 11 min.

Applying Concepts 4.1

45. Strategy

To find the part of the container filled at 3:39 P.M., write and solve an equation using x to represent the amount of liquid in the container at 3:39 P.M..

Solution

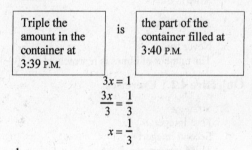

Triple the amount in the container at 3:39 P.M.	is	the part of the container filled at 3:40 P.M.

$$3x = 1$$
$$\frac{3x}{3} = \frac{1}{3}$$
$$x = \frac{1}{3}$$

$\frac{1}{3}$ of the container is filled at 3:39 P.M.

47. Strategy

To find how much the package weighs, write and solve an equation using x to represent the number of additional ounces the package weighs.

Solution

$.37 and $.27 for each additional ounce	is	$3.13

$$0.37 + 0.23x = 3.13$$
$$0.37 - 0.37 + 0.23x = 3.13 - 0.37$$
$$0.23x = 2.76$$
$$\frac{0.23x}{0.23} = \frac{2.76}{0.23}$$
$$x = 12$$

The package weighs more than 12 additional ounces but not over 13 ounces.

49. Strategy
To find how many coins are in the coin bank, write and solve an equation using x to represent the number of coins in the bank.

Solution

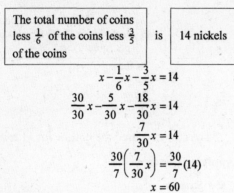

| The total number of coins less $\frac{1}{6}$ of the coins less $\frac{3}{5}$ of the coins | is | 14 nickels |

$$x - \frac{1}{6}x - \frac{3}{5}x = 14$$
$$\frac{30}{30}x - \frac{5}{30}x - \frac{18}{30}x = 14$$
$$\frac{7}{30}x = 14$$
$$\frac{30}{7}\left(\frac{7}{30}x\right) = \frac{30}{7}(14)$$
$$x = 60$$

There are 60 coins in the bank.

Section 4.2

Concept Review 4.2

1. Always true

3. Always true

5. Never true
The number of dimes is represented by $20 - n$.

Objective 4.2.1 Exercises

3. Strategy
First integer: n
Second integer: $n + 1$
Third integer: $n + 2$
The sum of the integers is 54.

Solution
$$n + (n + 1) + (n + 2) = 54$$
$$3n + 3 = 54$$
$$3n = 51$$
$$n = 17$$
$n + 1 = 17 + 1 = 18$
$n + 2 = 17 + 2 = 19$
The integers are 17, 18, and 19.

5. Strategy
First even integer: n
Second even integer: $n + 2$
Third even integer: $n + 4$
The sum of the three integers is 84.

Solution
$$n + (n + 2) + (n + 4) = 84$$
$$3n + 6 = 84$$
$$3n = 78$$
$$n = 26$$
$n + 2 = 26 + 2 = 28$
$n + 4 = 26 + 4 = 30$
The three integers are 26, 28, and 30.

7. Strategy
First odd integer: n
Second odd integer: $n + 2$
Third odd integer: $n + 4$
The sum of the three integers is 57.

Solution
$$n + (n + 2) + (n + 4) = 57$$
$$3n + 6 = 57$$
$$3n = 51$$
$$n = 17$$
$n + 2 = 17 + 2 = 19$
$n + 4 = 17 + 4 = 21$
The three integers are 17, 19, and 21.

9. Strategy
First even integer: n
Second even integer: $n + 2$
Five times the first equals four times the second.

Solution
$$5n = 4(n + 2)$$
$$5n = 4n + 8$$
$$n = 8$$
$n + 2 = 8 + 2 = 10$
The integers are 8 and 10.

11. Strategy
First odd integer: n
Second odd integer: $n + 2$
Nine times the first equals seven times the second.

Solution
$$9n = 7(n + 2)$$
$$9n = 7n + 14$$
$$2n = 14$$
$$n = 7$$
$n + 2 = 7 + 2 = 9$
The integers are 7 and 9.

13. Strategy
First integer: n
Next consecutive integer: $n + 1$
Next consecutive integer: $n + 2$
The sum of the integers is –24.

Solution
$$n + (n + 1) + (n + 2) = -24$$
$$3n + 3 = -24$$
$$3n = -27$$
$$n = -9$$
$n + 1 = -9 + 1 = -8$
$n + 2 = -9 + 2 = -7$
The integers are –9, –8, and –7.

15. Strategy
First even integer: n
Second even integer: $n + 2$
Third even integer: $n + 4$
Three times the smallest even integer equals two more than twice the largest.

Solution
$3n = 2(n + 4) + 2$
$3n = 2n + 8 + 2$
$n = 10$
$n + 2 = 10 + 2 = 12$
$n + 4 = 10 + 4 = 14$
The integers are 10, 12, and 14.

17. Strategy
First odd integer: n
Second odd integer: $n + 2$
Third odd integer: $n + 4$
Three times the middle number is six more than the sum of the first and third numbers.

Solution
$3(n + 2) = n + (n + 4) + 6$
$3n + 6 = 2n + 10$
$n + 6 = 10$
$n = 4$
$n + 2 = 4 + 2 = 6$
$n + 4 = 4 + 4 = 8$
There is no solution, since the solutions must be odd.

Objective 4.2.2 Exercises

21. Strategy
Number of dimes: x
Number of quarters: $27 - x$

Coin	Number	Value	Total value
Dime	x	10	$10x$
Quarter	$27 - x$	25	$25(27 - x)$

The sum of the total values of each type of coin equals the total value of all the coins (495 cents).

Solution
$10x + 25(27 - x) = 495$
$10x + 675 - 25x = 495$
$-15x + 675 = 495$
$-15x = -180$
$x = 12$
$27 - x = 27 - 12 = 15$
There are 12 dimes and 15 quarters in the bank.

23. Strategy
Number of 37¢ stamps: x
Number of 23¢ stamps: $40 - x$

Stamp	Number	Value	Total value
37¢	x	37	$37x$
23¢	$40 - x$	23	$23(40 - x)$

The sum of the total values of each type of stamp equals the total value of all the stamps (1368 cents).

Solution
$37x + 23(40 - x) = 1368$
$37x + 920 - 23x = 1368$
$920 + 14x = 1368$
$14x = 448$
$x = 32$
$40 - x = 40 - 32 = 8$
There were thirty-two 37¢ stamps and eight 23¢ stamps purchased.

25. Strategy
Number of 29¢ stamps: $3x - 4$
Number of 3¢ stamps: x

Stamp	Number	Value	Total value
29¢	$3x - 4$	29	$29(3x - 4)$
3¢	x	3	$3x$

The sum of the total values of each type of stamp equals the total value of all the stamps (154 cents).

Solution
$29(3x - 4) + 3x = 154$
$87x - 116 + 3x = 154$
$90x - 116 = 154$
$90x = 270$
$x = 3$
$3x - 4 = 3(3) - 4 = 9 - 4 = 5$
There were five 29¢ stamps in the drawer.

27. Strategy
Number of quarters: x
Number of dimes $44 - x$

Coin	Number	Value	Total value
Quarter	x	25	$25x$
Dime	$44 - x$	10	$10(44 - x)$

The sum of the total values of each type of coin equals the total value of all the coins (860 cents).

Solution
$25x + 10(44 - x) = 860$
$25x + 440 - 10x = 860$
$15x + 440 = 860$
$15x = 420$
$x = 28$
There are 28 quarters in the piggy bank.

29. Strategy
Number of one-dollar bills: x
Number of five-dollar bills: $26 - x$

Bill	Number	Value	Total value
$1	x	1	$1x$
$5	$26 - x$	5	$5(26 - x)$

The sum of the total values of each type of bill equals the total value of all the bills ($50).

Solution
$$1x + 5(26 - x) = 50$$
$$x + 130 - 5x = 50$$
$$130 - 4x = 50$$
$$-4x = -80$$
$$x = 20$$
$$26 - x = 26 - 20 = 6$$
There are 20 one-dollar bills and 6 five-dollar bills in the cash box.

31. Strategy
Number of pennies: x
Number of nickels: $6x$
Number of dimes: $4x$

Coin	Number	Value	Total value
Penny	x	1	$1x$
Nickel	$6x$	5	$5(6x)$
Dime	$4x$	10	$10(4x)$

The sum of the total values of each type of coin equals the total value of all the coins (781 cents).

Solution
$$1x + 5(6x) + 10(4x) = 781$$
$$x + 30x + 40x = 781$$
$$71x = 781$$
$$x = 11$$
There are 11 pennies in the bank.

33. Strategy
Number of 40¢ stamps: x
Number of 22¢ stamps: $4x + 3$

Stamp	Number	Value	Total value
40¢	x	40	$40x$
22¢	$4x + 3$	22	$22(4x + 3)$

The sum of the total values of each type of stamp equals the total value of all the stamps (834 cents).

Solution
$$40x + 22(4x + 3) = 834$$
$$40x + 88x + 66 = 834$$
$$128x + 66 = 834$$
$$128x = 768$$
$$x = 6$$
$$4x + 3 = 4(6) + 3 = 24 + 3 = 27$$
There are twenty-seven 22¢ stamps in the collection.

35. Strategy
Number of 3¢ stamps: $x - 5$
Number of 7¢ stamps: x
Number of 12¢ stamps: $\frac{1}{2}x$

Stamps	Number	Value	Total value
3¢	$x - 5$	3	$3(x - 5)$
7¢	x	7	$7x$
12¢	$\frac{1}{2}x$	12	$12\left(\frac{1}{2}x\right)$

The sum of the total values of each type of stamp equals the total value of all the stamps (273 cents).

Solution
$$3(x - 5) + 7x + 12\left(\frac{1}{2}x\right) = 273$$
$$3x - 15 + 7x + 6x = 273$$
$$16x - 15 = 273$$
$$16x = 288$$
$$x = 18$$
$$x - 5 = 18 - 5 = 13$$
$$\frac{1}{2}x = \frac{1}{2}(18) = 9$$
There are thirteen 3¢ stamps, eighteen 7¢ stamps, and nine 12¢ stamps in the collection.

37. Strategy
Number of 6¢ stamps: $3x$
Number of 8¢ stamps: x
Number of 15¢ stamps: $3x + 6$

Stamp	Number	Value	Total value
6¢	$3x$	6	$6(3x)$
8¢	x	8	$8x$
15¢	$3x + 6$	15	$15(3x + 6)$

The sum of the total values of each type of stamp equals the total value of all the stamps (516 cents).

Solution
$$6(3x) + 8x + 15(3x + 6) = 516$$
$$18x + 8x + 45x + 90 = 516$$
$$71x + 90 = 516$$
$$71x = 426$$
$$x = 6$$
$$3x = 3(6) = 18$$
$$3x + 6 = 3(6) + 6 = 18 + 6 = 24$$
There are eighteen 6¢ stamps, six 8¢ stamps, and twenty-four 15¢ stamps in the collection.

Applying Concepts 4.2

39. Strategy
First even integer: n
Second even integer: $n + 2$
Third even integer: $n + 4$
Fourth even integer: $n + 6$
The sum of the four integers is -36.

Solution
$$n + (n + 2) + (n + 4) + (n + 6) = -36$$
$$4n + 12 = -36$$
$$4n = -48$$
$$n = -12$$
$$n + 2 = -12 + 2 = -10$$
$$n + 4 = -12 + 4 = -8$$
$$n + 6 = -12 + 6 = -6$$
The four integers are -12, -10, -8, and -6.

41. Strategy
Number of dimes: x
Number of quarters: $2x - 2$

Coin	Number	Value	Total value
Dime	x	10	$10x$
Quarter	$2x - 2$	25	$25(2x - 2)$

The total number of coins is 34.

Solution
$$x + 2x - 2 = 34$$
$$3x - 2 = 34$$
$$3x = 36$$
$$x = 12$$
$$2x - 2 = 2(12) - 2 = 24 - 2 = 22$$
There are 12 dimes and 22 quarters in the bank.
Total value $= 10(12) + 25(22) = 120 + 550 = 670$
$6.70 is in the bank.

43. Strategy
First odd integer: n
Second odd integer: $n + 2$
Third odd integer: $n + 4$
The sum of the first and third is twice the second.

Solution
$$n + (n + 4) = 2(n + 2)$$
$$2n + 4 = 2n + 4$$
$$4 = 4$$
This is a true equation. For any three consecutive odd integers, the sum of the first and third is twice the second.

Section 4.3

Concept Review 4.3

1. Always true

3. Always true

5. Never true
The perimeter of a geometric figure measures the distance around the figure.

Objective 4.3.1 Exercises

3. Strategy
Each equal side: x
The third side: $0.5x$
Use the equation for the perimeter of a triangle.

Solution
$$a + b + c = P$$
$$x + x + 0.50x = 125$$
$$2.50x = 125$$
$$x = 50$$
$$0.50x = 0.50(50) = 25$$
The lengths of the sides are 25 ft, 50 ft, and 50 ft.

5. Strategy
Width: W
Length: $2W - 3$
Use the equation for the perimeter of
a rectangle.

Solution
$$2L + 2W = P$$
$$2(2W - 3) + 2W = 42$$
$$4W - 6 + 2W = 42$$
$$6W - 6 = 42$$
$$6W = 48$$
$$W = 8$$
$$2W - 3 = 2(8) - 3 = 16 - 3$$
$$= 13$$
The width is 8 m.
The length is 13 m.

7. Strategy
Width: W
Length: $2W$
Use the equation for the perimeter of
a rectangle.

Solution
$$2L + 2W = P$$
$$2(2W) + 2W = 120$$
$$4W + 2W = 120$$
$$6W = 120$$
$$W = 20$$
$$2W = 2(20) = 40$$
The width is 20 ft.
The length is 40 ft.

9. Strategy
First side: $2x$
Second side: x
Third side: $x + 30$
Use the equation for the perimeter of
a triangle.

Solution
$$a + b + c = P$$
$$2x + x + (x + 30) = 110$$
$$4x + 30 = 110$$
$$4x = 80$$
$$x = 20$$
$$2x = 2(20) = 40$$
$$x + 30 = 20 + 30 = 50$$
The sides of the triangle measure 20 cm, 40 cm,
and 50 cm.

11. Strategy
Length: L
Width: $0.30L$
Use the equation for the perimeter of
a rectangle.

Solution
$$2L + 2W = P$$
$$2L + 2(0.30L) = 338$$
$$2L + 0.60L = 338$$
$$2.60L = 338$$
$$L = 130$$
$$0.30L = 0.30(130) = 39$$
The length is 130 ft.
The width is 39 ft.

13. Strategy
To find the width, use the formula for the
perimeter of a rectangle. Substitute 64 for P and
20 for L. Solve for W.

Solution
$$P = 2L + 2W$$
$$64 = 2(20) + 2W$$
$$64 = 40 + 2W$$
$$24 = 2W$$
$$12 = W$$
The width is 12 ft.

15. Strategy
To find the length of each side, use the formula
for the perimeter of a square. Substitute 48 for P.
Solve for s.

Solution
$$P = 4s$$
$$48 = 4s$$
$$12 = s$$
The length of each side is 12 in.

Objective 4.3.2 Exercises

17. a. 90°
  b. 180°
  c. 360°
  d. between 0° and 90°
  e. between 90° and 180°
  f. 90°
  g. 180°

19. Strategy
    To find the complement, let x represent the complement of a 28° angle. Use the fact that complementary angles are two angles whose sum is 90° to write an equation. Solve for x.

    Solution
    $$x + 28° = 90°$$
    $$x = 62°$$
    The complement of a 28° angle is a 62° angle.

21. Strategy
    To find the supplement, let x represent the supplement of a 73° angle. Use the fact that supplementary angles are two angles whose sum is 180° to write an equation. Solve for x.

    Solution
    $$x + 73° = 180°$$
    $$x = 107°$$
    The supplement of a 73° angle is a 107° angle.

23. Strategy
    To find the measure of $\angle x$, write an equation using the fact that the sum of x and $x + 20°$ is 90°. Solve for $\angle x$.

    Solution
    $$x + 20° + x = 90°$$
    $$2x + 20° = 90°$$
    $$2x = 70°$$
    $$x = 35°$$
    The measure of $\angle x$. is 35°.

25. Strategy
    To find the measure of $\angle x$, write an equation using the fact that the sum of x and $3x + 10°$ is 90°. Solve for $\angle x$.

    Solution
    $$x + 3x + 10° = 90°$$
    $$4x + 10 = 90°$$
    $$4x = 80°$$
    $$x = 20°$$
    The measure of $\angle x$ is 20°.

27. Strategy
    To find the measure of $\angle a$, write an equation using the fact that the sum of the measure of $\angle a$ and 127° is 180°. Solve for $\angle a$.

    Solution
    $$\angle a + 127° = 180°$$
    $$\angle a = 53°$$
    The measure of $\angle a$ is 53°.

29. Strategy
    The sum of the measures of the three angles shown is 360°. To find $\angle a$, write an equation and solve for $\angle a$.

    Solution
    $$\angle a + 67° + 172° = 360°$$
    $$\angle a + 239° = 360°$$
    $$\angle a = 121°$$
    The measure of $\angle a$ is 121°.

31. Strategy
    The sum of the measures of the three angles shown is 180°. To find x, write an equation and solve for x.

    Solution
    $$4x + 6x + 2x = 180°$$
    $$12x = 180°$$
    $$x = 15°$$
    The measure of x is 15°.

33. Strategy
    The sum of the measures of the three angles shown is 180°. To find x, write an equation and solve for x.

    Solution
    $$3x + (x + 36°) + 4x = 180°$$
    $$8x + 36° = 180°$$
    $$8x = 144°$$
    $$x = 18°$$
    The measure of x is 18°.

35. Strategy
    The sum of the measures of the four angles shown is 360°. To find x, write an equation and solve for x.

    Solution
    $$x + 2x + 3x + 2x = 360°$$
    $$8x = 360°$$
    $$x = 45°$$
    The measure of x is 45°.

37. Strategy
    The angles labeled are adjacent angles of intersecting lines and are, therefore, supplementary angles. To find x, write an equation and solve for x.

    Solution
    $$x + 131° = 180°$$
    $$x = 49°$$
    The measure of x is 49°.

39. Strategy
The angles labeled are vertical angles and are, therefore, equal. To find x, write an equation and solve for x.

Solution
$7x = 4x + 36°$
$3x = 36°$
$x = 12°$
The measure of x is 12°.

41. Strategy
To find the measure of $\angle a$, use the fact that alternate interior angles of parallel lines are equal.
To find the measure of $\angle b$, use the fact that adjacent angles of intersecting lines are supplementary.

Solution
$\angle a = 122°$
$\angle b + \angle a = 180°$
$\angle b + 122° = 180°$
$\angle b = 58°$
The measure of $\angle a$ is 122°.
The measure of $\angle b$ is 58°.

43. Strategy
To find the measure of $\angle b$, use the fact that alternate interior angles of parallel lines are equal.
To find the measure of $\angle a$, use the fact that adjacent angles of intersecting lines are supplementary.

Solution
$\angle b = 136°$
$\angle a + \angle b = 180°$
$\angle a + 136° = 180°$
$\angle a = 44°$
The measure of $\angle a$ is 44°.
The measure of $\angle b$ is 136°.

45. Strategy

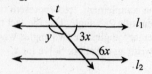

$6x = y$ because alternate interior angles have the same measure. $y + 3x = 180°$ because adjacent angles of intersecting lines are supplementary. Substitute $6x$ for y and solve for x.

Solution
$6x + 3x = 180°$
$9x = 180°$
$x = 20°$
The measure of x is 20°.

47. Strategy

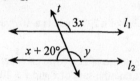

$y = 3x$ because corresponding angles have the same measure. $y + x + 20° = 180°$ because adjacent angles of intersecting lines are supplementary angles. Substitute $3x$ for y and solve for x.

Solution
$3x + x + 20° = 180°$
$4x + 20° = 180°$
$4x = 160°$
$x = 40°$
The measure of x is 40°.

49. Strategy

To find the measure of $\angle b$:
Use the fact that $\angle a$ and $\angle c$ are complementary angles.
Find $\angle b$ by using the fact that $\angle b$ and $\angle c$ are supplementary angles.

Solution
$\angle a + \angle c = 90°$
$38° + \angle c = 90°$
$\angle c = 52°$
$\angle b + \angle c = 180°$
$\angle b + 52° = 180°$
$\angle b = 128°$
The measure of $\angle b$ is 128°.

Objective 4.3.3 Exercises

51. Strategy

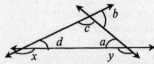

To find the measure of angle *y*, use the fact that sum of an interior angle and exterior angle is 180°.
To find the measure of angle *x*:
Find the measure of angle *c* by using the fact that the sum of an interior and exterior angle is 180°.
Find the measure of angle *d* by using the fact that the sum of the interior angles of a triangle is 180°.
Find the measure of angle *x* by using the fact that the sum of an interior and exterior angle is 180°.

Solution
$$\angle a + \angle y = 180°$$
$$35° + \angle y = 180°$$
$$\angle y = 145°$$
$$\angle b + \angle c = 180°$$
$$55° + \angle c = 180°$$
$$\angle c = 125°$$
$$\angle a + \angle c + \angle d = 180°$$
$$35° + 125° + \angle d = 180°$$
$$160° + \angle d = 180°$$
$$\angle d = 20°$$
$$\angle x + \angle d = 180°$$
$$\angle x + 20° = 180°$$
$$\angle x = 160°$$
The measure of $\angle x$ is 160°.
The measure of $\angle y$ is 145°.

53. Strategy

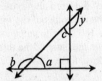

To find the measure of angle *a*:
Find the measure of angle *c* by using the fact that the sum of an interior and exterior angle is 180°.
Find the measure of angle *a* by using the fact that the sum of the measures of the interior angles of a triangle is 180°.
To find the measure of angle *b*, use the fact that the sum of an interior and exterior angle is 180°.

Solution
$$\angle c + \angle y = 180°$$
$$\angle c + 130° = 180°$$
$$\angle c = 50°$$
$$\angle a + \angle c + 90° = 180°$$
$$\angle a + 50° + 90° = 180°$$
$$\angle a + 140° = 180°$$
$$\angle a = 40°$$
$$\angle a + \angle b = 180°$$
$$40° + \angle b = 180°$$
$$\angle b = 140°$$
The measure of $\angle a$ is 40°.
The measure of $\angle b$ is 140°.

55. Strategy
To find the measure of $\angle AOC$, use the fact that the sum of the measures of the angles $x + 15°$, $\angle AOC$, and $\angle AOB$ is 180°. Since $AO \perp OB$, $\angle AOB$ is 90°.

Solution
$$(x + 15°) + \angle AOC + \angle AOB = 180°$$
$$x + 15° + \angle AOC + 90° = 180°$$
$$x + \angle AOC + 105° = 180°$$
$$\angle AOC = 75° - x$$
The measure of $\angle AOC$ is $75° - x$.

57. Strategy
To find the measure of the third angle, use the fact that the sum of the measures of the interior angles of a triangle is 180°. Write an equation using *x* to represent the measure of the third angle. Solve the equation of *x*.

Solution
$$x + 45° + 90° - 180°$$
$$x + 135° = 180°$$
$$x = 45°$$
The measure of the third angle is 45°.

59. Strategy
To find the measure of the third angle, use the fact that the sum of the measures of the interior angles of a triangle is 180°. Write an equation using x to represent the measure of the third angle. Solve the equation for x.

Solution
$x + 62° + 45° = 180°$
$x + 107° = 180°$
$x = 73°$
The measure of the third angle is 73°.

61. Strategy
To find the measure of the third angle, use the fact that the sum of the measures of interior angles of a triangle is 180°. Write an equation using x to represent the measure of the third angle. Solve the equation for x.

Solution
$x + 105° + 32° = 180°$
$x + 137° = 180°$
$x = 43°$
The measure of the third angle is 43°.

63. Strategy
Measure of one of the equal angles: x
Measure of the third angle: $3x - 10$
The sum of the measures of the angles of a triangle is 180°.

Solution
$x + x + 3x - 10 = 180$
$5x - 10 = 180$
$5x = 190$
$\dfrac{5x}{5} = \dfrac{190}{5}$
$x = 38$
$3x - 10 = 3(38) - 10 = 104$
The measures of the angles are 38°, 38°, and 104°.

65. Strategy
Measure of the first angle: $2x$
Measure of the second angle: x
Measure of the third angle: $3x$
The sum of the measures of the angles of triangle is 180°.

Solution
$2x + x + 3x = 180$
$6x = 180$
$\dfrac{6x}{6} = \dfrac{180}{6}$
$x = 30$
$2x = 60$
$3x = 90$
The measures of the angles are 60°, 30°, and 90°.

Applying Concepts 4.3

67. Strategy
Width of the original rectangle: W
Length of original rectangle: $2W + 1$
Width of new rectangle: $W - 1$
Length of new rectangle:
$2W + 1 - 2 = 2W - 1$
New perimeter is 20 cm.

Solution
$2(2W - 1) + 2(W - 1) = 20$
$4W - 2 + 2W - 2 = 20$
$6W - 4 = 20$
$6W = 24$
$W = 4$
$2W + 1 = 2(4) + 1 = 9$
The original length is 9 cm and the original width is 4 cm.

Section 4.4

Concept Review 4.4

1. Always true

3. Never true
The variable r represents the discount rate. The variable R represents the regular price.

5. Always true

Objective 4.4.1 Exercises

3. Strategy
Given: $r = 40\% = 0.40$, $C = \$40$
Unknown selling price: S
Use the equation $S = C + rC$

Solution
$S = C + rC$
$S = 40 + 0.40(40)$
$S = 40 + 16$
$S = 56$
The selling price is $56.

5. Strategy
Given: $r = 58\% = 0.58$, $C = \$358$
Unknown selling price: S
Use the equation $S = C + rC$.

Solution
$S = C + rC$
$S = 358 + 0.58(358)$
$S = 358 + 207.64$
$S = 565.64$
The selling price is $565.64.

7. Strategy

Given: $S = \$630$, $C = \$360$

Unknown markup rate: r

Use the equation $S = C + rC$.

Solution

$$S = C + rC$$
$$630 = 360 + 360r$$
$$270 = 360r$$
$$\frac{270}{360} = \frac{360r}{360}$$
$$0.75 = r$$

The markup rate is 75%.

9. Strategy

Given: $S = \$520$, $C = \$360$

Unknown markup rate: r

Use the equation $S = C + rC$.

Solution

$$S = C + rC$$
$$520 = 360 + 360r$$
$$160 = 360r$$
$$\frac{160}{360} = \frac{360r}{360}$$
$$0.444 \approx r$$

The markup rate is 44.4%.

11. Strategy

Given: $S = \$156.80$, $C = \$98$

Unknown markup rate: r

Use the equation $S = C + rC$.

Solution

$$S = C + rC$$
$$156.80 = 98 + 98r$$
$$58.80 = 98r$$
$$\frac{58.80}{98} = \frac{98r}{98}$$
$$0.60 = r$$

The markup rate is 60%.

13. Strategy

Given: $S = \$168$, $r = 40\% = 0.40$

Unknown cost: C

Use the equation $S = C + rC$.

Solution

$$S = C + rC$$
$$168 = C + 0.40C$$
$$168 = 1.40C$$
$$\frac{168}{1.40} = \frac{1.40C}{1.40}$$
$$120 = C$$

The cost is $120.

15. Strategy

Given: $S = \$82.60$, $r = 40\% = 0.40$

Unknown cost: C

Use the equation $S = C + rC$.

Solution

$$S = C + rC$$
$$82.60 = C + 0.40C$$
$$82.60 = 1.40C$$
$$\frac{82.60}{1.40} = \frac{1.40C}{1.40}$$
$$59 = C$$

The cost is $59.

17. Strategy

Given: $S = \$2187.50$, $r = 25\% = 0.25$

Unknown cost: C

Use the equation $S = C + rC$.

Solution

$$S = C + rC$$
$$2187.50 = C + 0.25C$$
$$2187.50 = 1.25C$$
$$\frac{2187.50}{1.25} = \frac{1.25C}{1.25}$$
$$1750 = C$$

The cost is $1750.

Objective 4.4.2 Exercises

21. Strategy

Given: $R = \$135$, $r = 25\% = 0.25$

Unknown sale price: S

Use the equation $S = R - rR$.

Solution

$$S = R - rR$$
$$S = 135 - 0.25(135)$$
$$S = 135 - 33.75$$
$$S = 101.25$$

The sale price is $101.25.

23. Strategy

Given: $R = \$230$, $r = 5\% = 0.05$

Unknown sale price: S

use the equation $S = R - rR$.

Solution

$$S = R - rR$$
$$S = 230 - 0.05(230)$$
$$S = 230 - 11.5$$
$$S = 218.5$$

The sale price is $218.50.

25. Strategy
Given: $R = \$425$, $S = \$318.75$
Unknown discount rate: r
Use the equation $S = R - rR$.

Solution
$$S = R - rR$$
$$318.75 = 425 - 425r$$
$$-106.25 = -425r$$
$$\frac{-106.25}{-425} = \frac{-425r}{-425}$$
$$0.25 = r$$
The discount rate is 25%.

27. Strategy
Given: $R = \$1295$, $S = \$995$
Unknown discount rate: r
Use the equation $S = R - rR$.

Solution
$$S = R - rR$$
$$995 = 1295 - 1295r$$
$$-300 = -1295r$$
$$\frac{-300}{-1295} = \frac{-1295r}{-1295}$$
$$0.232 \approx r$$
The discount rate is 23.2%.

29. Strategy
Given: $R = \$325$, $S = \$201.50$
Unknown discount rate: r
Use the equation $S = R - rR$.

Solution
$$S = R - rR$$
$$201.50 = 325 - 325r$$
$$-123.50 = -325r$$
$$\frac{-123.50}{-325} = \frac{-325r}{-325}$$
$$0.38 = r$$
The discount rate is 38%.

31. Strategy
Given: $S = \$408$, $r = 20\% = 0.20$
Unknown regular price: R
Use the equation $S = R - rR$.

Solution
$$S = R - rR$$
$$408 = R - 0.20R$$
$$408 = 0.80R$$
$$\frac{408}{0.80} = \frac{0.80R}{0.80}$$
$$510 = R$$
The regular price is $510.

33. Strategy
Given: $S = \$180$, $r = 40\% = 0.40$
Unknown regular price: R
Use the equation $S = R - rR$.

Solution
$$S = R - rR$$
$$180 = R - 0.40R$$
$$180 = 0.60R$$
$$\frac{180}{0.60} = \frac{0.60R}{0.60}$$
$$300 = R$$
The regular price is $300.

35. Strategy
Given: $S = \$165$, $r = 40\% = 0.40$
Unknown regular price: R
Use the equation $S = R - rR$.

Solution
$$S = R - rR$$
$$165 = R - 0.40R$$
$$165 = 0.60R$$
$$\frac{165}{0.60} = \frac{0.60R}{0.60}$$
$$275 = R$$
The regular price is $275.

Applying Concepts 4.4

37. Strategy
First find the cost.
Given: $r = 40\% = 0.40$, $S = \$63$
Unknown: C
Use the equation $S = C + rC$.
Then find the markup.
Given: $r = 40\% = 0.40$
C is known after the first equation is solved.
Unknown: M
Use the equation $M = r \cdot C$.

Solution
$$S = C + rC$$
$$63 = C + 0.40C$$
$$63 = 1.40C$$
$$\frac{63}{1.40} = \frac{1.40C}{1.40}$$
$$45 = C$$
The cost is $45.
$$M = r \cdot C$$
$$M = 0.40(45)$$
$$M = 18$$
The markup is $18.

39. Strategy

First find the cost.

Given: $S = \$770$, $M = \$220$

Unknown: C

Use the equation $S = C + M$.

Then find the markup rate:

Given: $M = \$220$

C is known after the first equation is solved.

Unknown: r

Use the equation $M = r \cdot C$.

Solution

$$S = C + M$$
$$770 = C + 220$$
$$770 - 220 = C + 220 - 220$$
$$550 = C$$

The cost is $550.

$$M = r \cdot C$$
$$220 = r(550)$$
$$\frac{220}{550} = \frac{550r}{550}$$
$$0.4 = r$$

The markup rate is 40%.

41. Strategy

Given: $r = 30\% = 0.30$, $S = \$299$

Unknown: C

Use the equation $S = C + rC$.

Solution

$$S = C + rC$$
$$299 = C + 0.30C$$
$$299 = 1.30C$$
$$\frac{299}{1.30} = \frac{1.30C}{1.30}$$
$$230 = C$$

The cost is $230.

43. Strategy

To find the regular price of a tire, write and solve an equation using x to represent the regular price of a tire. The fourth tire has a price of 80% of x or $0.80x$.

Solution

The sum of the prices of the 4 tires	is	$304

$$x + x + x + 0.80x = 304$$
$$3.80x = 304$$
$$\frac{3.80x}{3.80} = \frac{304}{3.80}$$
$$x = 80$$

The regular price of a tire is $80.

45. $S = (1 - 0.20)8500 = 6800$

The sale price after a 20% discount is $6800.

$S = (1 - 0.10)6800 = 6120$

Another 10% discount would give a sale price of $6120.

$S = (1 - 0.30)8500 = 5950$

A single discount of 30% would give a sale price of $5950.

Thus a 30% discount is not equivalent to successive discounts of 20% and 10%.

$$\frac{2380}{8500} = 0.28$$

The total discount of the successive discounts is $8500 - 6120 = 2380$. The single-discount equivalent is 28%.

Section 4.5

Concept Review 4.5

1. Always true

3. Always true

5. Never true

The correct equation is $0.05x = 0.08(9000 - x)$.

Objective 4.5.1 Exercises

1. Strategy
   Amount invested at 7%: x
   Amount invested at 6.5%: $15,000 - x$

	Principal	Rate	Interest
Amount at 7%	x	0.07	$0.07x$
Amount at 6.5%	$15,000 - x$	0.065	$0.065(15,000 - x)$

   The total annual earned is $1020.

   Solution
   $$0.07x + 0.065(15,000 - x) = 1020$$
   $$0.07x + 975 - 0.065x = 1020$$
   $$0.005x + 975 = 1020$$
   $$0.005x = 45$$
   $$x = 9000$$
   $15,000 - x = 15,000 - 9000 = 6000$
   The amount invested at 7% is $9000.
   The amount invested at 6.5% is $6000.

3. Amount invested in first deposit (mutual fund), at 13%: x
   Amount invested in second deposit, at 7%: $x + 2500$

	Principal	Rate	Interest
Amount at 13%	x	0.13	$0.13x$
Amount at 7%	$x + 2500$	0.07	$0.07(x + 2500)$

   The total annual interest earned by the two investments equals $475.

   Solution
   $$0.13x + 0.07(x + 2500) = 475$$
   $$0.13x + 0.07x + 175 = 475$$
   $$0.20x + 175 = 475$$
   $$0.20x = 300$$
   $$x = 1500$$
   The amount invested in the mutual fund is $1500.

5. Strategy
   Amount invested at 10%: x
   Amount invested at 8.5%: $300,000 - x$

	Principal	Rate	Interest
Amount at 10%	x	0.10	$0.10x$
Amount at 8.5%	$300,000 - x$	0.085	$0.085(300,000 - x)$

   The total annual interest earned by the two investments equals $28,500.

   Solution
   $$0.10x + 0.085(300,000 - x) = 28,500$$
   $$0.10x + 25,500 - 0.085x = 28,500$$
   $$0.015x + 25,500 = 28,500$$
   $$0.015x = 3000$$
   $$x = 200,000$$
   $200,000 - x = 300,000 - 200,000 = 100,000$
   The amount invested at 10% is $200,000.
   The amount invested at 8.5% is $100,000.

7. Strategy
   Amount invested at 7.5%: $5000
   Amount invested at 8%: x

	Principal	Rate	Interest
Amount at 7.5%	5000	0.075	0.075(5000)
Amount at 8%	x	0.08	0.08x

   The total annual interest earned by the two investments equals $615.

   Solution
   $$0.075(5000)+0.08x = 615$$
   $$375+0.08x = 615$$
   $$0.08x = 240$$
   $$x = 3000$$
   An additional $3000 must be invested in bonds.

9. Strategy
   Amount invested at 11%: x
   Amount invested at 7%: $2500

	Principal	Rate	Interest
Amount at 11%	x	0.11	0.11x
Amount at 7%	2500	0.07	0.07(2500)
Amount at 9%	2500 + x	0.09	0.09(2500 + x)

   The sum of the interest earned by the two investments equals the interest earned on the total investment.

   Solution
   $$0.07(2500)+0.11x = 0.09(2500+x)$$
   $$175+0.11x = 225+0.09x$$
   $$175+0.02x = 225$$
   $$0.02x = 50$$
   $$x = 2500$$
   An additional $2500 must be invested at 11%.

11. Amount invested at 8%: x
    Amount invested at 12%: $54,000 - x$

	Principal	Rate	Interest
Amount at 8%	x	0.08	0.08x
Amount at 12%	$54,000 - x$	0.12	0.12(54,000 − x)
Amount at 9%	54,000	0.09	0.09(54,000)

    The sum of the interest earned by the two investments equals the interest earned on the total investment.

    Solution
    $$0.08x + 0.12(54,000 - x) = 0.09(54,000)$$
    $$0.08x +6480 - 0.12x = 4860$$
    $$-0.04x +6480 = 4860$$
    $$-0.04x = -1620$$
    $$x = 40,500$$
    $54,000 - x = 54,000 - 40,500 = 13,500$
    The amount invested at 8% is $40,500.
    The amount invested at 12% is $13,500.

13. Strategy

Total amount invested: x

Amount invested at 8.25%: $55\%x = 0.55x$

Amount invested at 10%: $45\%x = 0.45x$

	Principal	Rate	Interest
Amount at 8.25%	$0.55x$	0.0825	$0.0825(0.55x)$
Amount at 10%	$0.45x$	0.10	$0.10(0.45x)$

The total annual interest earned by the two investments equals $58,743.75.

Solution

$$0.0825(0.55x) + 0.10(0.45x) = 58,743.75$$
$$0.045375x + 0.045x = 58,743.75$$
$$0.090375x = 58,743.75$$
$$x = 650,000$$

The total amount invested is $650,000.

15. Strategy

Total amount invested: x

Amount invested at 6%: $30\%x = 0.30x$

Amount invested at 8%: $25\%x = 0.25x$

Amount invested at 7.5%: $45\%x = 0.45x$

	Principal	Rate	Interest
Amount at 6%	$0.30x$	0.06	$0.06(0.30x)$
Amount at 8%	$0.25x$	0.08	$0.08(0.25x)$
Amount at 7.5%	$0.45x$	0.075	$0.075(0.45x)$

The total annual interest earned by the three investments equals $35,875.

Solution

$$0.06(0.30x) + 0.08(0.25x) + 0.075(0.45x) = 35,875$$
$$0.018x + 0.02x + 0.03375x = 35,875$$
$$0.07175x = 35,875$$
$$x = 500,000$$

The total amount invested is $500,000.

Applying Concepts 4.5

17. Strategy
Amount sales representative invests at 9%: x
Amount research consultant invests at 8%: $x + 5000$

	Principal	Rate	Interest
Amount invested by sales representative	x	0.09	$0.09x$
Amount invested by research consultant	$x + 5000$	0.08	$0.08(x + 5000)$

The research consultant's income is equal to the sales representative's.

Solution
$0.09x = 0.08(x + 5000)$
$0.09x = 0.08x + 400$
$0.01x = 400$
$x = 40,000$
$x + 5000 = 40,000 + 5000 = 45,000$
The research consultant invests $45,000.

19. Strategy
Amount invested in 9.5% bonds: x
Amount invested in 8% stocks: $x + 3000$

	Principal	Rate	Interest
Amount invested in bonds	x	0.095	$0.095x$
Amount invested in stocks	$x + 3000$	0.08	$0.08(x + 3000)$

Each investment yielded the same income.

Solution
$0.095x = 0.08(x + 3000)$
$0.095x = 0.08x + 240$
$0.015x = 240$
$x = 16,000$
$0.095x = 0.095(16,000) = 1520$
$0.08(x + 3000) = 0.08(16,000 + 3000) = 0.08(19,000) = 1520$
The income from each is $1520.
$2(1520) = 3040$
The total annual interest received on both investments is $3040.

21. Strategy
The value of the investment after one year is equal to the principal plus 8.5% of the principal $(P + 0.085P = 1.085P)$. The value of the investment after two years is equal to the value of the investment after one year $(1.085P)$ plus 8.5% of the value of the investment after one year
$$[1.085P + 0.085(1.085P) = 1.085P + 0.092225P$$
$$= 1.177225P].$$
The value of the investment after three years is equal to the value of the investment after two years $(1.177225P)$ plus 8.5% of the value of the investment after two years
$$[1.177225P + 0.085(1.177225P)].$$

Solution
$$1.177225P + 0.085(1.177225P)$$
$$= 1.177225P + 0.1000641P$$
$$= 1.2772891P$$
$$= 1.2772891(3000)$$
$$\approx 3831.867$$
The value of the investment in 3 years is $3831.87.

23. a. According to the table, a couple earning $75,000 should save $170,000 for retirement by age 55.

Section 4.6

Concept Review 4.6

1. Always true

3. Always true

5. Never true
The resulting solution will be between 9% and 4% acid.

7. Always true

Objective 4.6.1 Exercises

3. Strategy
Amount of dog food: x
Amount of supplement: $5 - x$

	Amount	Cost	Value
Dog food	x	6.75	$6.75x$
Supplement	$5 - x$	3.25	$3.25(5 - x)$
Mixture	5	4.65	$4.65(5)$

The sum of the values before mixing equals the value after mixing.

Solution
$$6.75x + 3.25(5 - x) = 4.65(5)$$
$$6.75x + 16.25 - 3.25x = 23.25$$
$$3.50x = 7$$
$$\frac{3.50x}{3.50} = \frac{7}{3.50}$$
$$x = 2$$
$$5 - x = 3$$
To make the mixture, 2 lb of dog food and 3 lb of vitamin supplement are needed.

5. Strategy
Cost of mixture: x

	Amount	Cost	Value
$9.20 type	8	9.20	$9.20(8)$
$5.50 type	12	5.50	$5.50(12)$
Mixture	20	x	$x(20)$

The sum of the values before mixing equals the value after mixing.

Solution
$$9.20(8) + 5.50(12) = 20x$$
$$73.60 + 66.00 = 20x$$
$$139.60 = 20x$$
$$6.98 = x$$
The cost of the mixture is $6.98 per pound.

7. Strategy
Amount of $4.30 alloy: x
Amount of $1.80 alloy: $200 - x$

	Amount	Cost	Value
$4.30 alloy	x	4.30	$4.30x$
$1.80 alloy	$200 - x$	1.80	$1.80(200 - x)$
Mixture	200	2.50	$2.50(200)$

The sum of the values before mixing equals the value after mixing.

Solution
$$4.30x + 1.80(200 - x) = 2.50(200)$$
$$4.30x + 360.00 - 1.80x = 500.00$$
$$2.50x + 360 = 500$$
$$2.50x = 140$$
$$x = 56$$
$$200 - x = 200 - 56 = 144$$
The amount of the $4.30 alloy is 56 oz.
The amount of the $1.80 alloy is 144 oz.

9. Strategy
Amount of blueberries: x

	Amount	Cost	Value
Blueberries	x	3.75	$3.75x$
Cherries	2	4.50	$4.50(2)$
Mixture	$x + 2$	4.25	$4.25(x + 2)$

The sum of the values before mixing equals the value after mixing.

Solution
$$3.75x + 4.50(2) = 4.25(x + 2)$$
$$3.75x + 9 = 4.25x + 8.50$$
$$0.50 = 0.50x$$
$$\frac{0.50}{0.50} = \frac{0.50x}{0.50}$$
$$1 = x$$
1 lb of blueberries is needed.

11. Strategy
Amount of hard candy at $7.50 per kilogram: x

	Amount	Cost	Value
Hard candy	x	7.50	$7.50x$
Jelly beans	24	3.25	$3.25(24)$
Mixture	$x + 24$	4.50	$4.50(x + 24)$

The sum of the values before mixing equals the value after mixing.

Solution
$$7.50x + 3.25(24) = 4.50(x + 24)$$
$$7.50x + 78 = 4.50x + 108$$
$$3x + 78 = 108$$
$$3x = 30$$
$$x = 10$$
The amount of hard candy needed is 10 kg.

13. Strategy
Amount of caramel: x

	Amount	Cost	Value
Caramel	x	2.40	$2.40x$
Popcorn	5	0.80	$0.80(5)$
Mixture	$x + 5$	1.40	$1.40(x + 5)$

The sum of the values before mixing equals the value after mixing.

Solution
$$2.40x + 0.80(5) = 1.40(x + 5)$$
$$2.40x + 4 = 1.40x + 7$$
$$x = 3$$
3 lb of caramel is needed.

15. Strategy
Amount of meal: x

	Amount	Cost	Value
Meal	x	0.80	$0.80x$
Grain	500	1.20	$1.20(500)$
Mixture	$x + 500$	1.05	$1.05(x + 500)$

The sum of the values before mixing equals the value after mixing.

Solution
$$0.80x + 1.20(500) = 1.05(x + 500)$$
$$0.80x + 600 = 1.05x + 525$$
$$75 = 0.25x$$
$$\frac{75}{0.25} = \frac{0.25x}{0.25}$$
$$300 = x$$
300 lb of meal are needed.

17. Strategy
Amount of almonds: x
Amount of walnuts: $100 - x$

	Amount	Cost	Value
Almonds	x	4.50	$4.50(x)$
Walnuts	$100 - x$	2.50	$2.50(100 - x)$
Mixture	100	3.24	$3.24(100)$

The sum of the values before mixing equals the value after mixing.

Solution
$$4.50x + 2.50(100 - x) = 3.24(100)$$
$$4.50x + 250 - 2.50x = 324$$
$$2.00x + 250 = 324$$
$$2.00x = 74$$
$$x = 37$$
$$100 - x = 100 - 37 = 63$$
The amount of almonds is 37 lb.
The amount of walnuts is 63 lb.

19. Strategy
Cost per pound of the sugar-coated cereal: x

	Amount	Cost	Value
Sugar	40	2.00	2.00(40)
Corn flakes	120	1.20	1.20(120)
Mixture	160	x	$160x$

The sum of the values before mixing equals the value after mixing.

Solution
$$2.00(40)+1.20(120)=160x$$
$$80+144=160x$$
$$224=160x$$
$$1.40=x$$
The cost per pound of the sugar-coated cereal is $1.40.

21. Strategy
Cost per ounce of sunscreen: x

	Amount	Cost	Value
$2.50 lotion	100	2.50	2.50(100)
$4.00 lotion	50	4.00	4.00(50)
Sunscreen	150	x	$150x$

The sum of the values before mixing equals the value after mixing.

Solution
$$2.50(100)+4.00(50)=150x$$
$$250+200=150x$$
$$450=150x$$
$$\frac{450}{150}=\frac{150x}{150}$$
$$3=x$$
The sunscreen costs $3 per ounce.

Objective 4.6.2 Exercises

25. Strategy
The percent concentration of gold in the combined alloy: x

	Amount	Percent	Quantity
30% gold alloy	40	0.30	0.30(40)
20% gold alloy	60	0.20	0.20(60)
Combined alloy	100	x	$100x$

The sum of the quantities before mixing is equal to the quantity after mixing.

Solution
$$0.30(40)+0.20(60)=100x$$
$$12+12=100x$$
$$24=100x$$
$$\frac{24}{100}=\frac{100x}{100}$$
$$0.24=x$$
The combined gold alloy is 24% gold.

27. Strategy
Amount of 15% solution: x

	Amount	Percent	Quantity
15% solution	x	0.15	$0.15x$
20% solution	5	0.20	0.20(5)
16% solution	$x+5$	0.16	$0.16(x+5)$

The sum of the quantities before mixing is equal to the quantity after mixing.

Solution
$$0.15x+0.20(5)=0.16(x+5)$$
$$0.15x+1=0.16x+0.8$$
$$0.2=0.01x$$
$$\frac{0.2}{0.01}=\frac{0.01x}{0.01}$$
$$20=x$$
20 gal of the 15% acid solution must be used.

29. Strategy
Number of pounds of 25% wool yarn: x

	Amount	Percent	Quantity
25% wool	x	0.25	$0.25x$
50% wool	20	0.50	$0.50(20)$
35% wool	$x + 20$	0.35	$0.35(x + 20)$

The sum of the quantities before mixing is equal to the quantity after mixing.

Solution
$$0.25x + 0.50(20) = 0.35(x + 20)$$
$$0.25x + 10 = 0.35x + 7$$
$$3 = 0.10x$$
$$\frac{3}{0.10} = \frac{0.10x}{0.10}$$
$$30 = x$$
30 lb of yarn that is 25% wool must be used.

31. Strategy
Number of gallons of 9% solution: x

	Amount	Percent	Quantity
9% solution	x	0.09	$0.09x$
25% solution	$10 - x$	0.25	$0.25(10 - x)$
15% solution	10	0.15	$0.15(10)$

The sum of the quantities before mixing is equal to the quantity after mixing.

Solution
$$0.09x + 0.25(10 - x) = 0.15(10)$$
$$0.09x + 2.5 - 0.25x = 1.5$$
$$-0.16x = -1$$
$$\frac{-0.16x}{-0.16} = \frac{-1}{-0.16}$$
$$x = 6.25$$
6.25 gal of plant food that is 9% nitrogen must be used.

33. Strategy
The percent concentration of sugar in the mix: x

	Amount	Percent	Quantity
Sugar	5	1.00	$(1.00)5$
Cereal	45	0.10	$0.10(40)$
Mixture	50	x	$50x$

The sum of the quantities before mixing is equal to the quantity after mixing.

Solution
$$1.00(5) + 0.10(45) = 50x$$
$$5 + 4.5 = 50x$$
$$9.5 = 50x$$
$$0.19 = x$$
The percent concentration of sugar in the mix is 19%.

35. Strategy
Amount of 60% lavender: x

	Amount	Percent	Quantity
60% lavender	x	0.60	$0.60x$
80% lavender	70	0.80	$0.80(70)$
74% lavender	$x + 70$	0.74	$0.74(x + 70)$

The sum of the quantities before mixing is equal to the quantity after mixing.

Solution
$$0.60x + 0.80(70) = 0.74(x + 70)$$
$$0.60x + 56 = 0.74x + 51.8$$
$$4.2 = 0.14x$$
$$\frac{4.2}{0.14} = \frac{0.14x}{0.14}$$
$$30 = x$$
30 oz of 60% lavender are needed.

37. Strategy
Amount of 7% hydrogen peroxide: x
Amount of 4% hydrogen peroxide: $300 - x$

	Amount	Percent	Quantity
7% hydrogen peroxide	x	0.07	$0.07x$
4% hydrogen peroxide	$300 - x$	0.04	$0.04(300 - x)$
5% hydrogen peroxide	300	0.05	$0.05(300)$

The sum of the quantities before mixing is equal to the quantity after mixing.

Solution
$$0.07x + 0.04(300 - x) = 0.05(300)$$
$$0.07x + 12 - 0.04x = 15$$
$$0.03x + 12 = 15$$
$$0.03x = 3$$
$$x = 100$$
$$300 - x = 300 - 100 = 200$$
100 ml of 7% hydrogen peroxide and 200 ml of 4% hydrogen peroxide are needed.

39. Strategy
Amount of pure chocolate: x

	Amount	Percent	Quantity
Pure chocolate	x	1	x
50% chocolate	150	0.50	0.50(150)
75% chocolate	$x + 150$	0.75	0.75(x + 150)

The sum of the quantities before mixing is equal to the quantity after mixing.

Solution
$$x + 0.50(150) = 0.75(x + 150)$$
$$x + 75 = 0.75x + 112.50$$
$$0.25x = 37.5$$
$$\frac{0.25x}{0.25} = \frac{37.5}{0.25}$$
$$x = 150$$
150 oz of pure chocolate are needed.

41. Strategy
The percent concentration of conditioner in the mixture: x

	Amount	Percent	Quantity
Pure shampoo	8	0	0
20% conditioner	12	0.20	0.20(12)
Mixture	20	x	$20x$

The sum of the quantities before mixing is equal to the quantity after mixing.

Solution
$$0 + 0.20(12) = 20x$$
$$2.4 = 20x$$
$$\frac{2.4}{20} = \frac{20x}{20}$$
$$0.12 = x$$
The percent concentration of conditioner in the mixture is 12%.

43. Strategy
Amount of dried apricots: x

	Amount	Percent	Quantity
Dried apricots	x	1.00	1.00(x)
Snack	18	0.20	0.20(18)
Mixture	$x + 18$	0.25	0.25(x + 18)

The sum of the quantities before mixing is equal to the quantity after mixing.

Solution
$$1.00(x) + 0.20(18) = 0.25(x + 18)$$
$$x + 3.6 = 0.25x + 4.5$$
$$0.75x + 3.6 = 4.5$$
$$0.75x = 0.9$$
$$x = 1.2$$
1.2 oz of dried apricots must be added.

Applying Concepts 4.6

45. Strategy
Cost per ounce of mixture: x

	Amount	Cost	Value
$4.50 alloy	30	4.50	4.50(30)
$3.50 alloy	40	3.50	3.50(40)
$3.00 alloy	30	3.00	3.00(30)
Mixture	100	x	$100x$

The sum of the values before mixing equals the value after mixing.

Solution
$$4.50(30) + 3.50(40) + 3.00(30) = 100x$$
$$135 + 140 + 90 = 100x$$
$$365 = 100x$$
$$3.65 = x$$
The mixture costs $3.65 per ounce.

47. Strategy
Ounces of water: x

	Amount	Percent	Quantity
12% salt solution	50	0.12	0.12(50)
Water	x	0	$0x$
15% salt solution	$50 - x$	0.15	0.15(50 − x)

The quantity of 12% salt solution less the evaporated water is equal to the quantity of 15% salt solution.

Solution
$$0.12(50) - 0x = 0.15(50 - x)$$
$$6 = 7.5 - 0.15x$$
$$-1.5 = -0.15x$$
$$10 = x$$
10 oz of water must be evaporated.

49. Strategy
Amount of pure water: x

	Amount	Percent	Quantity
Water	x	0	$0x$
Acid	50	1.00	1.00(50)
40% solution	$50 + x$	0.40	0.40(50 + x)

The sum of the quantities before mixing is equal to the quantity after mixing.

Solution
$$0x + 1.00(50) = 0.40(50 + x)$$
$$60 = 20 + 0.40x$$
$$30 = 0.40x$$
$$75 = x$$
75 g of pure water must be added.

51. Strategy

Number of adult tickets: x

Number of children's tickets: $120 - x$

	Amount	Cost	Value
Adult	x	5.50	$5.50x$
Child	$120 - x$	2.75	$2.75(120 - x)$

The total cost of the tickets sold was $563.75.

Solution

$$5.50x + 2.75(120 - x) = 563.75$$
$$5.50x + 330 - 2.75x = 563.75$$
$$2.75x + 330 = 563.75$$
$$2.75x = 233.75$$
$$x = 85$$
$$120 - x = 120 - 85 = 35$$

85 adults and 35 children attended the performance.

Section 4.7

Concept Review 4.7

1. Never true

In the uniform motion equation $d = rt$, r is the rate of travel.

3. Never true

$5 - t$ is used to represent the time spent on the return trip.

5. Always true

Objective 4.7.1 Exercises

1. Strategy

Rate of the second plane: r

Rate of the first plane: $r - 25$

	Rate	Time	Distance
First plane	$r - 25$	2	$2(r - 25)$
Second plane	r	2	$2r$

The total distance traveled by the two planes is 470 mi.

Solution

$$2(r - 25) + 2r = 470$$
$$2r - 50 + 2r = 470$$
$$4r - 50 = 470$$
$$4r = 520$$
$$r = 130$$
$$r - 25 = 130 - 25 = 105$$

The first plane is traveling at a rate of 105 mph.
The second plane is traveling at a rate of 130 mph.

3. Strategy

Time for the second skater: t

Time for the first skater: $t + 10$

	Rate	Time	Distance
First skater	8	$t + 10$	$8(t + 10)$
Second skater	10	t	$10t$

The skaters travel the same distance.

Solution

$$8(t + 10) = 10t$$
$$8t + 80 = 10t$$
$$80 = 2t$$
$$40 = t$$

The second skater passes the first skater in 40 s.

5. Strategy

Time for the tour boat: t

Time for the motorboat: $t + 2$

	Rate	Time	Distance
Motorboat	9	$t + 2$	$9(t + 2)$
Tour boat	18	t	$18t$

The boats travel the same distance.

Solution

$$9(t + 2) = 18t$$
$$9t + 18 = 18t$$
$$18 = 9t$$
$$2 = t$$

In 2 h the tour boat will be alongside the motorboat.

7. Strategy

Time spent driving: x

Time spent flying: $3 - x$

	Rate	Time	Distance
Driving	30	x	$30x$
Flying	60	$3 - x$	$60(3 - x)$

The entire distance was 150 mi.

Solution

$$30x + 60(3 - x) = 150$$
$$30x + 180 - 60x = 150$$
$$-30x = -30$$
$$\frac{-30x}{-30} = \frac{-30}{-30}$$
$$x = 1$$
$$60(3 - x) = 60(3 - 1) = 60(2) = 120$$

The airport is 120 mi from the corporate offices.

9. Strategy
Original speed of boat: x

	Rate	Time	Distance
First 3 h	x	3	$3x$
Second 3 h	$x - 5$	3	$3(x - 5)$

The total distance traveled was 57 mi.

Solution
$$3x + 3(x - 5) = 57$$
$$3x + 3x - 15 = 57$$
$$6x - 15 = 57$$
$$6x = 72$$
$$\frac{6x}{6} = \frac{72}{6}$$
$$x = 12$$
$$3x = 36$$
In the first 3 h the boat traveled 36 mi.

11. Strategy
Rate of the passenger train: x
Rate of the freight train: $x - 20$

	Rate	Time	Distance
Passenger train	x	3	$3x$
Freight train	$x - 20$	5	$5(x - 20)$

The distance traveled is the same.

Solution
$$3x = 5(x - 20)$$
$$3x = 5x - 100$$
$$-2x = -100$$
$$\frac{-2x}{-2} = \frac{-100}{-2}$$
$$x = 50$$
$$x - 20 = 30$$
The passenger train is traveling 50 mph and the freight train is traveling 30 mph.

13. Strategy
Time for the second ship to catch up: x

	Rate	Time	Distance
First ship	25	x	$25x + 10$
Second ship	35	x	$35x$

The distances are the same.

Solution
$$25x + 10 = 35x$$
$$10 = 10x$$
$$\frac{10}{10} = \frac{10x}{10}$$
$$1 = x$$
It takes the second ship 1 h to catch up to the first ship.

15. Strategy
Time for trains to pass each other: x

	Rate	Time	Distance
Train leaving Washington, D.C.	60	$x + 1$	$60(x + 1)$
Train leaving Charleston	50	x	$50x$

The sum of the distances is 500 mi.

Solution
$$60(x + 1) + 50x = 500$$
$$60x + 60 + 50x = 500$$
$$110x = 440$$
$$\frac{110x}{110} = \frac{440}{110}$$
$$x = 4$$
The trains will pass each other 4 h after the train leaves Charleston.

17. Strategy
Average speed on winding road: x
Average speed on straight road: $x + 20$

	Rate	Time	Distance
Winding road	x	3	$3x$
Straight road	$x + 20$	2	$2(x + 20)$

The total distance traveled was 210 mi.

Solution
$$3x + 2(x + 20) = 210$$
$$3x + 2x + 40 = 210$$
$$5x = 170$$
$$\frac{5x}{5} = \frac{170}{5}$$
$$x = 34$$
The average speed on the winding road is 34 mph.

19. Strategy
Time for the car to overtake the cyclist: x

	Rate	Time	Distance
Cyclist	12	$x + 3$	$12(x + 3)$
Car	48	x	$48x$

The distances are the same.

Solution
$$12(x + 3) = 48x$$
$$12x + 36 = 48x$$
$$36 = 36x$$
$$\frac{36}{36} = \frac{36x}{36}$$
$$1 = x$$
It takes 1 h for the car to overtake the cyclist. The rate of the car is 48 mph, so after 1 h the car has gone $1 \cdot 48 = 48$ mi.

Applying Concepts 4.7

21. Strategy
Speed of cyclist: r
Speed of the car: $3r + 5$

	Rate	Time	Distance
Cyclist	r	1.5	$1.5r$
Car	$3r + 5$	1.5	$1.5(3r + 5)$

In 1.5 h, the car is 46.5 mi ahead of the cyclist.

Solution
$$1.5(3r + 5) - 1.5r = 46.5$$
$$4.5r + 7.5 - 1.5r = 46.5$$
$$3r + 7.5 = 46.5$$
$$3r = 39$$
$$r = 13$$
The cyclist is traveling at 13 mph.

23. Strategy
Speed of the bus: r
Speed of the car: $2r - 30$

	Rate	Time	Distance
Bus	r	2	$2r$
Car	$2r - 30$	2	$2(2r - 30)$

In two hours the car is 30 mi ahead of the bus.

Solution
$$2(2r - 30) - 2r = 30$$
$$4r - 60 - 2r = 30$$
$$2r - 60 = 30$$
$$2r = 90$$
$$r = 45$$
$$2r - 30 = 2(45) - 30 = 90 - 30 = 60$$
The car is traveling at 60 mph.

25. Strategy
Time spent jogging: t

	Rate	Time	Distance
1st jogger	4	t	$4t$
2nd jogger	6	t	$6t$

The total distance is 8 mi.

Solution
$$4t + 6t = 8$$
$$10t = 8$$
$$t = \frac{4}{5} \quad \left(\frac{4}{5} \text{ of 1 h} = 48 \text{ min}\right)$$
The joggers will meet 48 min after 7 A.M., or at 7:48 A.M.

27. Strategy
Time for return trip: t

	Rate	Time	Distance
At 10 mph	10	2	$10(2)$
At 20 mph	20	t	$20t$

The distance covered is the same in both cases.

Solution
$$10(2) = 20t$$
$$20 = 20t$$
$$1 = t$$
The time for the return trip is 1 h.
The time for the whole trip is 3 h.
$$10(2) + 20t = 40$$
The distance covered is 40 mi.
$$\text{Average speed} = \frac{\text{distance}}{\text{time}} = \frac{40}{3} = 13\frac{1}{3}$$
The cyclist's average speed is $13\frac{1}{3}$ mph.

Section 4.8

Concept Review 4.8

1. Never true
"Is greater than" and "is more than" are indicated by the symbol >.

3. Always true

5. Never true
The expression $7(15) + 0.20m$ represents the cost to rent the car for one week.

Objective 4.8.1 Exercises

1. Strategy
The unknown number: x
Three-fifths of the number: $\frac{3}{5}x$

Solution

Three-fifths of the number	is greater than	two-thirds

$$\frac{3}{5}x > \frac{2}{3}$$
$$\frac{5}{3}\left(\frac{3}{5}x\right) > \frac{5}{3}\left(\frac{2}{3}\right)$$
$$x > \frac{10}{9}$$

The smallest integer that satisfies the inequality is 2.

3. **Strategy**
To find the amount of income tax that must be paid, write and solve an inequality using T to represent the amount of income tax.

Solution

| The amount of income tax that must be paid | is greater than or equal to | 90% of the annual income tax liability |

$$T \geq 0.90(3500)$$
$$T \geq 3150$$

The person must pay $3150 or more in income tax.

5. **Strategy**
To find the number of pounds of cans, write and solve an inequality using N to represent the number of pounds.

Solution

| The total number of pounds | is greater than | 1850 lb |

$$505 + 493 + 412 + N > 1850$$
$$1410 + N > 1850$$
$$1410 - 1410 + N > 1850 - 1410$$
$$N > 440$$

The organization must collect more than 440 pounds of cans.

7. **Strategy**
To find the scores, write and solve an inequality using N to represent the score on the last test.

Solution

| The average of the five scores | is greater than or equal to | 80 |

$$\frac{75 + 83 + 86 + 78 + N}{5} \geq 80$$
$$\frac{322 + N}{5} \geq 80$$
$$5\left(\frac{322 + N}{5}\right) \geq 5(80)$$
$$322 + N \geq 400$$
$$322 - 322 + N \geq 400 - 322$$
$$N \geq 78$$

The student must earn a grade of 78 or better on the last test.

9. **Strategy**
To find the dollar amounts, write and solve an inequality using N to represent the amount of sales.

Solution

| 35% of the sales | is greater than | $2000 |

$$0.35N > 2000$$
$$\frac{0.35N}{0.35} > \frac{2000}{0.35}$$
$$N > 5714.2857$$

The commission offer is more attractive if the sales are more than $5714.

11. **Strategy**
To find the dollar amount, write and solve an inequality using S to represent the amount of sales.

Solution

| $3200 | is greater than or equal to | $1000 plus 11% of sales |

$$3200 \geq 1000 + 0.11S$$
$$3200 - 1000 \geq 1000 - 1000 + 0.11S$$
$$2200 \geq 0.11S$$
$$\frac{2200}{0.11} \geq \frac{0.11S}{0.11}$$
$$20,000 \geq S$$

The agent expects to make sales totaling $20,000 or less.

13. Strategy

To find the number of minutes, write and solve an inequality using N to represent the number of minutes.

Solution

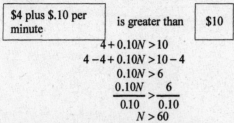

| $4 plus $.10 per minute | is greater than | $10 |

$$4 + 0.10N > 10$$
$$4 - 4 + 0.10N > 10 - 4$$
$$0.10N > 6$$
$$\frac{0.10N}{0.10} > \frac{6}{0.10}$$
$$N > 60$$

A person must use the service for more than 60 min.

15. Strategy

To find the number of ounces of artificial flavors that can be added, write and solve an inequality using N to represent the number of ounces.

Solution

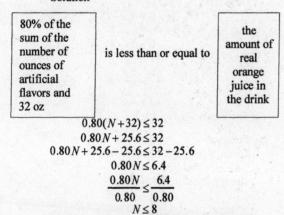

| 80% of the sum of the number of ounces of artificial flavors and 32 oz | is less than or equal to | the amount of real orange juice in the drink |

$$0.80(N + 32) \le 32$$
$$0.80N + 25.6 \le 32$$
$$0.80N + 25.6 - 25.6 \le 32 - 25.6$$
$$0.80N \le 6.4$$
$$\frac{0.80N}{0.80} \le \frac{6.4}{0.80}$$
$$N \le 8$$

If the drink is to be labeled orange juice, 8 or fewer ounces of artificial flavors can be added.

17. Strategy

To find the number of miles, write and solve an inequality using N to represent the number of miles to the ski area. Then $2N$ is the number of miles in the round trip.

Solution

| $8 per person each way | is less than | $45 plus $.25 per mile |

$$8(4)(2) < 45 + 0.25(2N)$$
$$64 < 45 + 0.5N$$
$$64 - 45 < 45 - 45 + 0.5N$$
$$19 < 0.5N$$
$$\frac{19}{0.5} < \frac{0.5N}{0.5}$$
$$38 < N$$

The ski area is more than 38 mi away.

19. Strategy

To find the maximum number of miles, write and solve an inequality using N to represent the number of miles.

Solution

| $15 plus 14¢ per mile | is less than | $25 plus 8¢ per mile |

$$15 + 0.14N < 25 + 0.08N$$
$$15 + 0.14N - 0.08N < 25 + 0.08N - 0.08N$$
$$15 + 0.06N < 25$$
$$15 - 15 + 0.06N < 25 - 15$$
$$0.06N < 10$$
$$\frac{0.06N}{0.06} < \frac{10}{0.06}$$
$$N < 166.\overline{6}$$

The maximum number of miles is 166.

Applying Concepts 4.8

21. Strategy

The first odd integer: x
The second consecutive odd integer: $x + 2$
The third consecutive odd integer: $x + 4$

Solution

| Three times the sum of the first two | is less than | four times the third integer |

$$3(x + x + 2) < 4(x + 4)$$
$$3(2x + 2) < 4(x + 4)$$
$$6x + 6 < 4x + 16$$
$$2x + 6 < 16$$
$$2x < 10$$
$$x < 5$$

The odd positive integers that will satisfy the inequality are 1 and 3.
$$x = 1 \qquad x = 3$$
$$x + 2 = 1 + 2 = 3 \qquad x + 2 = 3 + 2 = 5$$
$$x + 4 = 1 + 4 = 5 \qquad x + 4 = 3 + 4 = 7$$
The integers are 1, 3, and 5 or 3, 5, and 7.

23. Strategy

To find the number of aircraft, write and solve an inequality using N to represent the number of aircraft.

Solution

$$6(60) = 360 \text{ min}$$

There are 360 min in 6 h.

$$\frac{360}{45} < N < \frac{360}{30}$$
$$8 < N < 12$$

The crew can prepare between 8 and 12 aircraft in a 6-h period of time.

25. Strategy
To find the minimum number of calendars, write and solve an inequality using x to represent the number of calendars.

Solution

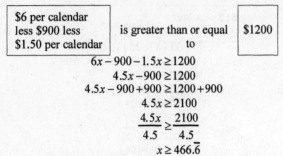

$$6x - 900 - 1.5x \geq 1200$$
$$4.5x - 900 \geq 1200$$
$$4.5x - 900 + 900 \geq 1200 + 900$$
$$4.5x \geq 2100$$
$$\frac{4.5x}{4.5} \geq \frac{2100}{4.5}$$
$$x \geq 466.\overline{6}$$

To make a profit of $1200, at least 467 calendars must be sold.

Chapter Review Exercises

1. the unknown number: x

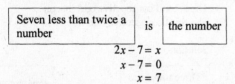

$$2x - 7 = x$$
$$x - 7 = 0$$
$$x = 7$$

The number is 7.

2. Strategy
To find the length of the shorter piece, write and solve an equation using x to represent the shorter piece and $35 - x$ to represent the longer piece.

Solution

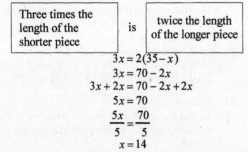

$$3x = 2(35 - x)$$
$$3x = 70 - 2x$$
$$3x + 2x = 70 - 2x + 2x$$
$$5x = 70$$
$$\frac{5x}{5} = \frac{70}{5}$$
$$x = 14$$

The length of the shorter piece is 14 in.

3. the smaller number: n
the larger number: $21 - n$

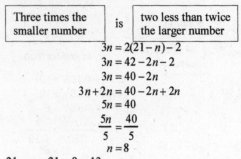

$$3n = 2(21 - n) - 2$$
$$3n = 42 - 2n - 2$$
$$3n = 40 - 2n$$
$$3n + 2n = 40 - 2n + 2n$$
$$5n = 40$$
$$\frac{5n}{5} = \frac{40}{5}$$
$$n = 8$$

$21 - n = 21 - 8 = 13$
The smaller number is 8. The larger number is 13.

4. the unknown number: x

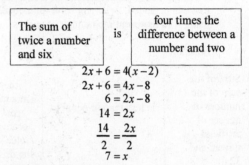

$$2x + 6 = 4(x - 2)$$
$$2x + 6 = 4x - 8$$
$$6 = 2x - 8$$
$$14 = 2x$$
$$\frac{14}{2} = \frac{2x}{2}$$
$$7 = x$$

The number is 7.

5. Strategy
Given: $R = \$90$, $S = \$60$
Unknown discount rate: r
Use the equation $S = R - rR$.

Solution
$$S = R - rR$$
$$60 = 90 - 90r$$
$$60 - 90 = 90 - 90 - 90r$$
$$-30 = -90r$$
$$\frac{-30}{-90} = \frac{-90r}{-90}$$
$$\frac{1}{3} = r$$

The discount rate is $33\frac{1}{3}\%$.

6. Strategy
Amount invested at 6%: x
Amount invested at 7%: $15,000

	Principal	Rate	Interest
Amount at 6%	x	0.06	0.06x
Amount at 7%	$15,000 - x$	0.07	$0.07(15,000 - x)$

The sum of the interest earned by the two investments equals the total annual interest earned ($970).

Solution
$$0.06x + 0.07(15,000 - x) = 970$$
$$0.06x + 1050 - 0.07x = 970$$
$$-0.01x + 1050 = 970$$
$$-0.01x = -80$$
$$x = 8000$$
$$15,000 - x = 15,000 - 8000 = 7000$$
The amount invested at 6% is $8000.
The amount invested at 7% is $7000.

7. Strategy
Measure of one of the equal angles: x
Measure of the third angle: $x - 15$
The sum of the measures of the angles of a triangle is 180°.

Solution
$$x + x + x - 15 = 180$$
$$3x - 15 = 180$$
$$3x = 195$$
$$\frac{3x}{3} = \frac{195}{3}$$
$$x = 65$$
$x - 15 = 50$
The measures of the angles are 65°, 65° and 50°.

8. Strategy
Speed of the bicyclist: x
Speed of the motorcyclist: $3x$

	Rate	Time	Distance
Bicyclist	x	2	$2(x)$
Motorcyclist	$3x$	2	$2(3x)$

In 2 h the motorcyclist is 60 mi ahead of the bicyclist.

Solution
$$2(3x) - 2x = 60$$
$$6x - 2x = 60$$
$$4x = 60$$
$$x = 15$$
$3x = 3(15) = 45$
The speed of the motorcyclist is 45 mph.

9. Strategy
Each equal angle: x
Third angle: $\frac{1}{2}x - 25$
The sum of the angles of a triangle is 180°.

Solution
$$x + x + \frac{1}{2}x - 25 = 180$$
$$2\frac{1}{2}x - 25 = 180$$
$$2.5x = 205$$
$$x = 82$$
$$\frac{1}{2}x - 25 = \frac{1}{2}(82) - 25 = 16$$
The angles measure 16°, 82°, and 82°.

10. Strategy
Given: $C = \$5.25$, $S = \$9.45$
Unknown markup rate: r
Use the equation $S = C + rC$.

Solution
$$S = C + rC$$
$$9.45 = 5.25 + 5.25r$$
$$4.20 = 5.25r$$
$$\frac{4.20}{5.25} = \frac{5.25r}{5.25}$$
$$0.80 = r$$
The markup rate is 80%.

11. Strategy

Percent of butterfat in the mixture: x

	Amount	Percent	Quantity
Cream	5	0.30	0.30(5)
Milk	8	0.04	0.04(8)
Mixture	13	x	$13x$

The sum of the quantities before mixing is equal to the quantity after mixing.

Solution

$$0.30(5) + 0.04(8) = 13x$$
$$1.5 + 0.32 = 13x$$
$$1.82 = 13x$$
$$0.14 = x$$

The mixture is 14% butterfat.

12. Strategy

Width: x

Length: $4x$

Use the equation for the perimeter of a rectangle.

Solution

$$2L + 2W = P$$
$$2(4x) + 2(x) = 200$$
$$8x + 2x = 200$$
$$10x = 200$$
$$x = 20$$
$$4x = 4(20) = 80$$

The width is 20 ft and the length is 80 ft.

13. Strategy

To find the minimum copies per week:

Write an expression for the cost of ordering from each company using x to represent the number of copies.

Write and solve an inequality.

Solution

Cost of ordering from Copy Center	is less than	Cost of ordering from Prints 4U

$$50 + 0.03x < 27 + 0.05x$$
$$50 + 0.03x - 0.03x < 27 + 0.05x - 0.03x$$
$$50 < 27 + 0.02x$$
$$-27 + 50 < -27 + 27 + 0.02x$$
$$23 < 0.02x$$
$$\frac{23}{0.02} < \frac{0.02x}{0.02}$$
$$1150 < x$$

The minimum number of copies that could be ordered is 1151.

14. Strategy

Given: $S = \$26.56$, $r = 17\% = 0.17$

Unknown regular price: R

Use the equation $S = R - rR$.

Solution

$$S = R - rR$$
$$26.56 = R - 0.17R$$
$$26.56 = 0.83R$$
$$\frac{26.56}{0.83} = \frac{0.83R}{0.83}$$
$$32 = R$$

The regular price is $32.

15. Strategy

Amount of cranberry juice: x

Amount of apple juice: $10 - x$

	Amount	Cost	Value
Cranberry juice	x	1.79	$1.79x$
Apple juice	$10 - x$	1.19	$1.19(10 - x)$
Mixture	10	1.61	$1.61(10)$

The sum of the values before mixing equals the value after mixing.

Solution

$$1.79x + 1.19(10 - x) = 1.61(10)$$
$$1.79x + 11.90 - 1.19x = 16.10$$
$$0.60x + 11.90 = 16.10$$
$$0.60x = 4.2$$
$$x = 7$$
$$10 - x = 10 - 7 = 3$$

7 qts of cranberry juice and 3 qts of apple juice are needed.

16. Strategy

Given: $S = \$33.74$, $r = 30\% = 0.30$

Unknown cost: C

Use the equation $S = C + rC$.

Solution

$$S = C + rC$$
$$33.74 = C + 0.30C$$
$$33.74 = 1.30C$$
$$\frac{33.74}{1.30} = \frac{1.30C}{1.30}$$
$$25.95 \approx C$$

The cost of the book is $25.95.

17. the smaller number: n
the larger number: $36 - n$

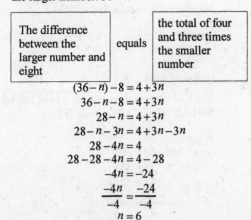

| The difference between the larger number and eight | equals | the total of four and three times the smaller number |

$$(36 - n) - 8 = 4 + 3n$$
$$36 - n - 8 = 4 + 3n$$
$$28 - n = 4 + 3n$$
$$28 - n - 3n = 4 + 3n - 3n$$
$$28 - 4n = 4$$
$$28 - 28 - 4n = 4 - 28$$
$$-4n = -24$$
$$\frac{-4n}{-4} = \frac{-24}{-4}$$
$$n = 6$$

$36 - n = 36 - 6 = 30$
The smaller number is 6.
The larger number is 30.

18. Strategy
Amount invested at 12%: x

	Principal	Rate	Interest
Amount at 8.15%	14,000	0.0815	0.0815(14,000)
Amount at 12%	x	0.12	$0.12x$
Amount at 9.25%	$14,000 + x$	0.0925	$0.0925(14,000 + x)$

The sum of the interest earned on the two accounts is equal to the interest earned at 9.25%.

Solution
$$0.0815(14,000) + 0.12x = 0.0925(14,000 + x)$$
$$1141 + 0.12x = 1295 + 0.0925x$$
$$1141 + 0.0275x = 1295$$
$$0.0275x = 154$$
$$x = 5600$$

$5600 must be deposited into an account paying 12% annual simple interest.

19. Strategy
Value of first card: x
Value of second card: $x + 1$
Assume that the values of an ace, jack, queen, and king are 1, 11, 12, and 13, respectively.
The sum of the values is 15.

Solution
$$x + x + 1 = 15$$
$$2x + 1 = 15$$
$$2x = 14$$
$$\frac{2x}{2} = \frac{14}{2}$$
$$x = 7$$
$$x + 1 = 8$$

The two cards are the 7 and 8 of hearts.

20. Strategy
Amount of pure water: x

	Amount	Percent	Quantity
Water	x	0.00	$0.00x$
Alcohol solution	15	0.80	0.80(15)
Mixture	$x + 15$	0.75	$0.75(x + 15)$

The sum of the quantities before mixing is equal to the quantity after mixing.

Solution
$$0.00x + 0.80(15) = 0.75(x + 15)$$
$$12 = 0.75x + 11.25$$
$$0.75 = 0.75x$$
$$1 = x$$

1 liter of pure water must be added.

21. Strategy
To find the lowest score, write and solve an inequality using N to represent the score on the sixth exam.

Solution

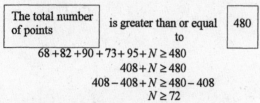

| The total number of points | is greater than or equal to | 480 |

$$68 + 82 + 90 + 73 + 95 + N \geq 480$$
$$408 + N \geq 480$$
$$408 - 408 + N \geq 480 - 408$$
$$N \geq 72$$

The lowest score on the sixth exam must be 72 or greater.

22. the unknown number: x

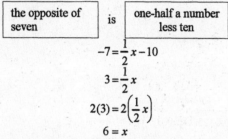

| the opposite of seven | is | one-half a number less ten |

$$-7 = \frac{1}{2}x - 10$$
$$3 = \frac{1}{2}x$$
$$2(3) = 2\left(\frac{1}{2}x\right)$$
$$6 = x$$

The number is 6.

23. Strategy
To find the height of the Eiffel Tower, write and solve an equation using h to represent the height.

Solution

| 1472 | is | 654 ft less than twice the height of the Eiffel Tower |

$$1472 = 2h - 654$$
$$1472 + 654 = 2h - 654 + 654$$
$$2126 = 2h$$
$$\frac{2126}{2} = \frac{2h}{2}$$
$$1063 = h$$

The height is 1063 ft.

24. Strategy
Measure of the first angle: $x + 15°$
Measure of the second angle: x
Measure of the third angle: $x - 15°$
Use the equation $A + B + C = 180°$.

Solution
$$A + B + C = 180$$
$$(x + 15) + x + (x - 15) = 180$$
$$3x = 180$$
$$x = 60$$
$$x + 15 = 60 + 15 = 75$$
$$x - 15 = 60 - 15 = 45$$
The measures of the angles are 75°, 60°, and 45°.

25. Strategy
Number of one-dollar bills: x
Number of five-dollar bills: $155 - x$

Bill	Number	Value	Total value
$1	x	1	$1x$
$5	$155 - x$	5	$5(155 - x)$

The sum of the values of each type of bill equals the total value of all the bills ($263).

Solution
$$1x + 5(155 - x) = 263$$
$$x + 775 - 5x = 263$$
$$-4x = -512$$
$$\frac{-4x}{-4} = \frac{-512}{-4}$$
$$x = 128$$
There are 128 one-dollar bills.

26. Strategy
To find the length of the longer piece, write and solve an equation using x to represent the length of the longer piece and $10 - x$ to represent the length of the shorter piece.

Solution

| Four times the length of the shorter piece | is | 2 ft less than two times the longer piece |

$$4(10 - x) = 2x - 2$$
$$40 - 4x = 2x - 2$$
$$40 - 4x - 2x = 2x - 2x - 2$$
$$40 - 6x = -2$$
$$40 - 40 - 6x = -2 - 40$$
$$-6x = -42$$
$$\frac{-6x}{-6} = \frac{-42}{-6}$$
$$x = 7$$

The length of the longer piece is 7 ft.

27. Strategy
To find the number of hours, write and solve an equation using n to represent the number of hours.

Solution

| $600 | equals | $80 and $65 for each hour of consultation |

$$600 = 80 + 65n$$
$$600 - 80 = 80 - 80 + 65n$$
$$520 = 65n$$
$$\frac{520}{65} = \frac{65n}{65}$$
$$8 = n$$

The number of hours of consultation was 8.

28. Strategy
Number of 17¢ stamps: x
Number of 12¢ stamps: $2x + 3$

Stamp	Number	Value	Total value
17¢	x	17	$17x$
12¢	$2x + 3$	12	$12(2x + 3)$

The sum of the total values of the stamps is equal to the given total face value of the stamps ($\$15.53 = 1553$¢).

Solution
$$17x + 12(2x + 3) = 1553$$
$$17x + 24x + 36 = 1553$$
$$41x = 1517$$
$$\frac{41x}{41} = \frac{1517}{41}$$
$$x = 37$$
$$2x + 3 = 2(37) + 3 = 74 + 3 = 77$$
There are 77 12¢ stamps and 37 17¢ stamps.

29. Strategy
Given: $r = 60\% = 0.60$, $S = \$1074$
Unknown cost: C
Use the equation $S = C + rC$.

Solution
$$S = C + rC$$
$$1074 = C + 0.60C$$
$$1074 = 1.60C$$
$$\frac{1074}{1.60} = \frac{1.60C}{1.60}$$
$$671.25 = C$$
The cost is $\$671.25$.

30. Strategy
Amount of time: x

	Rate	Time	Distance
Slower swimmer	55	x	$55x$
Faster swimmer	65	x	$65x$

The total distance is 480 m.

Solution
$$55x + 65x = 480$$
$$120x = 480$$
$$x = 4$$
They will meet in 4 min.

31. Strategy
First side: x
Second side: $x + 4$
Third side: $2x - 1$
Use the equation for the perimeter of a triangle.

Solution
$$a + b + c = P$$
$$x + x + 4 + 2x - 1 = 35$$
$$4x + 3 = 35$$
$$4x = 32$$
$$x = 8$$
$$x + 4 = 8 + 4 = 12$$
$$2x - 1 = 2(8) - 1 = 15$$
The lengths of the sides are 8 in., 12 in., and 15 in.

32. Strategy
First integer: n
Second integer: $n + 2$
Third integer: $n + 4$
The sum of the three integers is –45.

Solution
$$n + (n + 2) + (n + 4) = -45$$
$$3n + 6 = -45$$
$$3n = -51$$
$$n = -17$$
$$n + 2 = -17 + 2 = -15$$
$$n + 4 = -17 + 4 = -13$$
The integers are –17, –15, and –13.

33. Strategy
To find the maximum width:
Replace the variables in the area formula by the given values and solve for x.
Replace the variable in the expression $2x - 4$ with the value found for x.

Solution

Length times width	is less than	180 ft^2

$$15(2x - 4) < 180$$
$$30x - 60 < 180$$
$$30x - 60 + 60 < 180 + 60$$
$$30x < 240$$
$$\frac{30x}{30} < \frac{240}{30}$$
$$x < 8$$
$$2x - 4 < 12$$
The maximum width of the rectangle is 11 ft.

34. Strategy

$a = 138°$ because alternate interior angles have the same measure. $a + b = 180°$ because adjacent angles of intersecting lines are supplementary angles.

Substitute 138 for a and solve for b.

Solution

$$a + b = 180$$
$$138 + b = 180$$
$$b = 42$$

The measure of angle a is 138°, and the measure of angle b is 42°.

Chapter Test

1. the unknown number: n

The sum of six times a number and thirteen	is	five less than the product of three and the number

$$6n + 13 = 3n - 5$$
$$6n - 3n + 13 = 3n - 3n - 5$$
$$3n + 13 = -5$$
$$3n + 13 - 13 = -5 - 13$$
$$3n = -18$$
$$\frac{3n}{3} = \frac{-18}{3}$$
$$n = -6$$

The number is –6.

2. the unknown number: n

The difference between three times a number and fifteen	is	27

$$3n - 15 = 27$$
$$3n - 15 + 15 = 27 + 15$$
$$3n = 42$$
$$\frac{3n}{3} = \frac{42}{3}$$
$$n = 14$$

The number is 14.

3. the larger number: n

the smaller number: $18 - n$

The difference between four times the smaller number and seven	is equal to	the sum of two times the larger number and five

$$4(18 - n) - 7 = 2n + 5$$
$$72 - 4n - 7 = 2n + 5$$
$$65 - 4n = 2n + 5$$
$$65 - 4n - 2n = 2n - 2n + 5$$
$$65 - 6n = 5$$
$$65 - 65 - 6n = 5 - 65$$
$$-6n = -60$$
$$\frac{-6n}{-6} = \frac{-60}{-6}$$
$$n = 10$$

$18 - n = 18 - 10 = 8$

The larger number is 10.

The smaller number is 8.

4. Strategy

To find the length of each piece, write and solve an equation using x to represent the length of the longer piece and $18 - x$ to represent the length of the shorter piece.

Solution

Two feet less than the product of five and the length of the shorter piece	is equal to	the difference between three times the length of the longer piece and eight

$$5(18 - x) - 2 = 3x - 8$$
$$90 - 5x - 2 = 3x - 8$$
$$88 - 5x = 3x - 8$$
$$88 - 5x - 3x = 3x - 3x - 8$$
$$88 - 8x = -8$$
$$88 - 88 - 8x = -8 - 88$$
$$-8x = -96$$
$$\frac{-8x}{-8} = \frac{-96}{-8}$$
$$x = 12$$

$18 - x = 18 - 12 = 6$

The pieces measure 6 ft and 12 ft.

5. Strategy

Given: $r = 50\% = 0.50$, $S = \$300$

Unknown: C

Use the equation $S = C + rC$.

Solution

$$S = C + rC$$
$$300 = C + 0.50$$
$$300 = 1.50C$$
$$200 = C$$

The cost is $200.

6. Strategy
Given: $R = \$100$, $S = \$80$
Unknown: r
Use the equation $S = R - rR$.

Solution
$$S = R - rR$$
$$80 = 100 - r(100)$$
$$-20 = -100r$$
$$0.2 = r$$
The discount rate is 20%.

7. Strategy
Amount of 15% acid solution: x

	Amount	Percent	Quantity
15% acid solution	x	0.15	$0.15x$
20% acid solution	5	0.20	$0.20(5)$
16% acid solution	$x + 5$	0.16	$0.16(x + 5)$

The sum of the quantities before mixing is equal to the quantity after mixing.

Solution
$$0.15x + 0.20(5) = 0.16(x + 5)$$
$$0.15x + 1 = 0.16x + 0.8$$
$$-0.1x + 1 = 0.8$$
$$-0.01x = -0.2$$
$$x = 20$$
20 gal of the 15% acid solution should be used.

8. Strategy
Length: $3W - 1$
Width: W
Use the equation for the perimeter of a rectangle.

Solution
$$2L + 2W = P$$
$$2(3W - 1) + 2W = 38$$
$$6W - 2 + 2W = 38$$
$$8W - 2 = 38$$
$$8W = 40$$
$$W = 5$$
$$L = 3W - 1 = 3(5) - 1 = 15 - 1 = 14$$
The length is 14 m, and the width is 5 m.

9. Strategy
First odd integer: n
Second odd integer: $n + 2$
Third odd integer: $n + 4$
Three times the first integer is one less than the sum of the second and third integers.

Solution
$$3n = (n + 2) + (n + 4) - 1$$
$$3n = 2n + 5$$
$$n = 5$$
$$n + 2 = 5 + 2 = 7$$
$$n + 4 = 5 + 4 = 9$$
The three integers are 5, 7, and 9.

10. Strategy
To find the number of residents, write and solve an inequality using N to represent the number of residents.

Solution

$15 plus $18 per bouquet	is less than	$3 plus $21 per bouquet

$$15 + 18N < 3 + 21N$$
$$15 + 18N - 21N < 3 + 21N - 21N$$
$$15 - 3N < 3$$
$$15 - 15 - 3N < 3 - 15$$
$$-3N < -12$$
$$\frac{-3N}{-3} > \frac{-12}{-3}$$
$$N > 4$$
For florists B to be more economical, there must be 5 or more residents in the nursing home.

11. Strategy
Amount invested at 10%: x
Amount invested at 15%: $\$7000 - x$

	Principal	Rate	Interest
Amount at 10%	x	0.10	0.10
Amount at 15%	$7000 - x$	0.15	$0.15(7000 - x)$

The total annual interest earned is $800.

Solution
$$0.10x + 0.15(7000 - x) = 800$$
$$0.10x + 1050 - 0.15x = 800$$
$$-0.05x + 1050 = 800$$
$$-0.05x = -250$$
$$x = 5000$$
$$7000 - x = 7000 - 5000 = 2000$$
The amount invested at 10% is $5000.
The amount invested at 15% is $2000.

12. Strategy
Amount of $7 grade: x
Amount of $4 grade: $12 - x$

	Amount	Cost	Value
$7 grade	x	7.00	$7.00x$
$4 grade	$12 - x$	4.00	$4.00(12 - x)$
Mixture	12	6.00	$6.00(12)$

The sum of the values before mixing equals the value after mixing.

Solution
$$7.00x + 4.00(12 - x) = 6.00(12)$$
$$7.00x + 48 - 4.00x = 72$$
$$3.00x + 48 = 72$$
$$3.00x = 24$$
$$x = 8$$
$$12 - x = 12 - 8 = 4$$
To make the mixture, 8 lb of the $7 grade and 4 lb of the $4 grade should be used.

13. Strategy
Rate of first plane: $r + 100$
Rate of second plane: r

	Rate	Time	Distance
1st plane	$r + 100$	3	$3(r + 100)$
2nd plane	r	3	$3r$

In two hours, the two planes are 1050 mi apart.

Solution
$$3(r + 100) + 3r = 1050$$
$$3r + 300 + 3r = 1050$$
$$6r + 300 = 1050$$
$$6r = 750$$
$$r = 125$$
$$r + 100 = 125 + 100 = 225$$
The rate of the first plane is 225 mph.
The rate of the second plane is 125 mph.

14. Strategy
Measure of one angle: $15 + x$
Measure of second angle: x
Measure of third angle: $3x$
Use the equation $A + B + C = 180°$.

Solution
$$A + B + C = 180$$
$$(15 + x) + x + 3x = 180$$
$$5x + 15 = 180$$
$$5x = 165$$
$$x = 33$$
$$15 + x = 15 + 33 = 48$$
$$3x = 3(33) = 99$$
The measures of the angles are 33°, 48°, and 99°.

15. Strategy
Number of nickels: x
Number of quarters: $50 - x$

Coin	Number	Value	Total Value
Nickel	x	5	$5x$
Quarter	$50 - x$	25	$25(50 - x)$

The sum of the total values of each type of coin equals the total value of all the coins (950 cents).

Solution
$$5x + 25(50 - x) = 950$$
$$5x + 1250 - 25x = 950$$
$$-20x + 1250 = 950$$
$$-20x = -300$$
$$x = 15$$
$$50 - x = 50 - 15 = 35$$
The bank contains 15 nickels and 35 quarters.

16. Strategy
Amount invested at 6.75%: x
Amount invested at 9.45%: $2400 - x$

	Principal	Rate	Interest
Amount at 6.75%	x	0.0675	$0.0675x$
Amount at 9.45%	$2400 - x$	0.0945	$0.0945(2400 - x)$

The interest earned on each account is the same.

Solution
$0.0675x = 0.0945(2400 - x)$
$0.0675x = 226.8 - 0.0945x$
$0.162x = 226.8$
$x = 1400$
$2400 - x = 2400 - 1400 = 1000$
The amount deposited at 6.75% is $1400.
The amount deposited at 9.45% is $1000.

17. Strategy
To find the minimum length:
Replace the variables in the area formula by the given values and solve for x.
Replace the variable in the expression $3x + 5$ with value found for x

Solution
$$LW > A$$
$$(3x + 5)12 > 276$$
$$36x + 60 > 276$$
$$36x + 60 - 60 > 276 - 60$$
$$36x > 216$$
$$\frac{36x}{36} > \frac{216}{36}$$
$$x > 6$$
$3x + 5 > 23$
The minimum length of the rectangle is 24 ft.

18. Strategy
Given: $R = \$99$, $r = 20\% = 0.20$
Unknown sale price: S
Use the equation: $S = R - rR$

Solution
$S = R - rR$
$S = 99 - 0.20(99)$
$S = 99 - 19.80$
$S = 79.20$
The sale price is $79.20.

Cumulative Review Exercises

1. $-12 < -4$ true
$-6 < -4$ true
$-3 < -4$ false
$-1 < -4$ false
-12 and -6 are less than -4.

2. $-2 + (-8) - (-16) = -2 + (-8) + 16 = -10 + 16 = 6$

3. $\left(-\dfrac{2}{3}\right)^3 \left(-\dfrac{3}{4}\right)^2 = -\dfrac{8}{27}\left(\dfrac{9}{16}\right) = -\dfrac{1}{6}$

4. $\dfrac{5}{6} - \left(\dfrac{2}{3}\right)^2 \div \left(\dfrac{1}{2} - \dfrac{1}{3}\right) = \dfrac{5}{6} - \left(\dfrac{2}{3}\right)^2 \div \left(\dfrac{3}{6} - \dfrac{2}{6}\right)$

$= \dfrac{5}{6} - \left(\dfrac{2}{3}\right)^2 \div \dfrac{1}{6}$

$= \dfrac{5}{6} - \dfrac{4}{9} \cdot \dfrac{6}{1}$

$= \dfrac{5}{6} - \dfrac{8}{3}$

$= \dfrac{5}{6} - \dfrac{16}{6}$

$= -\dfrac{11}{6}$

5. $-|-18| = -18$

6. $b^2 - (a - b)^2$
$(-1)^2 - [4 - (-1)]^2 = (-1)^2 - (4 + 1)^2$
$= (-1)^2 - 5^2$
$= 1 - 25$
$= -24$

7. $5x - 3y - (-4x) + 7y = 5x - 3y + 4x + 7y$
$\qquad = 9x + 4y$

8. $-4(3 - 2x - 5x^3) = -12 + 8x + 20x^3$

9. $-2[x - 3(x - 1) - 5] = -2(x - 3x + 3 - 5)$
$\qquad = -2(-2x - 2)$
$\qquad = 4x + 4$

10. $-3x^2 - (-5x^2) + 4x^2 = -3x^2 + 5x^2 + 4x^2$
$\qquad = 2x^2 + 4x^2$
$\qquad = 6x^2$

11.

$4 - 2x - x^2 = 2 - 4x$	
$4 - 2(2) - (2)^2$	$2 - 4(2)$
$4 - 4 - 4$	$2 - 8$
$-4 \neq -6$	

No, 2 is not a solution.

12. $9 - x = 12$
$-9 + 9 - x = -9 + 12$
$\qquad -x = 3$
$\qquad -1(-x) = -1(3)$
$\qquad x = -3$
The solution is -3.

13. $-\dfrac{4}{5}x = 12$
$-\dfrac{5}{4}\left(-\dfrac{4}{5}x\right) = -\dfrac{5}{4}(12)$
$\qquad x = -15$
The solution is -15.

14. $8 - 5x = -7$
$8 - 8 - 5x = -7 - 8$
$\qquad -5x = -15$
$\qquad \dfrac{-5x}{-5} = \dfrac{-15}{-5}$
$\qquad x = 3$

15. $-6x - 4(3 - 2x) = 4x + 8$
$-6x - 12 + 8x = 4x + 8$
$\qquad -12 + 2x = 4x + 8$
$-12 + 2x - 4x = 4x - 4x + 8$
$\qquad -12 - 2x = 8$
$-12 + 12 - 2x = 8 + 12$
$\qquad -2x = 20$
$\qquad \dfrac{-2x}{-2} = \dfrac{20}{-2}$
$\qquad x = -10$
The solution is -10.

16. $40\% = 40\left(\dfrac{1}{100}\right) = \dfrac{2}{5}$

17. $4x \geq 16$
$\dfrac{4x}{4} \geq \dfrac{16}{4}$
$\quad x \geq 4$

18. $-15x \leq 45$
$\dfrac{-15x}{-15} \geq \dfrac{45}{-15}$
$\qquad x \geq -3$

19. $2x - 3 > x + 15$
$2x - x - 3 > x - x + 15$
$\qquad x - 3 > 15$
$x - 3 + 3 > 15 + 3$
$\qquad x > 18$

20. $12 - 4(x - 1) \leq 5(x - 4)$
$12 - 4x + 4 \leq 5x - 20$
$\qquad -4x + 16 \leq 5x - 20$
$-4x - 5x + 16 \leq 5x - 5x - 20$
$\qquad -9x + 16 \leq -20$
$-9x + 16 - 16 \leq -20 - 16$
$\qquad -9x \leq -36$
$\qquad \dfrac{-9x}{-9} \geq \dfrac{-36}{-9}$
$\qquad x \geq 4$

21. $0.025 = 0.025(100\%) = 2.5\%$

22. $\dfrac{3}{25} = \dfrac{3}{25}(100\%) = 12\%$

23. $16\dfrac{2}{3}\% = 16\dfrac{2}{3}\left(\dfrac{1}{100}\right) = \dfrac{50}{3}\left(\dfrac{1}{100}\right) = \dfrac{1}{6}$
$\qquad PB = A$
$\qquad \dfrac{1}{6}(18) = A$
$\qquad 3 = A$
$16\dfrac{2}{3}\%$ of 18 is 3.

24. $40\% = 0.40$
$\qquad PB = A$
$\qquad 0.40B = 18$
$\qquad \dfrac{0.40B}{0.40} = \dfrac{18}{0.40}$
$\qquad B = 5$
40% of 45 is 18.

25. the unknown number: n

The sum of eight times a number and twelve	is equal to	the product of four and the number

$\qquad 8n + 12 = 4n$
$8n - 8n + 12 = 4n - 8n$
$\qquad 12 = -4n$
$\qquad \dfrac{12}{-4} = \dfrac{-4n}{-4}$
$\qquad -3 = n$
The number is -3.

26. Strategy
To find the area of the garage, write and solve an equation using x to represent the area of the garage.

Solution

2000 ft^2	is	200 ft^2 more than three times the area of the garage

$$2000 = 200 + 3x$$
$$2000 - 200 = 200 - 200 + 3x$$
$$1800 = 3x$$
$$\frac{1800}{3} = \frac{3x}{3}$$
$$600 = x$$

The area of the garage is 600 ft^2.

27. Strategy
To find the number of hours of labor, write and solve an equation using x to represent the number of hours of labor.

Solution

$313	is	$88 for parts plus $45 for each hour of labor

$$313 = 88 + 45x$$
$$225 = 45x$$
$$5 = x$$

There were 5 h of labor.

28. Strategy
To find the percent of the libraries that had the reference book, solve the percent equation using $B = 250$ and $A = 50$. The percent is unknown.

Solution
$$PB = A$$
$$P(250) = 50$$
$$\frac{250P}{250} = \frac{50}{250}$$
$$P = 0.20$$

20% of the libraries had the reference book.

29. Strategy
Additional amount: x

Principal	Rate	Interest
4000	0.11	0.11(4000)
x	0.14	0.14x
4000 + x	0.12	0.12(4000 + x)

The total interest earned is 12% of the total investment.

Solution
$$0.11(4000) + 0.14x = 0.12(4000 + x)$$
$$440 + 0.14x = 480 + 0.12x$$
$$440 + 0.02x = 480$$
$$0.02x = 40$$
$$x = 2000$$

$2000 more must be invested at 14%.

30. Strategy
Given: $C = \$80$, $S = \$140$
Unknown: r
Use the equation $S = C + rC$.

Solution
$$S = C + rC$$
$$140 = 80 + r(80)$$
$$60 = 80r$$
$$0.75 = r$$

The markup rate is 75%.

31. Strategy
Amount of $4 gold alloy: x

	Amount	Cost	Value
$4 alloy	x	4.00	4.00x
$7 alloy	30	7.00	7.00(30)
$5 alloy	$x + 30$	5.00	5.00($x + 30$)

The sum of the values before mixing equals the value after mixing.

Solution
$$4.00x + 7.00(30) = 5.00(x + 30)$$
$$4.00x + 210 = 5.00x + 150$$
$$210 = 1.00x + 150$$
$$60 = x$$

60 g of the $4 gold alloy should be used.

32. Strategy
Amount of water: x

	Amount	Percent	Quantity
Water	x	0	$0x$
10% salt solution	70	0.10	0.10(70)
7% salt solution	$x + 70$	0.07	$0.07(x + 70)$

The sum of the quantities before mixing is equal to the quantity after mixing.

Solution
$$0x + 0.10(70) = 0.07(x + 70)$$
$$7 = 0.07x + 4.9$$
$$2.1 = 0.07x$$
$$30 = x$$
30 oz of water should be used.

33. Strategy
Measure of the equal angles. x
Measure of the third angle: $2x - 8$
Use the equation $A + B + C = 180°$.

Solution
$$A + B + C = 180$$
$$x + x + (2x - 8) = 180$$
$$4x - 8 = 180$$
$$4x = 188$$
$$x = 47$$
The measure of each of the equal angles is 47°.

34. Strategy
First integer: n
Second integer: $n + 2$
Third integer: $n + 4$
Three times the second integer is 14 more than the sum of the first and third integers.

Solution
$$3(n + 2) = 14 + (n + n + 4)$$
$$3n + 6 = 2n + 18$$
$$n + 6 = 18$$
$$n = 12$$
$$n + 2 = 12 + 2 = 14$$
The middle even integer is 14.

35. Strategy
Number of dimes: x
Number of quarters: $4x - 5$

Coin	Number	Value	Total value
Dime	x	10	$10x$
Quarters	$4x - 5$	25	$25(4x - 5)$

The sum of the total values of each type of coin equals the total value of all the coins (645 cents).

Solution
$$10x + 25(4x - 5) = 645$$
$$10x + 100x - 125 = 645$$
$$110x - 125 = 645$$
$$110x = 770$$
$$x = 7$$
There are 7 dimes in the bank.

Chapter 5: Linear Equations and Inequalities

1. $\dfrac{5-(-7)}{4-8} = \dfrac{5+7}{-4} = \dfrac{12}{-4} = -3$ [1.4.2]

2. $\dfrac{3-(-2)}{-3-2} = \dfrac{3+2}{-5} = \dfrac{5}{-5} = -1$ [2.1.1]

3. $-3x + 12$ [2.2.4]

4. $3x + 6 - 6 = 0 - 6$ [3.2.1]
   $\quad 3x = -6$
   $\quad\ x = -2$

5. $4x + 5(0) = 20$ [3.2.1]
   $\quad\ 4x = 20$
   $\quad\ \ x = 5$

6. $3(-1) - 7y = 11$ [3.2.1]
   $\quad -3 - 7y = 11$
   $\qquad\ -7y = 14$
   $\qquad\quad y = -2$

7. $y = -4(-2) + 5$ [2.1.1]
   $y = 8 + 5$
   $y = 13$

8. $\dfrac{1}{4}(3x - 16) = \dfrac{3}{4}x - \dfrac{16}{4} = \dfrac{3}{4}x - 4$ [2.2.4]

9.
   $\quad -4 < x + 3$
   $-4 - 3 < x$
   $\quad\ -7 < x$
   $\qquad\ x > -7$
   The solutions are a, b, and c.   [3.3.1]

Go Figure

If the ratio of $\overline{AB}$ to $\overline{AC}$ is 1:4, then $\overline{AB}$ is one unit and $\overline{AC}$ is four units. Then the length of $\overline{BC}$ is three units.

If the ratio of $\overline{BC}$ to $\overline{CD}$ is 1:2 and $\overline{BC}$ is three units, then $\overline{CD}$ is two times the length of $\overline{BC}$, or six units.

The ratio of $\overline{AB}$ to $\overline{CD}$ then is 1:6.

Section 5.1

Concept Review 5.1

1. Always true

3. Never true
   In an ordered pair, the first number is the *x*-coordinate and the second number is the *y*-coordinate.

5. Never true
   Any point plotted in quadrant III has a negative *y* value.

7. Always true

Objective 5.1.1 Exercises

3.

5.

7.

9. $A(2, 3)$, $B(4, 0)$, $C(-4, 1)$, $D(-2, -2)$

11. $A(-2, 5)$, $B(3, 4)$, $C(0, 0)$, $D(-3, -2)$

13. **a.** Abscissa of point A: 2
    Abscissa of point C: -4

    b. Ordinate of point B: 1
    Ordinate of point D: -3

Objective 5.1.2 Exercises

15.
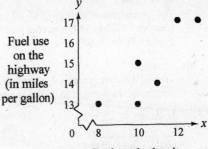

Fuel use on the highway (in miles per gallon)

Fuel use in the city (in miles per gallon)

17.
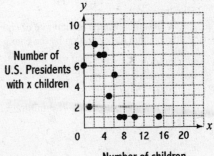

Number of U.S. Presidents with x children

Number of children

19.

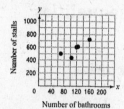

Number of stalls (y-axis): 1000, 800, 600, 400, 200
Number of bathrooms (x-axis): 0, 40, 80, 120, 160, 200

Objective 5.1.3 Exercises

21. Strategy
To find the average rate of change, divide the change in temperature (77 – 45) by the change in hours (2PM – 6AM).

Solution
$$\frac{77-45}{2PM-6AM} = \frac{32}{8} = 4$$
The rate of change is 4° per hour.

23a. Strategy
To find the average annual rate of change, divide the change in wage by the change in years.

Solution
$$\frac{3.80-2.65}{1990-1978} = \frac{1.15}{12} \approx 0.10$$
The annual rate of change was $0.10 per year.

23b. Strategy
To find the average annual rate of change, divide the change in wage by the change in years.

Solution
$$\frac{3.80-3.10}{1990-1980} = \frac{0.7}{10} = 0.07$$
The annual rate of change of $0.07 in wage was less from 1980 to 1990 than the rate of change from 1978 to 1990.

25a. Strategy
To find the average annual rate of change, divide the change in population by the change in years.

Solution
$$\frac{19-51}{2046-2031} = \frac{-32}{15} = -2.13$$
The rate of change was –2.13 million people per year.

25b. Strategy
To find the average annual rate of change, divide the change in population by the change in years.

Solution
$$\frac{19-79}{2046-2000} = \frac{-60}{46} = -1.3$$
The rate of change was –1.3 million people per year.

27a. Strategy
To find the average annual rate of change, divide the change in price by the change in years.

Solution
$$\frac{2300-42}{2001-1967} = \frac{2258}{34} \approx 66$$
The average rate of change was $66 thousand per year.

27b. Strategy
To find the average annual rate of change, divide the change in price by the change in years.

Solution
For 1971 to 1991,
$$\frac{800-72}{1991-1971} = \frac{728}{20} = 36.4$$
For 1981 to 2001,
$$\frac{2300-275}{2001-1981} = \frac{2025}{20} = 101.25$$
$$101.25 - 36.4 = 64.85$$
The average rate of change from 1981 to 2001 was greater than the rate of change from 1971 to 1991 by $64.85 thousand.

Applying Concepts 5.1

29. 1 unit

31. 0 units

33. 1 unit

35. The coordinates of a point plotted at the origin of the rectangular coordinate system is (0, 0).

Section 5.2

Concept Review 5.2

1. Never true
The equation $y = 3x^2 - 6$ is a quadratic equation. The equation $y = 3x + 2$ is a linear equation.

3. Always true

5. Never true
To find the x-intercept of a linear equation, let $y = 0$.

7. Never true
The graph of the equation $x = -4$ has an x-intercept of (–4, 0).

9. Always true

Objective 5.2.1 Exercises

1. $y = 4x + 1$; $m = 4$, $b = 1$

3. $y = \frac{5}{6}x + \frac{1}{6}$; $m = \frac{5}{6}$, $b = \frac{1}{6}$

5. $y = -x + 7$

$$\begin{array}{c|c} 4 & -(3)+7 \\ & -3+7 \\ & 4 \end{array}$$

$4 = 4$

Yes, (3, 4) is a solution of $y = -x + 7$.

7. $y = \dfrac{1}{2}x - 1$

$$\begin{array}{c|c} 2 & \frac{1}{2}(-1)-1 \\ & -\frac{1}{2}-1 \\ & -\frac{3}{2} \end{array}$$

$2 \neq -\dfrac{3}{2}$

No, (–1, 2) is not a solution of $y = \dfrac{1}{2}x - 1$.

9. $y = \dfrac{1}{4}x + 1$

$$\begin{array}{c|c} 1 & \frac{1}{4}(4)+1 \\ & 1+1 \\ & 2 \end{array}$$

$1 \neq 2$

No, (4, 1) is not a solution of $y = \dfrac{1}{4}x + 1$.

11. $y = \dfrac{3}{4}x + 4$

$$\begin{array}{c|c} 4 & \frac{3}{4}(0)+4 \\ & 0+4 \\ & 4 \end{array}$$

$4 = 4$

Yes, (0, 4) is a solution of $y = \dfrac{3}{4}x + 4$.

13. $y = 3x + 2$

$$\begin{array}{c|c} 0 & 3(0)+2 \\ & 0+2 \\ & 2 \end{array}$$

$0 \neq 2$

No, (0, 0) is not a solution of $y = 3x + 2$.

15. $y = 3x - 2$

$= 3(3) - 2$

$= 9 - 2$

$= 7$

The ordered-pair solution is (3, 7).

17. $y = \dfrac{2}{3}x - 1$

$= \dfrac{2}{3}(6) - 1$

$= 4 - 1$

$= 3$

The ordered-pair solution is (6, 3).

19. $y = -3x + 1$

$= -3(0) + 1$

$= 0 + 1$

$= 1$

The ordered-pair solution is (0, 1).

21. $y = \dfrac{2}{5}x + 2$

$= \dfrac{2}{5}(-5) + 2$

$= -2 + 2$

$= 0$

The ordered-pair solution is (–5, 0).

Objective 5.2.2 Exercises

23.

25.

27.

29.

31.

33.

35.

37.

39.

41.

43.

45.

47.

49.

51.

Objective 5.2.3 Exercises

53.
$$3x + y = 10$$
$$3x - 3x + y = -3x + 10$$
$$y = -3x + 10$$

55.
$$4x - y = 3$$
$$4x - 4x - y = -4x + 3$$
$$-y = -4x + 3$$
$$(-1)(-y) = (-1)(-4x + 3)$$
$$y = 4x - 3$$

57.
$$3x + 2y = 6$$
$$3x - 3x + 2y = -3x + 6$$
$$2y = -3x + 6$$
$$\frac{2y}{2} = \frac{-3x + 6}{2}$$
$$y = -\frac{3}{2}x + 3$$

59.
$$2x - 5y = 10$$
$$2x - 2x - 5y = -2x + 10$$
$$-5y = -2x + 10$$
$$\frac{-5y}{-5} = \frac{-2x + 10}{-5}$$
$$y = \frac{2}{5}x - 2$$

61.
$$2x + 7y = 14$$
$$2x - 2x + 7y = -2x + 14$$
$$7y = -2x + 14$$
$$\frac{7y}{7} = \frac{-2x + 14}{7}$$
$$y = -\frac{2}{7}x + 2$$

63.
$$x + 3y = 6$$
$$x - x + 3y = -x + 6$$
$$3y = -x + 6$$
$$\frac{3y}{3} = \frac{-x + 6}{3}$$
$$y = -\frac{1}{3}x + 2$$

65.

67.

69.

71.

73.

75.

77. *x*-intercept: *y*-intercept:

$x - y = 3$ $x - y = 3$

$x - 0 = 3$ $0 - y = 3$

$x = 3$ $-y = 3$

$y = -3$

(3, 0)   (0, −3)

79. *x*-intercept: *y*-intercept:

$y = 2x - 6$ $(0, b)$

$0 = 2x - 6$ $b = -6$

$6 = 2x$ $(0, -6)$

$3 = x$

(3, 0)

81. *x*-intercept: *y*-intercept:

$x - 5y = 10$ $x - 5y = 10$

$x - 5(0) = 10$ $0 - 5y = 10$

$x = 10$ $-5y = 10$

$y = -2$

(10, 0)   (0, −2)

83. *x*-intercept: *y*-intercept:

$y = 3x + 12$ $(0, b)$

$0 = 3x + 12$ $b = 12$

$-12 = 3x$ $(0, 12)$

$-4 = x$

(−4, 0)

85. *x*-intercept: *y*-intercept:

$2x - 3y = 0$ $2x - 3y = 0$

$2x - 3(0) = 0$ $2(0) - 3y = 0$

$2x = 0$ $-3y = 0$

$x = 0$ $y = 0$

(0, 0)   (0, 0)

87. *x*-intercept: *y*-intercept:

$y = \frac{1}{2}x + 3$ $(0, b)$

$b = 3$

$0 = \frac{1}{2}x + 3$ $(0, 3)$

$-3 = \frac{1}{2}x$

$-6 = x$

(−6, 0)

89. *x*-intercept: *y*-intercept:

$5x + 2y = 10$ $5x + 2y = 10$

$5x + 2(0) = 10$ $5(0) + 2y = 10$

$5x = 10$ $2y = 10$

$x = 2$ $y = 5$

(2, 0)   (0, 5)

91. *x*-intercept: *y*-intercept:

$3x - 4y = 12$ $3x - 4y = 12$

$3x - 4(0) = 12$ $3(0) - 4y = 12$

$3x = 12$ $-4y = 12$

$x = 4$ $y = -3$

(4, 0)   (0, −3)

93. *x*-intercept: *y*-intercept:

$2x + 3y = 6$ $2x + 3y = 6$

$2x + 3(0) = 6$ $2(0) + 3y = 6$

$2x = 6$ $3y = 6$

$x = 3$ $y = 2$

(3, 0)   (0, 2)

95. *x*-intercept: *y*-intercept:

$2x + 5y = 10$ $2x + 5y = 10$

$2x + 5(0) = 10$ $2(0) + 5y = 10$

$2x = 10$ $5y = 10$

$x = 5$ $y = 2$

(5, 0)   (0, 2)

97. *x*-intercept: *y*-intercept:

$x - 3y = 6$ $x - 3y = 6$

$x - 3(0) = 6$ $(0) - 3y = 6$

$x = 6$ $-3y = 6$

$y = -2$

(6, 0)   (0, −2)

99. *x*-intercept: *y*-intercept:

$x - 4y = 8$ $x - 4y = 8$

$x - 4(0) = 8$ $(0) - 4y = 8$

$x = 8$ $-4y = 8$

$y = -2$

(8, 0)   (0, −2)

101. x-intercept:

$$2x + y = 3$$
$$2x + 0 = 3$$
$$2x = 3$$
$$x = \frac{3}{2}$$
$$\left(\frac{3}{2}, 0\right)$$

y-intercept:

$$2x + y = 3$$
$$2(0) + y = 3$$
$$y = 3$$
$$(0, 3)$$

103. x-intercept:

$$4x - 3y = 6$$
$$4x - 3(0) = 6$$
$$4x = 6$$
$$x = \frac{3}{2}$$
$$\left(\frac{3}{2}, 0\right)$$

y-intercept:

$$4x - 3y = 6$$
$$4(0) - 3y = 6$$
$$-3y = 6$$
$$y = -2$$
$$(0, -2)$$

105. x-intercept:

$$3x + 4y = -12$$
$$3x + 4(0) = -12$$
$$3x = -12$$
$$x = -4$$
$$(-4, 0)$$

y-intercept:

$$3x + 4y = -12$$
$$3(0) + 4y = -12$$
$$4y = -12$$
$$y = -3$$
$$(0, -3)$$

107. x-intercept:

$$2x - y = 4$$
$$2x - 0 = 4$$
$$2x = 4$$
$$x = 2$$
$$(2, 0)$$

y-intercept:

$$2x - y = 4$$
$$2(0) - y = 4$$
$$-y = 4$$
$$y = -4$$
$$(0, -4)$$

109. x-intercept:

$$2x - 3y = 9$$
$$2x - 3(0) = 9$$
$$2x = 9$$
$$x = \frac{9}{2}$$
$$\left(\frac{9}{2}, 0\right)$$

y-intercept:

$$2x - 3y = 9$$
$$2(0) - 3y = 9$$
$$-3y = 9$$
$$y = -3$$
$$(0, -3)$$

Applying Concepts 5.2

111. **a.**
$$y + 4 = -\frac{1}{2}(x + 2)$$
$$y + 4 = -\frac{1}{2}x - 1$$
$$y = -\frac{1}{2}x - 5$$

b.
$$y = -\frac{1}{2}x - 5$$
$$y = -\frac{1}{2}(-2) - 5$$
$$y = 1 - 5$$
$$y = -4$$
$$(-2, -4)$$

113. $y = -x - 4$
Increase x by 1 unit.
$$y = -(x + 1) - 4$$
$$y = -x - 1 - 4$$
$$y = (-x - 4) - 1$$
y decreases by 1 unit.

Section 5.3

Concept Review 5.3

1. Always true

3. Never true
   A horizontal line has zero slope, and the slope of a vertical line is undefined.

5. Never true
   The line slopes upward to the right; it has a positive slope.

Objective 5.3.1 Exercises

3. $P_1(4, 2)$, $P_2(3, 4)$
$$m = \frac{y_2 - y_1}{x_2 - x_1} = \frac{4 - 2}{3 - 4} = \frac{2}{-1} = -2$$
The slope is -2.

5. $P_1(-1, 3)$, $P_2(2, 4)$

$$m = \frac{y_2 - y_1}{x_2 - x_1} = \frac{4 - 3}{2 - (-1)} = \frac{1}{3}$$

The slope is $\frac{1}{3}$.

7. $P_1(2, 4)$, $P_2(4, -1)$

$$m = \frac{y_2 - y_1}{x_2 - x_1} = \frac{-1 - 4}{4 - 2} = \frac{-5}{2} = -\frac{5}{2}$$

The slope is $-\frac{5}{2}$.

9. $P_1(-2, 3)$, $P_2(2, 1)$

$$m = \frac{y_2 - y_1}{x_2 - x_1} = \frac{1 - 3}{2 - (-2)} = \frac{-2}{4} = -\frac{1}{2}$$

The slope is $-\frac{1}{2}$.

11. $P_1(8, -3)$, $P_2(4, 1)$

$$m = \frac{y_2 - y_1}{x_2 - x_1} = \frac{1 - (-3)}{4 - 8} = \frac{4}{-4} = -1$$

The slope is -1.

13. $P_1(3, -4)$, $P_2(3, 5)$

$$m = \frac{y_2 - y_1}{x_2 - x_1} = \frac{5 - (-4)}{3 - 3} = \frac{9}{0}$$

The slope is undefined.

15. $P_1(4, -2)$, $P_2(3, -2)$

$$m = \frac{y_2 - y_1}{x_2 - x_1} = \frac{-2 - (-2)}{3 - 4} = \frac{0}{-1} = 0$$

The line has zero slope.

17. $P_1(0, -1)$, $P_2(3, -2)$

$$m = \frac{y_2 - y_1}{x_2 - x_1} = \frac{-2 - (-1)}{3 - 0} = \frac{-1}{3} = -\frac{1}{3}$$

The slope is $-\frac{1}{3}$.

19. $P_1(-2, 3)$, $P_2(1, 3)$

$$m = \frac{y_2 - y_1}{x_2 - x_1} = \frac{3 - 3}{1 - (-2)} = \frac{0}{3} = 0$$

The line has zero slope.

21. $P_1(-2, 4)$, $P_2(-1, -1)$

$$m = \frac{y_2 - y_1}{x_2 - x_1} = \frac{-1 - 4}{-1 - (-2)} = \frac{-5}{1} = -5$$

The slope is -5.

23. $P_1(2, -3)$, $P_2(-2, 1)$

$$m = \frac{y_2 - y_1}{x_2 - x_1} = \frac{1 - (-3)}{-2 - (-2)} = \frac{4}{0}$$

The slope is undefined.

25. $P_1(-1, 5)$, $P_2(5, 1)$

$$m = \frac{y_2 - y_1}{x_2 - x_1} = \frac{1 - 5}{5 - (-1)} = \frac{-4}{6} = -\frac{2}{3}$$

The slope is $-\frac{2}{3}$.

27. The slope of the line through $(-4, 2)$ and $(1, 6)$

$$m = \frac{y_2 - y_1}{x_2 - x_1} = \frac{6 - 2}{1 - (-4)} = \frac{4}{5}$$

The slope of the line through $(2, -4)$ and $(7, 0)$

$$m = \frac{y_2 - y_1}{x_2 - x_1} = \frac{0 - (-4)}{7 - 2} = \frac{4}{5}$$

Yes, the lines are parallel.

29. The slope of the line through $(-2, -3)$ and $(7, 1)$

$$m = \frac{y_2 - y_1}{x_2 - x_1} = \frac{1 - (-3)}{7 - (-2)} = \frac{4}{9}$$

The slope of the line through $(6, -5)$ and $(4, 1)$

$$m = \frac{y_2 - y_1}{x_2 - x_1} = \frac{1 - (-5)}{4 - 6} = \frac{6}{-2} = -3$$

No, the lines are not parallel.

37. $y = 3x + 1$

$m = 3$, $b = 1$

39. $y = \frac{2}{5}x - 2$

$m = \frac{2}{5}$, $b = -2$

41. $2x + y = 3$

$y = -2x + 3$

$m = -2$, $b = 3$

43. $x - 2y = 4$

$-2y = -x + 4$

$y = \frac{1}{2}x - 2$

$m = \frac{1}{2}$, $b = -2$

45. $y = \dfrac{2}{3}x$

$m = \dfrac{2}{3}, \; b = 0$

47. $y = -x + 1$

$m = -1, \; b = 1$

49. $3x - 4y = 12$

$-4y = -3x + 12$

$y = \dfrac{3}{4}x - 3$

$m = \dfrac{3}{4}, \; b = -3$

51. $y = -4x + 2$

$m = -4, \; b = 2$

53. $4x - 5y = 20$

$-5y = -4x + 20$

$y = \dfrac{4}{5}x - 4$

$m = \dfrac{4}{5}, \; b = -4$

Applying Concepts 5.3

55. Increasing the coefficient of x increases the slope of $y = mx + b$.

57. Increasing the constant term increases the y-intercept of $y = mx + b$. (moves the graph upward)

59. i. D slope $= -2$, y-intercept $(0, 4)$

ii. C slope $= 2$, y-intercept $(0, -4)$

iii. B slope $= 0$, y-intercept $(0, 2)$

iv. F slope $= -\dfrac{1}{2}$, y-intercept $(0, 0)$

v. E slope $= \dfrac{1}{2}$, y-intercept $(0, 4)$

vi. A slope $= -\dfrac{1}{4}$, y-intercept $(0, -2)$

61. No, not all straight lines have a y-intercept. Any line that is parallel to the y-axis does not have a y-intercept; an example is $x = 2$.

Section 5.4

Concept Review 5.4

1. Always true

3. Sometimes true
   If a line contains $(-3, 1)$, y is -3 when x is 1 only when the equation of the line is $y = -x - 2$.

5. Always true

Objective 5.4.1 Exercises

1. $y = 2x + b$

$2 = 2(0) + b$

$2 = b$

$y = 2x + 2$

3. $y = -3x + b$

$2 = -3(-1) + b$

$2 = 3 + b$

$-1 = b$

$y = -3x - 1$

5. $y = \dfrac{1}{3}x + b$

$1 = \dfrac{1}{3}(3) + b$

$1 = 1 + b$

$0 = b$

$y = \dfrac{1}{3}x$

7. $y = \dfrac{3}{4}x + b$

$-2 = \dfrac{3}{4}(4) + b$

$-2 = 3 + b$

$-5 = b$

$y = \dfrac{3}{4}x - 5$

9.
$$y = -\frac{3}{5}x + b$$
$$-3 = -\frac{3}{5}(5) + b$$
$$-3 = -3 + b$$
$$0 = b$$
$$y = -\frac{3}{5}x$$

11.
$$y = \frac{1}{4}x + b$$
$$3 = \frac{1}{4}(2) + b$$
$$3 = \frac{1}{2} + b$$
$$\frac{5}{2} = b$$
$$y = \frac{1}{4}x + \frac{5}{2}$$

13.
$$y = -\frac{2}{3}x + b$$
$$-5 = -\frac{2}{3}(-3) + b$$
$$-5 = 2 + b$$
$$-7 = b$$
$$y = -\frac{2}{3}x - 7$$

Objective 5.4.2 Exercises

17. $m = 2,\ (x_1, y_1) = (1, -1)$
$$y - y_1 = m(x - x_1)$$
$$y - (-1) = 2(x - 1)$$
$$y + 1 = 2x - 2$$
$$y = 2x - 3$$
The equation of the line is $y = 2x - 3$.

19. $m = -2,\ (x_1, y_1) = (-2, 1)$
$$y - y_1 = m(x - x_1)$$
$$y - 1 = -2[x - (-2)]$$
$$y - 1 = -2[x + 2]$$
$$y - 1 = -2x - 4$$
$$y = -2x - 3$$
The equation of the line is $y = -2x - 3$.

21. $m = \frac{2}{3},\ (x_1, y_1) = (0, 0)$
$$y - y_1 = m(x - x_1)$$
$$y - 0 = \frac{2}{3}(x - 0)$$
$$y = \frac{2}{3}x$$
The equation of the line is $y = \frac{2}{3}x$.

23. $m = \frac{1}{2},\ (x_1, y_1) = (2, 3)$
$$y - y_1 = m(x - x_1)$$
$$y - 3 = \frac{1}{2}(x - 2)$$
$$y - 3 = \frac{1}{2}x - 1$$
$$y = \frac{1}{2}x + 2$$
The equation of the line is $y = \frac{1}{2}x + 2$.

25. $m = -\frac{3}{4},\ (x_1, y_1) = (-4, 1)$
$$y - y_1 = m(x - x_1)$$
$$y - 1 = -\frac{3}{4}[x - (-4)]$$
$$y - 1 = -\frac{3}{4}[x + 4]$$
$$y - 1 = -\frac{3}{4}x - 3$$
$$y = -\frac{3}{4}x - 2$$
The equation of the line is $y = -\frac{3}{4}x - 2$.

27. $m = \frac{3}{4},\ (x_1, y_1) = (-2, 1)$
$$y - y_1 = m(x - x_1)$$
$$y - 1 = \frac{3}{4}[x - (-2)]$$
$$y - 1 = \frac{3}{4}[x + 2]$$
$$y - 1 = \frac{3}{4}x + \frac{3}{2}$$
$$y = \frac{3}{4}x + \frac{5}{2}$$
The equation of the line is $y = \frac{3}{4}x + \frac{5}{2}$.

29. $m = -\frac{4}{3},\ (x_1, y_1) = (-3, -5)$
$$y - y_1 = m(x - x_1)$$
$$y - (-5) = -\frac{4}{3}[x - (-3)]$$
$$y + 5 = -\frac{4}{3}(x + 3)$$
$$y + 5 = -\frac{4}{3}x - 4$$
$$y = -\frac{4}{3}x - 9$$

Applying Concepts 5.4

31. Find the equation of the line that contains (5, 1) and (4, 2).

$m = \dfrac{2-1}{4-5} = \dfrac{1}{-1} = -1$

$y - y_1 = m(x - x_1)$
$y - 1 = -1(x - 5)$
$y - 1 = -x + 5$
$\qquad y = -x + 6$

Does (0, 6) satisfy this equation?

$y = -x + 6$
$6 = -0 + 6$
$6 = 6$

(0, 6) satisfies $y = -x + 6$.
$y = -x + 6$ contains all three ordered pairs.

33. Find the equation of the line that contains (−1, −5) and (2, 4).

$m = \dfrac{4-(-5)}{2-(-1)} = \dfrac{4+5}{2+1} = \dfrac{9}{3} = 3$

$y - y_1 = m(x - x_1)$
$y - (-5) = 3[x - (-1)]$
$\quad y + 5 = 3(x + 1)$
$\quad y + 5 = 3x + 3$
$\qquad y = 3x - 2$

Does (0, 2) satisfy this equation?

$y = 3x - 2$
$2 \neq 3(0) - 2$
$2 \neq -2$

(0, 2) does not satisfy $y = 3x - 2$.
There is no linear equation that contains all of the given ordered pairs.

35. Find the equation of the line that contains (0, 1) and (4, 9).

$m = \dfrac{9-1}{4-0} = \dfrac{8}{4} = 2$

$y - y_1 = m(x - x_1)$
$y - 1 = 2(x - 0)$
$y - 1 = 2x$
$\quad y = 2x + 1$

Substitute $x = 3$, $y = n$ into this equation, and solve for n.

$n = 2(3) + 1$
$n = 6 + 1$
$n = 7$

37. Find the equation of the line that contains (2, −2) and (−2, −4).

$m = \dfrac{-4-(-2)}{-2-2} = \dfrac{-4+2}{-4} = \dfrac{-2}{-4} = \dfrac{1}{2}$

$y - y_1 = m(x - x_1)$
$y - (-2) = \dfrac{1}{2}(x - 2)$
$\quad y + 2 = \dfrac{1}{2}x - 1$
$\qquad y = \dfrac{1}{2}x - 3$

Substitute $x = 4$, $y = n$ into this equation and solve for n.

$n = \dfrac{1}{2}(4) - 3$
$n = 2 - 3$
$n = -1$

39. Strategy

Water freezers at 0°C, or 32°F.
Water boils at 100°C, or 212°F.
Use this information to form the ordered pairs (0, 32) and (100, 212).
Use these ordered pairs to write a linear equation.

Solution

$m = \dfrac{212-32}{100-0} = \dfrac{180}{100} = \dfrac{9}{5}$

$F - F_1 = m(C - C_1)$
$F - 32 = \dfrac{9}{5}(C - 0)$
$F - 32 = \dfrac{9}{5}C$
$\qquad F = \dfrac{9}{5}C + 32$

The equation expressing Fahrenheit temperature in terms of Celsius temperature is $F = \dfrac{9}{5}C + 32$.

41.

$y - y_1 = \dfrac{y_2 - y_1}{x_2 - x_1}(x - x_1)$

$y - 3 = \dfrac{-1-3}{4-(-2)}[x - (-2)]$

$y - 3 = \dfrac{-4}{6}(x + 2)$

$y - 3 = -\dfrac{2}{3}x - \dfrac{4}{3}$

$\quad y = -\dfrac{2}{3}x + \dfrac{5}{3}$

Section 5.5

Concept Review 5.5

1. Always true

3. Never true
$m = -\dfrac{2}{3}$; the slope is negative

5. Always true

7. Always true

Objective 5.5.1 Exercises

3. {(44, 40), (3, 3), (16, 28), (13, 13), (15, 13), (20, 17)}
Yes, the relation is a function.

5. {(197, 125), (205, 122), (257, 498), (226, 108), (205, 150)}
No, the relation is not a function because 205 is paired with two different second coordinates.

7. {(55, 273), (65, 355), (75, 447), (85, 549)}
Yes, the relation is a function.

9. The domain is {0, 2, 4, 6}.
The range is {0}.
No two ordered pairs have the same first component.
The relation is a function.

11. The domain is {2}.
The range is {2, 4, 6, 8}.
The ordered pairs (2, 2), (2, 4), (2, 6), and (2, 8) have the same first component and different second components.
The relation is not a function.

13. The domain is {0, 1, 2, 3}.
The range is {0, 1, 2, 3}.
No two ordered pairs have the same first component.
The relation is a function.

15. The domain is {−3, −2, −1, 1, 2, 3}.
The range is {−3, 2, 3, 4}.
No two ordered pairs have the same first component.
The relation is a function.

17. $f(x) = 4x$
$f(10) = 4(10)$
$f(10) = 40$
The ordered pair (10, 40) is an element of the function.

19. $f(x) = x - 5$
$f(-6) = -6 - 5$
$f(-6) = -11$
The ordered pair (−6, −11) is an element of the function.

21. $f(x) = 3x^2$
$f(-2) = 3(-2)^2$
$f(-2) = 3(4)$
$f(-2) = 12$
The ordered pair (−2, 12) is an element of the function.

23. $f(x) = 5x + 1$
$f\left(\dfrac{1}{2}\right) = 5\left(\dfrac{1}{2}\right) + 1$
$f\left(\dfrac{1}{2}\right) = \dfrac{5}{2} + 1$
$f\left(\dfrac{1}{2}\right) = \dfrac{7}{2}$
The ordered pair $\left(\dfrac{1}{2}, \dfrac{7}{2}\right)$ is an element of the function.

25. $f(x) = \dfrac{2}{5}x + 4$
$f(-5) = \dfrac{2}{5}(-5) + 4$
$f(-5) = -2 + 4$
$f(-5) = 2$
The ordered pair (−5, 2) is an element of the function.

27. $f(x) = 2x^2$
$f(-4) = 2(-4)^2$
$f(-4) = 2(16)$
$f(-4) = 32$
The ordered pair (−4, 32) is an element of the function.

29. $f(x) = 3x - 4$
$f(-5) = 3(-5) - 4 = -15 - 4 = -19$
$f(-3) = 3(-3) - 4 = -9 - 4 = -13$
$f(-1) = 3(-1) - 4 = -3 - 4 = -7$
$f(1) = 3(1) - 4 = 3 - 4 = -1$
$f(3) = 3(3) - 4 = 9 - 4 = 5$
The range is {−19, −13, −7, −1, 5}.
The ordered pairs (−5, −19), (−3, −13), (−1, −7), (1, −1), and (3, 5) belong to the function.

31. $f(x) = \dfrac{1}{2}x + 3$
$f(-4) = \dfrac{1}{2}(-4) + 3 = -2 + 3 = 1$
$f(-2) = \dfrac{1}{2}(-2) + 3 = -1 + 3 = 2$
$f(0) = \dfrac{1}{2}(0) + 3 = 0 + 3 = 3$
$f(2) = \dfrac{1}{2}(2) + 3 = 1 + 3 = 4$
$f(4) = \dfrac{1}{2}(4) + 3 = 2 + 3 = 5$
The range is {1, 2, 3, 4, 5}.
The ordered pairs (−4, 1), (−2, 2), (0, 3), (2, 4), and (4, 5) belong to the function.

33. $f(x) = x^2 + 6$

$f(-3) = (-3)^2 + 6 = 9 + 6 = 15$

$f(-1) = (-1)^2 + 6 = 1 + 6 = 7$

$f(0) = (0)^2 + 6 = 0 + 6 = 6$

$f(1) = (1)^2 + 6 = 1 + 6 = 7$

$f(3) = (3)^2 + 6 = 9 + 6 = 15$

The range is {6, 7, 15}.

The ordered pairs (–3, 15), (–1, 7), (0, 6), (1, 7), and (3, 15) belong to the function.

Objective 5.5.2 Exercises

35.

37.

39.

41.

43.

45.

47.

49.

51.

53.

55.

57. **a.** $V = 30,000 - 5000x$

$f(x) = 30,000 - 5000x$

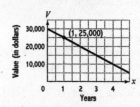

c. The value of the computer after 1 year is $25,000.

59. **a.** $C = 0.18m + 50$

$f(m) = 0.18m + 50$

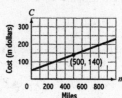

c. The cost to drive this car 500 miles is $140.

Section 5.6

Concept Review 5.6

1. Never true

$y \geq \frac{1}{2}x - 5$ is a linear inequality in two variables.

A linear equality must have an equals sign.

3. Always true

5. Always true

Objective 5.6.1 Exercises

3. $y > 2x + 3$

Graph $y = 2x + 3$ as a dashed line.

Shade the upper half-plane.

5. $y > \dfrac{3}{2}x - 4$

Graph $y = \dfrac{3}{2}x - 4$ as a dashed line.

Shade the upper half-plane.

7. $y \le -\dfrac{3}{4}x - 1$

Graph $y = -\dfrac{3}{4}x - 1$ as a solid line.

Shade the lower half-plane.

9. $y \le -\dfrac{6}{5}x - 2$

Graph $y = -\dfrac{6}{5}x - 2$ as a solid line.

Shade the lower half-plane.

11. $\quad x + y > 4$

$\quad x - x + y > -x + 4$

$\qquad\quad y > -x + 4$

Graph $y = -x + 4$ as a dashed line.

Shade the upper half-plane.

13. $\quad 2x + y \ge 4$

$\quad 2x - 2x + y \ge -2x + 4$

$\qquad\quad y \ge -2x + 4$

Graph $y = -2x + 4$ as a solid line.

Shade the upper half-plane.

15. $y \le -2$

Graph $y = -2$ as a solid line.

Shade the lower half-plane.

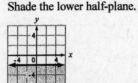

17. $\qquad 2x + 2y \le -4$

$\quad 2x - 2x + 2y \le -2x - 4$

$\qquad\qquad 2y \le -2x - 4$

$\qquad\qquad \dfrac{2y}{2} \le \dfrac{-2x - 4}{2}$

$\qquad\qquad\; y \le -x - 2$

Graph $y = -x - 2$ as a solid line.

Shade the lower half-plane.

Applying Concepts 5.6

19. First find the equation of the line with x-intercept $(-1, 0)$ and y-intercept $(0, 2)$.
Find the slope of the line.
$P_1(-1, 0), \; P_2(0, 2)$

$m = \dfrac{y_2 - y_1}{x_2 - x_1} = \dfrac{2 - 0}{0 - (-1)} = \dfrac{2}{0 + 1} = \dfrac{2}{1} = 2$

The line has slope 2.
The equation of the line is $y = 2x + 2$.
The line is a solid line, and the upper half-plane is shaded.
The inequality is $y \ge 2x + 2$.

21. First find the equation of the horizontal line with y-intercept $(0, 2)$.
The equation of the line is $y = 2$.
The line is dashed, and the upper half-plane is shaded.
The inequality is $y > 2$.

23. $y - 5 < 4(x - 2)$
$\quad y - 5 < 4x - 8$
$\qquad\; y < 4x - 3$

25.
$$3x - 2(y+1) \le y - (5-x)$$
$$3x - 2y - 2 \le y - 5 + x$$
$$3x - 3y - 2 \le -5 + x$$
$$-3y - 2 \le -2x - 5$$
$$-3y \le -2x - 3$$
$$y \ge \frac{2}{3}x + 1$$

Chapter Review Exercises

1.
$$y = -\frac{2}{3}x + 2$$
$$y = -\frac{2}{3}(3) + 2$$
$$y = -2 + 2$$
$$y = 0$$
The ordered-pair solution is (3, 0).

2. $m = 3, \; b = -1$
$$y = mx + b$$
$$y = 3x - 1$$

3.

4.

5. $f(x) = 3x^2 + 4$
$$f(-5) = 3(-5)^2 + 4$$
$$f(-5) = 3(25) + 4$$
$$f(-5) = 75 + 4$$
$$f(-5) = 79$$
The value of $f(-5)$ is 79.

6. $(x_1, y_1) = (3, -4), \; (x_2, y_2) = (1, -4)$
$$m = \frac{y_2 - y_1}{x_2 - x_1} = \frac{-4 - (-4)}{1 - 3} = \frac{0}{-2} = 0$$
The line has zero slope.

7.

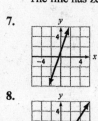

8.

9. $m = -\frac{2}{3}, \; (x_1, y_1) = (-1, 2)$
$$y - y_1 = m(x - x_1)$$
$$y - 2 = -\frac{2}{3}[x - (-1)]$$
$$y - 2 = -\frac{2}{3}[x + 1]$$
$$y - 2 = -\frac{2}{3}x - \frac{2}{3}$$
$$y = -\frac{2}{3}x - \frac{2}{3} + 2$$
$$y = -\frac{2}{3}x - \frac{2}{3} + \frac{6}{3}$$
$$y = -\frac{2}{3}x + \frac{4}{3}$$

10. $6x - 4y = 12$

x-intercept:	y-intercept:
$6x - 4(0) = 12$	$6(0) - 4y = 12$
$6x = 12$	$-4y = 12$
$x = 2$	$y = -3$
(2, 0)	(0, -3)

11.

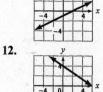

12.

13. $m = \frac{2}{3}, \; (x_1, y_1) = (-3, 1)$
$$y - y_1 = m(x - x_1)$$
$$y - 1 = \frac{2}{3}[x - (-3)]$$
$$y - 1 = \frac{2}{3}[x + 3]$$
$$y - 1 = \frac{2}{3}x + 2$$
$$y = \frac{2}{3}x + 3$$

14. $(x_1, y_1) = (2, -3), \; (x_2, y_2) = (4, 1)$
$$m = \frac{y_2 - y_1}{x_2 - x_1} = \frac{1 - (-3)}{4 - 2} = \frac{4}{2} = 2$$
The slope of the line is 2.

15. $f(x) = \frac{3}{5}x + 2$
$$f(-10) = \frac{3}{5}(-10) + 2$$
$$f(-10) = -6 + 2$$
$$f(-10) = -4$$

16. The domain is $\{-20, -10, 0, 10\}$.
The range is $\{-10, -5, 0, 5\}$.
No two ordered pairs have the same first coordinate. The relation is a function.

17.

18.

19.

20.

21. $2x - 3y = 12$

x-intercept: y-intercept:

$2x - 3(0) = 12$ $2(0) - 3y = 12$

$2x = 12$ $-3y = 12$

$x = 6$ $y = -4$

$(6, 0)$ $(0, -4)$

22. $m = 2,\ (x_1, y_1) = (-1, 0)$

$y - y_1 = m(x - x_1)$

$y - 0 = 2[x - (-1)]$

$y = 2[x + 1]$

$y = 2x + 2$

23.

24.

25. The slope of the line through $(-2, 1)$ and $(3, 5)$

$m = \dfrac{y_2 - y_1}{x_2 - x_1} = \dfrac{5 - 1}{3 - (-2)} = \dfrac{4}{5}$

The slope of the line through $(4, -5)$ and $(9, -1)$

$m = \dfrac{y_2 - y_1}{x_2 - x_1} = \dfrac{-1 - (-5)}{9 - 4} = \dfrac{4}{5}$

Yes, the lines are parallel.

26. $m = \dfrac{1}{2},\ (x_1, y_1) = (2, 3)$

$y - y_1 = m(x - x_1)$

$y - 3 = \dfrac{1}{2}(x - 2)$

$y - 3 = \dfrac{1}{2}x - 1$

$y = \dfrac{1}{2}x + 2$

27.

28.

29. $y = 2x - 1$

$y = 2(-2) - 1$

$y = -4 - 1$

$y = -5$

The ordered-pair solution is $(-2, -5)$.

30. $m = -3,\ b = 2$

$y = mx + b$

$y = -3x + 2$

31.

32. $f(x) = 3x - 5$

$f\left(\dfrac{5}{3}\right) = 3\left(\dfrac{5}{3}\right) - 5$

$f\left(\dfrac{5}{3}\right) = 5 - 5 = 0$

The value of $f\left(\dfrac{5}{3}\right) = 0$.

33. $(x_1, y_1) = (3, -2),\ (x_2, y_2) = (3, 5)$

$m = \dfrac{y_2 - y_1}{x_2 - x_1} = \dfrac{5 - (-2)}{3 - 3} = \dfrac{7}{0}$

The slope of the line is undefined.

34. $m = \dfrac{1}{2},\ (x_1, y_1) = (2, -1)$

$y - y_1 = m(x - x_1)$

$y - (-1) = \dfrac{1}{2}(x - 2)$

$y + 1 = \dfrac{1}{2}x - 1$

$y = \dfrac{1}{2}x - 2$

35. $y = \frac{1}{5}x + 2$

$$\begin{array}{c|c} 0 & \frac{1}{5}(-10)+2 \\ & -2+2 \end{array}$$

$0 = 0$

Yes, $(-10, 0)$ is a solution of $y = \frac{1}{5}x + 2$.

36. $y = 4x - 9$
$y = 4(2) - 9$
$y = 8 - 9$
$y = -1$
The ordered-pair solution is $(2, -1)$.

37. $4x - 3y = 0$

x-intercept:	y-intercept:
$4x - 3(0) = 0$	$4(0) - 3y = 0$
$4x = 0$	$-3y = 0$
$x = 0$	$y = 0$
$(0, 0)$	$(0, 0)$

38. $m = 0$, $(x_1, y_1) = (-2, 3)$
$y - y_1 = m(x - x_1)$
$y - 3 = 0[x - (-2)]$
$y - 3 = 0$
$y = 3$

39.

40.

41. $m = 3$, $b = -4$
$y = mx + b$
$y = 3x - 4$

42.

43. The domain is $\{-10, -5, 5\}$.
The range is $\{-5, 0\}$.
The ordered pairs $(-10, -5)$ and $(-10, 0)$ have the same first coordinate and different second coordinates. The relation is not a function.

44. $f(x) = 3x + 7$
$f(-20) = 3(-20) + 7 = -60 + 7 = -53$
$f(-10) = 3(-10) + 7 = -30 + 7 = -23$
$f(0) = 3(0) + 7 = 7$
$f(10) = 3(10) + 7 = 30 + 7 = 37$
$f(20) = 3(20) + 7 = 60 + 7 = 67$
The range is $\{-53, -23, 7, 37, 67\}$.

45. $f(x) = \frac{1}{3}x + 4$
$f(-6) = \frac{1}{3}(-6) + 4 = -2 + 4 = 2$
$f(-3) = \frac{1}{3}(-3) + 4 = -1 + 4 = 3$
$f(0) = \frac{1}{3}(0) + 4 = 0 + 4 = 4$
$f(3) = \frac{1}{3}(3) + 4 = 1 + 4 = 5$
$f(6) = \frac{1}{3}(6) + 4 = 2 + 4 = 6$
The range is $\{2, 3, 4, 5, 6\}$.

46.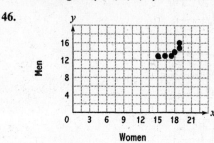

47.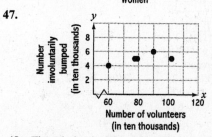

48. The relation is $\{(25, 4.8), (35, 3.5), (40, 2.1), (20, 5.5), (45, 1.0)\}$
No two ordered pairs have the same first coordinate.
The relation is a function.

49. **a.** $C = 70s + 40,000$
$f(s) = 70s + 40,000$

c. The contractor's cost to build a house of 1500 ft^2 is $145,000.

50. Strategy
To find the average annual rate of change, divide the change in salary ($44,047 - 23,600$) by the change in years ($2001 - 1985$).

Solution
$$\frac{44,047 - 23,600}{2001 - 1985} = \frac{20,447}{16} \approx 1277.94$$
The annual rate of change was $1277.94.

Chapter Test

1. $m = -\dfrac{1}{3}, \; (x_1, y_1) = (9, -3)$

$y - y_1 = m(x - x_1)$

$y - (-3) = -\dfrac{1}{3}(x - 9)$

$y + 3 = -\dfrac{1}{3}x + 3$

$y = -\dfrac{1}{3}x$

2. $(x_1, y_1) = (9, 8), \; (x_2, y_2) = (-2, 1)$

$m = \dfrac{y_2 - y_1}{x_2 - x_1} = \dfrac{1 - 8}{-2 - 9} = \dfrac{-7}{-11} = \dfrac{7}{11}$

The slope of the line is $\dfrac{7}{11}$.

3. $3x - 2y = 24$

x-intercept: y-intercept:

$3x - 2(0) = 24$ $3(0) - 2y = 24$

$3x = 24$ $-2y = 24$

$x = 8$ $y = -12$

$(8, 0)$ $(0, -12)$

4. $y = -\dfrac{4}{3}x - 1$

$y = -\dfrac{4}{3}(9) - 1$

$y = -12 - 1$

$y = -13$

The ordered-pair solution is $(9, -13)$.

5.

6.

7. $f(x) = 4x + 7$

$f\left(\dfrac{3}{4}\right) = 4\left(\dfrac{3}{4}\right) + 7$

$f\left(\dfrac{3}{4}\right) = 3 + 7$

$f\left(\dfrac{3}{4}\right) = 10$

The value of $f\left(\dfrac{3}{4}\right)$ is 10.

8. $y = \dfrac{2}{3}x + 1$

$\begin{array}{c|c} 3 & \dfrac{2}{3}(6) + 1 \\ 3 & 4 + 1 \\ 3 \ne 5 \end{array}$

No, $(6, 3)$ is not a solution of $y = \dfrac{2}{3}x + 1$.

9.

10.

11.

12.

13. $(x_1, y_1) = (4, -3), \; (x_2, y_2) = (-2, -3)$

$m = \dfrac{y_2 - y_1}{x_2 - x_1} = \dfrac{-3 - (-3)}{-2 - 4} = \dfrac{0}{-6} = 0$

The line has zero slope.

14. $m = -\dfrac{2}{5}, \; b = 7$

$y = mx + b$

$y = -\dfrac{2}{5}x + 7$

15. $m = 4, \; (x_1, y_1) = (2, 1)$

$y - y_1 = m(x - x_1)$

$y - 1 = 4(x - 2)$

$y - 1 = 4x - 8$

$y = 4x - 7$

16. $f(x) = 4x^2 - 3$

$f(-2) = 4(-2)^2 - 3$

$f(-2) = 4(4) - 3 = 16 - 3$

$f(-2) = 13$

The value of $f(-2)$ is 13.

17.

18.

19.

20.

21.

22.

23. $f(x) = -3x + 7$
$f(-6) = -3(-6) + 7$
$f(-6) = 18 + 7$
$f(-6) = 25$
The value of $f(-6)$ is 25.

24. $f(x) = -3x + 5$
$f(-7) = -3(-7) + 5 = 21 + 5 = 26$
$f(-2) = -3(-2) + 5 = 6 + 5 = 11$
$f(0) = -3(0) + 5 = 0 + 5 = 5$
$f(4) = -3(4) + 5 = -12 + 5 = -7$
$f(9) = -3(9) + 5 = -27 + 5 = -22$
The range is $\{-22, -7, 5, 11, 26\}$.

25. The slope of the line through $(-3, -2)$ and $(6, 0)$
$m = \dfrac{y_2 - y_1}{x_2 - x_1} = \dfrac{0 - (-2)}{6 - (-3)} = \dfrac{2}{9}$
The slope of the line through $(6, -4)$ and $(3, 2)$
$m = \dfrac{y_2 - y_1}{x_2 - x_1} = \dfrac{2 - (-4)}{3 - 6} = \dfrac{6}{-3} = -2$
No, the lines are not parallel.

26. Strategy
To find the average annual rate of change, divide the change in drivers $(18 - 13)$ in millions, by the change in years (10).

Solution
$\dfrac{18 - 13}{10} = \dfrac{5}{10} = 0.5$
The annual rate of change was 0.5 million.

27.

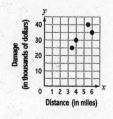

28. a. $C = 8t + 1000$
$f(t) = 8t + 1000$

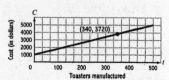

c. The cost to manufacture 340 toasters is $3720.

29. The relation is $\{(8.5, 64), (9.4, 68), (10.1, 76), (11.4, 87), (12.0, 92)\}$
No two ordered pairs have the same first coordinate. The relation is a function.

30. $(x_1, y_1) = (1, 300)$, $(x_2, y_2) = (6, 480)$
$m = \dfrac{y_2 - y_1}{x_2 - x_1} = \dfrac{480 - 300}{6 - 1} = \dfrac{180}{5} = 36$
The cost of the lumber has been increasing by $36 per month.

Cumulative Review Exercises

1. $12 - 18 \div 3 \cdot (-2)^2 = 12 - 18 \div 3 \cdot 4$
$= 12 - 6 \cdot 4$
$= 12 - 24$
$= -12$

2. $\dfrac{a - b}{a^2 - c}$
$\dfrac{-2 - 3}{(-2)^2 - (-4)} = \dfrac{-2 - 3}{4 - (-4)}$
$= \dfrac{-5}{4 + 4}$
$= \dfrac{-5}{8}$
$= -\dfrac{5}{8}$

3. $4(2 - 3x) - 5(x - 4) = 8 - 12x - 5x + 20$
$= -12x - 5x + 8 + 20$
$= -17x + 28$

4. $2x - \dfrac{2}{3} = \dfrac{7}{3}$
$2x - \dfrac{2}{3} + \dfrac{2}{3} = \dfrac{7}{3} + \dfrac{2}{3}$
$2x = \dfrac{9}{3}$
$2x = 3$
$\dfrac{2x}{2} = \dfrac{3}{2}$
$x = \dfrac{3}{2}$
The solution is $\dfrac{3}{2}$.

5.
$$3x - 2[x - 3(2 - 3x)] = x - 6$$
$$3x - 2[x - 6 + 9x] = x - 6$$
$$3x - 2[10x - 6] = x - 6$$
$$3x - 20x + 12 = x - 6$$
$$-17x + 12 = x - 6$$
$$-17x - x + 12 = x - x - 6$$
$$-18x + 12 = -6$$
$$-18x + 12 - 12 = -6 - 12$$
$$-18x = -18$$
$$\frac{-18x}{-18} = \frac{-18}{-18}$$
$$x = 1$$

The solution is 1.

6.
$$6\frac{2}{3}\% = 6\frac{2}{3} \cdot \frac{1}{100} = \frac{20}{3} \cdot \frac{1}{100}$$
$$= \frac{20}{300} = \frac{1}{15}$$

7. $\{1, 2, 3, 4, 5, 6, 7, 8\}$

8.
$-23 > -16$	false
$-18 > -16$	false
$-4 > -16$	true
$0 > -16$	true
$5 > -16$	true

9.
$$8a - 3 \geq 5a - 6$$
$$8a - 5a - 3 \geq 5a - 5a - 6$$
$$3a - 3 \geq -6$$
$$3a - 3 + 3 \geq -6 + 3$$
$$3a \geq -3$$
$$\frac{3a}{3} \geq \frac{-3}{3}$$
$$a \geq -1$$

10.
$$4x - 5y = 15$$
$$4x - 4x - 5y = -4x + 15$$
$$-5y = -4x + 15$$
$$\frac{-5y}{-5} = \frac{-4x + 15}{-5}$$
$$y = \frac{4}{5}x - 3$$

11.
$$y = 3x - 1$$
$$y = 3(-2) - 1$$
$$y = -6 - 1$$
$$y = -7$$
The ordered-pair solution is $(-2, -7)$.

12. $(x_1, y_1) = (2, 3)$, $(x_2, y_2) = (-2, 3)$
$$m = \frac{y_2 - y_1}{x_2 - x_1} = \frac{3 - 3}{-2 - 2} = \frac{0}{-4} = 0$$
The line has zero slope.

13. $5x + 2y = 20$

x-intercept:	*y*-intercept:
$5x + 2(0) = 20$	$5(0) + 2y = 20$
$5x = 20$	$2y = 20$
$x = 4$	$y = 10$
$(4, 0)$	$(0, 10)$

14. $m = -1$, $(x_1, y_1) = (3, 2)$
$$y - y_1 = m(x - x_1)$$
$$y - 2 = -1(x - 3)$$
$$y - 2 = -x + 3$$
$$y = -x + 5$$

15.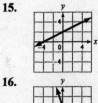

16.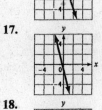

17.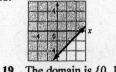

18.

19. The domain is $\{0, 1, 2, 3, 4\}$.
The range is $\{0, 1, 2, 3, 4\}$.
No two ordered pairs have the same first coordinate. The relation is a function.

20.
$$f(x) = -4x + 9$$
$$f(5) = -4(5) + 9$$
$$f(5) = -20 + 9$$
$$f(5) = -11$$
The value of $f(5)$ is 11.

21.
$$f(x) = -\frac{5}{3}x + 3$$
$$f(-9) = -\frac{5}{3}(-9) + 3 = 15 + 3 = 18$$
$$f(-6) = -\frac{5}{3}(-6) + 3 = 10 + 3 = 13$$
$$f(-3) = -\frac{5}{3}(-3) + 3 = 5 + 3 = 8$$
$$f(0) = -\frac{5}{3}(0) + 3 = 0 + 3 = 3$$
$$f(3) = -\frac{5}{3}(3) + 3 = -5 + 3 = -2$$
$$f(6) = -\frac{5}{3}(6) + 3 = -10 + 3 = -7$$

The range is $\{-7, -2, 3, 8, 13, 18\}$.

22. Strategy
- Smaller number: x
  Larger number: $24 - x$
- Twice the smaller number is three less than the large number.

Solution
$$2x = (24 - x) - 3$$
$$2x = 21 - x$$
$$2x + x = 21 - x + x$$
$$3x = 21$$
$$\frac{3x}{3} = \frac{21}{3}$$
$$x = 7$$
$$24 - x = 24 - 7 = 17$$
The two numbers are 7 and 17.

23. Strategy
To find the location of the fulcrum when the system balances, replace the variables F_1, F_2, and d in the lever system equation with the given values, and solve for x.

Solution
$$F_1 x = F_2(d - x)$$
$$80x = 560(8 - x)$$
$$80x = 4480 - 560x$$
$$80x + 560x = 4480 - 560x + 560x$$
$$640x = 4480$$
$$\frac{640x}{640} = \frac{4480}{640}$$
$$x = 7$$
The fulcrum is 7 ft from the 80-lb force.

24. Strategy
- Length of the third side: x
  Length of the first side: $2x$
  Length of the second side: $x + 5$
- Use the equation $P = a + b + c$.

Solution
$$P = a + b + c$$
$$49 = x + 2x + (x + 5)$$
$$49 = 4x + 5$$
$$44 = 4x$$
$$11 = x$$
$$2x = 2(11) = 22$$
The length of the first side is 22 ft.

25. Strategy
Given: $R = \$89$
$\qquad r = 30\% = 0.30$
Unknown: S

Solution
$$S = R - rR$$
$$S = 89 - 0.30(89)$$
$$S = 89 - 26.70$$
$$S = 62.30$$
The sale price is $62.30.

Chapter 6: Systems of Linear Equations

Prep Test

1. $3x - 4y = 24$ [5.2.3]
$$-4y = -3x + 24$$
$$y = \frac{3}{4}x - 6$$

2. $50 + 0.07x = 0.05(x + 1400)$ [3.2.3]
$$50 + 0.07x = 0.05x + 70$$
$$0.02x = 20$$
$$x = 1000$$

3. $-3(2x - 7y) + 3(2x + 4y)$ [2.2.5]
$$= -6x + 21y + 6x + 12y$$
$$= 33y$$

4. $4x + 2(3x - 5) = 4x + 6x - 10$ [2.2.5]
$$= 10x - 10$$

5.

$3x - 5y = -22$	
$3(-4) - 5(2)$	-22
$-12 - 10$	-22
$-22 = -22$	

[3.1.1[

Yes, $(-4, 2)$ is a solution.

6. x-intercept: y-intercept: [5.2.3]
$$3x - 4y = 12 \qquad 3x - 4y = 12$$
$$3x - 4(0) = 12 \qquad 3(0) - 4y = 12$$
$$3x = 12 \qquad\qquad -4y = 12$$
$$x = 4 \qquad\qquad\quad y = -3$$
$(4, 0)$ $(0, -3)$

7. Yes, the slopes are both -3.   [5.3.1]

8.   [5.2.2]

9. **Strategy**
Time for the second hiker to catch up: x

	Rate	Time	Distance
First hiker	3	$x + 0.5$	$3(x + 0.5)$
Second hiker	4	x	$4x$

The distances are the same.

Solution
$$3x + 1.5 = 4x$$
$$1.5 = x$$
It takes the second hiker 1.5 h to catch up to the first hiker.

Go Figure

Carla will arrive home first.
Carla runs for half the *time* and walk for half the *time*.
James runs for half the *distance* and walk for half the *distance*.
Since they run and walk at the same rate, by running for half the time, Carla will go farther than James when he runs for half the distance.

Section 6.1

Concept Review 6.1

1. Always true

3. Sometimes true
An ordered pair can be a solution of one equation, but not of the other equation, in a system of linear equations.

5. Never true
The graph could be the solution of two equations having the same solution.

Objective 6.1.1 Exercises

3.

$3x + 4y = 18$		$2x - y = 1$	
$3(2) + (4)3$	18	$2(3) - 3$	1
$6 + 12$	18	$4 - 3$	1
$18 = 18$		1	

Yes, $(2, 3)$ is a solution of the system of equations.

5.

$3x - y = 5$		$2x + 5y = -8$	
$3(1) - (-2)$	5	$2(1) + 5(-2)$	-8
$3 + 2$	5	$2 - 10$	-8
$5 = 5$		$-8 = -8$	

Yes, $(1, -2)$ is a solution of the system of equations.

7.

$5x - 2y = 14$		$x + y = 8$
$5(4) - 2(3)$	14	$4 + 3$ 8
$20 - 6$	14	$7 \neq 8$
$14 = 14$		

No, $(4, 3)$ is not a solution of the system of equations.

9.

$4x - y = -5$		$2x + 5y = 13$	
$4(-1) - 3$	5	$2(-1) + 5(3)$	13
$-4 - 3$	5	$-2 + 15$	13
$-7 \neq -5$		$13 = 13$	

No, $(-1, 3)$ is not a solution of the system of equations.

11.

$4x + 3y = 0$		$2x - y = 1$	
$4(0) + 3(0)$	0	$2(0) - 0$	1
$0 + 0$	0	$0 - 0$	1
$0 = 0$		$0 \neq 1$	

No, $(0, 0)$ is not a solution of the system of equations.

13. $y = 2x - 7$ $3x - y = 9$

$-3 \mid 2(2) - 7$ $3(2) - (-3) \mid 9$

$-3 \mid 4 - 7$ $6 + 3 \mid 9$

$-3 = -3$ $9 = 9$

Yes, $(2, -3)$ is a solution of the system of equations.

15. $y = 2x - 8$ $y = 3x - 13$

$2 \mid 2(5) - 8$ $2 \mid 3(5) - 13$

$2 \mid 10 - 8$ $2 \mid 15 - 13$

$2 = 2$ $2 = 2$

Yes, $(5, 2)$ is a solution of the system of equations.

19. independent

21. dependent

23. inconsistent

25. $(2, -1)$

27. no solution

29. $(-2, 4)$

31.

The solution is $(4, 1)$.

33.

The solution is $(4, 1)$.

35.

The solution is $(4, 3)$.

37.

The solution is $(3, -2)$

39.

The solution is $(2, -2)$.

41.

The system of equations is inconsistent and has no solution.

43.

The system of equations is dependent. The solutions are the ordered pairs that satisfy the equation $y = 2x - 2$.

45.

The solution is $(1, -4)$.

47.

The solution is $(0, 0)$.

49.

The system of equations is inconsistent and has no solution.

51.

The solution is $(0, -2)$.

53.

The solution is $(1, -1)$.

55.

The solution is $(2, 0)$.

57.

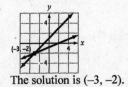

The solution is $(-3, -2)$.

59.

The system of equations is inconsistent and has no solution.

Applying Concepts 6.1

61. Find the equation of the line that contains the points $(-2, 2)$ and $(0, 4)$.
Find the slope of the line.
$P_1(-2, 2)$, $P_2(0, 4)$
$$m = \frac{y_2 - y_1}{x_2 - x_1} = \frac{4 - 2}{0 - (-2)} = \frac{2}{2} = 1$$
The line has slope 1.
$y = x + b$
$4 = 0 + b$
$4 = b$
The equation of the line is $y = x + 4$.
The horizontal line with y-intercept $(0, 2)$ has the equation $y = 2$.
The system of equations is $y = x + 4$
$\qquad\qquad\qquad\qquad\qquad\quad y = 2$.

63. Find the equation of the line that contains the points $(2, 0)$ and $(1, 1)$.
Find the slope of the line.
$P_1(2, 0)$, $P_2(1, 1)$
$$m = \frac{y_2 - y_1}{x_2 - y_1} = \frac{1 - 0}{1 - 2} = \frac{1}{-1} = -1$$
The line has slope -1.
$y = -x + b$
$1 = -1 + b$
$2 = b$
The equation of the line is $y = -x + 2$.
Find the equation of the line that contains the points $(0, 0)$ and $(1, 1)$.
Find the slope of the line.
$P_1(0, 0)$, $P_2(1, 1)$
$$m = \frac{y_2 - y_1}{x_2 - x_1} = \frac{1 - 0}{1 - 0} = 1$$
The line has slope 1.
$y = x + b$
$0 = 0 + b$
$0 = b$
The equation of the line is $y = x + 0$ or $y = x$.
The system of equations is $y = x$
$\qquad\qquad\qquad\qquad\qquad\quad y = -x + 2$.

65. a. Always true

b. Never true

c. Always true

Section 6.2

Concept Review 6.2

1. Always true

3. Never true
The solution of a system of two linear equations is an ordered pair.

5. Never true
The system of equations is inconsistent. The system does not have a solution.

Objective 6.2.1 Exercises

1. (1) $2x + 3y = 7$
(2) $\qquad x = 2$
Substitute in equation (1).
$2x + 3y = 7$
$2(2) + 3y = 7$
$4 + 3y = 7$
$3y = 3$
$y = 1$
The solution is $(2, 1)$.

3. (1) $y = x - 3$
(2) $x + y = 5$
Substitute into equation (2).
$x + y = 5$
$x + (x - 3) = 5$
$2x - 3 = 5$
$2x = 8$
$x = 4$
Substitute x in equation (1).
$y = x - 3$
$y = 4 - 3$
$y = 1$
The solution is $(4, 1)$.

5. (1) $\qquad x = y - 2$
(2) $x + 3y = 2$
Substitute in equation (2).
$x + 3y = 2$
$(y - 2) + 3y = 2$
$4y - 2 = 2$
$4y = 4$
$y = 1$
Substitute y in equation (1).
$x = y - 2$
$x = 1 - 2$
$x = -1$
The solution is $(-1, 1)$.

7. (1) $2x+3y=9$
(2) $y=x-2$
Substitute in equation (1)
$2x+3y=9$
$2x+3(x-2)=9$
$2x+3x-6=9$
$5x-6=9$
$5x=15$
$x=3$
Substitute x into equation (2).
$y=x-2$
$y=3-2$
$y=1$
The solution is (3, 1).

9. (1) $3x-y=2$
(2) $y=2x-1$
Substitute into equation (1).
$3x-y=2$
$3x-(2x-1)=1$
$3x-2x+1=2$
$x+1=2$
$x=1$
Substitute x in equation (2).
$y=2x-1$
$y=2(1)-1$
$y=2-1$
$y=1$
The solution is (1, 1).

11. (1) $x=2y-3$
(2) $2x-3y=-5$
Substitute in equation (2).
$2x-3y=-5$
$2(2y-3)-3y=-5$
$4y-6-3y=-5$
$y-6=-5$
$y=1$
Substitute y in equation (1).
$x=2y-3$
$x=2(1)-3$
$x=2-3$
$x=-1$
The solution is (−1, 1).

13. (1) $y=4-3x$
(2) $3x+y=5$
Substitute in equation (2).
$3x+y=5$
$3x+(4-3x)=5$
$0+4=5$
$4=5$
The system is inconsistent and has no solution.

15. (1) $x=3y+3$
(2) $2x-6y=12$
Substitute in equation (2).
$2x-6y=12$
$2(3y+3)-6y=12$
$6y+6-6y=12$
$0+6=12$
$6=12$
The system is inconsistent and has no solution.

17. (1) $3x+5y=-6$
(2) $x=5y+3$
Substitute in equation (1).
$3x+5y=-6$
$3(5y+3)+5y=-6$
$15y+9+5y=-6$
$20y+9=-6$
$20y=-15$
$y=-\dfrac{3}{4}$
Substitute y in equation (2).
$x=5y+4$
$x=5\left(-\dfrac{3}{4}\right)+3$
$x=-\dfrac{15}{4}+3$
$x=-\dfrac{15}{4}+\dfrac{12}{4}$
$x=-\dfrac{3}{4}$
The solution is $\left(-\dfrac{3}{4},-\dfrac{3}{4}\right)$.

19. (1) $x=4y-3$
(2) $2x-3y=0$
Substitute in equation (2).
$2x-3y=0$
$2(4y-3)-3y=0$
$8y-6-3y=0$
$5y-6=0$
$5y=6$
$y=\dfrac{6}{5}$
Substitute y in equation (1).
$x=4y-3$
$x=4\left(\dfrac{6}{5}\right)-3$
$x=\dfrac{24}{5}-3$
$x=\dfrac{24}{5}-\dfrac{15}{5}$
$x=\dfrac{9}{5}$
The solution is $\left(\dfrac{9}{5},\dfrac{6}{5}\right)$.

21. (1) $y = 2x - 9$
   (2) $3x - y = 2$
Substitute in equation (2).
$$3x - y = 2$$
$$3x - (2x - 9) = 2$$
$$3x - 2x + 9 = 2$$
$$x + 9 = 2$$
$$x = -7$$
Substitute x in equation (1).
$$y = 2x - 9$$
$$y = 2(-7) - 9$$
$$y = -14 - 9$$
$$y = -23$$
The solution is $(-7, -23)$.

23. (1) $3x + 4y = 7$
   (2) $x = 3 - 2y$
Substitute in equation (1).
$$3x + 4y = 7$$
$$3(3 - 2y) + 4y = 7$$
$$9 - 6y + 4y = 7$$
$$9 - 2y = 7$$
$$-2y = -2$$
$$y = 1$$
Substitute y in equation (2).
$$x = 3 - 2y$$
$$x = 3 - 2(1)$$
$$x = 3 - 2$$
$$x = 1$$
The solution is $(1, 1)$.

25. (1) $2x - y = 4$
   (2) $3x + 2y = 6$
Solve equation (1) for y.
$$2x - y = 4$$
$$-y = -2x + 4$$
$$y = 2x - 4$$
Substitute in equation (2).
$$3x + 2y = 6$$
$$3x + 2(2x - 4) = 6$$
$$3x + 4y - 8 = 6$$
$$7x - 8 = 6$$
$$7x = 14$$
$$x = 2$$
Substitute x in equation (1).
$$2x - y = 4$$
$$2(2) - y = 4$$
$$4 - y = 4$$
$$-y = 0$$
$$y = 0$$
The solution $(2, 0)$.

27. (1) $4x - 3y = 5$
   (2) $x + 2y = 4$
Solve equation (2) for x.
$$x + 2y = 4$$
$$x = -2y + 4$$
Substitute in equation (1).
$$4x - 3y = 5$$
$$4(-2y + 4) - 3y = 5$$
$$-8y + 16 - 3y = 5$$
$$-11y + 16 = 5$$
$$-11y = -11$$
$$y = 1$$
Substitute y in equation (1).
$$4x - 3y = 5$$
$$4x - 3(1) = 5$$
$$4x - 3 = 5$$
$$4x = 8$$
$$x = 2$$
The solution is $(2, 1)$.

29. (1) $7x - y = 4$
   (2) $5x + 2y = 1$
Solve equation (1) for y.
$$7x - y = 4$$
$$-y = -7x + 4$$
$$y = 7x - 4$$
Substitute in equation (2).
$$5x + 2y = 1$$
$$5x + 2(7x - 4) = 1$$
$$5x + 14x - 8 = 1$$
$$19x - 8 = 1$$
$$19x = 9$$
$$x = \frac{9}{19}$$
Substitute x in equation (1).
$$7x - y = 4$$
$$7\left(\frac{9}{19}\right) - y = 4$$
$$\frac{63}{19} - y = 4$$
$$-y = 4 - \frac{63}{19}$$
$$-y = \frac{76}{19} - \frac{63}{19}$$
$$-y = \frac{13}{19}$$
$$y = -\frac{13}{19}$$
The solution is $\left(\frac{9}{19}, -\frac{13}{19}\right)$.

31. (1) $4x - 3y = -1$
 (2) $y = 2x - 3$

Substitute in equation (1).
$$4x - 3y = -1$$
$$4x - 3(2x - 3) = -1$$
$$4x - 6x + 9 = -1$$
$$-2x + 9 = -1$$
$$-2x = -10$$
$$x = 5$$

Substitute x in equation (2).
$$y = 2x - 3$$
$$y = 2(5) - 3$$
$$y = 10 - 3$$
$$y = 7$$

The solution is (5, 7).

33. (1) $7x + y = 14$
 (2) $2x - 5y = -33$

Solve equation (1) for y.
$$7x + y = 14$$
$$y = -7x + 14$$

Substitute in equation (2).
$$2x - 5y = -33$$
$$2x - 5(-7x + 14) = -33$$
$$2x + 35x - 70 = -33$$
$$37x - 70 = -33$$
$$37x = 37$$
$$x = 1$$

Substitute x in equation (1).
$$7x + y = 14$$
$$7(1) + y = 14$$
$$7 + y = 14$$
$$y = 7$$

The solution is (1, 7).

35. (1) $x - 4y = 9$
 (2) $2x - 3y = 11$

Solve equation (1) for x.
$$x - 4y = 9$$
$$x = 4y + 9$$

Substitute in equation (2).
$$2x - 3y = 11$$
$$2(4y + 9) - 3y = 11$$
$$8y + 18 - 3y = 11$$
$$5y + 18 = 11$$
$$5y = -7$$
$$y = -\frac{7}{5}$$

Substitute y in equation (1).
$$x - 4y = 9$$
$$x - 4\left(-\frac{7}{5}\right) = 9$$
$$x + \frac{28}{5} = 9$$
$$x = 9 - \frac{28}{5}$$
$$x = \frac{45}{5} - \frac{28}{5}$$
$$x = \frac{17}{5}$$

The solution is $\left(\frac{17}{5}, -\frac{7}{5}\right)$.

37. (1) $4x - y = -5$
 (2) $2x + 5y = 13$

Solve equation (1) for y.
$$4x - y = -5$$
$$-y = -4x - 5$$
$$y = 4x + 5$$

Substitute in equation (2).
$$2x + 5y = 13$$
$$2x + 5(4x + 5) = 13$$
$$2x + 20x + 25 = 13$$
$$22x + 25 = 13$$
$$22x = -12$$
$$x = -\frac{12}{22} = -\frac{6}{11}$$

Substitute x in equation (1).
$$4x - y = -5$$
$$4\left(-\frac{6}{11}\right) - y = -5$$
$$-\frac{24}{11} - y = -5$$
$$-y = -5 + \frac{24}{11}$$
$$-y = \frac{-55}{11} + \frac{24}{11}$$
$$-y = -\frac{31}{11}$$
$$y = \frac{31}{11}$$

The solution is $\left(-\frac{6}{11}, \frac{31}{11}\right)$.

39. (1) $3x + 4y = 18$
(2) $2x - y = 1$
Solve equation (2) for y.
$2x - y = 1$
$-y = -2x + 1$
$y = 2x - 1$
Substitute in equation (1).
$3x + 4y = 18$
$3x + 4(2x - 1) = 18$
$3x + 8x - 4 = 18$
$11x - 4 = 18$
$11x = 22$
$x = 2$
Substitute x in equation (2).
$2x - y = 1$
$2(2) - y = 1$
$4 - y = 1$
$-y = -3$
$y = 3$
The solution is (2, 3).

41. (1) $5x + 2y = 0$
(2) $x - 3y = 0$
Solve equation (2) for x.
$x - 3y = 0$
$x = 3y$
Substitute in equation (1).
$5x + 2y = 0$
$5(3x) + 2y = 0$
$15y + 2y = 0$
$17y = 0$
$y = 0$
Substitute y in equation (2).
$x - 3y = 0$
$x - 3(0) = 0$
$x - 0 = 0$
$x = 0$
The solution is (0, 0).

43. (1) $3x + y = 4$
(2) $9x + 3y = 12$
Solve equation (1) for y.
$3x + y = 4$
$y = -3x + 4$
Substitute in equation (2).
$9x + 3y = 12$
$9x + 3(-3x + 4) = 12$
$9x - 9x + 12 = 12$
$12 = 12$
The system is dependent. The solutions are the ordered pairs that satisfy the equation $3x + y = 4$.

45. (1) $x + 7y = -5$
(2) $2x - 3y = 5$
Solve equation (1) for x.
$x + 7y = -5$
$x = -7y - 5$
Substitute in equation (2).
$2x - 3y = 5$
$2(-7y - 5) - 3y = 5$
$-14y - 10 - 3y = 5$
$-17y - 10 = 5$
$-17y = 15$
$y = -\dfrac{15}{17}$
Substitute y in equation (1).
$x + 7y = -5$
$x + 7\left(-\dfrac{15}{17}\right) = -5$
$x = -5 + \dfrac{105}{17}$
$x = -\dfrac{85}{17} + \dfrac{105}{17}$
$x = \dfrac{20}{17}$
The solution is $\left(\dfrac{20}{17}, -\dfrac{15}{17}\right)$.

47. (1) $y = 2x - 8$
(2) $y = 3x - 13$
Substitute in equation (2).
$y = 3x - 13$
$2x - 8 = 3x - 13$
$-x - 8 = -13$
$-x = -5$
$x = 5$
Substitute x in equation (1).
$y = 2x - 8$
$y = 2(5) - 8$
$y = 10 - 8$
$y = 2$
The solution is (5, 2).

49. (1) $x = 3y + 7$
(2) $x = 2y - 1$
Substitute in equation (2).
$x = 2y - 1$
$3y + 7 = 2y - 1$
$y + 7 = -1$
$y = -8$
Substitute y in equation (1).
$x = 3y + 7$
$x = 3(-8) + 7$
$x = -24 + 7$
$x = -17$
The solution is (17, -8).

51. (1) $x = 3 - 2y$

(2) $x = 5y - 10$

Substitute in equation (2).

$x = 5y - 10$

$3 - 2y = 5y - 10$

$3 - 7y = -10$

$-7y = -13$

$y = -\dfrac{13}{7}$

Substitute y in equation (1).

$x = 3 - 2y$

$x = 3 - 2\left(\dfrac{13}{7}\right)$

$x = 3 - \dfrac{26}{7}$

$x = \dfrac{21}{7} - \dfrac{26}{7}$

$x = -\dfrac{5}{7}$

The solution is $\left(-\dfrac{5}{7}, \dfrac{13}{7}\right)$.

53. (1) $3x - y = 11$

(2) $2x + 5y = -4$

Solve equation (1) for y.

$3x - y = 11$

$-y = -3x + 11$

$y = 3x - 11$

Substitute in equation (2).

$2x + 5y = -4$

$2x + 5(3x - 11) = -4$

$2x + 15x - 55 = -4$

$17x - 55 = -4$

$17x = 51$

$x = 3$

Substitute x in equation (1).

$3x - y = 11$

$3(3) - y = 11$

$9 - y = 11$

$-y = 2$

$y = -2$

The solution $(3, -2)$.

Applying Concepts 6.2

55. (1) $0.1x - 0.6y = -0.4$

$x - 6y = -4$

(2) $-0.7x + 0.2y = 0.5$

$-7x + 2y = 5$

Solve equation (1) for x.

$x = 6y - 4$

Substitute in equation (2).

$-7(6y - 4) + 2y = 5$

$-42y + 28 + 2y = 5$

$-40y + 28 = 5$

$-40y = -23$

$y = \dfrac{23}{40}$

Substitute y in equation (1).

$x - 6\left(\dfrac{23}{40}\right) = -4$

$x = -4 + \dfrac{69}{20}$

$x = -\dfrac{80}{20} + \dfrac{69}{20}$

$x \approx -\dfrac{11}{20}$

The solution is $\left(-\dfrac{11}{20}, \dfrac{23}{40}\right)$.

57. (1) $0.4x + 0.5y = 0.2$

$4x + 5y = 2$

(2) $0.3x - 0.1y = 1.1$

$3x - y = 11$

Solve equation (2) for y.

$-y = -3x + 11$

$y = 3x - 11$

Substitute in equation (1).

$4x + 5(3x - 11) = 2$

$4x + 15x - 55 = 2$

$19x - 55 = 2$

$19x = 57$

$x = 3$

Substitute x in equation (2).

$3(3) - y = 11$

$9 - y = 11$

$-y = 2$

$y = -2$

The solution is $(3, -2)$.

59. (1) $1.2x + 0.1y = 1.9$

$\quad\quad 12x + y = 19$

(2) $0.1x + 0.3y = 2.2$

$\quad\quad x + 3y = 22$

Solve equation (2) for x.

$x = 22 - 3y$

Substitute in equation (1).

$12(22 - 3y) + y = 19$

$264 - 36y + y = 19$

$264 - 35y = 19$

$-35y = -245$

$y = 7$

Substitute y in equation (2).

$x = 22 - 3y$

$x = 22 - 3(7)$

$x = 22 - 21$

$x = 1$

The solution is $(1, 7)$.

61. (1) $2x - 3y = 7$

$\quad\quad -3y = -2x + 7$

$\quad\quad 3y = 2x - 7$

$\quad\quad y = \dfrac{2x - 7}{3}$

$\quad\quad y = \dfrac{2}{3}x - \dfrac{7}{3}$

(2) $kx - 3y = 4$

$\quad\quad -3y = -kx + 4$

$\quad\quad y = \dfrac{k}{3}x - \dfrac{4}{3}$

For the system of equations to have no solution, the slopes of both lines must be equal.

$\dfrac{k}{3} = \dfrac{2}{3}$

$k = 2$

63. (1) $x = 4y + 4$

$\quad\quad -4y = -x + 4$

$\quad\quad y = \dfrac{1}{4}x - 1$

(2) $kx - 8y = 4$

$\quad\quad -8y = -kx + 4$

$\quad\quad y = \dfrac{k}{8}x - \dfrac{1}{2}$

For the system of equations to have no solution, the slopes of both lines must be equal.

$\dfrac{k}{8} = \dfrac{1}{4}$

$k = 2$

Section 6.3

Concept Review 6.3

1. Always true

3. Never true

If two equations in a system of linear equations have the same slope and different y-intercepts, then the graphs of the two equations are parallel and never intersect.

5. Never true

The solution of a system of equations is an ordered pair. The y-coordinate of that ordered pair is equal to -2.

Objective 6.3.1 Exercises

1. (1) $x + y = 4$

(2) $x - y = 6$

Add the equations

$2x = 10$

$\quad x = 5$

Replace x in equation (1).

$x + y = 4$

$5 + y = 4$

$\quad\quad y = -1$

The solution is $(5, -1)$.

3. (1) $x + y = 4$

(2) $2x + y = 5$

Eliminate y.

$-1(x + y) = -1 \cdot 4$

$\quad\quad 2x + y = 5$

$\quad\quad -x - y = -4$

$\quad\quad 2x + y = 5$

Add the equations

$x = 1$

Replace x in equation (2).

$2x + y = 5$

$2(1) + y = 5$

$\quad 2 + y = 5$

$\quad\quad y = 3$

The solution is $(1, 3)$.

5. (1) $2x - y = 1$

(2) $x + 3y = 4$

Eliminate y

$3(2x - y) = 3 \cdot 1$

$\quad\quad x + 3y = 4$

$\quad 6x - 3y = 3$

$\quad x + 3y = 4$

Add the equations.

$7x = 7$

$\quad x = 1$

Replace x in equation (2).

$x + 3y = 4$

$1 + 3y = 4$

$\quad 3y = 3$

$\quad\quad y = 1$

The solution is $(1, 1)$.

7. (1) $\quad 4x - 5y = 22$
   (2) $\quad x + 2y = -1$
   Eliminate x.
   $$4x - 5y = 22$$
   $$-4(x + 2y) = -4(-1)$$

   $$4x - 5y = 22$$
   $$-4x - 8y = 4$$
   Add the equations.
   $$-13y = 26$$
   $$y = -2$$
   Replace y in equation (1).
   $$4x - 5y = 22$$
   $$4x - 5(-2) = 22$$
   $$4x + 10 = 22$$
   $$4x = 12$$
   $$x = 3$$
   The solution is $(3, -2)$.

9. (1) $\quad 2x - y = 1$
   (2) $\quad 4x - 2y = 2$
   Eliminate y.
   $$-2(2x - y) = -2 \cdot 1$$
   $$4x - 2y = 2$$

   $$-4x + 2y = -2$$
   $$4x - 2y = 2$$
   Add the equations.
   $$0 = 0$$
   The system is dependent. The solutions are the ordered pairs that satisfy the equation $2x - y = 1$.

11. (1) $\quad 4x + 3y = 15$
    (2) $\quad 2x - 5y = 1$
    Eliminate y.
    $$5(4x + 3y) = 5 \cdot 15$$
    $$3(2x - 5y) = 3 \cdot 1$$

    $$20x + 15y = 75$$
    $$6x - 15y = 3$$
    Add the equations.
    $$26x = 78$$
    $$x = 3$$
    Replace x in equation (1).
    $$4x + 3y = 15$$
    $$4(3) + 3y = 15$$
    $$12 + 3y = 15$$
    $$3y = 3$$
    $$y = 1$$
    The solution is $(3, 1)$.

13. (1) $\quad 2x - 3y = 5$
    (2) $\quad 4x - 6y = 3$
    Eliminate x.
    $$-2(2x - 3y) = -2 \cdot 5$$
    $$4x - 6y = 3$$

    $$-4x + 6y = -10$$
    $$4x - 6y = 3$$
    Add the equations.
    $$0 = -7$$
    The system is inconsistent and has no solution.

15. (1) $\quad 5x - 2y = -1$
    (2) $\quad x + y = 4$
    Eliminate x.
    $$5x - 2y = -1$$
    $$-5(x + y) = -5(4)$$

    $$5x - 2y = -1$$
    $$-5x - 5y = -20$$
    Add the equations.
    $$-7y = -21$$
    $$y = 3$$
    Replace y in equation (1).
    $$5x - 2y = -1$$
    $$5x - 2(3) = -1$$
    $$5x - 6 = -1$$
    $$5x = -1 + 6$$
    $$5x = 5$$
    $$\frac{5x}{5} = \frac{5}{5}$$
    $$x = 1$$
    The solution is $(1, 3)$.

17. (1) $\quad 5x + 7y = 10$
    (2) $\quad 3x - 14y = 6$
    Eliminate y.
    $$2(5x + 7y) = 2 \cdot 10$$
    $$3x - 14y = 6$$

    $$10x + 14y = 20$$
    $$3x - 14y = 6$$
    Add the equations.
    $$13x = 26$$
    $$x = 2$$
    Replace x in equation (2).
    $$3x - 14y = 6$$
    $$3(2) - 14y = 6$$
    $$6 - 14y = 6$$
    $$-14y = 0$$
    $$y = 0$$
    The solution is $(2, 0)$.

19. (1) $3x - 2y = 0$
(2) $6x + 5y = 0$
Eliminate x.
$-2(3x - 2y) = -2(0)$
$6x + 5y = 0$

$-6x + 4y = 0$
$6x + 5y = 0$
Add the equations.
$9y = 0$
$y = 0$
Replace y in equation (2).
$6x + 5y = 0$
$6x + 5(0) = 0$
$6x = 0$
$x = 0$
The solution is $(0, 0)$.

21. (1) $2x - 3y = 16$
(2) $3x + 4y = 7$
Eliminate y.
$4(2x - 3y) = 4 \cdot 16$
$3(3x + 4y) = 3 \cdot 7$

$8x - 12y = 64$
$9x + 12y = 21$
Add the equations.
$17x = 85$
$x = 5$
Replace x in equation (1).
$2x - 3y = 16$
$2(5) - 3y = 16$
$10 - 3y = 16$
$-3y = 6$
$y = -2$
The solution is $(5, -2)$.

23. (1) $x + 3y = 4$
(2) $2x + 5y = 1$
Eliminate x.
$-2(x + 3y) = -2 \cdot 4$
$2x + 5y = 1$

$-2x - 6y = -8$
$2x + 5y = 1$
Add the equations.
$-y = -7$
$y = 7$
Replace y in equation (2).

$2x + 5y = 1$
$2x + 5(7) = 1$
$2x + 35 = 1$
$2x = 1 - 35$
$2x = -34$
$x = -17$
The solution is $(-17, 7)$.

25. (1) $7x - 2y = 13$
(2) $5x + 3y = 27$
Eliminate y.
$3(7x - 2y) = 3 \cdot 13$
$2(5x + 3y) = 2 \cdot 27$

$21x - 6y = 39$
$10x + 6y = 54$
Add the equations.
$31x = 93$
$x = 3$
Replace x in equation (1).
$7x - 2y = 13$
$7(3) - 2y = 13$
$21 - 2y = 13$
$-2y = -8$
$y = 4$
The solution is $(3, 4)$.

27. (1) $8x - 3y = 11$
(2) $6x - 5y = 11$
Eliminate y.
$5(8x - 3y) = 5 \cdot 11$
$-3(6x - 5y) = -3 \cdot 11$

$40x - 15y = 55$
$-18x + 15y = -33$
Add the equations.
$22x = 22$
$x = 1$
Replace x in equation (1).
$8x - 3y = 11$
$8(1) - 3y = 11$
$8 - 3y = 11$
$-3y = 3$
$y = -1$
The solution is $(1, -1)$.

29. (1) $5x + 15y = 20$
(2) $2x + 6y = 8$
Eliminate x.
$-2(5x + 15y) = -2 \cdot 20$
$5(2x + 6y) = 5 \cdot 8$

$-10x - 30y = -40$
$10x + 30y = 40$
Add the equations.
$0 = 0$
The system is dependent. The solutions are the ordered pairs that satisfy the equation
$5x + 15y = 20$.

31. (1) $\quad 3x = 2y + 7$
(2) $\quad 5x - 2y = 13$
Write equation (1) in the form $Ax + By = C$.
$$3x = 2y + 7$$
$$3x - 2y = 7$$
Eliminate y.
$$-1(3x - 2y) = -1 \cdot 7$$
$$5x - 2y = 13$$

$$-3x + 2y = -7$$
$$5x - 2y = 13$$
Add the equations.
$$2x = 6$$
$$x = 3$$
Replace x in equation (2).
$$5x - 2y = 13$$
$$5(3) - 2y = 13$$
$$15 - 2y = 13$$
$$-2y = -2$$
$$y = 1$$
The solution is (3, 1).

33. (1) $\quad 2x + 9y = 5$
(2) $\quad 5x = 6 - 3y$
Write equation (2) in the form $Ax + By = C$.
$$5x = 6 - 3y$$
$$5x + 3y = 6$$
Eliminate y.
$$2x + 9y = 5$$
$$-3(5x + 3y) = -3 \cdot 6$$

$$2x + 9y = 5$$
$$-15x - 9y = -18$$
Add the equations.
$$-13x = -13$$
$$x = 1$$
Replace x in equation (1).
$$2x + 9y = 5$$
$$2(1) + 9y = 5$$
$$2 + 9y = 5$$
$$9y = 3$$
$$y = \frac{1}{3}$$
The solution is $\left(1, \frac{1}{3}\right)$.

35. (1) $\quad 2x + 3y = 7 - 2x$
(2) $\quad 7x + 2y = 9$
Write equation (1) in the form $Ax + By = C$.
$$2x + 3y = 7 - 2x$$
$$4x + 3y = 7$$
Eliminate y.
$$2(4x + 3y) = 2 \cdot 7$$
$$-3(7x + 2y) = -3 \cdot 9$$

$$8x + 6y = 14$$
$$-21x - 6y = -27$$
Add the equations.
$$-13x = -13$$
$$x = 1$$
Replace x in equation (2).
$$7x + 2y = 9$$
$$7(1) + 2y = 9$$
$$7 + 2y = 9$$
$$2y = 2$$
$$y = 1$$
The solution is (1, 1).

37. (1) $\quad 3x + y = 1$
(2) $\quad 5x + y = 2$
Eliminate y.
$$3x + y = 1$$
$$-1(5x + y) = -1 \cdot 2$$

$$3x + y = 1$$
$$-5x - y = -2$$
Add the equations.
$$-2x = -1$$
$$x = \frac{1}{2}$$
Replace x in equation (1).
$$3x + y = 1$$
$$3\left(\frac{1}{2}\right) + y = 1$$
$$\frac{3}{2} + y = 1$$
$$y = 1 - \frac{3}{2}$$
$$y = -\frac{1}{2}$$
The solution is $\left(\frac{1}{2}, -\frac{1}{2}\right)$.

39. (1) $4x + 3y = 3$
(2) $x + 3y = 1$
Eliminate y.
$4x + 3y = 3$
$-1(x + 3y) = -1 \cdot 1$

$4x + 3y = 3$
$-x - 3y = -1$
Add the equations.
$3x = 2$
$x = \dfrac{2}{3}$
Replace x in equation (2).
$x + 3y = 1$
$\dfrac{2}{3} + 3y = 1$
$3y = 1 - \dfrac{2}{3}$
$3y = \dfrac{1}{3}$
$y = \dfrac{1}{9}$
The solution is $\left(\dfrac{2}{3}, \dfrac{1}{9} \right)$.

41. (1) $3x - 4y = 1$
(2) $4x + 3y = 1$
Eliminate y.
$3(3x - 4y) = 3 \cdot 1$
$4(4x + 3y) = 4 \cdot 1$

$9x - 12y = 3$
$6x + 12y = 4$
Add the equations.
$25x = 7$
$x = \dfrac{7}{25}$
Replace x in equation (1).
$3x - 4y = 1$
$3\left(\dfrac{7}{25} \right) - 4y = 1$
$\dfrac{21}{25} - 4y = 1$
$-4y = 1 - \dfrac{21}{25}$
$-4y = \dfrac{25}{25} - \dfrac{21}{25}$
$-4y = \dfrac{4}{25}$
$y = -\dfrac{1}{25}$
The solution is $\left(\dfrac{7}{25}, -\dfrac{1}{25} \right)$.

43. (1) $y = 2x - 3$
(2) $4x + 4y = -1$
Write equation (1) in the form $Ax + By = C$.
$y = 2x - 3$
$-2x + y = -3$
Eliminate y.
$-4(-2x + y) = -4(-3)$
$4x + 4y = -1$

$8x - 4y = 12$
$4x + 4y = -1$
Add the equations.
$12x = 11$
$x = \dfrac{11}{12}$
Replace x in equation (1).
$y = 2x - 3$
$y = 2\left(\dfrac{11}{12} \right) - 3$
$y = \dfrac{11}{6} - 3$
$y = \dfrac{11}{6} - \dfrac{18}{6}$
$y = -\dfrac{7}{6}$
The solutions is $\left(\dfrac{11}{12}, -\dfrac{7}{6} \right)$.

Applying Concepts 6.3

45. (1) $x - 0.2y = 0.2$
$10x - 2y = 2$
(2) $0.2x + 0.5y = 2.2$
$2x + 5y = 22$
Eliminate y.
$5(10x - 2y) = 5(2)$
$2(2x + 5y) = 2(22)$

$50x - 10y = 10$
$4x + 10y = 44$
Add the equations.
$54x = 54$
$x = 1$
Replace x in equation (1).
$10x - 2y = 2$
$10(1) - 2y = 2$
$10 - 2y = 2$
$-2y = -8$
$y = 4$
The solution is $(1, 4)$.

47. (1) $1.25x - 1.5y = -1.75$
$125x - 150y = -175$
$5x - 6y = -7$
(2) $2.5x - 1.75y = -1$
$250x - 175y = -100$
$10x - 7y = -4$

Eliminate x.
$-2(5x - 6y) = -2(-7)$
$10x - 7y = -4$

$-10x + 12y = 14$
$10x - 7y = -4$
Add the equations.
$5y = 10$
$y = 2$
Replace y in equation (1).
$5x - 6y = -7$
$5x - 6(2) = -7$
$5x - 12 = -7$
$5x = 5$
$x = 1$
The solution is (1, 2).

49.
$Ax - 4y = 9$ $4x + By = -1$
$A(-1) - 4(-3) = 9$ $4(-1) + B(-3) = -1$
$-A + 12 = 9$ $-4 - 3B = -1$
$-A = -3$ $-3B = 3$
$A = 3$ $B = -1$

51. (1) $3x - 2y = -2$
(2) $2x - y = 0$
Eliminate y.
$3x - 2y = -2$
$-2(2x - y) = -2 \cdot 0$

$3x - 2y = -2$
$-4x + 2y = 0$
Add the equations.
$-x = -2$
$x = 2$
Replace x in equation (2).
$2x - y = 0$
$2(2) - y = 0$
$4 - y = 0$
$-y = -4$
$y = 4$
The solution is (2, 4).
$Ax + y = 8$
$A(2) + 4 = 8$
$2A + 4 = 8$
$2A = 4$
$A = 2$

53. For the system to be independent, the slopes must not be the same. Write each equation in slope-intercept form and solve for k.

a. (1) $x + y = 7 \Rightarrow y = -x + 7$
(2) $kx + y = 3 \Rightarrow y = -kx + 3$

$-1 \neq -k$
$1 \neq k$ k cannot equal 1.

b. (1) $x + 2y = 4 \Rightarrow 2y = -x + 4 \Rightarrow y = -\dfrac{1}{2}x + 2$
(2) $kx + 3y = 2 \Rightarrow 3y = -kx + 2 \Rightarrow y = -\dfrac{k}{3}x + \dfrac{2}{3}$

$-\dfrac{1}{2} \neq -\dfrac{k}{3}$
$2k \neq 3$
$k \neq \dfrac{3}{2}$ k cannot equal $\dfrac{3}{2}$.

c. (1) $2x + ky = 1 \Rightarrow ky = -2x + 1 \Rightarrow y = -\dfrac{2}{k}x + \dfrac{1}{k}$
(2) $x + 2y = 2 \Rightarrow 2y = -x + 2 \Rightarrow y = -\dfrac{1}{2}x + 1$

$-\dfrac{2}{k} \neq -\dfrac{1}{2}$
$k \neq 4$ k cannot equal 4.

Section 6.4

Concept Review 6.4

1. Always true

3. Never true
   The rate of the boat traveling against the current is $b - c$. The rate of the boat traveling with the current is $b + c$.

5. Never true
   The rate of the plane is 100 mph.

Objective 6.4.1 Exercises

1. Strategy
   Rate of the whale in calm water: w
   Rate of the ocean current: c

	Rate	Time	Distance
With current	$w + c$	1.5	$1.5(w + c)$
Without current	$w - c$	2	$2(w - c)$

   The distance traveled in each direction is 60 mi.

   Solution
   $$1.5(w + c) = 60 \qquad \frac{1}{1.5} \cdot 1.5(w + c) = 60 \cdot \frac{1}{1.5}$$
   $$2(w - c) = 60 \qquad \frac{1}{2} \cdot 2(w - c) = 60 \cdot \frac{1}{2}$$

   $$w + c = 40$$
   $$w - c = 30$$

   $$2w = 70$$
   $$w = 35$$

   $$w + c = 40$$
   $$35 + c = 40$$
   $$c = 5$$
   The rate of the whale in calm water is 35 mph.
   The rate of the ocean current is 5 mph.

3. Strategy
   Rate of rowing in calm water: r
   Rate of the current: c

	Rate	Time	Distance
With current	$r + c$	2	$2(r + c)$
Against current	$r - c$	2	$2(r - c)$

   The distance traveled with the current is 40 km. The distance traveled against the current is 16 km.

   Solution
   $$2(r + c) = 40 \qquad \frac{1}{2} \cdot 2(r + c) = 40 \cdot \frac{1}{2}$$
   $$2(r - c) = 16 \qquad \frac{1}{2} \cdot 2(r - c) = 16 \cdot \frac{1}{2}$$

   $$r + c = 20$$
   $$r - c = 8$$

   $$2r = 28$$
   $$r = 14$$

   $$r + c = 20$$
   $$14 + c = 20$$
   $$c = 6$$
   The rate of rowing in calm water is 14 km/h.
   The rate of the current is 6 km/h.

5. Strategy
   Rate of the boat in calm water: r
   Rate of the current: c

	Rate	Time	Distance
With current	$r + c$	3.5	$3.5(r + c)$
Against current	$r - c$	3	$3(r - c)$

   The distance traveled with the current is 35 mi. The distance traveled against the current is 12 mi.

   Solution
   $$3.5(r + c) = 35 \qquad \frac{1}{3.5} \cdot 3.5(r + c) = 35 \cdot \frac{1}{3.5}$$
   $$3(r - c) = 12 \qquad \frac{1}{3} \cdot 3(r - c) = 12 \cdot \frac{1}{3}$$

   $$r + c = 10$$
   $$r - c = 4$$

   $$2r = 14$$
   $$r = 7$$

   $$r + c = 10$$
   $$7 + c = 10$$
   $$c = 3$$
   The rate of the boat in calm water is 7 mph.
   The rate of the current is 3 mph.

7. Strategy

Rate of jet in calm air: j
Rate of wind: r

	Rate	Time	Distance
With wind	$j+w$	2	$2(j+w)$
Against wind	$j-w$	2	$2(j-w)$

The distance flown with the wind is 1120 mi.
The distance flown against the wind is 980 mi.

Solution

$$2(j+w)=1120 \qquad \frac{1}{2}\cdot 2(j+w)=1120\cdot\frac{1}{2}$$
$$2(j-w)=980 \qquad \frac{1}{2}\cdot 2(j-w)=980\cdot\frac{1}{2}$$

$$j+w=560$$
$$j-w=490$$

$$2j=1050$$
$$j=525$$

$$j+w=560$$
$$525+w=560$$
$$w=35$$

The rate of the jet in calm air is 525 mph. and the rate of the wind is 35 mph.

9. Strategy

Rate of the plane in calm air: p
Rate of the wind: w

	Rate	Time	Distance
With wind	$p+w$	2	$2(p+w)$
Against wind	$p-w$	3	$3(p-w)$

The distance with the wind and against the wind is 240 mi.

Solution

$$2(p+w)=240 \qquad \frac{1}{2}\cdot 2(p+w)=240\cdot\frac{1}{2}$$
$$3(p-w)=240 \qquad \frac{1}{3}\cdot 3(p-w)=240\cdot\frac{1}{3}$$

$$p+w=120$$
$$p-w=80$$

$$2p=200$$
$$p=100$$

$$p+w=120$$
$$100+w=120$$
$$w=20$$

The rate of the plane in calm water is 100 mph.
The rate of the wind is 20 mph.

11. Strategy

Rate of the helicopter in calm air: h
Rate of wind: w

	Rate	Time	Distance
With wind	$h+w$	$1\frac{2}{3}$	$\frac{5}{3}(p+w)$
Against wind	$h-w$	2.5	$2.5(p-w)$

The distance traveled in each direction is 450 mi.

Solution

$$2.5(h-w)=450 \qquad \frac{1}{2.5}\cdot 2.5(h-w)=450\cdot\frac{1}{2.5}$$
$$\frac{5}{3}(h+w)=450 \qquad \frac{3}{5}\cdot\frac{5}{3}(h+w)=450\cdot\frac{3}{5}$$

$$h-w=180$$
$$h+w=270$$

$$2h=450$$
$$h=225$$

$$h+w=270$$
$$225+w=270$$
$$w=45$$

The rate of the helicopter in calm air is 225 mph.
The rate of the wind is 45 mph.

Objective 6.4.2 Exercises

13. Strategy
Cost per copy for a word processing program: x
Cost per copy for a spreadsheet program: x
First Shipment

	Number	Cost	Total Cost
Word processing	12	x	$12x$
Spreadsheet	10	y	$10y$

Second Shipment

	Number	Cost	Total Cost
Word processing	5	x	$5x$
Spreadsheet	8	y	$8y$

The first shipment costs $6190.
The second shipment costs $3825.

Solution
$$12x + 10y = 6190 \qquad 5(12x + 10y) = 6190(5)$$
$$5x + 8y = 3825 \qquad -12(5x + 8y) = 3825(-12)$$

$$60x + 50y = 30{,}950$$
$$-60x - 96y = -45{,}900$$

$$-46y = -14{,}950$$
$$y = 325$$

$$5x + 8y = 3825$$
$$5x + 8(325) = 3825$$
$$5x + 2600 = 3825$$
$$5x = 1225$$
$$x = 245$$

The cost per copy of a word processing program
is $245.

15. Strategy
Cost per pound of the wheat flour: x
Cost per pound of the rye flour: y
First Purchase

	Number	Cost	Total Cost
Wheat	12	x	$12x$
Rye	15	y	$15y$

Second Purchase

	Number	Cost	Total Cost
Wheat	15	x	$15x$
Rye	10	y	$10y$

The first purchase costs $18.30.
The second purchase costs $16.75.

Solution
$$12x + 15y = 18.30 \qquad 5(12x + 15y) = 18.30(5)$$
$$15x + 10y = 16.75 \qquad -4(15x + 10y) = 16.75(-4)$$

$$60x + 75y = 91.50$$
$$-60x - 40y = -67.00$$

$$35y = 24.50$$
$$y = 0.70$$

$$12x + 15y = 18.30$$
$$12x + 15(0.70) = 18.30$$
$$12x + 10.50 = 18.30$$
$$12x = 7.80$$
$$x = 0.65$$

The cost per pound of the wheat flour is $.65.
The cost per pound of the rye flour is $.70.

17. Strategy
Rate for daytime hours: d
Rate for nighttime hours; n
First week

	Rate	Hours	Total Pay
Daytime	d	17	$17d$
Nighttime	n	8	$8n$

Second Week

	Rate	Hours	Total Pay
Daytime	d	12	$12d$
Nighttime	n	15	$15n$

The first week Shelly earned \$191.
The second week she earned \$219.

Solution
$$17d + 8n = 191 \qquad 15(17d + 8n) = 191(15)$$
$$12d + 15n = 219 \qquad -8(12d + 15n) = 219(-8)$$

$$255d + 120n = 2865$$
$$-96d - 120n = -1752$$

$$159d = 1113$$
$$d = 7$$

$$17d + 8n = 191$$
$$17(7) + 8n = 191$$
$$119 + 8n = 191$$
$$8n = 72$$
$$n = 9$$

Shelly earns \$7 an hour for daytime hours and \$9 an hour for nighttime hours.

19. Strategy
Number of two-point baskets: x
Number of three-point baskets: y
First Game

	Point	Number	Total Points
Two-point baskets	2	x	$2x$
Three-point baskets	3	y	$3y$

Second Game

	Point	Number	Total Points
Two-point baskets	2	y	$2y$
Three-point baskets	3	x	$3x$

The score for the first game was 87 points.
The score for the second game was 93 points.

Solution
$$2x + 3y = 87 \qquad 3(2x + 3y) = 87(3)$$
$$3x + 2y = 93 \qquad -2(3x + 2y) = 93(-2)$$

$$6x + 9y = 261$$
$$-6x - 4y = -186$$

$$5y = 75$$
$$y = 15$$

$$2x + 3y = 87$$
$$2x + 3(15) = 87$$
$$2x + 45 = 87$$
$$2x = 42$$
$$x = 21$$

The number of two-point baskets was 21.
The number of three-point baskets was 15.

21. Strategy
Number of nickels in the first bank: n
Number of quarters in the first bank: q
First Bank

	Number	Value	Total Value
Nickels	n	5	$5n$
Quarters	q	25	$25q$

Second Bank

	Number	Value	Total Value
Nickels	$2n$	5	$5(2n)$
Quarters	$q+2$	25	$25(q+2)$

The total value of the coins in the first bank is $2.90.
The total value of the coins in the second bank is $3.80.

Solution
$$5n+25q=290$$
$$5(2n)+25(q+2)=380$$

$$5n+25q=290$$
$$10n+25q+50=380$$

$$5n+25q=290 \quad -1(5n+25q)=-1(290)$$
$$10n+25q=330 \quad 10n+25q=330$$

$$-5n-25q=-290$$
$$10n+25q=330$$

$$5n=40$$
$$n=8$$

$$5n+25q=290$$
$$5(8)+25q=290$$
$$40+25q=290$$
$$25q=250$$
$$q=10$$

There are 8 nickels and 10 quarters in the first bank.

23. Strategy
Number of dimes in the bank: n
Number of quarters in the bank: q

	Number	Value	Total Value
Dimes	d	10	$10d$
Quarters	q	25	$25q$

	Number	Value	Total Value
Dimes	q	10	$10q$
Quarters	d	25	$25d$

The total value of the coins is $3.30.
If the quarters were dimes and the dimes were quarters, the total value would be $3.00.

Solution
$$10d+25q=330 \quad 5(10d+25q)=5(330)$$
$$25d+10q=300 \quad -2(25q+10q)=-2(300)$$

$$50d+125q=1650$$
$$-50d-20q=-600$$

$$105q=1050$$
$$q=10$$

$$10d+25q=330$$
$$10d+25(10)=330$$
$$10d+250=330$$
$$10d=80$$
$$d=8$$

There are 8 dimes and 10 quarters in the bank.

Applying Concepts 6.4

25. Strategy
Smaller angle: a
Larger angle: c
The angles a and c are supplementary.
Angle c is 15° more than twice angle a.

Solution
(1) $\quad a+c=180$
(2) $\qquad c=2a+15$
Substitute in equation (1).
$$a+(2a+15)=180$$
$$3a+15=180$$
$$3a=165$$
$$a=55$$
Substitute in equation (2).
$$c=2a+15$$
$$c=2(55)+15$$
$$c=110+15$$
$$c=125$$
The two supplementary angles are 55° and 125°.

27. Strategy
    Amount invested at 9%: x
    Amount invested at 8%: y

	Principal	Rate	Interest
Amount at 9%	x	0.09	$0.09x$
Amount at 8%	y	0.08	$0.08y$

	Principal	Rate	Interest
Amount at 9%	y	0.09	$0.09y$
Amount at 8%	x	0.08	$0.08x$

Investing x at 9% and y at 8% yields $860 in interest.
Investing x at 8% and y at 9% yields $840 in interest.

Solution

$$0.09x + 0.08y = 860 \qquad 9x + 8y = 86,000$$
$$0.08x + 0.09y = 840 \qquad 8x + 9y = 84,000$$

$$8(9x + 8y) = 8(86,000)$$
$$-9(8x + 9y) = -9(84,000)$$

$$72x + 64y = 688,000$$
$$-72x - 81y = -756,000$$
$$-17y = -68,000$$
$$y = 4000$$

$$9x + 8y = 86,000$$
$$9x + 8(4000) = 86,000$$
$$9x + 32,000 = 86,000$$
$$9x = 54,000$$
$$x = 6000$$

The amount invested at 9% was $6000.
The amount invested at 8% was $4000.

29. Strategy
Number of nickels in the bank: n
Number of dimes in the bank: d
In the bank

	Number	Value	Total Value
Nickels	n	5	$5n$
Dimes	d	10	$10d$

	Number	Value	Total Value
Nickels	$2n$	5	$5(2n)$
Dimes	$2d$	10	$10(2d)$

The value of the nickels and dimes in the bank is $.25. If the number of nickels and the number of dimes were doubled, the value of the coins would be $.50.

Solution
$$5n + 10d = 25$$
$$5(2n) + 10(2d) = 50$$

$$\begin{array}{ll} 5n + 10d = 25 & -2(5n + 10d) = -2(25) \\ 10n + 20d = 50 & 10n + 20d = 50 \end{array}$$

$$\begin{array}{l} -10n - 20d = -50 \\ 10n + 20d = 50 \end{array}$$

$$0 = 0$$

This is a true equation. There is more than one solution.
The numbers of dimes and nickels must be whole numbers.
Substitute 0, 1, and 2 for the number of dimes in the bank. There cannot be 3 dimes, as 3 dimes = $.30 > $.25.

0 dimes:

$$5n + 10d = 25$$
$$5n + 10(0) = 25$$
$$5n + 0 = 25$$
$$5n = 25$$
$$n = 5$$

1 dime:

$$5n + 10d = 25$$
$$5n + 10(1) = 25$$
$$5n + 10 = 25$$
$$5n = 15$$
$$n = 3$$

2 dimes:

$$5n + 10d = 25$$
$$5n + 10(2) = 25$$
$$5n + 20 = 25$$
$$5n = 5$$
$$n = 1$$

If there are 0 dimes, there are 5 nickels.   If there is 1 dime, there are 3 nickels.   If there are 2 dimes, there is 1 nickel.

There are 0 dimes and 5 nickels, 1 dime and 3 nickels, or 2 dimes and 1 nickel in the bank.

Chapter Review Exercises

1. (1) $4x + 7y = 3$
     (2) $x = y - 2$
Substitute in equation (1).
$$4(y - 2) + 7y = 3$$
$$4y - 8 + 7y = 3$$
$$11y - 8 = 3$$
$$11y = 11$$
$$y = 1$$
Substitute y in equation (2).
$$x = y - 2$$
$$x = 1 - 2$$
$$x = -1$$
The solution is $(-1, 1)$.

2.

The solution is $(2, -3)$.

3. (1) $3x + 8y = -1$
(2) $x - 2y = -5$

Eliminate y.   Add the equations.
$3x + 8y = -1$ $3x + 8y = -1$
$4(x - 2y) = -5(4)$ $4x - 8y = -20$

$7x = -21$
$x = -3$

Replace x in equation (2).
$x - 2y = -5$
$-3 - 2y = -5$
$-2y = -2$
$y = 1$
The solution is $(-3, 1)$.

4. (1) $8x - y = 2$
(2) $y = 5x + 1$
Substitute in equation (1).
$8x - (5x + 1) = 2$
$8x - 5x - 1 = 2$
$3x - 1 = 2$
$3x = 3$
$x = 1$
Substitute x in equation (2).
$y = 5x + 1$
$y = 5(1) + 1$
$y = 5 + 1$
$y = 6$
The solution is $(1, 6)$.

5.

The solution is $(1, 1)$.

6. (1) $4x - y = 9$
(2) $2x + 3y = -13$

Eliminate y.   Add the equations.
$3(4x - y) = 9(3)$ $12x - 3y = 27$
$2x + 3y = -13$ $2x + 3y = -13$

$14x = 14$
$x = 1$

Replace x in equation (1).
$4x - y = 9$
$4(1) - y = 9$
$4 - y = 9$
$-y = 5$
$y = -5$
The solution is $(1, -5)$.

7. (1) $-2x + y = -4$
(2) $x = y + 1$
Substitute in equation (1).
$-2x + y = -4$
$-2(y + 1) + y = -4$
$-2y - 2 + y = -4$
$-y - 2 = -4$
$-y = -2$
$y = 2$
Replace y in equation (2).
$x = y + 1$
$x = 2 + 1$
$x = 3$
The solution is $(3, 2)$.

8. (1) $8x - y = 25$
(2) $32x - 4y = 100$
Eliminate y.
$-4(8x - y) = -4(25)$
$32x - 4y = 100$

$-32x + 4y = -100$
$32x - 4y = 100$
Add the equations.
$0 = 0$
The system of equations is dependent. The solutions are the ordered pairs that satisfy the equation $8x - y = 25$.

9. (1) $5x - 15y = 30$
(2) $2x + 6y = 0$
Eliminate x.
$2(5x - 15y) = 2(30)$
$-5(2x + 6y) = -5(0)$

$10x - 30y = 60$
$-10x - 30y = 0$
Add the equations.
$-60y = 60$
$y = -1$
Replace y in equation (2).
$2x + 6y = 0$
$2x + 6(-1) = 0$
$2x - 6 = 0$
$2x = 6$
$x = 3$
The solution is $(3, -1)$.

10.

The solution is $(1, -3)$.

11. (1) $7x - 2y = 0$
   (2) $2x + y = -11$

Eliminate y.
$$7x - 2y = 0$$
$$2(2x + y) = 2(-11)$$

$$7x - 2y = 0$$
$$4x + 2y = -22$$
Add the equations.
$$11x = -22$$
$$x = -2$$
Replace x in equation (1).
$$7x - 2y = 0$$
$$7(-2) - 2y = 0$$
$$-14 - 2y = 0$$
$$-2y = 14$$
$$y = -7$$
The solution is $(-2, -7)$.

12. (1) $x - 5y = 4$
   (2) $y = x - 4$

Substitute in equation (1).
$$x - 5y = 4$$
$$x - 5(x - 4) = 4$$
$$x - 5x + 20 = 4$$
$$-4x + 20 = 4$$
$$-4x = -16$$
$$x = 4$$
Replace x in equation (2).
$$y = x - 4$$
$$y = 4 - 4$$
$$y = 0$$
The solution is $(4, 0)$.

13.

$5x + 4y = -17$		$2x - y = 1$	
$5(-1) + 4(-3)$	-17	$2(-1) - (-3)$	1
$-5 - 12$	-17	$-2 + 3$	1
$-17 = -17$		$1 = 1$	

Yes, $(-1, -3)$ is a solution of the system of equations.

14. (1) $6x + 4y = -3$
   (2) $12x - 10y = -15$

Eliminate x.   Add the equations.
$$-2(6x + 4y) = -3(-2) \quad -12x - 8y = 6$$
$$12x - 10y = -15 \quad\quad 12x - 10y = -15$$
$$-18y = -9$$
$$y = \frac{1}{2}$$

Replace y in equation (1).
$$6x + 4y = -3$$
$$6x + 4\left(\frac{1}{2}\right) = -3$$
$$6x + 2 = -3$$
$$6x = -5$$
$$x = -\frac{5}{6}$$
The solution is $\left(-\frac{5}{6}, \frac{1}{2}\right)$.

15. (1) $5x + 2y = -9$
   (2) $12x - 7y = 2$

Eliminate y.   Add the equations.
$$7(5x + 2y) = -9(7) \quad 35x + 14y = -63$$
$$2(12x - 7y) = 2(2) \quad 24x - 14y = 4$$

$$59x = -59$$
$$x = -1$$
Replace x in equation (1).
$$5x + 2y = -9$$
$$5(-1) + 2y = -9$$
$$-5 + 2y = -9$$
$$2y = -4$$
$$y = -2$$
The solution is $(-1, -2)$.

16.

The solution is $(3, -3)$.

17. (1) $5x + 7y = 21$
   (2) $20x + 28y = 63$

Eliminate x.   Add the equations.
$$-4(5x + 7y) = 21(-4) \quad -20x - 28y = -84$$
$$20x + 28y = 63 \quad\quad 20x + 28y = 63$$

$$0 \neq -21$$

The system is inconsistent and has no solution.

18. (1) $9x + 12y = -1$
   (2) $x - 4y = -1$
Solve equation (2) for x.
$$x - 4y = -1$$
$$x = 4y - 1$$
Substitute in equation (1).
$$9x + 12y = -1$$
$$9(4y - 1) + 12y = -1$$
$$36y - 9 + 12y = -1$$
$$48y - 9 = -1$$
$$48y = 8$$
$$y = \frac{1}{6}$$
Replace y in equation (2).
$$x = 4y - 1$$
$$x = 4\left(\frac{1}{6}\right) - 1$$
$$x = \frac{2}{3} - 1$$
$$x = -\frac{1}{3}$$
The solution is $\left(-\frac{1}{3}, \frac{1}{6}\right)$.

19.

The system of equations is dependent. The solutions are the ordered pairs that satisfy the equation $y = 2x - 4$.

20. (1) $3x + y = -2$
(2) $-9x - 3y = 6$

Eliminate x.   Add the equations.
$3(3x + y) = -2(3)$ $9x + 3y = -6$
$9x - 3y = 6$ $-9x - 3y = 6$
                        $0 = 0$

The system is dependent. The solutions are the ordered pairs that satisfy the equation $3x + y = -2$.

21. (1) $11x - 2y = 4$
(2) $25x - 4y = 2$
Eliminate y.
$-2(11x - 2y) = -2(4)$
$25x - 4y = 2$

$-22x + 4y = -8$
$25x - 4y = 2$
Add the equations.
$3x = -6$
$x = -2$
Replace x in equation (1).
$11x - 2y = 4$
$11(-2) - 2y = 4$
$-22 - 2y = 4$
$-2y = 26$
$y = -13$
The solution is $(-2, -13)$.

22. (1) $4x + 3y = 12$
(2) $y = -\dfrac{4}{3}x + 4$

Substitute in equation (1).
$4x + 3\left(-\dfrac{4}{3}x + 4\right) = 12$
$4x - 4x + 12 = 12$
$12 = 12$
The system of equations is dependent. The solutions are the ordered pairs that satisfy the equation $y = -\dfrac{4}{3}x + 4$.

23.

The solution is $(4, 2)$.

24. (1) $2x - y = 5$
(2) $10x - 5y = 20$
Eliminate y.
$-5(2x - y) = -5(5)$
$10x - 5y = 20$

$-10x + 5y = -25$
$10 - 5y = 20$
Add the equations.
$0 \neq -5$.
The system of equations is inconsistent and has no solution.

25. (1) $6x + 5y = -2$
(2) $y = 2x - 2$
Substitute in equation (1).
$6x + 5y = -2$
$6x + 5(2x - 2) = -2$
$6x + 10x - 10 = -2$
$16x - 10 = -2$
$16x = 8$
$x = \dfrac{1}{2}$
Replace x in equation (2).
$y = 2x - 2$
$y = 2\left(\dfrac{1}{2}\right) - 2$
$y = 1 - 2$
$y = -1$
The solution is $\left(\dfrac{1}{2}, -1\right)$.

26.

$$\begin{array}{c|c} -x + 9y = 2 \\ \hline -(-2) + 9(0) & 2 \\ -(-2) + 0 & 2 \\ 2 = 2 \end{array} \qquad \begin{array}{c|c} 6x - 4y = 12 \\ \hline 6(-2) - 4(0) & 12 \\ 6(-2) - 0 & 12 \\ -12 \neq 12 \end{array}$$

No, $(-2, 0)$ is not a solution to the system of equations.

27. (1) $6x - 18y = 7$
(2) $9x + 24y = 2$
Eliminate x.
$9(6x - 18y) = 7(9)$
$-6(9x + 24y) = 2(-6)$

$54x - 162y = 63$
$-54x - 144y = -12$
Add the equations.
$-306y = 51$
$y = -\dfrac{1}{6}$
Replace y in equation (1).
$6x - 18y = 7$
$6x - 18\left(-\dfrac{1}{6}\right) = 7$
$6x + 3 = 7$
$6x = 4$
$x = \dfrac{2}{3}$
The solution is $\left(\dfrac{2}{3}, -\dfrac{1}{6}\right)$.

28. (1) $12x - 9y = 18$
(2) $y = \dfrac{4}{3}x - 3$
Substitute in equation (1).
$12x - 9\left(\dfrac{4}{3}x - 3\right) = 18$
$12x - 12x + 27 = 18$
$27 \neq 18$
The system is inconsistent and has no solution.

29. (1) $9x - y = -3$
(2) $18x - y = 0$
Solve equation (2) for y.
$18x - y = 0$
$18x = y$
Substitute in equation (1).
$9x - y = -3$
$9x - 18x = -3$
$-9x = -3$
$x = \dfrac{1}{3}$
Replace x in equation (2).
$18x - y = 0$
$18\left(\dfrac{1}{3}\right) - y = 0$
$6 - y = 0$
$6 = y$
The solution is $\left(\dfrac{1}{3}, 6\right)$.

30.

The solution of equations is inconsistent and has no solution.

31. (1) $7x - 9y = 9$
(2) $3x - y = 1$
Eliminate y.
$7x - 9y = 9$
$-9(3x - y) = -9(1)$

$7x - 9y = 9$
$-27x + 9y = -9$
Add the equations.
$-20x = 0$
$x = 0$
Replace x in equation (2).
$3x - y = 1$
$3(0) - y = 1$
$0 - y = 1$
$-y = 1$
$y = -1$
The solution is $(0, -1)$.

32. (1) $7x + 3y = -16$
(2) $x - 2y = 5$
Solve equation (2) for x.
$x - 2y = 5$
$x = 2y + 5$
Substitute in equation (1).
$7x + 3y = -16$
$7(2y + 5) + 3y = -16$
$14y + 35 + 3y = -16$
$17y + 35 = -16$
$17y = -51$
$y = -3$
Replace y in equation (2).
$x - 2y = 5$
$x - 2(-3) = 5$
$x + 6 = 5$
$x = -1$
The solution is $(-1, -3)$.

33. (1) $5x - 3y = 6$
    (2) $x - y = 2$
Solve equation (2) for x.
$x - y = 2$
$x = y + 2$
Substitute in equation (1).
$5x - 3y = 6$
$5(y + 2) - 3y = 6$
$5y + 10 - 3y = 6$
$2y + 10 = 6$
$2y = -4$
$y = -2$
Replace y in equation (2).
$x - y = 2$
$x - (-2) = 2$
$x + 2 = 2$
$x = 0$
The solution is $(0, -2)$.

34. (1) $6x + y = 12$
    (2) $9x + 2y = 18$
Eliminate y.
$-2(6x + y) = -2(12)$
$9x + 2y = 18$

$-12x - 2y = -24$
$9x + 2y = 18$
Add the equations.
$-3x = -6$
$x = 2$
Replace x in equation (2).
$9x + 2y = 18$
$9(2) + 2y = 18$
$18 + 2y = 18$
$2y = 0$
$y = 0$
The solution is $(2, 0)$.

35. (1) $5x + 12y = 4$
    (2) $x + 6y = 8$
Eliminate y.
$5x + 12y = 4$
$-2(x + 6y) = -2(8)$

$5x + 12y = 4$
$-2x - 12y = -16$
Add the equations.
$3x = -12$
$x = -4$
Replace x in equation (2).
$x + 6y = 8$
$-4 + 6y = 8$
$6y = 12$
$y = 2$
The solution is $(-4, 2)$.

36. (1) $6x - y = 0$
    (2) $7x - y = 1$
Solve equation (1) for y.
$6x - y = 0$
$6x = y$
Substitute in equation (2).
$7x - y = 1$
$7x - 6y = 1$
$x = 1$
Replace x in equation (1).
$6x - y = 0$
$6(1) - y = 0$
$6 - y = 0$
$6 = y$
The solution is $(1, 6)$.

37. Strategy
Rate of the plane in calm air: r
Rate of the wind: r

	Rate	Time	Distance
With wind	$r + w$	4	$4(r + w)$
Against wind	$r - w$	5	$5(r - w)$

The distance traveled with and against the wind is 800 mi.

Solution
$4(r + w) = 800$ $\quad \frac{1}{4} \cdot 4(r + w) = \frac{1}{4} \cdot 800$
$5(r - w) = 800$ $\quad \frac{1}{5} \cdot 5(r - w) = \frac{1}{5} \cdot 800$

$r + w = 200$
$r - w = 160$

$2r = 360$
$r = 180$

$r + w = 200$
$180 + w = 200$
$w = 20$
The rate of the plane in calm air is 180 mph.
The rate of the wind is 20 mph.

38. Strategy

Number of adult tickets sold: a

Number of children's tickets sold: c

	Number	Price	Total
Adult tickets	a	7	$7a$
Children's tickets	c	5	$5c$

A total of 200 tickets were sold.
The total receipts were $120.

Solution

$a + c = 200$
$7a + 5c = 1120$

$-5(a+c) = -5(200)$
$7a + 5c = 1120$

$-5a - 5c = -1000$
$7a + 5c = 1120$

$2a = 120$
$a = 60$
$a + c = 200$
$60 + c = 200$
$c = 140$

There were 60 adult tickets and 140 children's tickets sold.

39. Strategy

Rate of the canoeist in still water: r

Rate of the current: c

	Rate	Time	Distance
With current	$r + c$	3	$3(r+c)$
Against current	$r - c$	5	$5(r-c)$

The distance with the current and against the current is 30 mi.

Solution

$3(r+c) = 30 \quad \frac{1}{3} \cdot 3(r+c) = 30 \cdot \frac{1}{3}$
$5(r-c) = 30 \quad \frac{1}{5} \cdot 5(r-c) = 30 \cdot \frac{1}{5}$

$r + c = 10$
$r - c = 6$
$2r = 16$
$r = 8$

$r + c = 10$
$8 + c = 10$
$c = 2$

The rate of the canoeist in still water is 8 mph.
The rate of the current is 2 mph.

40. Strategy

Number of advertisements requiring 33¢ postage: x

Number of advertisements requiring 55¢ postage: y

A total number of 190 advertisements were mailed.

33¢ · number requiring 33¢ postage
+ 55¢ · number requiring 55¢ postage
= total cost
Total cost was $75.90.

Solution

$x + y = 190 \qquad -33(x+y) = 190(-33)$
$33x + 55y = 7590 \qquad 33x + 55y = 7590$

$-33x - 33y = -6270$
$33x + 55y = 7590$

$22y = 1320$
$y = 60$

$x + y = 190$
$x + 60 = 190$
$x = 130$

The number of advertisements requiring 33¢ postage is 130.

41. Strategy

Rate of the boat in calm water: r

Rate of the current: c

	Rate	Time	Distance
With current	$r + c$	3	$3(r+c)$
Against current	$r - c$	2	$2(r-c)$

The distance traveled with the current is 48 km.
The distance traveled against the current is 24 km.

Solution

$3(r+c) = 48 \quad \frac{1}{3} \cdot 3(r+c) = \frac{1}{3} \cdot 48$
$2(r-c) = 24 \quad \frac{1}{2} \cdot 2(r-c) = \frac{1}{2} \cdot 24$

$r + c = 16$
$r - c = 12$

$2r = 28$
$r = 14$

$r + c = 16$
$14 + c = 16$
$c = 2$

The rate the boat in calm water is 14 km/h.
The rate of the current is 2 km/h.

42. Strategy
Number of $15 discs purchased: x
Number of $10 discs purchased: y
The total number of discs purchased is 10.
$15 · number of $15 discs purchased
+ $10 · number of $10 discs purchased
= total spent
Total spent was $120.

Solution
$$x + y = 10 \qquad -10(x + y) = -10(10)$$
$$15x + 10y = 120 \qquad 15x + 10y = 120$$

$$-10x - 10y = -100$$
$$15x + 10y = 120$$

$$5x = 20$$
$$x = 4$$

$$x + y = 10$$
$$4 + y = 10$$
$$y = 6$$

The customer purchased four $15 compact discs and six $10 compact discs.

43. Rate of the plane in calm air: p
Rate of the wind: w

	Rate	Time	Distance
With wind	$p + w$	3	$3(p + w)$
Against wind	$p - w$	4	$4(p - w)$

The distance traveled with the wind is 420 km.
The distance traveled against the wind is 440 km.

Solution
$$3(p + w) = 420 \qquad \frac{1}{3} \cdot 3(p + w) = 420 \cdot \frac{1}{3}$$
$$4(p - w) = 440 \qquad \frac{1}{4} \cdot 4(p - w) = 440 \cdot \frac{1}{4}$$

$$p + w = 140$$
$$p - w = 110$$

$$2p = 250$$
$$p = 125$$

$$p + w = 140$$
$$125 + w = 140$$
$$w = 15$$

The rate of the plane in calm air is 125 km/h.
The rate of the wind is 15 km/h.

44. Strategy
Rate of the boat in calm water: r
Rate of the current: c

	Rate	Time	Distance
With current	$r + c$	1	$r + c$
Against current	$r - c$	1	$r - c$

With the current, the boat traveled 4 mi.
Against the current, the boat traveled 2 mi.

Solution
$$r + c = 4$$
$$r - c = 2$$

$$2r = 6$$
$$r = 3$$

$$r + c = 4$$
$$3 + c = 4$$
$$c = 1$$

The rate of the boat in calm water is 3 mph.
The rate of the current is 1 mph.

45. Strategy
Number of bushels of lentils: x
Number of bushels of corn: y
The number of bushels of lentils plus 50 equals twice the number of bushels of corn.
The number of bushels of corn plus 150 equals the number of bushels of lentils.

Solution
$$x + 50 = 2y \qquad x - 2y = -50$$
$$y + 150 = x \qquad -x + y = -150$$

$$-y = -200$$
$$y = 200$$

$$y + 150 = x$$
$$200 + 150 = x$$
$$350 = x$$

There are 350 bushels of lentils and 200 bushels of corn in the silo.

46. Strategy

Rate of the plane in calm air: p

Rate of the wind: w

	Rate	Time	Distance
With wind	$p+w$	3	$3(p+w)$
Against wind	$p-w$	4	$4(p-w)$

The distance traveled with the wind and against the wind is 360 mi.

Solution

$$3(p+w)=360 \qquad \frac{1}{3}\cdot 3(p+w)=360\cdot\frac{1}{3}$$
$$4(p-w)=360 \qquad \frac{1}{4}\cdot 4(p-w)=360\cdot\frac{1}{4}$$

$$p+w=120$$
$$p-w=90$$

$$2p=210$$
$$p=105$$

$$p+w=120$$
$$105+w=120$$
$$w=15$$

The rate of the plane in calm air is 105 mph.
The rate of the wind is 15 mph.

47. Strategy

Number of nickels: n

Number of dimes: d

	Number	Value	Total Value
Nickels	n	5	$5n$
Dimes	d	10	$10d$

There are 12 coins in all.
The total value of the nickels and dimes is 80¢.

Solution

$$n+d=12 \qquad -5(n+d)=-5(12)$$
$$5n+10d=80 \qquad 5n+10d=80$$

$$-5-5d=-60$$
$$5n+10d=80$$

$$5d=20$$
$$d=4$$

There are 4 dimes in the coin purse.

48. Strategy

Rate of the plane in calm air: r

Rate of the wind: w

	Rate	Time	Distance
With wind	$r+w$	6	$6(r+w)$
Against wind	$r+w$	7	$7(r-w)$

The distance traveled with and against the wind is 2100 mi.

Solution

$$6(r+w)=2100 \qquad \frac{1}{6}\cdot 6(r+w)=\frac{1}{6}\cdot 2100$$
$$7(r-w)=2100 \qquad \frac{1}{7}\cdot 7(r-w)=\frac{1}{7}\cdot 2100$$

$$r+w=350$$
$$r-w=300$$

$$2r=650$$
$$r=325$$

$$r+w=350$$
$$325+w=350$$
$$w=25$$

The rate of the plane in calm air is 325 mph.
The rate of the wind is 25 mph.

49. Strategy

Number of shares at $6 per share: x

Number of shares at $25 per share: y

The total number of shares is 1500.

$6 · number of $6 shares
+ $25 · number of $25 shares
= total cost

Total cost is $12,800.

Solution

$$x+y=1500 \qquad -6(x+y)=1500(-6)$$
$$6x+25y=12,800 \qquad 6x+25y=12,800$$

$$-6x-6y=-9000$$
$$6x+25y=12,800$$

$$19y=3800$$
$$y=200$$

$$x+y=1500$$
$$x+200=1500$$
$$x=1300$$

The number of $6 shares purchased is 1300.
The number of $25 shares purchased is 200.

50. Strategy

Rate of the sculling team in calm water: r

Rate of the current: c

	Rate	Time	Distance
With current	$r + c$	2	$2(r + c)$
Against current	$r - c$	3	$3(r - c)$

The distance with the current is 24 mi.
The distance against the current is 18 mi.

Solution

$2(r + c) = 24 \qquad \frac{1}{2} \cdot 2(r + c) = 24 \cdot \frac{1}{2}$

$3(r - c) = 18 \qquad \frac{1}{3} \cdot 3(r - c) = 18 \cdot \frac{1}{3}$

$$r + c = 12$$
$$r - c = 6$$

$$2r = 18$$
$$r = 9$$

$r + c = 12$
$9 + c = 12$
$c = 3$

The rate of the sculling team in calm water is 9 mph.

The rate of the current is 3 mph.

Chapter Test

1. (1) $\quad 4x - y = 11$
  (2) $\qquad y = 2x - 5$

Substitute in equation (1)
$4x - (2x - 5) = 11$
$\quad 4x - 2x + 5 = 11$
$\qquad 2x + 5 = 11$
$\qquad\qquad 2x = 6$
$\qquad\qquad\; x = 3$

Substitute x in equation (2).
$y = 2x - 5$
$y = 2(3) - 5$
$y = 6 - 5$
$y = 1$

The solution is (3, 1).

2. (1) $\quad 4x + 3y = 11$
  (2) $\quad 5x - 3y = 7$

Add the equations.
$9x = 18$
$\; x = 2$

Replace x in equation (1).
$\quad 4x + 3y = 11$
$4(2) + 3y = 11$
$\quad 8 + 3y = 11$
$\qquad\; 3y = 3$
$\qquad\;\; y = 1$

The solution is (2, 1).

3.

$\begin{array}{c|c} 2x + 5y = 11 \\ \hline 2(-2) + 5(3) & 11 \\ -4 + 15 & 11 \\ 11 = 11 \end{array}$ $\qquad \begin{array}{c|c} x + 3y = 7 \\ \hline (-2) + 3(3) & 7 \\ -2 + 9 & 7 \\ 7 = 7 \end{array}$

Yes, $(-2, 3)$ is a solution.

4. (1) $\qquad x = 2y + 3$
  (2) $\quad 3x - 2y = 5$

Substitute in equation (2).
$\qquad 3x - 2y = 5$
$3(2y + 3) - 2y = 5$
$\quad 6y + 9 - 2y = 5$
$\qquad\quad 4y + 9 = 5$
$\qquad\qquad 4y = -4$
$\qquad\qquad\; y = -1$

Replace y in equation (1).
$x = 2y + 3$
$x = 2(-1) + 3$
$x = -2 + 3$
$x = 1$

The solution is (1, −1).

5. (1) $\quad 2x - 5y = 6$
  (2) $\quad 4x + 3y = -1$

Eliminate x
$-2(2x - 5y) = 6(-2)$
$\qquad 4x + 3y = -1$

$-4x + 10y = -12$
$\quad\; 4x + 3y = -1$

Add the equations.
$13y = -13$
$\;\; y = -1$

Replace y in equation (1).
$\quad 2x - 5y = 6$
$2x - 5(-1) = 6$
$\quad 2x + 5 = 6$
$\qquad 2x = 1$
$\qquad\; x = \dfrac{1}{2}$

The solution is $\left(\dfrac{1}{2}, -1\right)$.

6

The solution is (−2, 6).

7. (1) $\quad 4x + 2y = 3$
  (2) $\qquad y = -2x + 1$

Substitute in equation (1)
$\qquad 4x + 2y = 3$
$4x + 2(-2x + 1) = 3$
$\quad 4x - 4x + 2 = 3$
$\qquad\qquad\quad 2 \neq 3$

The system of equations is inconsistent and has no solution.

8. (1) $3x + 5y = 1$

(2) $2x - y = 5$

Solve equation (2) for y.

$2x - y = 5$

$-y = -2x + 5$

$y = 2x - 5$

Substitute in equation (1).

$3x + 5y = 1$

$3x + 5(2x - 5) = 1$

$3x + 10x - 25 = 1$

$13x - 25 = 1$

$13x = 26$

$x = 2$

Substitute x in equation (2).

$2x - y = 5$

$2(2) - y = 5$

$4 - y = 5$

$-y = 1$

$y = -1$

The solution is $(2, -1)$.

9. (1) $7x + 3y = 11$

(2) $2x - 5y = 9$

Eliminate y.

$5(7x + 3y) = 11(5)$

$3(2x - 5y) = 9(3)$

Add the equations.

$35x + 15y = 55$

$6x - 15y = 27$

$41x = 82$

$x = 2$

Replace x in equation (1).

$7x + 3y = 11$

$7(2) + 3y = 11$

$14 + 3y = 11$

$3y = -3$

$y = -1$

The solution is $(2, -1)$.

10. (1) $3x - 5y = 13$

(2) $x + 3y = 1$

Solve equation (2) for x.

$x + 3y = 1$

$x = -3y + 1$

Substitute in equation (1).

$3x - 5y = 13$

$3(-3y + 1) - 5y = 13$

$-9y + 3 - 5y = 13$

$-14y + 3 = 13$

$-14y = 10$

$y = -\dfrac{5}{7}$

Substitute y in equation (2).

$x + 3y = 1$

$x + 3\left(-\dfrac{5}{7}\right) = 1$

$x - \dfrac{15}{7} = 1$

$x = 1 + \dfrac{15}{7}$

$x = \dfrac{7}{7} + \dfrac{15}{7}$

$x = \dfrac{22}{7}$

The solution is $\left(\dfrac{22}{7}, -\dfrac{5}{7}\right)$.

11. (1) $5x + 6y = -7$

(2) $3x + 4y = -5$

Eliminate x.

$3(5x + 6y) = -7(3)$

$-5(3x + 4y) = -5(-5)$

Add the equations.

$15x + 18y = -21$

$-15x - 20y = 25$

$-2y = 4$

$y = -2$

Replace y in equation (1).

$5x + 6y = -7$

$5x + 6(-2) = -7$

$5x - 12 = -7$

$5x = 5$

$x = 1$

The solution is $(1, -2)$.

12.

$\begin{array}{c|c} 3x - 2y = 8 \\ \hline 3(2) - 2(1) & 8 \\ 6 - 2 & 8 \\ 4 \neq 8 \end{array}$ $\begin{array}{c|c} 4x + 5y = 3 \\ \hline 4(2) + 5(1) & 3 \\ 8 + 5 & 3 \\ 13 \neq 3 \end{array}$

No, $(2, 1)$ is not a solution to the system of equations.

13. (1) $3x - y = 5$
(2) $y = 2x - 3$

Substitute in equation (1).
$3x - y = 5$
$3x - (2x - 3) = 5$
$3x - 2x + 3 = 5$
$x + 3 = 5$
$x = 2$
Substitute x in equation (2).
$y = 2x - 3$
$y = 2(2) - 3$
$y = 4 - 3$
$y = 1$
The solution is (2, 1).

14. (1) $3x + 2y = 2$
(2) $5x - 2y = 14$

Add the equations.
$8x = 16$
$x = 2$
Substitute x in equation (1).
$3x + 2y = 2$
$3(2) + 2y = 2$
$6 + 2y = 2$
$2y = -4$
$y = -2$
The solution is (2, −2).

15.

The solution is (2, 0).

16. (1) $x = 3y + 1$
(2) $2x + 5y = 13$

Substitute in equation (2).
$2x + 5y = 13$
$2(3y + 1) + 5y = 13$
$(6y + 2) + 5y = 13$
$11y + 2 = 13$
$11y = 11$
$y = 1$
Substitute in equation (1).
$x = 3y + 1$
$x = 3(1) + 1$
$x = 3 + 1$
$x = 4$
The solution is (4, 1).

17. (1) $5x + 4y = 7$
(2) $3x - 2y = 13$

Eliminate y.
$5x + 4y = 7$
$2(3x - 2y) = 2(13)$

$5x + 4y = 7$
$6x - 4y = 26$
Add the equations.
$11x = 33$
$x = 3$
Replace x in equation (2).
$3x - 2y = 13$
$3(3) - 2y = 13$
$9 - 2y = 13$
$-2y = 4$
$y = -2$
The solution is (3, −2).

18.

The system of equations is dependent. The solutions are the ordered pairs that satisfy the equation $y = -\frac{1}{2} + \frac{1}{3}$.

19. (1) $4x - 3y = 1$
(2) $2x + y = 3$

Solve equation (2) for y.
$2x + y = 3$
$y = 3 - 2x$
Substitute in equation (1).
$4x - 3y = 1$
$4x - 3(3 - 2x) = 1$
$4x - 9 + 6x = 1$
$10x - 9 = 1$
$10x = 10$
$x = 1$
Substitute x in equation (2).
$2x + y = 3$
$2(1) + y = 3$
$2 + y = 3$
$y = 1$
The solution is (1, 1).

20. (1) $5x - 3y = 29$
   (2) $4x + 7y = -5$

Eliminate y.
$7(5x - 3y) = 7(29)$
$3(4x + 7y) = 3(-5)$

$35x - 21y = 203$
$12x + 21y = -15$
Add the equations.
$47x = 188$
$x = 4$
Replace x in equation (2).
$4x + 7y = -5$
$4(4) + 7y = -5$
$16 + 7y = -5$
$7y = -21$
$y = -3$
The solution is $(4, -3)$.

21. (1) $3x - 5y = -23$
   (2) $x + 2y = -4$
Solve equation (2) for x.
$x + 2y = -4$
$x = -2y - 4$
Substitute in equation (1).
$3x - 5y = -23$
$3(-2y - 4) - 5y = -23$
$-6y - 12 - 5y = -23$
$-12 - 11y = -23$
$-11y = -11$
$y = 1$
Substitute y in equation (2).
$x = -2y - 4$
$x = -2(1) - 4$
$x = -2 - 4$
$x = -6$
The solution is $(-6, 1)$.

22. (1) $9x - 2y = 17$
   (2) $5x + 3y = -7$

Eliminate y.
$3(9x - 2y) = 3(17)$
$2(5x + 3y) = 2(-7)$

$27x - 6y = 51$
$10x + 6y = -14$
Add the equations.
$37x = 37$
$x = 1$
Replace x in equation (2).
$5x + 3y = -7$
$5(1) + 3y = -7$
$5 + 3y = -7$
$3y = -12$
$y = -4$
The solution is $(1, -4)$.

23. Strategy
Rate of the plane in calm air: p
Rate of the wind: w

	Rate	Time	Distance
With wind	$p + w$	2	$2(p + w)$
Against wind	$p - w$	3	$3(p - w)$

The distance traveled with the wind is 240 mi.
The distance traveled against the wind is 240 mi.

Solution
$2(p + w) = 240 \qquad \dfrac{1}{2} \cdot 2(p + w) = 240 \cdot \dfrac{1}{2}$
$3(p - w) = 240 \qquad \dfrac{1}{3} \cdot 3(p - w) = 240 \cdot \dfrac{1}{3}$

$p + w = 120$
$p - w = 80$

$2p = 200$
$p = 100$

$p + w = 120$
$100 + w = 120$
$w = 20$
The rate of the plane in calm air is 100 mph.
The rate of the wind is 20 mph.

24. Strategy
Rate of the boat in calm water: r
Rate of the current: c

	Rate	Time	Distance
With current	$r + c$	3	$3(r + c)$
Against current	$r - c$	4	$4(r - c)$

The distance traveled with the current is 48 mi.
The distance traveled against the current is 48 mi.

Solution
$3(r + c) = 48 \qquad \dfrac{1}{3} \cdot 3(r + c) = \dfrac{1}{3} \cdot 48$
$4(r - c) = 48 \qquad \dfrac{1}{4} \cdot 4(r - c) = \dfrac{1}{4} \cdot 48$

$r + c = 16$
$r - c = 12$

$2r = 28$
$r = 14$

$r + c = 16$
$14 + c = 16$
$c = 2$
The rate of the boat in calm water is 14 mph.
The rate of the current is 2 mph.

25. The number of nickels in the first bank: x
The number of dimes in the first bank: y
First Bank

	Number	Value	Total Value
Nickels	x	5	$5x$
Dimes	y	10	$10y$

Second Bank

	Number	Value	Total Value
Nickels	$x-10$	5	$5(x-10)$
Dimes	$\frac{1}{2}y$	10	$5y$

The total value of the coins in the first bank is $5.50.
The total value of the coins in the second bank is $3.00.

Solution
$$5x+10y=550 \quad 5x+10y=550$$
$$5(x-10)+5y=300 \quad 5x+5y=350$$

$$5x+10y=550$$
$$-5x-5y=-350$$

$$5y=200$$
$$y=40$$

$$5x+5y=350$$
$$5x+5(40)=350$$
$$5x+200=350$$
$$5x=150$$
$$x=30$$

There are 30 nickels and 40 dimes in the first bank.

Cumulative Review Exercises

1. $-8 \le -4$ true
$-4 \le -4$ true
$0 \le -4$ false
-8 and -4 are less than or equal to -4.

2. $\{1, 2, 3, 4, 5, 6, 7, 8, 9, 10\}$

3. $12-2(7-5)^2 \div 4 = 12-2(2)^2 \div 4$
$= 12-2(4) \div 4$
$= 12-(-8) \div 4$
$= 12-2$
$= 10$

4. $2[5a-3(2-5a)-8] = 2[5a-6+15a-8]$
$= 2[20a-14]$
$= 40a-28$

5. $\dfrac{a^2-b^2}{2a} = \dfrac{4^2-(-2)^2}{2(4)}$
$= \dfrac{16-4}{8}$
$= \dfrac{12}{8}$
$= \dfrac{3}{2}$

6. $-\dfrac{3}{4}x = \dfrac{9}{8}$
$-\dfrac{4}{3}\left(-\dfrac{3}{4}x\right) = -\dfrac{4}{3}\left(\dfrac{9}{8}\right)$
$x = -\dfrac{3}{2}$
The solution is $-\dfrac{3}{2}$.

7. $4-3(2-3x) = 7x-9$
$4-6+9x = 7x-9$
$-2+9x = 7x-9$
$-2+9x-7x = 7x-7x-9$
$-2+2x = -9$
$-2+2+2x = -9+2$
$2x = -7$
$\dfrac{2x}{2} = \dfrac{-7}{2}$
$x = -\dfrac{7}{2}$
The solution is $-\dfrac{7}{2}$.

8. $3[2-4(x+1))] = 6x-2$
$3[2-4x-4] = 6x-2$
$3[-2-4x] = 6x-2$
$-6-12x = 6x-2$
$-6-12x-6x = 6x-6x-2$
$-6-18x = -2$
$-6+6-18x = -2+6$
$-18x = 4$
$\dfrac{-18x}{-18} = \dfrac{4}{-18}$
$x = -\dfrac{2}{9}$
The solution is $-\dfrac{2}{9}$.

9. $-7x-5 > 4x+50$
$-7x-4x-5 > 4x-4x+50$
$-11x-5 > 50$
$-11x-5+5 > 50+5$
$-11x > 55$
$\dfrac{-11x}{-11} < \dfrac{55}{-11}$
$x < -5$

10. $5+2(x+1)\le13$
$5+2x+2\le13$
$2x+7\le13$
$2x+7-7\le13-7$
$2x\le6$
$\dfrac{2x}{2}\le\dfrac{6}{2}$
$x\le3$

11. $PB=A$
$P(50)=12$
$\dfrac{50P}{50}=\dfrac{12}{50}$
$P=0.24$
12 is 24% of 50.

12.

x-intercept:	y-intercept
$3x-6y=12$	$3x-6y=12$
$3x-6(0)=12$	$3(0)-6y=12$
$3x=12$	$-6y=12$
$x=4$	$y=-2$
$(4, 0)$	$(0, -2)$

13. $(x_1, y_1)=(2, -3),\ (x_2, y_2)=(-3, 4)$
$m=\dfrac{y_2-y_1}{x_2-x_1}=\dfrac{4-(-3)}{-3-2}=\dfrac{7}{-5}=-\dfrac{7}{5}$
The slope is $-\dfrac{7}{5}$.

14. $m=-\dfrac{3}{2},\ (x_1, y_1)=(-2, 3)$
$y-y_1=m(x-x_1)$
$y-3=-\dfrac{3}{2}[x-(-2)]$
$y-3=-\dfrac{3}{2}(x+2)$
$y-3=-\dfrac{3}{2}x-3$
$y=-\dfrac{3}{2}x$

15.

16.

17.

18.

19. The domain is {–5, 0, 1, 5}
The range is {5}.

No two ordered pairs have the same first component. The relation is a function.

20. $f(x)=-2x-5$
$f(-4)=-2(-4)-5$
$f(-4)=8-5$
$f(-4)=3$
The value of $f(-4)$ is 3.

21.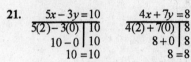

$5x-3y=10$	$4x+7y=8$
$5(2)-3(0)\ \vert\ 10$	$4(2)+7(0)\ \vert\ 8$
$10-0\ \vert\ 10$	$8+0\ \vert\ 8$
$10=10$	$8=8$

Yes, (2, 0) is a solution of the system.

22. (1) $2x-3y=-7$
(2) $x+4y=2$

Solve equation (2) for x.
$x+4y=2$
$x=-4y+2$
Replace x in equation (1).
$2x-3y=-7$
$2(2-4y)-3y=-7$
$4-8y-3y=-7$
$4-11y=-7$
$-11y=-11$
$y=1$
Substitute y in equation (2).
$x+4y=2$
$x+4(1)=2$
$x+4=2$
$x=-2$
The solution is (–2, 1).

23.

24. (1) $5x - 2y = 8$
(2) $4x + 3y = 11$

Eliminate y.
$3(5x - 2y) = 3(8)$
$2(4x + 3y) = 2(11)$

$15x - 6y = 24$
$8x + 6y = 22$
Add the equations.
$23x = 46$
$x = 2$
Substitute x in equation (2).
$4x + 3y = 11$
$4(2) + 3y = 11$
$8 + 3y = 11$
$3y = 3$
$y = 1$
The solution is (2, 1).

25. $f(x) = 5x - 2$
$f(-4) = 5(-4) - 2 = -20 - 2 = -22$
$f(-2) = 5(-2) - 2 = -10 - 2 = -12$
$f(0) = 5(0) - 2 = 0 - 2 = -2$
$f(2) = 5(2) - 2 = 10 - 2 = 8$
$f(4) = 5(4) - 2 = 20 - 2 = 18$
The range is $\{-22, -12, -2, 8, 18\}$.

26. Strategy
To find the number of cameras produced, replace each of the variables by its given value and solve for N.

Solution
$$T = U \cdot N + F$$
$$21,500 = 90N + 3500$$
$$21,500 - 3500 = 90N + 3500 - 3500$$
$$18,000 = 90N$$
$$\frac{18,000}{90} = \frac{90N}{90}$$
$$200 = N$$
200 cameras were produced.

27. Strategy
Amount invested at 9.6%: x
Amount invested at 7.2%: $8750 - x$

	Principal	Rate	Interest
Amount at 9.6%	x	0.096	$0.096(x)$
Amount at 7.2%	$8750 - x$	0.072	$0.072(8750 - x)$

Both investments earn the same annual interest.

Solution
$0.096x = 0.072(8750 - x)$
$0.096x = 630 - 0.072x$
$0.168x = 630$
$x = 3750$
$8750 - x = 8750 - 3750 = 5000$

The amount invested at 9.6% is $3750.
The amount invested at 7.2% is $5000.

28. Strategy
Rate of the plane: p
Rate of the wind: w

	Rate	Time	Distance
With wind	$p+w$	3	$3(p+w)$
Against wind	$p-w$	3	$3(p-w)$

The distance traveling with the wind is 570 miles.
The distance traveling against the wind is 390 miles.

Solution
$$3(p+w)=570$$
$$3(p-w)=390$$

$$\frac{1}{3}\cdot 3(p+w)=570\cdot\frac{1}{3}$$
$$\frac{1}{3}\cdot 3(p-w)=390\cdot\frac{1}{3}$$

$$p+w=190$$
$$p-w=130$$

$$2p=320$$
$$p=160$$

$$p+w=190$$
$$160+w=190$$
$$w=30$$

The rate of the plane is 160 mph. The rate of the wind is 30 mph.

29. Rate of the boat in calm water: r
Rate of the current: c

	Rate	Time	Distance
With current	$r+c$	2	$2(r+c)$
Against current	$r-c$	3	$3(r-c)$

The distance traveled with the current is 24 mi.
The distance traveled against the current is 24 mi.

Solution
$$2(r+c)=24 \qquad \frac{1}{2}\cdot 2(r+c)=\frac{1}{2}\cdot 24$$
$$3(r-c)=24 \qquad \frac{1}{3}\cdot 3(r-c)=\frac{1}{3}(24)$$

$$r+c=12$$
$$r-c=8$$

$$2r=20$$
$$r=10$$

The rate of the boat in calm water is 10 mph.

30. Strategy
Number of nickels in the first bank: n
Number of dimes in the first bank: d
First Bank

	Number	Value	Total Value
Nickels	n	5	$5n$
Dimes	d	10	$10d$

Second Bank

	Number	Value	Total Value
Nickels	$n-15$	5	$5(n-15)$
Dimes	$\frac{1}{2}d$	10	$5d$

The total value of the coins in the first bank is $5.25.
The total value of the coins in the second bank is $2.50.

Solution
$$5n+10d=525$$
$$5(n-15)+5d=250$$

$$5n+10d=525$$
$$5n-75+5d=250$$

$$5n+10d=525 \qquad 5n+10d=525$$
$$5n+5d=325 \qquad -1(5n+5d)=-1(325)$$

$$5n+10d=525$$
$$-5n-5d=-325$$

$$5d=200$$
$$d=40$$

There are 40 dimes in the first bank.

Chapter 7: Polynomials

Prep Test

1. $-2 - (-3) = -2 + 3 = 1$ [1.2.2.]

2. -18 [1.2.3]

3. $\dfrac{2}{3}$ [1.2.4]

4. $3n^4$ when $n = -2$ [2.1.1]
   $3(-2)^4 = 3(-2)(-2)(-2)(-2) = 48$

5. 0 [1.2.4]

6. No   [2.2.2]

7. $3x^2 - 4x + 1 + 2x^2 - 5x - 7$ [2.2.2]
   $= (3x^2 + 2x^2) + (-4x - 5x) + 1 - 7$
   $= 5x^2 - 9x - 6$

8. 0 [2.2.2]

9. $-6x + 14$ [2.2.4]

10. $3xy - 4y - 2(5xy - 7y)$ [2.2.5]
    $= 3xy - 4y - 10xy + 14y$
    $= (3xy - 10xy) + (-4y + 14y)$
    $= -7xy + 10y$

Go Figure

If I have two more sisters than brothers and each of my sisters have two more sisters than brothers, then I must be a sister. If I have one brother, then I must have three sisters. So then my brother has zero brothers and four sisters. Similarly, if I have two brothers, then I must have four sisters. Then my youngest brother has one brother and five sisters. He has four more sisters than brothers.

Section 7.1

Concept Review 7.1

1. Always true

3. Never true
   The degree of the polynomial $7x^2 - 9x + 1$ is 2.

5. Always true

Objective 7.1.1 Exercises

1. binomial

3. monomial

5. No

7. Yes

9. Yes

---

11. $x^2 + 7x$
    $\underline{-3x^2 - 4x}$
    $-2x^2 + 3x$

13. $y^2 + 4y$
    $\underline{\quad -4y - 8}$
    $y^2 \quad -8$

15. $2x^2 + 6x + 12$
    $\underline{3x^2 + \ x + \ 8}$
    $5x^2 + 7x + 20$

17. $x^3 \qquad -7x + \ 4$
    $\underline{\quad\ 2x^2 + \ x - 10}$
    $x^3 + 2x^2 - 6x - \ 6$

19. $2a^3 \qquad -7a + 1$
    $\underline{\quad -3a^2 - \ 4a + 1}$
    $2a^3 - 3a^2 - 11a + 2$

21. $(4x^2 + 2x) + (x^2 + 6x)$
    $= (4x^2 + x^2) + (2x + 6x)$
    $= 5x^2 + 8x$

23. $(4x^2 - 5xy) + (3x^2 + 6xy - 4y^2)$
    $= (4x^2 + 3x^2) + (-5xy + 6xy) - 4y^2$
    $= 7x^2 + xy - 4y^2$

25. $(2a^2 - 7a + 10) + (a^2 + 4a + 7)$
    $= (2a^2 + a^2) + (-7a + 4a) + (10 + 7)$
    $= 3a^2 - 3a + 17$

27. $(5x^3 + 7x - 7) + (10x^2 - 8x + 3)$
    $= 5x^3 + 10x^2 + (7x - 8x) + (-7 + 3)$
    $= 5x^3 + 10x^2 - x - 4$

29. $(2r^2 - 5r + 7) + (3r^3 - 6r)$
    $= 3r^3 + 2r^2 + (-5r - 6r) + 7$
    $= 3r^3 + 2r^2 - 11r + 7$

31. $(3x^2 + 7x + 10) + (-2x^3 + 3x + 1)$
    $= -2x^3 + 3x^2 + (7x + 3x) + (10 + 1)$
    $= -2x^3 + 3x^2 + 10x + 11$

Objective 7.1.2 Exercises

33. $x^2 - 6x$ $x^2 - 6x$
    $\underline{-(x^2 - 10x)}$ $\underline{-x^2 + 10x}$
                     $4x$

35. $2y^2 - 4y$ $2y^2 - 4y$
    $\underline{-(-y^2 \quad + 2)}$ $\underline{\ y^2 \quad -2}$
                     $3y^2 - 4y - 2$

37.
$$\begin{array}{ll} x^2-2x+1 & x^2-2x+1 \\ -(x^2+5x+8) & \underline{-x^2-5x-8} \\ & -7x-7 \end{array}$$

39.
$$\begin{array}{ll} 4x^3+5x+2 & 4x^3 \qquad +5x+2 \\ -(-3x^2+2x+1) & \underline{3x^2-2x-1} \\ & 4x^3+3x^2+3x+1 \end{array}$$

41.
$$\begin{array}{ll} 2y^3 \quad +6y-2 & 2y^3 \qquad +6y-2 \\ -(y^3+y^2 \quad +4) & \underline{-y^3-y^2 \qquad -4} \\ & y^3-y^2+6y-6 \end{array}$$

43. $(y^2-10xy)-(2y^2+3xy)$
$\quad =(y^2-10xy)+(-2y^2-3xy)$
$\quad =-y^2-13xy$

45. $(3x^2+x-3)-(x^2+4x-2)$
$\quad =(3x^2+x-3)+(-x^2-4x+2)$
$\quad =2x^2-3x-1$

47. $(-2x^3+x-1)-(-x^2+x-3)$
$\quad =(-2x^3+x-1)+(x^2-x+3)$
$\quad =-2x^3+x^2+2$

49. $(4a^3-2a+1)-(a^3-2a+3)$
$\quad =(4a^3-2a+1)+(-a^3+2a-3)$
$\quad =3a^3-2$

51. $(4y^3-y-1)-(2y^2-3y+3)$
$\quad =(4y^3-y-1)+(-2y^2+3y-3)$
$\quad =4y^3-2y^2+2y-4$

Applying Concepts 7.1

53. $\left(\dfrac{2}{3}a^2+\dfrac{1}{2}a-\dfrac{3}{4}\right)-\left(\dfrac{5}{3}a^2+\dfrac{1}{2}a+\dfrac{1}{4}\right)$
$\quad =\left(\dfrac{2}{3}a^2+\dfrac{1}{2}a-\dfrac{3}{4}\right)+\left(-\dfrac{5}{3}a^2-\dfrac{1}{2}a-\dfrac{1}{4}\right)$
$\quad =-a^2-1$

55. $(-x^2+2x+1)-(3x^2-4x-2)$
$\quad =(-x^2+2x+1)+(-3x^2+4x+2)$
$\quad =-4x^2+6x+3$

57. $(6x^2-4x-2)-(2x^2+2x-5)$
$\quad =(6x^2-4x-2)+(-2x^2-2x+5)$
$\quad =4x^2-6x+3$

59. Yes, it is possible to subtract two polynomials, each of degree 3, and have the difference be a polynomial of degree 2. One example to show this is $(3x^3-2x^2+3x-4)-(3x^3+4x^2-6x+5)$
$\quad =-6x^2+9x-9$

Section 7.2

Concept Review 7.2

1. Never true
   $6x$ is the base and 4 is the exponent.

3. Always true

5. Never true
   $(4x^3y^5)^2=4^2(x^3)^2(y^5)^2=16x^6y^{10}$

Objective 7.2.1 Exercises

3. **a.** product
   b. power

5. **a.** power
   b. product

7. $(x)(2x)=2(x\cdot x)=2x^2$

9. $(3x)(4x)=(3\cdot4)(x\cdot x)=12x^2$

11. $(-2a^3)(-3a^4)=[(-2)(-3)](a^3\cdot a^4)=6a^7$

13. $(x^2y)(xy^4)=(x^2\cdot x)(y\cdot y^4)=x^3y^5$

15. $(-2x^4)(5x^5y)=[(-2)(5)](x^4\cdot x^5)(y)$
    $\qquad =-10x^9y$

17. $(x^2y^4)(x^5y^4)=(x^2\cdot x^5)(y^4\cdot y^4)=x^7y^8$

19. $(2xy)(-3x^2y^4)=[2(-3)](x\cdot x^2)(y\cdot y^4)$
    $\qquad =-6x^3y^5$

21. $(x^2yz)(x^2y^4)=(x^2\cdot x^2)(y\cdot y^4)(z)$
    $\qquad =x^4y^5z$

23. $(a^2b^3)(ab^2c^4)=(a^2\cdot a)(b^3\cdot b^2)(c^4)$
    $\qquad =a^3b^5c^4$

25. $(-a^2b^2)(a^3b^6)=-(a^2\cdot a^3)(b^2\cdot b^6)=-a^5b^8$

27. $(-6a^3)(a^2b)=(-6)(a^3\cdot a^2)(b)=-6a^5b$

29. $(-5y^4z)(-8y^6z^5)=[(-5)(-8)](y^4\cdot y^6)(z\cdot z^5)$
    $\qquad =40y^{10}z^6$

31. $(10ab^2)(-2ab)=[10(-2)](a\cdot a)(b^2\cdot b)=-20a^2b^3$

33. $(xy^2z)(x^2y)(z^2y^2)=(x\cdot x^2)(y^2\cdot y\cdot y^2)(z\cdot z^2)$
    $\qquad =x^3y^5z^3$

35. $(4a^2b)(-3a^3h^4)(a^5b^2)$
    $\quad =[4(-3)](a^2\cdot a^3\cdot a^5)(b\cdot b^4\cdot b^2)$
    $\quad =-12a^{10}b^7$

Objective 7.2.2 Exercises

37. $(2^2)^3=2^6=64$

39. $(-2)^2 = 4$

41. $(-2^2)^3 = (-4)^3 = -64$

43. $(x^3)^3 = x^9$

45. $(x^7)^2 = x^{14}$

47. $(-x^2)^2 = x^4$

49. $(2x)^2 = 2^2 \cdot x^2 = 4x^2$

51. $(-2x^2)^3 = (-2)^3 \cdot x^6 = -8x^6$

53. $(x^2 y^3)^2 = x^4 y^6$

55. $(3x^2 y)^2 = 3^2 x^4 y^2 = 9x^4 y^2$

57. $(a^2)(3a^2)^3 = (a^2)(3)^3 a^6 = 27(a^2 \cdot a^6) = 27a^8$

59. $(-2x)(2x^3)^2 = (-2x)(2)^2 x^6$
$$= [(-2)(4)](x \cdot x^6)$$
$$= -8x^7$$

61. $(x^2 y)(x^2 y)^3 = (x^2 y)x^6 y^3$
$$= (x^2 \cdot x^6)(y \cdot y^3)$$
$$= x^8 y^4$$

63. $(ab^2)^2 (ab)^2 = (a^2 b^4)(a^2 b^2)$
$$= (a^2 \cdot a^2)(b^4 \cdot b^2)$$
$$= a^4 b^6$$

65. $(-2x)(-2x^3 y)^3 = (-2x)(-2)^3 x^9 y^3$
$$= [(-2)(-8)](x \cdot x^9)y^3$$
$$= 16x^{10} y^3$$

67. $(-2x)(-3xy^2)^2 = (-2x)(-3)^2 x^2 y^4$
$$= [(-2)(9)](x \cdot x^2)y^4$$
$$= -18x^3 y^4$$

69. $(ab^2)(-2a^2 b)^3 = (ab^2)(-2)^3 a^6 b^3$
$$= -8(a \cdot a^6)(b^2 \cdot b^3)$$
$$= -8a^7 b^5$$

71. $(-2a^3)(3a^2 b)^3 = (-2a^3)(3)^3 a^6 b^3$
$$= [(-2)(27)](a^3 \cdot a^6)b^3$$
$$= -54a^9 b^3$$

73. $(-3ab)^2 (-2ab)^3 = (-3)^2 a^2 b^2 (-2)^3 a^3 b^3$
$$= [9(-8)](a^2 \cdot a^3)(b^2 \cdot b^3)$$
$$= -72a^5 b^5$$

Applying Concepts 7.2

75. $(6x)(2x^2) + (4x^2)(5x) = 12x^3 + 20x^3$
$$= 32x^3$$

77. $(3a^2 b^2)(2ab) - (9ab^2)(a^2 b) = (6a^3 b^3) - (9a^3 b^3)$
$$= -3a^3 b^3$$

79. $(5xy^3)(3x^4 y^2) - (2x^3 y)(x^2 y^4)$
$$= (15x^5 y^5) - (2x^5 y^5)$$
$$= 13x^5 y^5$$

81. $4a^2 (2ab)^3 - 5b^2 (a^5 b) = (4a^2)(8a^3 b^3) - (5a^5 b^3)$
$$= (32a^5 b^3) - (5a^5 b^3)$$
$$= 27a^5 b^3$$

83. $-2xy(x^2 y)^3 - 3x^5 (xy^2)^2$
$$= -2xy(x^6 y^3) - 3x^5 (x^2 y^4)$$
$$= (-2x^7 y^4) - (3x^7 y^4)$$
$$= -5x^7 y^4$$

85. $a^n \cdot a^n = a^{n+n}$
$$= a^{2n}$$

87. $(a^2)^n = a^{2 \cdot n}$
$$= a^{2n}$$

89. True.

91. False.
$$(x^2)^5 = x^{2 \cdot 5} = x^{10}$$

93. $(2^3)^2 = 2^{3 \cdot 2} = 2^6 = 64$
$2^{(3^2)} = 2^9 = 512$. The results are not the same.
$2^{(3^2)} > (2^3)^2$

95. If n is a positive integer and $x^n = y^n$, then x does not have to equal y, though sometimes it does equal y. For example, $2^4 = (-2)^4$, but $2 \neq -2$. $x = y$ when n is odd.

Section 7.3

Concept Review 7.3

1. Always true

3. Sometimes true
$x^2 + 2$ is a binomial of degree 2; $x^3 - 8$ is a binomial of degree 3.

5. Always true

7. Sometimes true
The FOIL method is used to multiply two binomials.

Objective 7.3.1 Exercises

1. $x(x - 2) = x^2 - 2x$

3. $-x(x + 7) = -x^2 - 7x$

5. $3a^2 (a - 2) = 3a^3 - 6a^2$

7. $-5x^2 (x^2 - x) = -5x^4 + 5x^3$

9. $-x^3 (3x^2 - 7) = -3x^5 + 7x^3$

11. $2x(6x^2 - 3x) = 12x^3 - 6x^2$

13. $(2x - 4)(3x) = 6x^2 - 12x$

15. $-xy(x^2 - y^2) = -x^3y + xy^3$

17. $x(2x^3 - 3x + 2) = 2x^4 - 3x^2 + 2x$

19. $-a(-2a^2 - 3a - 2) = 2a^3 + 3a^2 + 2a$

21. $x^2(3x^4 - 3x^2 - 2) = 3x^6 - 3x^4 - 2x^2$

23. $2y^2(-3y^2 - 6y + 7) = -6y^4 - 12y^3 + 14y^2$

25. $(a^2 + 3a - 4)(-2a) = -2a^3 - 6a^2 + 8a$

27. $-3y^2(-2y^2 + y - 2) = 6y^4 - 3y^3 + 6y^2$

29. $xy(x^2 - 3xy + y^2) = x^3y - 3x^2y^2 + xy^3$

Objective 7.3.2 Exercises

31.
$$\begin{array}{r} x^2 + 3x + 2 \\ \times \qquad x + 1 \\ \hline x^2 + 3x + 2 \\ x^3 + 3x^2 + 2x \\ \hline x^3 + 4x^2 + 5x + 2 \end{array}$$

33.
$$\begin{array}{r} a^2 - 3a + 4 \\ \times \qquad a - 3 \\ \hline -3a^2 + 9a - 12 \\ a^3 - 3a^2 + 4a \\ \hline a^3 - 6a^2 + 13a - 12 \end{array}$$

35.
$$\begin{array}{r} -2b^2 - 3b + 4 \\ \times \qquad b - 5 \\ \hline 10b^2 + 15b - 20 \\ -2b^3 - 3b^2 + 4b \\ \hline -2b^3 + 7b^2 + 19b - 20 \end{array}$$

37.
$$\begin{array}{r} -2x^2 + 7x - 2 \\ \times \qquad 3x - 5 \\ \hline 10x^2 - 35x + 10 \\ -6x^3 + 21x^2 - 6x \\ \hline -6x^3 + 31x^2 - 41x + 10 \end{array}$$

39.
$$\begin{array}{r} x^3 \qquad - 3x + 2 \\ \times \qquad x - 4 \\ \hline -4x^3 \qquad + 12x - 8 \\ x^4 \qquad - 3x^2 + 2x \\ \hline x^4 - 4x^3 - 3x^2 + 14x - 8 \end{array}$$

41.
$$\begin{array}{r} 5y^2 + 8y - 2 \\ \times \qquad 3y - 8 \\ \hline -40y^2 - 64y + 16 \\ 15y^3 + 24y^2 - 6y \\ \hline 15y^3 - 16y^2 - 70y + 16 \end{array}$$

43.
$$\begin{array}{r} 5a^3 \qquad - 15a + 2 \\ \times \qquad a - 4 \\ \hline -20a^3 \qquad + 60a - 8 \\ 5a^4 \qquad - 15a^2 + 2a \\ \hline 5a^4 - 20a^3 - 15a^2 + 62a - 8 \end{array}$$

45.
$$\begin{array}{r} y^3 + 2y^2 - 3y + 1 \\ \times \qquad y + 2 \\ \hline 2y^3 + 4y^2 - 6y + 2 \\ y^4 + 2y^3 - 3y^2 + y \\ \hline y^4 + 4y^3 + y^2 - 5y + 2 \end{array}$$

Objective 7.3.3 Exercises

47. a. $y, 2y$

b. $y, 3$

c. $-8, 2y$

d. $-8, 3$

49. $(x + 1)(x + 3) = x^2 + 3x + x + 3$
$$= x^2 + 4x + 3$$

51. $(a - 3)(a + 4) = a^2 + 4a - 3a - 12$
$$= a^2 + a - 12$$

53. $(y + 3)(y - 8) = y^2 - 8y + 3y - 24$
$$= y^2 - 5y - 24$$

55. $(y - 7)(y - 3) = y^2 - 3y - 7y + 21$
$$= y^2 - 10y + 21$$

57. $(2x + 1)(x + 7) = 2x^2 + 14x + x + 7$
$$= 2x^2 + 15x + 7$$

59. $(3x - 1)(x + 4) = 3x^2 + 12x - x - 4$
$$= 3x^2 + 11x - 4$$

61. $(4x - 3)(x - 7) = 4x^2 - 28x - 3x + 21$
$$= 4x^2 - 31x + 21$$

63. $(3y - 8)(y + 2) = 3y^2 + 6y - 8y - 16$
$$= 3y^2 - 2y - 16$$

65. $(3x + 7)(3x + 11) = 9x^2 + 33x + 21x + 77$
$$= 9x^2 + 54x + 77$$

67. $(7a - 16)(3a - 5) = 21a^2 - 35a - 48a + 80$
$$= 21a^2 - 83a + 80$$

69. $(3b+13)(5b-6) = 15b^2 - 18b + 65b - 78$
$\qquad\qquad\qquad = 15b^2 + 47b - 78$

71. $(2a+b)(a+3b) = 2a^2 + 6ab + ab + 3b^2$
$\qquad\qquad\qquad = 2a^2 + 7ab + 3b^2$

73. $(2a-b)(3a+2b) = 6a^2 + 4ab - 3ab - 2b^2$
$\qquad\qquad\qquad = 6a^2 + ab - 2b^2$

75. $(2x+y)(x-2y) = 2x^2 - 4xy + xy - 2y^2$
$\qquad\qquad\qquad = 2x^2 - 3xy - 2y^2$

77. $(2x+3y)(5x+7y) = 10x^2 + 14xy + 15xy + 21y^2$
$\qquad\qquad\qquad = 10x^2 + 29xy + 21y^2$

79. $(3a-2b)(2a-7b) = 6a^2 - 21ab - 4ab + 14b^2$
$\qquad\qquad\qquad = 6a^2 - 25ab + 14b^2$

81. $(a-9b)(2a+7b) = 2a^2 + 7ab - 18ab - 63b^2$
$\qquad\qquad\qquad = 2a^2 - 11ab - 63b^2$

83. $(10a-3b)(10a-7b)$
$\qquad = 100a^2 - 70ab - 30ab + 21b^2$
$\qquad = 100a^2 - 100ab + 21b^2$

85. $(5x+12y)(3x+4y) = 15x^2 + 20xy + 36xy + 48y^2$
$\qquad\qquad\qquad = 15x^2 + 56xy + 48y^2$

87. $(2x-15y)(7x+4y) = 14x^2 + 8xy - 105xy - 60y^2$
$\qquad\qquad\qquad = 14x^2 - 97xy - 60y^2$

89. $(8x-3y)(7x-5y) = 56x^2 - 40xy - 21xy + 15y^2$
$\qquad\qquad\qquad = 56x^2 - 61xy + 15y^2$

Objective 7.3.4 Exercises

91. $(y-5)(y+5) = y^2 - 25$

93. $(2x+3)(2x-3) = 4x^2 - 9$

95. $(x+1)^2 = x^2 + 2x + 1$

97. $(3a-5)^2 = 9a^2 - 30a + 25$

99. $(3x-7)(3x+7) = 9x^2 - 49$

101. $(2a+b)^2 = 4a^2 + 4ab + b^2$

103. $(x-2y)^2 = x^2 - 4xy + 4y^2$

105. $(4-3y)(4+3y) = 16 - 9y^2$

107. $(5x+2y)^2 = 25x^2 + 20xy + 4y^2$

Objective 7.3.5 Exercises

109. Strategy
To find the area, replace the variables L and W in the equation $A = L \cdot W$ by the given values and solve for A.

Solution
$A = L \cdot W$
$A = (5x)(2x - 7)$
$A = 10x^2 - 35x$
The area is $(10x^2 - 35x)$ ft^2.

111. Strategy
To find the area, replace the variables L and W in the equation $A = L \cdot W$ by the given values and solve for A. $W = 3x + 1$, and $L = 2W$.

Solution
$A = L \cdot W$
$A = 2W \cdot W = 2W^2$
$A = 2(3x+1)^2$
$A = 2(9x^2 + 6x + 1)$
$A = 18x^2 + 12x + 2$
The area is $(18x^2 + 12x + 2)$ in^2.

113. Strategy
To find the area, replace the variable s in the equation $A = s^2$ by the given value and solve for A.

Solution
$A = s^2$
$A = (2x+1)^2$
$A = 4x^2 + 4x + 1$
The area is $(4x^2 + 4x + 1)$ km^2.

115. Strategy
To find the area, replace the variables b and h in the equation $A = \frac{1}{2}bh$ by the given values and solve for A.

Solution
$A = \frac{1}{2}bh$
$A = \frac{1}{2}(4x)(2x + 5)$
$A = 2x(2x+5)$
$A = 4x^2 + 10x$
The area is $(4x^2 + 10x)$ m^2.

117. Strategy

To find the area, replace the variable r in the equation $A = \pi r^2$ by the given value and solve for A.

Solution

$A = \pi r^2$

$A = \pi(x+4)^2$

$A = \pi(x^2 + 8x + 16)$

$A = \pi x^2 + 8\pi x + 16\pi$

The area is $(\pi x^2 + 8\pi x + 16\pi)$ cm^2.

119. Strategy

Use the equation $A = s^2$ to find the area of the square.

Use the equation $A = L \cdot W$ to find the area of each of the base paths.

Add the areas.

Solution

$A = s^2 = 45^2 = 2025$

$A = L \cdot W = 45 \cdot x = 45x$

Total area = $2025 + 45x + 45x$

Total area = $90x + 2025$

The total area is $(90x + 2025)$ ft^2.

Applying Concepts 7.3

121. $(a+b)^2 - (a-b)^2 = (a^2 + 2ab + b^2) - (a^2 - 2ab + b^2) = (a^2 + 2ab + b^2) + (-a^2 + 2ab - b^2) = 4ab$

123. $(3a^2 - 4a + 2)^2 = (9a^4 - 12a^3 + 6a^2) + (-12a^3 + 16a^2 - 8a) + (6a^2 - 8a + 4) = 9a^4 - 24a^3 + 28a^2 - 16a + 4$

125. $3x^2(2x^3 + 4x - 1) - 6x^3(x^2 - 2) = (6x^5 + 12x^3 - 3x^2) - (6x^5 - 12x^3)$

$= (6x^5 + 12x^3 - 3x^2) + (-6x^5 + 12x^3)$

$= 24x^3 - 3x^2$

127. $x^n(x^n + 1) = x^{2n} + x^n$

129. $(x^n + 1)(x^n + 1) = x^{2n} + 2x^n + 1$

131. $(x+1)(x-1) = x^2 - 1$

133.
$$
\begin{array}{r}
x^3 - x^2 + x - 1 \\
\times \qquad\quad x + 1 \\
\hline
x^3 - x^2 + x - 1 \\
x^4 - x^3 + x^2 - x \qquad\quad \\
\hline
x^4 \qquad\qquad\qquad\quad -1
\end{array}
$$

135. Using the pattern reflected in Exercises 131–134:

$(x+1)(x^5 - x^4 + x^3 - x^2 + x - 1) = x^6 - 1$

137.
$$2n - 3 = 4n - 7$$
$$2n - 2n - 3 = 4n - 2n - 7$$
$$-3 = 2n - 7$$
$$-3 + 7 = 2n - 7 + 7$$
$$4 = 2n$$
$$2 = n$$
$$(4n^3)^2 = 16n^6 = 16(2)^6 = 16(64) = 1024$$

139. Multiply the quotient by the divisor.

$(3x-4)(4x+5) = 12x^2 + 15x - 16x - 20$

$= 12x^2 - x - 20$

The polynomial is $12x^2 - x - 20$.

141. $(2x-5)(3x+1) + (x^2 + 2x - 3)$

$= 6x^2 + 2x - 15x - 5 + x^2 + 2x - 3$

$= 7x^2 - 11x - 8$

Section 7.4

Concept Review 7.4

1. Never true

The bases must be the same for us to use the Rule for Dividing Exponential Expressions.

3. Always true

5. Never true

$4^{-3} = \dfrac{1}{4^3} = \dfrac{1}{64}$; a negative exponent does not indicate a negative number.

7. Never true

$2x^{-5} = 2(x^{-5}) = \dfrac{2}{x^5}$

Objective 7.4.1 Exercises

1. $5^{-2} = \dfrac{1}{5^2} = \dfrac{1}{25}$

3. $\dfrac{1}{8^{-2}} = 8^2 = 64$

5. $\dfrac{3^{-2}}{3} = 3^{-2-1} = 3^{-3} = \dfrac{1}{3^3} = \dfrac{1}{27}$

7. $\dfrac{2^3}{2^3} = 2^{3-3} = 2^0 = 1$

11. $\dfrac{y^7}{y^3} = y^{7-3} = y^4$

13. $\dfrac{a^8}{a^5} = a^{8-5} = a^3$

15. $\dfrac{p^5}{p} = p^{5-1} = p^4$

17. $\dfrac{4x^8}{2x^5} = 2x^{8-5} = 2x^3$

19. $\dfrac{22k^5}{11k^4} = 2k^{5-4} = 2k$

21. $\dfrac{m^9 n^7}{m^4 n^5} = m^{9-4}n^{7-5} = m^5 n^2$

23. $\dfrac{6r^4}{4r^2} = \dfrac{3r^{4-2}}{2} = \dfrac{3r^2}{2}$

25. $\dfrac{-16a^7}{24a^6} = \dfrac{-2a^{7-6}}{3} = \dfrac{-2a}{3} = -\dfrac{2a}{3}$

27. $x^{-2} = \dfrac{1}{x^2}$

29. $\dfrac{1}{a^{-6}} = a^6$

31. $4x^{-7} = 4 \cdot \dfrac{1}{x^7} = \dfrac{4}{x^7}$

33. $\dfrac{5}{b^{-8}} = 5 \cdot \dfrac{1}{b^{-8}} = 5b^8$

35. $\dfrac{1}{3x^{-2}} = \dfrac{1}{3} \cdot \dfrac{1}{x^{-2}} = \dfrac{1}{3} \cdot x^2 = \dfrac{x^2}{3}$

37. $(ab^5)^0 = 1$

39. $\dfrac{y^3}{y^8} = y^{3-8} = y^{-5} = \dfrac{1}{y^5}$

41. $\dfrac{a^5}{a^{11}} = a^{5-11} = a^{-6} = \dfrac{1}{a^6}$

43. $\dfrac{4x^2}{12x^5} = \dfrac{x^{2-5}}{3} = \dfrac{x^{-3}}{3} = \dfrac{1}{3x^3}$

45. $\dfrac{-12x}{-18x^6} = \dfrac{2x^{1-6}}{3} = \dfrac{2x^{-5}}{3} = \dfrac{2}{3x^5}$

47. $\dfrac{x^6 y^5}{x^8 y} = x^{6-8}y^{5-1} = x^{-2}y^4 = \dfrac{y^4}{x^2}$

49. $\dfrac{2m^6 n^2}{5m^9 n^{10}} = \dfrac{2m^{6-9}n^{2-10}}{5} = \dfrac{2m^{-3}n^{-8}}{5} = \dfrac{2}{5m^3 n^8}$

51. $\dfrac{pq^3}{p^4 q^4} = p^{1-4}q^{3-4} = p^{-3}q^{-1} = \dfrac{1}{p^3 q}$

53. $\dfrac{3x^4 y^5}{6x^4 y^8} = \dfrac{1x^{4-4}y^{5-8}}{2} = \dfrac{x^0 y^{-3}}{2} = \dfrac{1}{2y^3}$

55. $\dfrac{14x^4 y^6 z^2}{16x^3 y^9 z} = \dfrac{7x^{4-3}y^{6-9}z^{2-1}}{8} = \dfrac{7xy^{-3}z}{8} = \dfrac{7xz}{8y^3}$

57. $\dfrac{15mn^9 p^3}{30m^4 n^9 p} = \dfrac{1m^{1-4}n^{9-9}p^{3-1}}{2}$

$= \dfrac{m^{-3}n^0 p^2}{2}$

$= \dfrac{p^2}{2m^3}$

59. $(-2xy^{-2})^3 = (-2)^3 x^3 y^{-6} = \dfrac{-8x^3}{y^6} = -\dfrac{8x^3}{y^6}$

61. $(3x^{-1}y^{-2})^2 = 3^2 x^{-2} y^{-4} = \dfrac{9}{x^2 y^4}$

63. $(2x^{-1})(x^{-3}) = 2x^{-4} = \dfrac{2}{x^4}$

65. $(-5a^2)(a^{-5})^2 = (-5a^2)(a^{-10}) = -5a^{-8} = \dfrac{-5}{a^8}$

67. $(-2ab^{-2})(4a^{-2}b)^{-2} = (-2ab^{-2})(4^{-2}a^4 b^{-2})$

$= \left(\dfrac{-2a}{b^2}\right)\left(\dfrac{a^4}{4^2 b^2}\right)$

$= \dfrac{-2a^5}{4^2 b^4}$

$= \dfrac{-2a^5}{16b^4}$

$= -\dfrac{a^5}{8b^4}$

69. $(-5x^{-2}y)(-2x^{-2}y^2) = \left(\dfrac{-5y}{x^2}\right)\left(\dfrac{-2y^2}{x^2}\right)$

$= \dfrac{10y^3}{x^4}$

71. $\dfrac{3x^{-2}y^2}{6xy^2} = \dfrac{1}{2x^3}$

73. $\dfrac{3x^{-2}y}{xy} = \dfrac{3}{x^3}$

75. $\dfrac{2x^{-1}y^{-4}}{4xy^2} = \dfrac{1}{2x^2 y^6}$

77. $\dfrac{-16xy^4}{96x^4 y^4} = -\dfrac{1}{6x^3}$

81. No, 39.4 is not between 1 and 10.

83. No, 2.4 is not an integer.

Objective 7.4.2 Exercises

85. $2,370,000 = 2.37 \cdot 10^6$

87. $0.00045 = 4.5 \cdot 10^{-4}$

89. $309,000 = 3.09 \cdot 10^5$

91. $0.000000601 = 6.01 \cdot 10^{-7}$

93. $7.1 \cdot 10^5 = 710,000$

95. $4.3 \cdot 10^{-5} = 0.000043$

97. $6.71 \cdot 10^8 = 671,000,000$

99. $7.13 \cdot 10^{-6} = 0.00000713$

101. $16,000,000,000 = 1.6 \cdot 10^{10}$

103. $5,980,000,000,000,000,000,000,000 = 5.98 \cdot 10^{24}$

105. $0.0000000000000000000016 = 1.6 \cdot 10^{-19}$

107. $0.000000000001 = 1 \cdot 10^{-12}$

109. $(1.9 \cdot 10^{12})(3.5 \cdot 10^7) = (1.9)(3.5) \cdot 10^{12+7}$
$= 6.65 \cdot 10^{19}$

111. $(2.3 \cdot 10^{-8})(1.4 \cdot 10^{-6}) = (2.3)(1.4) \cdot 10^{-8-6}$
$= 3.22 \cdot 10^{-14}$

113. $\dfrac{6.12 \cdot 10^{14}}{1.7 \cdot 10^9} = \dfrac{6.12}{1.7} \cdot 10^{14-9}$
$= 3.6 \cdot 10^5$

115. $\dfrac{5.58 \cdot 10^{-7}}{3.1 \cdot 10^{11}} = \dfrac{5.58}{3.1} \cdot 10^{-7-11}$
$= 1.8 \cdot 10^{-18}$

Applying Concepts 7.4

117. $8^{-2} + 2^{-5} = \dfrac{1}{8^2} + \dfrac{1}{2^5} = \dfrac{1}{64} + \dfrac{1}{32} = \dfrac{1}{64} + \dfrac{2}{64} = \dfrac{3}{64}$

119.

x	2^x	2^{-x}
-2	$2^{-2} = \frac{1}{4}$	$2^{-(-2)} = 2^2 = 4$
-1	$2^{-1} = \frac{1}{2}$	$2^{-(-1)} = 2^1 = 2$
0	$2^0 = 1$	$2^0 = 1$
1	$2^1 = 2$	$2^{-1} = \frac{1}{2}$
2	$2^2 = 4$	$2^{-2} = \frac{1}{4}$

121. $2^{-4} = 0.0625$

123. $\left(\dfrac{9x^2y^4}{3xy^2}\right) - \left(\dfrac{12x^5y^6}{6x^4y^4}\right) = (3xy^2) - (2xy^2)$
$= xy^2$

125. $\left(\dfrac{6x^4yz^3}{2x^2y^3}\right)\left(\dfrac{2x^2z^3}{4y^2z}\right) \div \left(\dfrac{6x^2y^3}{x^4y^2z}\right)$
$= \left(\dfrac{3x^2z^3}{y^2}\right)\left(\dfrac{x^2z^2}{2y^2}\right) \div \left(\dfrac{6y}{x^2z}\right)$
$= \dfrac{3x^4z^5}{2y^4} \div \dfrac{6y}{x^2z} = \dfrac{3x^4z^5}{2y^4} \cdot \dfrac{x^2z}{6y} = \dfrac{x^6z^6}{4y^5}$

127. $\dfrac{a^m}{a^n} = a^{m-n} = a^{n-n} = a^0 = 1$

129. $(-4.8)^x = 1$
$x = 0$

131. False.
$(2a)^{-3} = \dfrac{1}{(2a)^3} = \dfrac{1}{8a^3}$

133. False.
$(2+3)^{-1} = (5)^{-1} = \dfrac{1}{5}$

Section 7.5

Concept Review 7.5

1. Always true

3. Never true
The remainder was not multiplied by $(3x + 2)$.

Objective 7.5.1 Exercises

1. $\dfrac{2x+2}{2} = \dfrac{2x}{2} + \dfrac{2}{2} = x + 1$

3. $\dfrac{10a-25}{5} = \dfrac{10a}{5} - \dfrac{25}{5} = 2a - 5$

5. $\dfrac{3a^2+2a}{a} = \dfrac{3a^2}{a} + \dfrac{2a}{a} = 3a + 2$

7. $\dfrac{4b^3-3b}{b} = \dfrac{4b^3}{b} - \dfrac{3b}{b} = 4b^2 - 3$

9. $\dfrac{3x^2-6x}{3x} = \dfrac{3x^2}{3x} - \dfrac{6x}{3x} = x - 2$

11. $\dfrac{5x^2-10x}{-5x} = \dfrac{5x^2}{-5x} + \dfrac{-10x}{-5x} = -x + 2$

13. $\dfrac{x^3+3x^2-5x}{x} = \dfrac{x^3}{x} + \dfrac{3x^2}{x} - \dfrac{5x}{x} = x^2 + 3x - 5$

15. $\dfrac{x^6-3x^4-x^2}{x^2} = \dfrac{x^6}{x^2} - \dfrac{3x^4}{x^2} - \dfrac{x^2}{x^2} = x^4 - 3x^2 - 1$

17. $\dfrac{5x^2y^2+10xy}{5xy}=\dfrac{5x^2y^2}{5xy}+\dfrac{10xy}{5xy}=xy+2$

19. $\dfrac{9y^6-15y^3}{-3y^3}=\dfrac{9y^6}{-3y^3}-\dfrac{15y^3}{-3y^3}=-3y^3+5$

21. $\dfrac{3x^2-2x+1}{x}=\dfrac{3x^2}{x}-\dfrac{2x}{x}+\dfrac{1}{x}=3x-2+\dfrac{1}{x}$

23. $\dfrac{-3x^2+7x-6}{x}=\dfrac{-3x^2}{x}+\dfrac{7x}{x}-\dfrac{6}{x}$

$\qquad\qquad =-3x+7-\dfrac{6}{x}$

25. $\dfrac{16a^2b-20ab+24ab^2}{4ab}=\dfrac{16a^2b}{4ab}-\dfrac{20ab}{4ab}+\dfrac{24ab^2}{4ab}$

$\qquad\qquad =4a-5+6b$

27. $\dfrac{9x^2y+6xy-3xy^2}{xy}=\dfrac{9x^2y}{xy}+\dfrac{6xy}{xy}-\dfrac{3xy^2}{xy}$

$\qquad\qquad =9x+6-3y$

29.
$$
\begin{array}{r}
x+2 \\
x-3\overline{)x^2-\ x-6} \\
\underline{x^2-3x} \\
2x-6 \\
\underline{2x-6} \\
0
\end{array}
$$
$(x^2-x-6)\div(x-3)=x+2$

31.
$$
\begin{array}{r}
2x+1 \\
x+2\overline{)2x^2+5x+2} \\
\underline{2x^2+4x} \\
x+2 \\
\underline{x+2} \\
0
\end{array}
$$
$(2x^2+5x+2)\div(x+2)=2x+1$

33.
$$
\begin{array}{r}
2x-\ 4 \\
2x+4\overline{)4x^2+0\ -16} \\
\underline{4x^2+8x} \\
-8x-16 \\
\underline{-8x-16} \\
0
\end{array}
$$
$(4x^2-16)\div(2x+4)=2x-4$

35.
$$
\begin{array}{r}
x+1 \\
x-1\overline{)x^2+0+1} \\
\underline{x^2-x} \\
x+1 \\
\underline{x-1} \\
2
\end{array}
$$
$(x^2+1)\div(x-1)=x+1+\dfrac{2}{x-1}$

37.
$$
\begin{array}{r}
2x-1 \\
3x-2\overline{)6x^2-7x+0} \\
\underline{6x^2-4x} \\
-3x+0 \\
\underline{-3x+2} \\
-2
\end{array}
$$
$(6x^2-7x)\div(3x-2)=2x-1-\dfrac{2}{3x-2}$

39.
$$
\begin{array}{r}
5x+12 \\
x-1\overline{)5x^2+7x+\ 0} \\
\underline{5x^2-5x} \\
12x+\ 0 \\
\underline{12x-12} \\
12
\end{array}
$$
$(5x^2+7x)\div(x-1)=5x+12+\dfrac{12}{x-1}$

41.
$$
\begin{array}{r}
a+\ 3 \\
a+2\overline{)a^2+5a+10} \\
\underline{a^2+2a} \\
3a+10 \\
\underline{3a+\ 6} \\
4
\end{array}
$$
$(a^2+5a+10)\div(a+2)=a+3+\dfrac{4}{a+2}$

43.
$$
\begin{array}{r}
y-\ 6 \\
2y+3\overline{)2y^2-9y+\ 8} \\
\underline{2y^2+3y} \\
-12y+\ 8 \\
\underline{-12y-18} \\
26
\end{array}
$$
$(2y^2-9y+8)\div(2y+3)=y-6+\dfrac{26}{2y+3}$

45.
$$
\begin{array}{r}
2x+5 \\
2x-1\overline{)4x^2+8x+3} \\
\underline{4x^2-2x} \\
10x+3 \\
\underline{10x-5} \\
8
\end{array}
$$
$(4x^2+8x+3)\div(2x-1)=2x+5+\dfrac{8}{2x-1}$

47.
$$
\begin{array}{r}
5a-6 \\
3a+2\overline{)15a^2-\ 8a-\ 8} \\
\underline{15a^2+10a} \\
-18a-\ 8 \\
\underline{-18a-12} \\
4
\end{array}
$$
$(15a^2-8a-8)\div(3a+2)=5a-6+\dfrac{4}{3a+2}$

49.

$$\begin{array}{r} 3x-5 \\ 4x-1{\overline{\smash{\big)}\,2x^2-23x+5}} \\ \underline{12x^2-\ 3x} \\ -20x+5 \\ \underline{-20x+5} \\ 0 \end{array}$$

$(12x^2-23x+4)\div(4x-1)=3x-5$

51.

$$\begin{array}{r} x^2+2x+3 \\ x+1{\overline{\smash{\big)}\,x^3+3x^2+5x+3}} \\ \underline{x^3+\ x^2} \\ 2x^2+5x \\ \underline{2x^2+2x} \\ 3x+3 \\ \underline{3x+3} \\ 0 \end{array}$$

$(x^3+3x^2+5x+3)\div(x+1)=x^2+2x+3$

53.

$$\begin{array}{r} x^2-3 \\ x^2+0+2{\overline{\smash{\big)}\,x^4+0-\ x^2+0-6}} \\ \underline{x^4+0+2x^2} \\ -3x^2+0-6 \\ \underline{-3x^2+0-6} \\ 0 \end{array}$$

$(x^4-x^2-6)\div(x^2+2)=x^2-3$

59. The unknown polynomial: p

$$\frac{p}{2x+1}=2x-4+\frac{7}{2x+1}$$

$$(2x+1)\left(\frac{p}{2x+1}\right)=(2x+1)\left(2x-4+\frac{7}{2x+1}\right)$$

$$p=(2x+1)2x+(2x+1)(-4)+(2x+1)\left(\frac{7}{2x+1}\right)$$

$$p=4x^2+2x-8x-4+7$$

$$p=4x^2-6x+3$$

The polynomial is $4x^2-6x+3$.

Chapter Review Exercises

1. $12y^2+17y-4$

$9y^2-13y+3$

$\overline{21y^2+\ 4y-1}$

2. $(5xy^2)(-4x^2y^3)=-20x^3y^5$

3. $-2x(4x^2+7x-9)=-2x(4x^2)-2x(7x)-2x(-9)$

$=-8x^3-14x^2+18x$

4. $(5a-7)(2a+9)=10a^2+45a-14a-63$

$=10a^2+31a-63$

Applying Concepts 7.5

55. To find the expression, multiply both sides by $(x-3)$:

$$\frac{x^2-x-6}{x-3}=(x+2)$$

$$(x-3)\frac{x^2-x-6}{x-3}=(x+2)(x-3)$$

$$x^2-x-6=(x+2)(x-3)$$

Then, $x^2-x-6=(x+2)(x-3)$.

57. The unknown monomial: m

$$m(4b)=12a^2b$$

$$\frac{m(4b)}{4b}=\frac{12a^2b}{4b}$$

$$m=3a^2$$

The monomial is $3a^2$.

5. $\dfrac{36x^2-42x+60}{6}=\dfrac{36x^2}{6}-\dfrac{42x}{6}+\dfrac{60}{6}$

$\phantom{\dfrac{36x^2-42x+60}{6}}=6x^2-7x+10$

6. $\begin{array}{l} 5x^2-2x-1 \\ \underline{-(3x^2-5x+7)} \end{array}$ $\begin{array}{l} 5x^2-2x-1 \\ \underline{-3x^2+5x-7} \\ 2x^2+3x-8 \end{array}$

7. $(-3^2)^3=-3^6=-729$

8. $(x^2-5x+2)(x-1)$

$=x^2(x-1)-5x(x-1)+2(x-1)$

$=x^3-x^2-5x^2+5x+2x-2$

$=x^3-6x^2+7x-2$

9. $(a+7)(a-7) = a^2 - 7a + 7a - 49$
$$= a^2 - 49$$

10. $\dfrac{6^2}{6^{-2}} = 6^{2-(-2)} = 6^4 = 1296$

11.
$$
\begin{array}{r}
x-6 \\
x+7\overline{)x^2+\ x-42} \\
\underline{x^2+7x} \\
-6x-42 \\
\underline{-6x-42} \\
0
\end{array}
$$
$(x^2 + x - 42) \div (x+7) = x - 6$

12. $\quad 2x^3 + 7x^2 +\ x$
$$\dfrac{2x^2 - 4x - 12}{2x^3 + 9x^2 - 3x - 12}$$

13. $(6a^2 b^5)(3a^6 b) = 18a^8 b^6$

14. $x^2 y(3x^2 - 2x + 12)$
$$= x^2 y(3x^2) - x^2 y(2x) + x^2 y(12)$$
$$= 3x^4 y - 2x^3 y + 12x^2 y$$

15. $(2b-3)(4b+5) = 8b^2 + 10b - 12b - 15$
$$= 8b^2 - 2b - 15$$

16. $\dfrac{16y^2 - 32y}{-4y} = \dfrac{16y^2}{-4y} + \dfrac{-32y}{-4y} = -4y + 8$

17.
$$
\begin{array}{ll}
13y^3 - 7y - 2 & 13y^3\quad\ -7y-2 \\
\underline{-(12y^2 - 2y - 1)} & \underline{-12y^2 + 2y + 1} \\
& 13y^3 - 12y^2 - 5y - 1
\end{array}
$$

18. $(2^3)^2 = 2^6 = 64$ or $(8)^2 = 64$

19.
$$
\begin{array}{r}
3y^2 + 4y - 7 \\
\times \qquad 2y + 3 \\
\hline
9y^2 + 12y - 21 \\
6y^3 + 8y^2 - 14y \\
\hline
6y^3 + 17y^2 - 2y - 21
\end{array}
$$

20. $(2b-9)(2b+9) = 4b^2 + 18b - 18b - 81$
$$= 4b^2 - 81$$

21. $(a^{-2}b^3 c)^2 = a^{-4}b^6 c^2 = \dfrac{b^6 c^2}{a^4}$

22.
$$
\begin{array}{r}
2y-9 \\
3y-4\overline{)6y^2-35y+36} \\
\underline{6y^2-\ 8y} \\
-27y+36 \\
\underline{-27y+36} \\
0
\end{array}
$$
$(6y^2 - 35y + 36) \div (3y - 4) = 2y - 9$

23. $0.00000397 = 3.97 \cdot 10^{-6}$

24. $(xy^5 z^3)(x^3 y^3 z) = x^4 y^8 z^4$

25. $(6y^2 - 2y + 9)(-2y^3)$
$$= 6y^2(-2y^3) - 2y(-2y^3) + 9(-2y^3)$$
$$= -12y^5 + 4y^4 - 18y^3$$

26. $(6x-12)(3x-2) = 18x^2 - 12x - 36x + 24$
$$= 18x^2 - 48x + 24$$

27. $6.23 \cdot 10^{-5} = 0.0000623$

28.
$$
\begin{array}{ll}
8a^2 - a & 8a^2 - a \\
\underline{-(15a^2\quad - 4)} & \underline{-15a^2\quad + 4} \\
& -7a^2 - a + 4
\end{array}
$$

29. $(-3x^2 y^3)^2 = (-3)^2 x^4 y^6 = 9x^4 y^6$

30. $(4a^2 - 3)(3a - 2) = 12a^3 - 8a^2 - 9a + 6$

31. $(5y-7)^2 = (5y-7)(5y-7)$
$$= 25y^2 - 35y - 35y + 49$$
$$= 25y^2 - 70y + 49$$

32. $(-3x^{-2} y^{-3})^{-2} = (-3)^{-2} x^4 y^6 = \dfrac{x^4 y^6}{(-3)^2} = \dfrac{x^4 y^6}{9}$

33.
$$
\begin{array}{r}
x+5 \\
x+12\overline{)x^2+17x+64} \\
\underline{x^2+12x} \\
5x+64 \\
\underline{5x+60} \\
4
\end{array}
$$
$(x^2 + 17x + 64) \div (x+12) = x + 5 + \dfrac{4}{x+12}$

34. $2.4 \cdot 10^5 = 240{,}000$

35. $(a^2 b^7 c^6)(ab^3 c)(a^3 bc^2) = a^6 b^{11} c^9$

36. $2ab^3(4a^2 - 2ab + 3b^2)$
$$= 2ab^3(4a^2) + 2ab^3(-2ab) + 2ab^3(3b^2)$$
$$= 8a^3 b^3 - 4a^2 b^4 + 6ab^5$$

37. $(3x+4y)(2x-5y) = 6x^2 - 15xy + 8xy - 20y^2$
$$= 6x^2 - 7xy - 20y^2$$

38. $\dfrac{12b^7 + 36b^5 - 3b^3}{3b^3} = \dfrac{12b^7}{3b^3} + \dfrac{36b^5}{3b^3} - \dfrac{3b^3}{3b^3}$
$$= 4b^4 + 12b^2 - 1$$

39.
$$
\begin{array}{ll}
b^2 - 11b + 19 & b^2 - 11b + 19 \\
\underline{-(5b^2 + 2b - 9)} & \underline{-5b^2 - 2b + 9} \\
& -4b^2 - 13b + 28
\end{array}
$$

40. $(5a^7 b^6)^2(4ab) = (5^2 a^{14} b^{12})(4ab)$
$$= (25a^{14} b^{12})(4ab)$$
$$= 100a^{15} b^{13}$$

41.
$$\begin{array}{r} 6b^3 - 2b^2 - 5 \\ \times \qquad 2b^2 - 1 \\ \hline -6b^3 + 2b^2 + 5 \\ 12b^5 - 4b^4 \qquad -10b^2 \\ \hline 12b^5 - 4b^4 - 6b^3 - 8b^2 + 5 \end{array}$$

42. $(6 - 5x)(6 + 5x) = 36 + 30x - 30x - 25x^2$
$$= 36 - 25x^2$$

43. $\dfrac{6x^{-2}y^4}{3xy} = 2x^{-2-1}y^{4-1}$
$$= 2x^{-3}y^3$$
$$= \dfrac{2y^3}{x^3}$$

44.
$$\begin{array}{r} a^2 - 2a + 6 \\ a+3 \overline{\smash{\big)}\ a^3 + a^2 + 0a + 18} \\ \underline{a^3 + 3a^2} \\ -2a^2 + 0a \\ \underline{-2a^2 - 6a} \\ 6a + 18 \\ \underline{6a + 18} \\ 0 \end{array}$$
$$(a^3 + a^2 + 18) \div (a+3)^2 = a^2 - 2a + 6$$

45.
$$\begin{array}{r} 4b^3 - 7b^2 \qquad + 10 \\ 2b^2 - 9b - 3 \\ \hline 4b^3 - 5b^2 - 9b + 7 \end{array}$$

46. $(2a^{12}b^3)(-9b^2c^6)(3ac)$
$$= [2(-9)(3)]a^{12+1}b^{3+2}c^{6+1}$$
$$= -54a^{13}b^5c^7$$

47. $-9x^2(2x^2 + 3x - 7)$
$$= -9x^2(2x^2) - 9x^2(3x) - 9x^2(-7)$$
$$= -18x^4 - 27x^3 + 63x^2$$

48. $(10y - 3)(3y - 10) = 30y^2 - 100y - 9y + 30$
$$= 30y^2 - 109y + 30$$

49. $9{,}176{,}000{,}000{,}000 = 9.176 \cdot 10^{12}$

50.
$$\begin{array}{rr} 6y^2 + 2y + 7 & 6y^2 + 2y + 7 \\ -(8y^2 + y + 12) & -8y^2 - y - 12 \\ \hline & -2y^2 + y - 5 \end{array}$$

51. $(6x^4y^7z^2)^2(-2x^3y^2z^6)^2$
$$= (6^2x^8y^{14}z^4)[(-2)^2x^6y^4z^{12}]$$
$$= (36x^8y^{14}z^4)(4x^6y^4z^{12})$$
$$= 144x^{14}y^{18}z^{16}$$

52.
$$\begin{array}{r} -3x^3 - 2x^2 + x - 9 \\ \times \qquad 4x + 3 \\ \hline -9x^3 - 6x^2 + 3x - 27 \\ -12x^4 - 8x^3 + 4x^2 - 36x \\ \hline -12x^4 - 17x^3 - 2x^2 - 33x - 27 \end{array}$$

53. $(8a + 1)^2 = (8a + 1)(8a + 1)$
$$= 64a^2 + 8a + 8a + 1$$
$$= 64a^2 + 16a + 1$$

54. $\dfrac{4a^{-2}b^{-8}}{2a^{-1}b^{-2}} = 2a^{-2-(-1)}b^{-8-(-2)}$
$$= 2a^{-1}b^{-6}$$
$$= \dfrac{2}{ab^6}$$

55.
$$\begin{array}{r} b^2 + 5b + 2 \\ b-7 \overline{\smash{\big)}\ b^3 - 2b^2 - 33b - 7} \\ \underline{b^3 - 7b^2} \\ 5b^2 - 33b \\ \underline{5b^2 - 35b} \\ 2b - 7 \\ \underline{2b - 14} \\ 7 \end{array}$$
$$(b^3 - 2b^2 - 33b - 7) \div (b - 7) = b^2 + 5b + 2 + \dfrac{7}{b-7}$$

56. Strategy
To find the area, replace the variables L and W in the equation $A = L \cdot W$ by the given values and solve for A.

Solution
$A = L \cdot W$
$A = 5x(4x - 7)$
$A = 20x^2 - 35x$
The area is $(20x^2 - 35x)\,\text{m}^2$.

57. Strategy
To find the area, replace the variable s in the equation $A = s^2$ by the given value and solve for A.

Solution
$A = s^2$
$A = (5x + 4)^2$
$A = 25x^2 + 40x + 16$
The area is $(25x^2 + 40x + 16)\,\text{in}^2$.

58. Strategy
To find the area, replace the variables b and h in the equation $A = \frac{1}{2}bh$ by the given values and solve for A.

Solution
$A = \frac{1}{2}bh$
$A = \frac{1}{2}(3x-2)(6x+4)$
$A = \frac{1}{2}(6x+4)(3x-2)$
$A = (3x+2)(3x-2)$
$A = 9x^2 - 4$
The area is $(9x^2 - 4)$ ft^2.

59. Strategy
To find the area, replace the variable r in the equation $A = \pi r^2$ by the given value and solve for A.

Solution
$A = \pi r^2$
$A = \pi(x-6)^2$
$A = \pi(x^2 - 12x + 36)$
$A = \pi x^2 - 12\pi x + 36\pi$
The area is $(\pi x^2 - 12\pi x + 36\pi)$ cm^2.

60. Strategy
To find the area, replace the variables L and W in the equation $A = L \cdot W$ by the given values and solve for A.

Solution
$A = L \cdot W$
$A = (5x+4)(3x-8)$
$A = 15x^2 - 40x + 12x - 32$
$A = 15x^2 - 28x - 32$
The area is $(15x^2 - 28x - 32)$ mi^2.

Chapter Test

1. $3x^3 - 2x^2 \qquad -4$
    $\underline{\qquad 8x^2 - 8x + 7}$
    $3x^3 + 6x^2 - 8x + 3$

2. $\qquad -2x^3 + x^2 \qquad\quad -7$
    $\underline{\times \qquad\qquad\quad 2x - 3}$
    $\qquad\quad 6x^3 - 3x^2 \qquad +21$
    $\underline{-4x^4 + 2x^3 \qquad -14x}$
    $-4x^4 + 8x^3 - 3x^2 - 14x + 21$

3. $2x(2x^2 - 3x) = 2x(2x^2) + 2x(-3x)$
    $\qquad\qquad\qquad = 4x^3 - 6x^2$

4. $(-2a^2b)^3 = (-2)^3 a^6b^3 = -8a^6b^3$

5. $\dfrac{12x^2}{-3x^{-4}} = -4x^{2-(-4)} = -4x^6$

6. $(2ab^{-3})(3a^{-2}b^4) = 6a^{-1}b = \dfrac{6b}{a}$

7. $\qquad 3a^2 - 2a - \ 7 \qquad\qquad 3a^2 - 2a - \ 7$
    $\underline{-(5a^3 \quad +2a - 10)} \quad \underline{-5a^3 \qquad\quad -2a +10}$
    $\qquad\qquad\qquad\qquad\qquad -5a^3 + 3a^2 - 4a + \ 3$

8. $(a - 2b)(a + 5b) = a^2 + 5ab - 2ab - 10b^2$
    $\qquad\qquad\qquad\quad = a^2 + 3ab - 10b^2$

9. $\dfrac{16x^5 - 8x^3 + 20x}{4x} = \dfrac{16x^5}{4x} - \dfrac{8x^3}{4x} + \dfrac{20x}{4x}$
    $\qquad\qquad\qquad\qquad = 4x^4 - 2x^2 + 5$

10.
    $$2x - 3 \overline{\smash{\big)}\ \begin{array}{r} 2x + 3 \\ 4x^2 + 0x - 7 \end{array}}$$
    $\qquad\quad \underline{4x^2 - 6x}$
    $\qquad\qquad\quad 6x - 7$
    $\qquad\qquad\quad \underline{6x - 9}$
    $\qquad\qquad\qquad\quad 2$

    $(4x^2 - 7) \div (2x - 3) = 2x + 3 + \dfrac{2}{2x-3}$

11. $(-2xy^2)(3x^2y^4) = -6x^3y^6$

12. $-3y^2(-2y^2 + 3y - 6)$
    $= -3y^2(-2y^2) - 3y^2(3y) - 3y^2(-6)$
    $= 6y^4 - 9y^3 + 18y^2$

13. $\dfrac{27xy^3}{3x^4y^3} = 9x^{1-4}y^{3-3} = 9x^{-3}y^0 = \dfrac{9}{x^3}$

14. $(2x - 5)^2 = (2x - 5)(2x - 5)$
    $\qquad\qquad = 4x^2 - 10x - 10x + 25$
    $\qquad\qquad = 4x^2 - 20x + 25$

15. $(2x - 7y)(5x - 4y) = 10x^2 - 8xy - 35xy + 28y^2$
    $\qquad\qquad\qquad\qquad = 10x^2 - 43xy + 28y^2$

16. $\qquad x^2 - \ 4x + \ 5$
    $\underline{\times \qquad\qquad x - \ 3}$
    $\qquad\quad -3x^2 + 12x - 15$
    $\underline{x^3 - 4x^2 + \ 5x}$
    $x^3 - 7x^2 + 17x - 15$

17. $(a^2b^{-3})^2 = a^4b^{-6} = \dfrac{a^4}{b^6}$

18. $0.000029 = 2.9 \cdot 10^{-5}$

19. $(4y - 3)(4y + 3) = 16y^2 + 12y - 12y - 9$
    $\qquad\qquad\qquad\quad = 16y^2 - 9$

20.
$$
\begin{array}{ll}
3y^3 - 5y + 8 & 3y^3 \quad\quad -5y + 8 \\
\underline{-(-2y^2 + 5y + 8)} & \underline{+2y^2 - 5y - 8} \\
& 3y^3 + 2y^2 - 10y
\end{array}
$$

21. $(-3a^3b^2)^2 = (-3)^2a^6b^4 = 9a^6b^4$

22.
$$
\begin{array}{r}
5a^2 - 2a + 3 \\
\times \quad\quad 2a - 7 \\
\hline
-35a^2 + 14a - 21 \\
10a^3 - 4a^2 + 6a \\
\hline
10a^3 - 39a^2 + 20a - 21
\end{array}
$$

23. $(3b+2)^2 = (3b+2)(3b+2)$
$$= 9b^2 + 6b + 6b + 4$$
$$= 9b^2 + 12b + 4$$

24. $\dfrac{-2a^2b^3}{8a^4b^8} = -\dfrac{a^{2-4}b^{3-8}}{4}$
$$= -\dfrac{a^{-2}b^{-5}}{4}$$
$$= -\dfrac{1}{4a^2b^5}$$

25.
$$
\begin{array}{r}
4x + 8 \\
2x-3 \overline{)8x^2 + 4x - 3} \\
\underline{8x^2 - 12x} \\
16x - 3 \\
\underline{16x - 24} \\
21
\end{array}
$$
$$(8x^2 + 4x - 3) \div (2x - 3) = 4x + 8 + \dfrac{21}{2x-3}$$

26. $(a^2b^5)(ab^2) = a^3b^7$

27. $(a-3b)(a+4b) = a^2 + 4ab - 3ab - 12b^2$
$$= a^2 + ab - 12b^2$$

28. $3.5 \cdot 10^{-8} = 0.000000035$

29. Strategy
To find the area, replace the variable s in the equation $A = s^2$ with the given value and solve for A.

Solution
$A = s^2$
$A = (2x + 3)^2$
$A = (2x + 3)(2x + 3)$
$A = 4x^2 + 12x + 9$
The area is $(4x^2 + 12x + 9)$ m^2.

30. Strategy
To find the area, replace the variable r in the equation $A = \pi r^2$ with the given value and solve for A.

Solution
$A = \pi r^2$
$A = \pi(x - 5)^2$
$A = \pi(x^2 - 10x + 25)$
$A = \pi x^2 - 10\pi x + 25\pi$
The area is $(\pi x^2 - 10\pi x + 25\pi)$ in^2.

Cumulative Review Exercises

1. $\dfrac{3}{16} - \left(-\dfrac{3}{8}\right) - \dfrac{5}{9} = \dfrac{3}{16} + \dfrac{3}{8} + \left(-\dfrac{5}{9}\right)$
$$= \dfrac{27}{144} + \dfrac{54}{144} + \left(-\dfrac{80}{144}\right)$$
$$= \dfrac{1}{144}$$

2. $-5^2 \cdot \left(\dfrac{2}{3}\right)^3 \cdot \left(-\dfrac{3}{8}\right) = -25 \cdot \dfrac{8}{27} \cdot \left(-\dfrac{3}{8}\right)$
$$= -\dfrac{200}{27} \cdot \left(-\dfrac{3}{8}\right)$$
$$= \dfrac{25}{9}$$

3. $\left(-\dfrac{1}{2}\right)^2 \div \left(\dfrac{5}{8} - \dfrac{5}{6}\right) + 2 = \left(-\dfrac{1}{2}\right)^2 \div \left(\dfrac{30}{48} - \dfrac{40}{48}\right) + 2$
$$= \left(-\dfrac{1}{2}\right)^2 \div \left(-\dfrac{10}{48}\right) + 2$$
$$= \dfrac{1}{4} \cdot \left(-\dfrac{48}{10}\right) + 2$$
$$= -\dfrac{6}{5} + 2$$
$$= -\dfrac{6}{5} + \dfrac{10}{5} = \dfrac{4}{5}$$

4. 87

5.
$$
\begin{array}{r}
0.775 \\
40 \overline{)31.000} \\
\underline{-280} \\
300 \\
\underline{-280} \\
200 \\
\underline{-200} \\
0
\end{array}
$$

6. $\dfrac{b-(a-b)^2}{b^2} = \dfrac{-2-[3-(-2)]^2}{(-2)^2}$

$= \dfrac{-2-(3+2)^2}{4}$

$= \dfrac{-2-5^2}{4}$

$= \dfrac{-2-25}{4}$

$= \dfrac{-27}{4} = -\dfrac{27}{4}$

7. $-3x-(-xy)+2x-5xy = -3x+xy+2x-5xy$
$\qquad\qquad\qquad\qquad = -x-4xy$

8. $(16x)\left(-\dfrac{3}{4}\right) = -12x$

9. $-2[3x-4(3-2x)+2] = -2[3x-12+8x+2]$
$\qquad\qquad\qquad\qquad = -2[11x-10]$
$\qquad\qquad\qquad\qquad = -22x+20$

10. $-8+8 = 0$

11. $\qquad 12 = -\dfrac{2}{3}x$

$-\dfrac{3}{2}(12) = -\dfrac{3}{2}\left(-\dfrac{2}{3}x\right)$

$\qquad -18 = x$

The solution is -18.

12. $\qquad 3x-7 = 2x+9$

$3x-2x-7 = 2x-2x+9$

$\qquad x-7 = 9$

$x-7+7 = 9+7$

$\qquad x = 16$

The solution is 16.

13. $\qquad 3-4(2-x) = 3x+7$

$\qquad 3-8+4x = 3x+7$

$\qquad -5+4x = 3x+7$

$-5+4x-3x = 3x-3x+7$

$\qquad -5+x = 7$

$-5+5+x = 5+7$

$\qquad x = 12$

The solution is 12.

14. $\qquad -\dfrac{4}{5}x = 16-x$

$-\dfrac{4}{5}x+x = 16-x+x$

$-\dfrac{4}{5}x+\dfrac{5}{5}x = 16$

$\qquad \dfrac{1}{5}x = 16$

$5\left(\dfrac{1}{5}x\right) = 5(16)$

$\qquad x = 80$

The solution is 80.

15. $\qquad PB = A$

$P(160) = 38.4$

$\dfrac{160P}{160} = \dfrac{38.4}{160}$

$\qquad P = 0.24$

The percent is 24%.

16. $\qquad 7x-8 \ge -29$

$7x-8+8 \ge -29+8$

$\qquad 7x \ge -21$

$\qquad \dfrac{7x}{7} \ge \dfrac{-21}{7}$

$\qquad x \ge -3$

17. $P_1 = (x_1,y_1) = (3,-4),\ P_2 = (x_2,y_2) = (-2,5)$

$m = \dfrac{y_2-y_1}{x_2-x_1} = \dfrac{5-(-4)}{-2-3} = \dfrac{9}{-5} = -\dfrac{9}{5}$

The slope of the line is $-\dfrac{9}{5}$.

18. $m = -\dfrac{3}{2},\ (x_1,y_1) = (1,-3)$

$y-y_1 = m(x-x_1)$

$y-(-3) = -\dfrac{3}{2}(x-1)$

$y+3 = -\dfrac{3}{2}x+\dfrac{3}{2}$

$y = -\dfrac{3}{2}x+\dfrac{3}{2}-3$

$y = -\dfrac{3}{2}x+\dfrac{3}{2}-\dfrac{6}{2}$

$y = -\dfrac{3}{2}x-\dfrac{3}{2}$

19. $3x-2y = -6$

$\quad -2y = -3x-6$

$\qquad y = \dfrac{3}{2}x+3$

20. $y \le \dfrac{4}{5}x-3$

Graph $y = \dfrac{4}{5}x-3$ as a solid line.

Shade the lower half-plane.

21. Domain: $\{-8, -6, -4, -2\}$
Range: $\{-7, -5, -2, 0\}$
No two ordered pairs have the same first coordinate. The relation is a function

22. $f(x) = -2x + 10$
$f(6) = -2(6) + 10$
$f(6) = -12 + 10$
$f(6) = -2$
The value of $f(6)$ is –2.

23. (1) $\quad x = 3y + 1$
(2) $2x + 5y = 13$
Substitute in equation (2).
$2(3y + 1) + 5y = 14$
$6y + 2 + 5y = 13$
$11y + 2 = 13$
$11y = 11$
$y = 1$
Substitute in equation (1).
$x = 3y + 1$
$x = 3(1) + 1$
$x = 3 + 1$
$x = 4$
The solution is (4, 1).

24. (1) $9x - 2y = 17$
(2) $5x + 3y = -7$
Eliminate y.
$3(9x - 2y) = 3(17)$
$2(5x + 3y) = 2(-7)$
$27x - 6y = 51$
$10x + 6y = -14$
Add the equations.
$37x = 37$
$x = 1$
Replace x in equation (2).
$5x + 3y = -7$
$5(1) + 3y = -7$
$5 + 3y = -7$
$3y = -12$
$y = -4$
The solution is (1, –4).

25. $(5b^3 - 4b^2 - 7) - (3b^2 - 8b + 3)$
$(5b^3 - 4b^2 - 7) + (-3b^2 + 8b - 3)$
$5b^3 - 7b^2 + 8b - 10$

26. $\quad\quad 5x^2 - 2x + 1$
$\underline{\times \quad\quad\quad\quad 3x - 4}$
$\quad -20x^2 + 8x - 4$
$\underline{15x^3 - 6x^2 + 3x}$
$15x^3 - 26x^2 + 11x - 4$

27. $(4b - 3)(5b - 8) = 20b^2 - 32b - 15b + 24$
$= 20b^2 - 47b + 24$

28. $(5b + 3)^2 = (5b + 3)(5b + 3)$
$= 25b^2 + 15b + 15b + 9$
$= 25b^2 + 30b + 9$

29. $\dfrac{-3a^3b^2}{12a^4b^{-2}} = -\dfrac{a^{3-4}b^{2-(-2)}}{4}$
$= -\dfrac{a^{-1}b^4}{4}$
$= -\dfrac{b^4}{4a}$

30. $\dfrac{-15y^2 + 12y - 3}{-3y} = \dfrac{-15y^2}{-3y} + \dfrac{12y}{-3y} - \dfrac{3}{-3y}$
$= 5y - 4 + \dfrac{1}{y}$

31.
$$a + 4 \,)\overline{\begin{array}{l} a - 7 \\ a^2 - 3a - 28 \\ \underline{a^2 + 4a} \\ -7a - 28 \\ \underline{-7a - 28} \\ 0 \end{array}}$$
$(a^2 - 3a - 28) \div (a + 4) = a - 7$

32. $(-3x^{-4}y)(-3x^{-2}y) = [(-3)(-3)]x^{-4-2}y^{1+1}$
$= 9x^{-6}y^2$
$= \dfrac{9y^2}{x^6}$

33. $\quad f(x) = -\dfrac{4}{3}x + 9$
$f(-12) = -\dfrac{4}{3}(-12) + 9 = 16 + 9 = 25$
$f(-9) = -\dfrac{4}{3}(-9) + 9 = 12 + 9 = 21$
$f(-6) = -\dfrac{4}{3}(-6) + 9 = 8 + 9 = 17$
$f(0) = -\dfrac{4}{3}(0) + 9 = 0 + 9 = 9$
$f(6) = -\dfrac{4}{3}(6) + 9 = -8 + 9 = 1$
The range is {1, 9, 17, 21, 25}.

34. The *product* of five and the *difference* between a number and twelve
$5(n - 12)$
$5n - 60$

35. The unknown number: n

The difference between eight times a number and twice a number	is	18

$8n - 2n = 18$
$6n = 18$
$n = 3$
The number is 3.

36. Strategy
Length: L
Width: $0.40L$
Use the equation for the perimeter of a rectangle.

Solution
$$2L + 2W = P$$
$$2L + 2(0.40L) = 42$$
$$2L + 0.8L = 42$$
$$2.8L = 42$$
$$L = 15$$
$$0.40L = 0.40(15) = 6$$
The length is 15 m. The width is 6 m.

37. Strategy
Given: $r = 80\% = 0.80$
$C = \$24$
Unknown: S
Use the equation $S = C + rC$.

Solution
$$S = C + rC$$
$$= 24 + 0.80(24)$$
$$= 24 + 19.20$$
$$= 43.20$$
The selling price is $43.20.

38. Strategy
% concentration orange juice: x

	Amount	Percent	Quantity
Pure orange juice	50	1.00	1.00(50)
Fruit punch	200	0.10	0.10(200)
Mixture	250	x	$250x$

The sum of the quantities before mixing is equal to the quantity after mixing.

Solution
$$1.00(50) + 0.10(200) = 250x$$
$$50 + 20 = 250x$$
$$70 = 250x$$
$$0.28 = x$$
The resulting mixture is 28% orange juice.

39. Strategy
Time of car: t
Time of cyclist: $t + 2$

	Rate	Time	Distance
Car	50	t	$50t$
Cyclist	10	$t + 2$	$10(t + 2)$

The car and the cyclist travel the same distance.

Solution
$$50t = 10(t + 2)$$
$$50t = 10t + 20$$
$$40t = 20$$
$$t = \frac{1}{2}$$
$$50t = 50\left(\frac{1}{2}\right) = 25$$

The car overtakes the cyclist 25 mi from the starting point.

40. Strategy
To find the area, replace the variable s in the equation $A = s^2$ by the given value and solve for A.

Solution
$$A = s^2$$
$$A = (3x + 2)^2$$
$$A = (3x + 2)(3x + 2)$$
$$A = 9x^2 + 12x + 4$$
The area is $(9x^2 + 12x + 4)$ ft^2.

Chapter 8: Factoring

Prep Test

1. $2 \cdot 3 \cdot 5$ [1.3.2]

2. $-12y + 15$ [2.2.4]

3. $-a + b$ [2.2.4]

4. $2(a - b) - 5(a - b)$ [2.2.5]
   $= 2a - 2b - 5a + 5b$
   $= -3a + 3b$

5. $x = 0$ [3.1.3]

6. $2x + 1 = 0$ [2.2.1]
   $2x = -1$
   $x = -\dfrac{1}{2}$

7. $(x + 4)(x - 6)$ [7.3.3]
   $= x^2 - 6x + 4x - 24$
   $= x^2 - 2x - 24$

8. $(2x - 5)(3x + 2)$ [7.3.3]
   $= 6x^2 + 4x - 15x - 10$
   $= 6x^2 - 11x - 10$

9. x^3 [7.4.1]

10. $3x^3 y$ [7.4.1]

Go Figure

We can rewrite $2^{150} = 2^{50} \cdot 2^{50} \cdot 2^{50} = (2 \cdot 2 \cdot 2)^{50}$,

and $3^{100} = 3^{50} \cdot 3^{50} = (3 \cdot 3)^{50}$, and we have 5^{50}.

Since they are each to the 50th power, we can compare the bases of these numbers.
Since $9 > 8 > 5$, then $3^{100} > 2^{150} > 5^{50}$.

Section 8.1

Concept Review 8.1

1. Always true

3. Always true

5. Never true
   $y(y + 4) + 6$ is not the product of factors.

7. Never true
   $a^2 - 8a - ab + 8b = a(a - 8) - b(a - 8)$
   $\qquad\qquad\qquad\qquad = (a - 8)(a - b)$

Objective 8.1.1 Exercises

3. The GCF is x^3

5. The GCF is xy^4

7. The GCF is $xy^4 z^2$

9. $14a^3 = 2 \cdot 7a^3$
   $49a^7 = 7 \cdot 7a^7$
   The GCF is $7a^3$.

11. $3x^2 y^2 = 3x^2 y^2$
    $5ab^2 = 5ab^2$
    There is no common factor other than 1.

13. $9a^2 b^4 = 3 \cdot 3a^2 b^4$
    $24a^4 b^2 = 2 \cdot 2 \cdot 2 \cdot 3a^4 b^2$
    The GCF is $3a^2 b^2$.

15. $ab^3 = ab^3$
    $4a^2 b = 2 \cdot 2a^2 b$
    $12a^2 b^3 = 2 \cdot 2 \cdot 3a^2 b^3$
    The GCF is ab.

17. $2x^2 y = 2x^2 y$
    $4xy = 2 \cdot 2xy$
    $8x = 2 \cdot 2 \cdot 2x$
    The GCF is $2x$.

19. $3x^2 y^2 = 3x^2 y^2$
    $6x = 2 \cdot 3x$
    $9x^3 y^3 = 3 \cdot 3x^3 y^3$
    The GCF is $3x$.

21. The GCF is 5.
    $5a + 5 = 5(a) + 5(1) = 5(a + 1)$

23. $16 = 2 \cdot 2 \cdot 2 \cdot 2$
    $8a^2 = 2 \cdot 2 \cdot 2a^2$
    The GCF is 8.
    $16 - 8a^2 = 8(2) + 8(-a^2)$
    $\qquad\qquad = 8(2 - a^2)$

25. $8x = 2 \cdot 2 \cdot 2x$
    $12 = 2 \cdot 2 \cdot 3$
    The GCF is 4.
    $8x + 12 = 4(2x) + 4(3)$
    $\qquad\qquad = 4(2x + 3)$

27. $30a = 2 \cdot 3 \cdot 5a$
    $6 = 2 \cdot 3$
    The GCF is 6.
    $30a - 6 = 6(5a) + 6(-1)$
    $\qquad\qquad = 6(5a - 1)$

29. $7x^2 = 7x^2$
    $3x = 3x$
    The GCF is x.
    $7x^2 - 3x = x(7x) + x(-3)$
    $\qquad\qquad = x(7x - 3)$

31. $3a^2 = 3a^2$
$5a^5 = 5a^5$
The GCF is a^2.
$3a^2 + 5a^5 = a^2(3) + a^2(5a^3)$
$\quad\quad\quad\quad = a^2(3 + 5a^3)$

33. $14y^2 = 2 \cdot 7y^2$
$11y = 11y$
The GCF is y.
$14y^2 + 11y = y(14y) + y(11)$
$\quad\quad\quad\quad = y(14y + 11)$

35. $2x^4 = 2x^4$
$4x = 2 \cdot 2x$
The GCF is $2x$.
$2x^4 - 4x = 2x(x^3) + 2x(-2)$
$\quad\quad\quad = 2x(x^3 - 2)$

37. $10x^4 = 2 \cdot 5x^4$
$12x^2 = 2 \cdot 2 \cdot 3x^2$
The GCF is $2x^2$.
$10x^4 - 12x^2 = 2x^2(5x^2) + 2x^2(-6)$
$\quad\quad\quad\quad = 2x^2(5x^2 - 6)$

39. The GCF is xy.
$x^2y - xy^3 = xy(x) + xy(-y^2)$
$\quad\quad\quad\quad = xy(x - y^2)$

41. There is no common factor other than 1.
$2a^5b + 3xy^3$

43. $6a^2b^3 = 2 \cdot 3a^2b^3$
$12b^2 = 2 \cdot 2 \cdot 3b^2$
The GCF is $6b^2$.
$6a^2b^3 - 12b^2 = 6b^2(a^2b) + 6b^2(-2)$
$\quad\quad\quad\quad = 6b^2(a^2b - 2)$

45. $6a^2bc = 2 \cdot 3a^2bc$
$4ab^2c = 2 \cdot 2ab^2c$
The GCF is $2abc$.
$6a^2bc + 4ab^2c = 2abc(3a) + 2abc(2b)$
$\quad\quad\quad\quad = 2abc(3a + 2b)$

47. $18x^2y^2 = 2 \cdot 3 \cdot 3x^2y^2$
$9a^2b^2 = 3 \cdot 3a^2b^2$
The GCF is 9.
$18x^2y^2 - 9a^2b^2 = 9(2x^2y^2) + 9(-a^2b^2)$
$\quad\quad\quad\quad = 9(2x^2y^2 - a^2b^2)$

49. $6x^3y^3 = 2 \cdot 3x^3y^3$
$12x^6y^6 = 2 \cdot 2 \cdot 3x^6y^6$
The GCF is $6x^3y^3$.
$6x^3y^3 - 12x^6y^6 = 6x^3y^3(1) + 6x^3y^3(-2x^3y^3)$
$\quad\quad\quad\quad = 6x^3y^3(1 - 2x^3y^3)$

51. The GCF is x.
$x^3 - 3x^2 - x = x(x^2) + x(-3x) + x(-1)$
$\quad\quad\quad\quad = x(x^2 - 3x - 1)$

53. $2x^2 = 2 \cdot x^2$
$8x = 2 \cdot 2 \cdot 2x$
$12 = 2 \cdot 2 \cdot 3$
The GCF is 2.
$2x^2 + 8x - 12 = 2(x^2) + 2(4x) + 2(-6)$
$\quad\quad\quad\quad = 2(x^2 + 4x - 6)$

55. The GCF is b.
$b^3 - 5b^2 - 7b = b(b^2) + b(-5b) + b(-7)$
$\quad\quad\quad\quad = b(b^2 - 5b - 7)$

57. $8y^2 = 2 \cdot 2 \cdot 2y^2$
$12y = 2 \cdot 2 \cdot 3y$
$32 = 2 \cdot 2 \cdot 2 \cdot 2 \cdot 2$
The GCF is 4.
$8y^2 - 12y + 32 = 4(2y^2) + 4(-3y) + 4(8)$
$\quad\quad\quad\quad = 4(2y^2 - 3y + 8)$

59. $5y^3 = 5y^3$
$20y^2 = 2 \cdot 2 \cdot 5y^2$
$10y = 2 \cdot 5y$
The GCF is $5y$.
$5y^3 - 20y^2 + 10y = 5y(y^2) + 5y(-4y) + 5y(2)$
$\quad\quad\quad\quad = 5y(y^2 - 4y + 2)$

61. $3y^4 = 3y^4$
$9y^3 = 3 \cdot 3y^3$
$6y^2 = 2 \cdot 3y^2$
The GCF is $3y^2$.
$3y^4 - 9y^3 - 6y^2 = 3y(y^2) + 3y^2(-3y) + 3y^2(-2)$
$\quad\quad\quad\quad = 3y^2(y^2 - 3y - 2)$

63. $3y^3 = 3y^3$
$9y^2 = 3 \cdot 3y^2$
$24y = 2 \cdot 2 \cdot 2 \cdot 3y$
The GCF is $3y$.
$3y^3 - 9y^2 + 24y = 3y(y^2) + 3y(-3y) + 3y(8)$
$\quad\quad\quad\quad = 3y(y^2 - 3y + 8)$

65. The GCF is a^2.
$6a^5 - 3a^3 - 2a^2 = a^2(6a^3) + a^2(-3a) + a^2(-2)$
$\quad\quad\quad\quad = a^2(6a^3 - 3a - 2)$

67. $8x^2y^2 = 2 \cdot 2 \cdot 2x^2y^2$
$4x^2y = 2 \cdot 2x^2y$
$x^2 = x^2$
The GCF is x^2.
$8x^2y^2 - 4x^2y + x^2$
$= x^2(8y^2) + x^2(-4y) + x^2(1)$
$= x^2(8y^2 - 4y + 1)$

69. The GCF is $x^3 y^3$.

$4x^5 y^5 - 8x^4 y^4 + x^3 y^3$

$= x^3 y^3 (4x^2 y^2) + x^3 y^3 (-8xy) + x^3 y^3 (1)$

$= x^3 y^3 (4x^2 y^2 - 8xy + 1)$

Objective 8.1.2 Exercises

71. $x(a+b) + 2(a+b) = (a+b)(x+2)$

73. $x(b+2) - y(b+2) = (b+2)(x-y)$

75. $a(y-4) - b(y-4) = (y-4)(a-b)$

77. $a(x-2) - b(2-x) = a(x-2) + b(x-2)$

79. $b(a-7) + 3(7-a) = b(a-7) - 3(a-7)$

81. $x(a-2b) - y(2b-a) = x(a-2b) + y(a-2b)$

83. $a(x-2) + 5(2-x) = a(x-2) - 5(x-2)$
$= (x-2)(a-5)$

85. $b(y-3) + 3(3-y) = b(y-3) - 3(y-3)$
$= (y-3)(b-3)$

87. $a(x-y) - 2(y-x) = a(x-y) + 2(x-y)$
$= (x-y)(a+2)$

89. $z(c+5) - 8(5+c) = z(c+5) - 8(c+5)$
$= (c+5)(z-8)$

91. $w(3x-4) - (4-3x) = w(3x-4) + (3x-4)$
$= (3x-4)(w+1)$

93. $x^3 + 4x^2 + 3x + 12 = (x^3 + 4x^2) + (3x+12)$
$= x^2(x+4) + 3(x+4)$
$= (x+4)(x^2+3)$

95. $2y^3 + 4y^2 + 3y + 6 = (2y^3 + 4y^2) + (3y+6)$
$= 2y^2(y+2) + 3(y+2)$
$= (y+2)(2y^2+3)$

97. $ab + 3b - 2a - 6 = (ab+3b) + (-2a-6)$
$= b(a+3) - 2(a+3)$
$= (a+3)(b-2)$

99. $x^2 a - 2x^2 - 3a + 6 = (x^2 a - 2x^2) + (-3a+6)$
$= x^2(a-2) - 3(a-2)$
$= (a-2)(x^2-3)$

101. $3ax - 3bx - 2ay + 2by$
$= (3ax - 3bx) + (-2ay + 2by)$
$= 3x(a-b) \quad 2y(a-b)$
$= (a-b)(3x-2y)$

103. $x^2 - 3x + 4ax - 12a = (x^2 - 3x) + (4ax - 12a)$
$= x(x-3) + 4a(x-3)$
$= (x-3)(x+4a)$

105. $xy - 5y - 2x + 10 = (xy - 5y) - (2x - 10)$
$= y(x-5) - 2(x-5)$
$= (x-5)(y-2)$

107. $21x^2 + 6xy - 49x - 14y$
$= (21x^2 + 6xy) - (49x + 14y)$
$= 3x(7x+2y) - 7(7x+2y)$
$= (7x+2y)(3x-7)$

109. $2ra + a^2 - 2r - a = (2ra + a^2) - (2r+a)$
$= a(2r+a) - 1(2r+a)$
$= (2r+a)(a-1)$

111. $4x^2 + 3xy - 12y - 16x = (4x^2 + 3xy) - (12y + 16x)$
$= x(4x+3y) - 4(3y+4x)$
$= (4x+3y)(x-4)$

113. $10xy^2 - 15xy + 6y - 9 = (10xy^2 - 15xy) + (6y - 9)$
$= 5xy(2y-3) + 3(2y-3)$
$= (2y-3)(5xy+3)$

Applying Concepts 8.1

115. a. $2x^2 + 6x + 5x + 15 = (2x^2 + 6x) + (5x+15)$
$= 2x(x+3) + 5(x+3)$
$= (x+3)(2x+5)$

b. $2x^2 + 5x + 6x + 15 = (2x^2 + 5x) + (6x+15)$
$= x(2x+5) + 3(2x+5)$
$= (2x+5)(x+3)$

117. a. $2a^2 - 2ab - 3ab + 3b^2$
$= (2a^2 - 2ab) - (3ab - 3b^2)$
$= 2a(a-b) - 3b(a-b)$
$= (a-b)(2a-3b)$

b. $2a^2 - 3ab - 2ab + 3b^2$
$= (2a^2 - 3ab) - (2ab - 3b^2)$
$= a(2b-3b) - b(2a-3b)$
$= (2a-3b)(a-b)$

119. 28 is a perfect number.
The factors of 28 are 1, 2, 4, 7, and 14,
and $1 + 2 + 4 + 7 + 14 = 28$.

121. $P = 2L + 2W = 2(L+W)$
When $L + W$ doubles, P also doubles.

Section 8.2

Concept Review 8.2

1. Never true
The value of b is -8.

3. Always true

5. Never true
When the constant term is negative, the constant terms of the binomials have opposite signs.

180 **Chapter 8:** *Factoring*

Objective 8.2.1 Exercises

1.

Factors of 18	Sum
1, 18	19
2, 9	11
3, 6	9

3.

Factors of –21	Sum
1, –21	–20
–1, 21	20
3, –7	–4
–3, 7	4

5.

Factors of –28	Sum
1, –28	–27
–1, 28	27
2, –14	–12
–2, 14	12
4, –7	–3
–4, 7	3

7.

Factors	Sum
+1, +2	3

$$x^2 + 3x + 2 = (x+1)(x+2)$$

9.

Factors	Sum
–1, +2	1
+1, –2	–1

$$x^2 - x - 2 = (x+1)(x-2)$$

11.

Factors	Sum
–1, +12	11
+1, –12	–11
–2, +6	4
+2, –6	–4
–3, +4	1
+3, –4	–1

$$a^2 + a - 12 = (a+4)(a-3)$$

13.

Factors	Sum
–1, –2	–3

$$a^2 - 3a + 2 = (a-1)(a-2)$$

15.

Factors	Sum
–1, +2	1
+1, –2	–1

$$a^2 + a - 2 = (a+2)(a-1)$$

17.

Factors	Sum
–1, –9	–10
–3, –3	–6

$$b^2 - 6b + 9 = (b-3)(b-3)$$

19.

Factors	Sum
–1, +8	7
+1, –8	–7
–2, +4	2
+2, –4	–2

$$b^2 + 7b - 8 = (b+8)(b-1)$$

21.

Factors	Sum
–1, +55	54
+1, –55	–54
–5, +11	6
+5, –11	–6

$$y^2 + 6y - 55 = (y+11)(y-5)$$

23.

Factors	Sum
–1, –6	–7
–2, –3	–5

$$y^2 - 5y + 6 = (y-2)(y-3)$$

25.

Factors	Sum
–1, –45	–46
–3, –15	–18
–5, –9	–14

$$z^2 - 14z + 45 = (z-5)(z-9)$$

27.

Factors	Sum
–1, +160	159
+1, –160	–159
–2, +80	78
+2, –80	–78
–4, +40	36
+4, –40	–36
–5, +32	27
+5, –32	–27
–8, +20	12
+8, –20	–12
–10, +16	6
+10, –16	–6

$$z^2 - 12z - 160 = (z+8)(z-20)$$

29.

Factors	Sum
+1, +27	28
+3, +9	12

$$p^2 + 12p + 27 = (p+3)(p+9)$$

31.

Factors	Sum
+1, +100	101
+2, +50	52
+4, +25	29
+5, +20	25
+10, +10	20

$$x^2 + 20x + 100 = (x+10)(x+10)$$

33.

Factors	Sum
+1, +20	21
+2, +10	12
+4, +5	9

$b^2 + 9b + 20 = (b+4)(b+5)$

35.

Factors	Sum
−1, +42	41
+1, −42	−41
−2, +21	19
+2, −21	−19
−3, +14	11
+3, −14	−11
−6, +7	1
+6, −7	−1

$x^2 - 11x - 42 = (x+3)(x-14)$

37.

Factors	Sum
−1, +20	19
+1, −20	−19
−2, +10	8
+2, −10	−8
−4, +5	1
+4, −5	−1

$b^2 - b - 20 = (b+4)(b-5)$

39.

Factors	Sum
−1, +51	50
+1, −51	−50
−3, +17	14
+3, −17	−14

$y^2 - 14y - 51 = (y+3)(y-17)$

41.

Factors	Sum
−1, +21	20
+1, −21	−20
−3, +7	4
+3, −7	−4

$p^2 - 4p - 21 = (p+3)(p-7)$

43.

Factors	Sum
−1, −32	−33
−2, −16	−18
−4, −8	−12

$y^2 - 8y + 32$ is nonfactorable over the integers.

45.

Factors	Sum
−1, −75	−76
−3, −25	−28
−5, −15	−20

$x^2 - 20x + 75 = (x-5)(x-15)$

47.

Factors	Sum
−1, −56	−57
−2, −28	−30
−4, −14	−18
−7, −8	−15

$x^2 - 15x + 56 = (x-7)(x-8)$

49.

Factors	Sum
−1, +56	55
+1, −56	−55
−2, +28	26
+2, −28	−26
−4, +14	10
+4, −14	−10
−7, +8	1
+7, −8	−1

$x^2 + x - 56 = (x+8)(x-7)$

51.

Factors	Sum
−1, +72	71
+1, −72	−71
−2, +36	34
+2, −36	−34
−3, +24	21
+3, −24	−21
−4, +18	14
+4, −18	−14
−6, +12	6
+6, −12	−6
−8, +9	1
+8, −9	−1

$a^2 - 21a - 72 = (a+3)(a-24)$

53.

Factors	Sum
−1, −36	−37
−2, −18	−20
−3, −12	−15
−4, −9	−13
−6, −6	−12

$a^2 - 15a + 36 = (a-3)(a-12)$

55.

Factors	Sum
−1, +136	135
+1, −136	−135
−2, +68	66
+2, −68	−66
−4, +34	30
+4, −34	−30
−8, +17	9
+8, −17	−9

$z^2 - 9z - 136 = (z+8)(z-17)$

57.

Factors	Sum
−1, +90	89
+1, −90	−89
−2, +45	43
+2, −45	−43
−3, +30	27
+3, −30	−27
−5, +18	13
+5, −18	−13
−6, +15	9
+6, −15	−9
−9, +10	1
+9, −10	−1

$$c^2 - c - 90 = (c + 9)(c - 10)$$

Objective 8.2.2 Exercises

59. The GCF is 2.
$$2x^2 + 6x + 4 = 2(x^2 + 3x + 2)$$
Factor the trinomial.

Factors	Sum
+1, +2	+3

$$2x^2 + 6x + 4 = 2(x + 1)(x + 2)$$

61. The GCF is 3.
$$3a^2 + 3a - 18 = 3(a^2 + a - 6)$$
Factor the trinomial.

Factors	Sum
−1, +6	+5
+1, −6	−5
−2, +3	+1
+2, −3	−1

$$3a^2 + 3a - 18 = 3(a + 3)(a - 2)$$

63. The GCF is a.
$$ab^2 + 2ab - 15a = a(b^2 + 2b - 15)$$
Factor the trinomial.

Factors	Sum
−1, +15	+14
+1, −15	−14
−3, +5	+2
+3, −5	−2

$$ab^2 + 2ab - 15a = a(b + 5)(b - 3)$$

65. The GCF is x.
$$xy^2 - 5xy + 6x = x(y^2 - 5y + 6)$$
Factor the trinomial.

Factors	Sum
−1, −6	−7
−2, −3	−5

$$xy^2 - 5xy + 6x = x(y - 2)(y - 3)$$

67. The GCF is z.
$$z^3 - 7z^2 + 12z = z(z^2 - 7z + 12)$$
Factor the trinomial.

Factors	Sum
−1, −12	−13
−2, −6	−8
−3, −4	−7

$$z^3 - 7z^2 + 12z = z(z - 3)(z - 4)$$

69. The GCF is $3y$.
$$3y^3 - 15y^2 + 18y = 3y(y^2 - 5y + 6)$$
Factor the trinomial.

Factors	Sum
−1, −6	−7
−2, −3	−5

$$3y^3 - 15y^2 + 18y = 3y(y - 2)(y - 3)$$

71. The GCF is 3.
$$3x^2 + 3x - 36 = 3(x^2 + x - 12)$$
Factor the trinomial.

Factors	Sum
−1, +12	+11
+1, −12	−11
−2, +6	+4
+2, −6	−4
−3, +4	+1
+3, −4	−1

$$3x^2 + 3x - 36 = 3(x + 4)(x - 3)$$

73. The GCF is 5.
$$5z^2 - 15z - 140 = 5(z^2 - 3z - 28)$$
Factor the trinomial.

Factors	Sum
−1, +28	+27
+1, −28	−27
−2, +14	+12
+2, −14	−12
−4, +7	+3
+4, −7	−3

$$5z^2 - 15z - 140 = 5(z + 4)(z - 7)$$

75. The GCF is $2a$.
$$2a^3 + 8a^2 - 64a = 2a(a^2 + 4a - 32)$$
Factor the trinomial.

Factors	Sum
−1, +32	+31
+1, −32	−31
−2, +16	+14
+2, −16	−14
−4, +8	+4
+4, −8	−4

$$2a^3 + 8a^2 - 64a = 2a(a + 8)(a - 4)$$

77. There is no common factor.
Factor the trinomial.

Factors	Sum
−1, −6	−7
−2, −3	−5

$x^2 - 5xy + 6y^2 = (x - 2y)(x - 3y)$

79. There is no common factor.
Factor the trinomial.

Factors	Sum
−1, −20	−21
−2, −10	−12
−4, −5	−9

$a^2 - 9ab + 20b^2 = (a - 4b)(a - 5b)$

81. There is no common factor.
Factor the trinomial.

Factors	Sum
−1, +28	+27
+1, −28	−27
−2, +14	+12
+2, −14	−12
−4, +7	+3
+4, −7	−3

$x^2 - 3xy - 28y^2 = (x + 4y)(x - 7y)$

83. There is no common factor.
Factor the trinomial.

Factors	Sum
−1, +41	+40
+1, −41	−40

$y^2 - 15yz - 41z^2$ is nonfactorable over the integers.

85. The GCF is z^2.
$z^4 - 12z^3 + 35z^2 = z^2(z^2 - 12z + 35)$
Factor the trinomial.

Factors	Sum
−1, −35	−36
−5, −7	−12

$z^4 - 12z^3 + 35z^2 = z^2(z - 5)(z - 7)$

87. The GCF is b^2.
$b^4 - 22b^3 + 120b^2 = b^2(b^2 - 22b + 120)$
Factor the trinomial.

Factors	Sum
−1, −120	−121
−2, −60	−62
−3, −40	−43
−4, −30	−34
−5, −24	−29
−6, −20	−26
−8, −15	−23
−10, −12	−22

$b^4 - 22b^3 + 120b^2 = b^2(b - 10)(b - 12)$

89. The GCF is $2y^2$.
$2y^4 - 26y^3 - 96y^2 = 2y^2(y^2 - 13y - 48)$
Factor the trinomial.

Factors	Sum
−1, +48	+47
+1, −48	−47
−2, +24	+22
+2, −24	−22
−3, +16	+13
+3, −16	−13
−4, +12	+8
+4, −12	−8
−6, +8	+2
+6, −8	−2

$2y^4 - 26y^3 - 96y^2 = 2y^2(y + 3)(y - 16)$

91. The GCF is x^2.
$x^4 + 7x^3 - 8x^2 = x^2(x^2 + 7x - 8)$
Factor the trinomial.

Factors	Sum
−1, +8	+7
+1, −8	−7
−2, +4	+2
+2, −4	−2

$x^4 + 7x^3 - 8x^2 = x^2(x + 8)(x - 1)$

93. The GCF is $4y$.
$4x^2y + 20xy - 56y = 4y(x^2 + 5x - 14)$
Factor the trinomial.

Factors	Sum
−1, +14	+13
+1, −14	−13
−2, +7	+5
+2, −7	−5

$4x^2y + 20xy - 56y = 4y(x + 7)(x - 2)$

95. The GCF is 8.
$8y^2 - 32y + 24 = 8(y^2 - 4y + 3)$
Factor the trinomial.

Factors	Sum
−1, −3	−4

$8y^2 - 32y + 24 = 8(y - 1)(y - 3)$

97. The GCF is c.
$c^3 + 13c^2 + 30c = c(c^2 + 13c + 30)$
Factor the trinomial.

Factors	Sum
+1, +30	+31
+2, +15	+17
+3, +10	+13
+5, +6	+11

$c^3 + 13c^2 + 30c = c(c + 3)(c + 10)$

99. The GCF is $3x$.

$3x^3 - 36x^2 + 81x = 3x(x^2 - 12x + 27)$

Factor the trinomial.

Factors	Sum
$-1, -27$	-28
$-3, -9$	-12

$3x^3 - 36x^2 + 81x = 3x(x - 3)(x - 9)$

101. There is no common factor

Factor the trinomial.

Factors	Sum
$-1, -15$	-16
$-3, -5$	-8

$x^2 - 8xy + 15y^2 = (x - 3y)(x - 5y)$

103. There is no common factor.

Factor the trinomial.

Factors	Sum
$-1, -42$	-43
$-2, -21$	-23
$-3, -14$	-17
$-6, -7$	-13

$a^2 - 13ab + 42b^2 = (a - 6b)(a - 7b)$

105. There is no common factor.

Factor the trinomial.

Factors	Sum
$+1, +7$	$+8$

$y^2 + 8yz + 7z^2 = (y + z)(y + 7z)$

107. The GCF is $3y$.

$3x^2y + 60xy - 63y = 3y(x^2 + 20x - 21)$

Factor the trinomial.

Factors	Sum
$-1, +21$	$+20$
$+1, -21$	-20
$-3, +7$	$+4$
$+3, -7$	-4

$3x^2y + 60xy - 63y = 3y(x + 21)(x - 1)$

109. The GCF is $3x$.

$3x^3 + 3x^2 - 36x = 3x(x^2 + x - 12)$

Factor the trinomial.

Factors	Sum
$-1, +12$	$+11$
$+1, -12$	-11
$-2, +6$	$+4$
$+2, -6$	-4
$-3, +4$	$+1$
$+3, -4$	-1

$3x^3 + 3x^2 - 36x = 3x(x + 4)(x - 3)$

111. The GCF is $4z$.

$4z^3 + 32z^2 - 132z = 4z(z^2 + 8z - 33)$

Factor the trinomial.

Factors	Sum
$-1, +33$	$+32$
$+1, -33$	-32
$-3, +11$	$+8$
$+3, -11$	-8

$4z^3 + 32z^2 - 132z = 4z(z + 11)(z - 3)$

Applying Concepts 8.2

113. $20 + c^2 + 9c = c^2 + 9c + 20$

Factors	Sum
$+1, +20$	$+21$
$+2, +10$	$+12$
$+4, +5$	$+9$

$20 + c^2 + 9c = (c + 4)(c + 5)$

115. The GCF is a^2.

$45a^2 + a^2b^2 - 14a^2b = a^2(45 + b^2 - 14b)$

Factor the trinomial.

$45 + b^2 - 14b = b^2 - 14b + 45$

Factors	Sum
$-1, -45$	-46
$-3, -15$	-18
$-5, -9$	-14

$45a^2 + a^2b^2 - 14a^2b = a^2(b - 5)(b - 9)$

117. $x^2 + kx + 35$

Factors	Sum (k)
$+7, +5$	$+12$
$-7, -5$	-12
$+35, +1$	$+36$
$-35, -1$	-36

$k = 12, -12, 36,$ or -36

119. $x^2 - kx + 21$

Factors	Sum (k)
$+1, +21$	$+22$
$-1, -21$	-22
$+3, +7$	$+10$
$-3, -7$	-10

$k = 22, -22, 10,$ or -10

121. $y^2 + 4k + k, \ k > 0$

Factor two positive integers that sum to 4. Their product is k.

Integers	Product
$+1, +3,$	3
$+2, +2$	4

k can be 3 or 4.

123. $a^2 - 6a + k, k > 0$

Find two negative integers that sum to -6. Their product is k.

Integers	Product
$-1, -5$	5
$-2, -4$	8
$-3, -3$	9

k can be 5, 8, or 9.

125. $x^2 - 3x + k, k > 0$

Find two negative integers that sum to -3. Their product is k.

Integers	Product
$-2, -1$	2

k can be 2.

Section 8.3

Concept Review 8.3

1. Never true
The value of a is 3.

3. Sometimes true
The factorization of the trinomial may have a common factor.

5. Always true

Objective 8.3.1 Exercises

1. Factors of 2: 1, 2   Factors of 1: $+1, +1$

Trial Factors	Middle Term
$(1x+1)(2x+1)$	$x + 2x = 3x$

$2x^2 + 3x + 1 = (x+1)(2x+1)$

3. Factors of 2: 1, 2   Factors of 3: $+1, +3$

Trial Factors	Middle Term
$(1y+1)(2y+3)$	$3y + 2y = 5y$
$(1y+3)(2y+1)$	$y + 6y = 7y$

$2y^2 + 7y + 3 = (y+3)(2y+1)$

5. Factors of 2: 1, 2   Factors of 1: $-1, -1$

Trial Factors	Middle Term
$(1a-1)(2a-1)$	$-a - 2a = -3a$

$2a^2 - 3a + 1 = (a-1)(2a-1)$

7. Factors of 2: 1, 2   Factors of 5: $-1, -5$

Trial Factors	Middle Term
$(1b-1)(2b-5)$	$-5b - 2b = -7b$
$(1b-5)(2b-1)$	$-b - 10b = -11b$

$2b^2 - 11b + 5 = (b-5)(2b-1)$

9. Factors of 2: 1, 2   Factors of -1: $-1, +1$

Trial Factors	Middle Term
$(1x-1)(2x+1)$	$x - 2x = -x$
$(1x+1)(2x-1)$	$-x + 2x = x$

$2x^2 + x - 1 = (x+1)(2x-1)$

11. Factors of 2: 1, 2   Factors of -3: $-1, +3$
                                          $+1, -3$

Trial Factors	Middle Term
$(1x-1)(2x+3)$	$3x - 2x = x$
$(1x+3)(2x-1)$	$-x + 6x = 5x$
$(1x+1)(2x-3)$	$-3x + 2x = -x$
$(1x-3)(2x+1)$	$x - 6x = -5x$

$2x^2 - 5x - 3 = (x-3)(2x+1)$

13. Factors of 6: 1, 6   Factors of 3: $-1, -3$
                     2, 3

Trial Factors	Middle Term
$(1z-1)(6z-3)$	Common factor
$(1z-3)(6z-1)$	$-z - 18z = -19z$
$(2z-1)(3z-3)$	Common factor
$(2z-3)(3z-1)$	$-2z - 9z = -11z$

$6z^2 - 7z + 3$ is nonfactorable over the integers.

15. Factors of 6: 1, 6   Factors of 4: $-1, -4$
                     2, 3 $-2, -2$

Trial Factors	Middle Term
$(1t-1)(6t-4)$	Common factor
$(1t-4)(6t-1)$	$-t - 24t = -25t$
$(1t-2)(6t-2)$	Common factor
$(2t-1)(3t-4)$	$-8t - 3t = -11t$
$(2t-4)(3t-1)$	Common factor
$(2t-2)(3t-2)$	Common factor

$6t^2 - 11t + 4 = (2t-1)(3t-4)$

17. Factors of 8: 1, 8   Factors of 4: $+1, +4$
                     2, 4 $+2, +2$

Trial Factors	Middle Term
$(1x+1)(8x+4)$	Common factor
$(1x+4)(8x+1)$	$x + 32x = 33x$
$(1x+2)(8x+2)$	Common factor
$(2x+1)(4x+4)$	Common factor
$(2x+4)(4x+1)$	Common factor
$(2x+2)(4x+2)$	Common factor

$8x^2 + 33x + 4 = (x+4)(8x+1)$

19. Factors of 3: 1, 3   Factors of 16: $-1, -16$
                                             $-2, -8$
                                             $-4, -4$

Trial Factors	Middle Term
$(1b-1)(3b-16)$	$-16b - 3b = -19b$
$(1b-16)(3b-1)$	$-b - 48b = -49b$
$(1b-2)(3b-8)$	$-8b - 6b = -14b$
$(1b-8)(3b-2)$	$-2b - 24b = -26b$
$(1b-4)(3b-4)$	$-4b - 12b = -16b$

$3b^2 - 16b + 16 = (b-4)(3b-4)$

21. Factors of 2: 1, 2   Factors of -14: $-1, +14$
                              $+1, -14$
                              $-2, +7$
                              $+2, -7$

Trial Factors	Middle Term
$(1z - 1)(2z + 14)$	Common factor
$(1z + 14)(2z - 1)$	$-z + 28z = 27z$
$(1z + 1)(2z - 14)$	Common factor
$(1z - 14)(2z + 1)$	$z - 28z = -27z$
$(1z - 2)(2z + 7)$	$7z - 4z = 3z$
$(1z + 7)(2z - 2)$	Common factor
$(1z + 2)(2z - 7)$	$-7z + 4z = -3z$
$(1z - 7)(2z + 2)$	Common factor

$2z^2 - 27z - 14 = (z - 14)(2z + 1)$

23. Factors of 3: 1, 3   Factors of -16: $-1, +16$
                              $+1, -16$
                              $-2, +8$
                              $+2, -8$
                              $-4, +4$

Trial Factors	Middle Term
$(1p - 1)(3p + 16)$	$16p - 3p = 13p$
$(1p + 16)(3p - 1)$	$-p + 48p = 47p$
$(1p + 1)(3p - 16)$	$-16p + 3p = -13p$
$(1p - 16)(3p + 1)$	$p - 48p = -47p$
$(1p - 2)(3p + 8)$	$8p - 6p = 2p$
$(1p + 8)(3p - 2)$	$-2p + 24p = 22p$
$(1p + 2)(3p - 8)$	$-8p + 6p = -2p$
$(1p - 8)(3p + 2)$	$2p - 24p = -22p$
$(1p - 4)(3p + 4)$	$4p - 12p = -8p$
$(1p + 4)(3p - 4)$	$-4p + 12p = 8p$

$3p^2 + 22p - 16 = (p + 8)(3p - 2)$

25. Factors of 6: 1, 6   Factors of 12: $-1, -12$
                    2, 3 $-2, -6$
                                            $-3, -4$

Trial Factors	Middle Term
$(1x - 1)(6x - 12)$	Common factor
$(1x - 12)(6x - 1)$	$-x - 72x = -73x$
$(1x - 2)(6x - 6)$	Common factor
$(1x - 6)(6x - 2)$	Common factor
$(1x - 3)(6x - 4)$	Common factor
$(1x - 4)(6x - 3)$	Common factor
$(2x - 1)(3x - 12)$	Common factor
$(2x - 12)(3x - 1)$	Common factor
$(2x - 2)(3x - 6)$	Common factor
$(2x - 6)(3x - 2)$	Common factor
$(2x - 3)(3x - 4)$	$-8x - 9x = -17x$
$(2x - 4)(3x - 3)$	Common factor

$6x^2 - 17x + 12 = (2x - 3)(3x - 4)$

27. Factors of 5: 1, 5   Factors of -14: $-1, +14$
                              $+1, -14$
                              $-2, 7$
                              $+2, -7$

Trial Factors	Middle Term
$(1b - 1)(5b + 14)$	$14b - 5b = 9b$
$(1b + 14)(5b - 1)$	$-b + 70b = 69b$
$(1b + 1)(5b - 14)$	$-14b + 5b = -9b$
$(1b - 14)(5b + 1)$	$b - 70b = -69b$
$(1b - 2)(5b + 7)$	$7b - 10b = -3b$
$(1b + 7)(5b - 2)$	$-2b + 35b = 33b$
$(1b + 2)(5b - 7)$	$-7b + 10b = 3b$
$(1b - 7)(5b + 2)$	$2b - 35b = -33b$

$5b^2 + 33b - 14 = (b + 7)(5b - 2)$

29. Factors of 6: 1, 6 Factors of −24: −1, +24
　　　　　　　 2, 3　　　　　　　　　　 +1, −24
　　　　　　　　　　　　　　　　　　 −2, +12
　　　　　　　　　　　　　　　　　　 +2, −12
　　　　　　　　　　　　　　　　　　 −3, +8
　　　　　　　　　　　　　　　　　　 +3, −8
　　　　　　　　　　　　　　　　　　 −4, +6
　　　　　　　　　　　　　　　　　　 +4, −6

Trial Factors	Middle Term
$(1a-1)(6a+24)$	Common factor
$(1a+24)(6a-1)$	$-a+144a=143a$
$(1a+1)(6a-24)$	Common factor
$(1a-24)(6a+1)$	$a-144a=-143a$
$(1a-2)(6a+12)$	Common factor
$(1a+12)(6a-2)$	Common factor
$(1a+2)(6a-12)$	Common factor
$(1a-12)(6a+2)$	Common factor
$(1a-3)(6a+8)$	Common factor
$(1a+8)(6a-3)$	Common factor
$(1a+3)(6a-8)$	Common factor
$(1a-8)(6a+3)$	Common factor
$(1a-4)(6a+6)$	Common factor
$(1a+6)(6a-4)$	Common factor
$(1a+4)(6a-6)$	Common factor
$(1a-6)(6a+4)$	Common factor
$(2a-1)(3a+24)$	Common factor
$(2a+24)(3a-1)$	Common factor
$(2a+1)(3a-24)$	Common factor
$(2a-24)(3a+1)$	Common factor
$(2a-2)(3a+12)$	Common factor
$(2a+12)(3a-2)$	Common factor
$(2a+2)(3a-12)$	Common factor
$(2a-12)(3a+2)$	Common factor
$(2a-3)(3a+8)$	$16a-9a=7a$
$(2a+8)(3a-3)$	Common factor
$(2a+3)(3a-8)$	$-16a+9a=-7a$
$(2a-8)(3a+3)$	Common factor
$(2a-4)(3a+6)$	Common factor
$(2a+6)(3a-4)$	Common factor
$(2a+4)(3a-6)$	Common factor
$(2a-6)(3a+4)$	Common factor

$6a^2+7a-24=(2a-3)(3a+8)$

31. Factors of 18: 1, 18 Factors of −5: −1, +5
　　　　　　　 2, 9　　　　　　　　　 +1, −5
　　　　　　　 3, 6

Trial Factors	Middle Term
$(1t-1)(18t+5)$	$5t-18t=-13t$
$(1t+5)(18t-1)$	$-t+90t=89t$
$(1t+1)(18t-5)$	$-5t+18t=13t$
$(1t-5)(18t+1)$	$t-90t=-89t$
$(2t-1)(9t+5)$	$10t-9t=t$
$(2t+5)(9t-1)$	$-2t+45t=43t$
$(2t+1)(9t-5)$	$-10t+9t=-t$
$(2t-5)(9t+1)$	$2t-45t=-43t$
$(3t-1)(6t+5)$	$15t-6t=9t$
$(3t+5)(6t-1)$	$-3t+30t=27t$
$(3t+1)(6t-5)$	$-15t+6t=-9t$
$(3t-5)(6t+1)$	$3t-30t=-27t$

$18t^2-9t-5=(3t+1)(6t-5)$

33. Factors of 15: 1, 15 Factors of −21: −1, +21
　　　　　　　 3, 5　　　　　　　　　　 +1, −21
　　　　　　　　　　　　　　　　　　 −3, +7
　　　　　　　　　　　　　　　　　　 +3, −7

Trial Factors	Middle Term
$(1a-1)(15a+21)$	Common factor
$(1a+21)(15a-1)$	$-a+315a=314a$
$(1a+1)(15a-21)$	Common factor
$(1a-21)(15a+1)$	$a-315a=-314a$
$(1a-3)(15a+7)$	$7a-45a=-38a$
$(1a+7)(15a-3)$	Common factor
$(1a+3)(15a-7)$	$-7a+45a=38a$
$(1a-7)(15a+3)$	Common factor
$(3a-1)(5a+21)$	$63a-5a=58a$
$(3a+21)(5a-1)$	Common factor
$(3a+1)(5a-21)$	$-63a+5a=-58a$
$(3a-21)(5a+1)$	Common factor
$(3a-3)(5a+7)$	Common factor
$(3a+7)(5a-3)$	$-9a+35a=26a$
$(3a+3)(5a-7)$	Common factor
$(3a-7)(5a+3)$	$9a-35a=-26a$

$15a^2+26a-21=(3a+7)(5a-3)$

35. Factors of 8: 1, 8 Factors of 15: −1, −15
　　　　　　　 2, 4　　　　　　　　　 −3, −5

Trial Factors	Middle Term
$(1y-1)(8y-15)$	$-15y-8y=-23y$
$(1y-15)(8y-1)$	$-y-120y=-121y$
$(1y-3)(8y-5)$	$-5y-24y=-29y$
$(1y-5)(8y-3)$	$-3y-40y=-43y$
$(2y-1)(4y-15)$	$-30y-4y=-34y$
$(2y-15)(4y-1)$	$-2y-60y=-62y$
$(2y-3)(4y-5)$	$-10y-12y=-22y$
$(2y-5)(4y-3)$	$-6y-20y=-26y$

$8y^2-26y+15=(2y-5)(4y-3)$

37. Factors of 8: 1, 8   Factors of −15: −1, +15
                      2, 4   +1, −15
                                                     −3, +5
                                                     +3, −5

Trial Factors	Middle Term
$(1z - 1)(8z + 15)$	$15z - 8z = 7z$
$(1z + 15)(8z - 1)$	$-z + 120z = 119z$
$(1z + 1)(8z - 15)$	$-15z + 8z = -7z$
$(1z - 15)(8z + 1)$	$Z - 120z = -119z$
$(1z - 3)(8z + 5)$	$5z - 24z = -19z$
$(1z + 5)(8z - 3)$	$-3z + 40z = 37z$
$(1z + 3)(8z - 5)$	$-5z + 24z = 19z$
$(1z - 5)(8z + 3)$	$3z - 40z = -37z$
$(2z - 1)(4z + 15)$	$30z - 4z = 26z$
$(2z + 15)(4z - 1)$	$-2z + 60z = 58z$
$(2z + 1)(4z - 15)$	$-30z + 4z = -26z$
$(2z - 15)(4z + 1)$	$2z - 60z = -58z$
$(2z - 3)(4z + 5)$	$10z - 12z = -2z$
$(2z + 5)(4z - 3)$	$-6z + 20z = 14z$
$(2z + 3)(4z - 5)$	$-10z + 12z = 2z$
$(2z - 5)(4z + 3)$	$6z - 20z = -14z$

$8z^2 + 2z - 15 = (2z + 3)(4z - 5)$

39. Factors of 3: 3, 1   Factors of −5: −1, +5
                                                  +1, −5

Trial Factors	Middle Term
$(3x + 5)(1x - 1)$	$-3x + 5x = 2x$
$(3x - 5)(1x + 1)$	$3x - 5x = -2x$
$(3x + 1)(1x - 5)$	$-15x + x = -14x$
$(3x - 1)(1x + 5)$	$15x - x = 14x$

$3x^2 + 14x - 5 = (3x - 1)(x + 5)$

41. Factors of 12: 1, 12   Factors of 12: +1, +12
                        2, 6   +2, +6
                        3, 4   +3, +4

Trial Factors	Middle Term
$(1x + 1)(12x + 12)$	Common factor
$(1x + 12)(12x + 1)$	$x + 144x = 145x$
$(1x + 2)(12x + 6)$	Common factor
$(1x + 6)(12x + 2)$	Common factor
$(1x + 3)(12x + 4)$	Common factor
$(1x + 4)(12x + 3)$	Common factor
$(2x + 1)(6x + 12)$	Common factor
$(2x + 12)(6x + 1)$	$2x + 72x = 74x$
$(2x + 2)(6x + 6)$	Common factor
$(2x + 6)(6x + 2)$	Common factor
$(2x + 3)(6x + 4)$	Common factor
$(2x + 4)(6x + 3)$	Common factor
$(3x + 1)(4x + 12)$	Common factor
$(3x + 12)(4x + 1)$	Common factor
$(3x + 2)(4x + 6)$	Common factor
$(3x + 6)(4x + 2)$	Common factor
$(3x + 3)(4x + 4)$	Common factor
$(3x + 4)(4x + 3)$	$9x + 16x = 25x$

$12x^2 + 25x + 12 = (3x + 4)(4x + 3)$

43. Factors of 3: 1, 3   Factors of 10: +1, +10
                                                    +2, +5

Trial Factors	Middle Term
$(1z + 1)(3z + 10)$	$10z + 3z = 13z$
$(1z + 10)(3z + 1)$	$z + 30z = 31z$
$(1z + 2)(3z + 5)$	$5z + 6z = 11z$
$(1z + 5)(3z + 2)$	$2z + 15z = 17z$

$3z^2 + 95z + 10$ is nonfactorable over the integers.

45. Factors of 3: 1, 3   Factors of −2: −1, +2
                                                    +1, −2

Trial Factors	Middle Term
$(1x - 1y)(3x + 2y)$	$2xy - 3xy = -xy$
$(1x + 2y)(3x - 1y)$	$-xy + 6xy = 5xy$
$(1x + 1y)(3x - 2y)$	$-2xy + 3xy = xy$
$(1x - 2y)(3x + 1y)$	$xy - 6xy = -5xy$

$3x^2 + xy - 2y^2 = (x + y)(3x - 2y)$

47.   Factors of 28: 1, 28   Factors of -1: $-1, +1$
                2, 14
                4, 7

Trial Factors	Middle Term
$(1-1z)(28+1z)$	$z-28z=-27z$
$(1+1z)(28-1z)$	$-z+28z=27z$
$(2-1z)(14+1z)$	$2z-14z=-12z$
$(2+1z)(14-1z)$	$-2z+14z=12z$
$(4-1z)(7+1z)$	$4z-7z=-3z$
$(4+1z)(7-1z)$	$-4z+7z=3z$

$$28+3z-z^2 = (4+z)(7-z)$$

49.   Factors of 8: 1, 8   Factors of -1: $-1, +1$
              2, 4

Trial Factors	Middle Term
$(1-1x)(8+1x)$	$x-8x=-7x$
$(1+1x)(8-1x)$	$-x+8x=7x$
$(2-1x)(4+1x)$	$2x-4x=-2x$
$(2+1x)(4-1x)$	$-2x+4x=2x$

$$8-7x-x^2 = (1-x)(8+x)$$

51. The GCF is 3.
$$9x^2+33x-60 = 3(3x^2+11x-20)$$
Factor the trinomial.

Factors of 3: 1, 3   Factors of -20: $-1, +20$
                                $+1, -20$
                                $-2, +10$
                                $+2, -10$
                                $-4, +5$
                                $+4, -5$

Trial Factors	Middle Term
$(1x-1)(3x+20)$	$20x-3x=17x$
$(1x+20)(3x-1)$	$-x+60x=59x$
$(1x+1)(3x-20)$	$-20x+3x=-17x$
$(1x-20)(3x+1)$	$x-60x=-59x$
$(1x-2)(3x+10)$	$10x-6x=4x$
$(1x+10)(3x-2)$	$-2x+30x=28x$
$(1x+2)(3x-10)$	$-10x+6x=-4x$
$(1x-10)(3x+2)$	$2x-30x=-28x$
$(1x-4)(3x+5)$	$5x-12x=-7x$
$(1x+5)(3x-4)$	$-4x+15x=11x$
$(1x+4)(3x-5)$	$-5x+12x=7x$
$(1x-5)(3x+4)$	$4x-15x=-11x$

$$9x^2+33x-60 = 3(x+5)(3x-4)$$

53. The GCF is 4.
$$24x^2-52x+24 = 4(6x^2-13x+6)$$
Factor the trinomial.

Factors of 6: 1, 6   Factors of 6: $-1, -6$
             2, 3 $-2, -3$

Trial Factors	Middle Term
$(1x-1)(6x-6)$	Common factor
$(1x-6)(6x-1)$	$-x-36x=-37x$
$(1x-2)(6x-3)$	Common factor
$(1x-3)(6x-2)$	Common factor
$(2x-1)(3x-6)$	Common factor
$(2x-6)(3x-1)$	Common factor
$(2x-2)(3x-3)$	Common factor
$(2x-3)(3x-2)$	$-4x-9x=-13x$

$$24x^2-52x+24 = 4(2x-3)(3x-2)$$

55. The GCF is a^2.
$$35a^4+9a^3-2a^2 = a^2(35a^2+9a-2)$$
Factor the trinomial.

Factors of 35: 1, 35   Factors of -2: $-1, +2$
                 5, 7 $+1, -2$

Trial Factors	Middle Term
$(1a-1)(35a+2)$	$2a-35a=-33a$
$(1a+2)(35a-1)$	$-a+70a=69a$
$(1a+1)(35a-2)$	$-2a+35a=33a$
$(1a-2)(35a+1)$	$a-70a=-69a$
$(5a-1)(7a+2)$	$10a-7a=3a$
$(5a+2)(7a-1)$	$-5a+14a=9a$
$(5a+1)(7a-2)$	$-10a+7a=-3a$
$(5a-2)(7a+1)$	$5a-14a=-9a$

$$35a^4+9a^3-2a^2 = a^2(5a+2)(7a-1)$$

57. The GCF is 5.
$$15b^2-115b+70 = 5(3b^2-23b+14)$$
Factor the trinomial.

Factors of 3: 1, 3   Factors of 14: $-1, -14$
                                 $-2, -7$

Trial Factors	Middle Term
$(1b-1)(3b-14)$	$-14b-3b=-17b$
$(1b-14)(3b-1)$	$-b-42b=-43b$
$(1b-2)(3b-7)$	$-7b-6b=-13b$
$(1b-7)(3b-2)$	$-2b-21b=-23b$

$$15b^2-115b+70 = 5(b-7)(3b-2)$$

59. The GCF is $2x$.
$$10x^3+12x^2+2x = 2x(5x^2+6x+1)$$
Factor the trinomial.

Factors of 5: 1, 5   Factors of 1: $+1, +1$

Trial Factors	Middle Term
$(1x+1)(5x+1)$	$x+5x=6x$

$$10x^3+12x^2+2x = 2x(x+1)(5x+1)$$

61. The GCF is $2y$.

$$10y^3 - 44y^2 + 16y = 2y(5y^2 - 22y + 8)$$

Factor the trinomial.

| Factors of 5: 1, 5 | Factors of 8: −1, −8 |
| | −2, −4 |

Trial Factors	Middle Term
$(1y-1)(5y-8)$	$-8y-5y=-13y$
$(1y-8)(5y-1)$	$-y-40y=-41y$
$(1y-2)(5y-4)$	$-4y-10y=-14y$
$(1y-4)(5y-2)$	$-2y-20y=-22y$

$$10y^3 - 44y^2 + 16y = 2y(y-4)(5y-2)$$

63. The GCF is yz.

$$4yz^3 + 5yz^2 - 6yz = yz(4z^2 + 5z - 6)$$

Factor the trinomial.

Factors of 4: 1, 4	Factors of −6: −1, +6
	2, 2	+1, −6
		−2, +3
		+2, −3

Trial Factors	Middle Term
$(1z-1)(4z+6)$	Common factor
$(1z+6)(4z-1)$	$-z+24z=23z$
$(1z+1)(4z-6)$	Common factor
$(1z-6)(4z+1)$	$z-24z=-23z$
$(1z-2)(4z+3)$	$3z-8z=-5z$
$(1z+3)(4z-2)$	Common factor
$(1z+2)(4z-3)$	$-3z+8z=5z$
$(1z-3)(4z+2)$	Common factor
$(2z-1)(2z+6)$	Common factor
$(2z+1)(2z-6)$	Common factor
$(2z-2)(2z+3)$	Common factor
$(2z+2)(2z-3)$	Common factor

$$4yz^3 + 5yz^2 - 6yz = yz(z+2)(4z-3)$$

65. The GCF is b^2.

$$20b^4 + 41b^3 + 20b^2 = b^2(20b^2 + 41b + 20)$$

Factor the trinomial.

Factors of 20: 1, 20	Factors of 20: +1, +20
	2, 10	+2, +10
	4, 5	+4, +5

Trial Factors	Middle Term
$(1b+1)(20b+20)$	Common factor
$(1b+20)(20b+1)$	$b+400b=401b$
$(1b+2)(20b+10)$	Common factor
$(1b+10)(20b+2)$	Common factor
$(1b+4)(20b+5)$	Common factor
$(1b+5)(20b+4)$	Common factor
$(2b+1)(10b+20)$	Common factor
$(2b+20)(10b+1)$	Common factor
$(2b+2)(10b+10)$	Common factor
$(2b+10)(10b+2)$	Common factor
$(2b+4)(10b+5)$	Common factor
$(2b+5)(10b+4)$	Common factor
$(4b+1)(5b+20)$	Common factor
$(4b+20)(5b+1)$	Common factor
$(4b+2)(5b+10)$	Common factor
$(4b+10)(5b+2)$	Common factor
$(4b+4)(5b+5)$	Common factor
$(4b+5)(5b+4)$	$16b+25b=41b$

$$20b^4 + 41b^3 + 20b^2 = b^2(4b+5)(5b+4)$$

67. The GCF is xy.

$$9x^3y + 12x^2y + 4xy = xy(9x^2 + 12x + 4)$$

Factor the trinomial.

| Factors of 9: 1, 9 | Factors of 4: +1, +4 |
| | 3, 3 | +2, +2 |

Trial Factors	Middle Term
$(1x+1)(9x+4)$	$4x+9x=13x$
$(1x+4)(9x+1)$	$x+36x=37x$
$(1x+2)(9x+2)$	$2x+18x=20x$
$(3x+1)(3+4)$	$12x+3x=15x$
$(3x+2)(3x+2)$	$6x+6x=12x$

$$9x^3y + 12x^2y + 4xy = xy(3x+2)(3x+2)$$

Objective 8.3.2 Exercises

69. $2t^2 - t - 10$

$2(-10) = -20$

Factors of −20 whose sum is −1: 4 and −5

$$2t^2 - t - 10 = 2t^2 + 4t - 5t - 10$$
$$= (2t^2 + 4t) - (5t + 10)$$
$$= 2t(t+2) - 5(t+2)$$
$$= (t+2)(2t-5)$$

71. $3p^2 - 16p + 5$
$3 \cdot 5 = 15$
Factors of 15 whose sum is -16: -1 and -15
$3p^2 - 16p + 5 = 3p^2 - p - 15p + 5$
$\qquad = (3p^2 - p) - (15p - 5)$
$\qquad = p(3p - 1) - 5(3p - 1)$
$\qquad = (3p - 1)(p - 5)$

73. $12y^2 - 7y + 1$
$12 \cdot 1 = 12$
Factors of 12 whose sum is -7: -3 and -4
$12y^2 - 7y + 1 = 12y^2 - 3y - 4y + 1$
$\qquad = (12y^2 - 3y) - (4y - 1)$
$\qquad = 3y(4y - 1) - 1(4y - 1)$
$\qquad = (4y - 1)(3y - 1)$

75. $5x^2 - 62x - 7$
$5(-7) = -35$
There are no factors of -35 whose sum is -62.
$5x^2 - 62x - 7$ is nonfactorable over the integers.

77. $12y^2 + 19y + 5$
$12 \cdot 5 = 60$
Factors of 60 whose sum is 19: 4 and 15
$12y^2 + 19y + 5 = 12y^2 + 4y + 15y + 5$
$\qquad = (12y^2 + 4y) + (15y + 5)$
$\qquad = 4y(3y + 1) + 5(3y + 1)$
$\qquad = (3y + 1)(4y + 5)$

79. $7a^2 + 47a - 14$
$7(-14) = -98$
Factors of -98 whose sum is 47: 49 and -2
$7a^2 + 47a - 14 = 7a^2 + 49a - 2a - 14$
$\qquad = (7a^2 + 49a) - (2a + 14)$
$\qquad = 7a(a + 7) - 2(a + 7)$
$\qquad = (a + 7)(7a - 2)$

81. $4z^2 + 11z + 6$
$4 \cdot 6 = 24$
Factors of 24 whose sum is 11: 3 and 8
$4z^2 + 11z + 6 = 4z^2 + 3z + 8z + 6$
$\qquad = (4z^2 + 3z) + (8z + 6)$
$\qquad = z(4z + 3) + 2(4z + 3)$
$\qquad = (4z + 3)(z + 2)$

83. $22p^2 + 51p - 10$
$22(-10) = -220$
Factors of -220 whose sum is 51: 55 and -4
$22p^2 + 51p - 10 = 22p^2 + 55p - 4p - 10$
$\qquad = (22p^2 + 55p) - (4p + 10)$
$\qquad = 11p(2p + 5) - 2(2p + 5)$
$\qquad = (2p + 5)(11p - 2)$

85. $8y^2 + 17y + 9$
$8 \cdot 9 = 72$
Factors of 72 whose sum is 17: 8 and 9
$8y^2 + 17y + 9 = 8y^2 + 8y + 9y + 9$
$\qquad = (8y^2 + 8y) + (9y + 9)$
$\qquad = 8y(y + 1) + 9(y + 1)$
$\qquad = (y + 1)(8y + 9)$

87. $6b^2 - 13b + 6$
$6 \cdot 6 = 36$
Factors of 36 whose sum is -13: -9 and -4
$6b^2 - 13b + 6 = 6b^2 - 9b - 4b + 6$
$\qquad = (6b^2 - 9b) - (4b - 6)$
$\qquad = 3b(2b - 3) - 2(2b - 3)$
$\qquad = (2b - 3)(3b - 2)$

89. $33b^2 + 34b - 35$
$33(-35) = -1155$
Factors of -1155 whose sum is 34: 55 and -21
$33b^2 + 34b - 35 = 33b^2 + 55b - 21b - 35$
$\qquad = (33b^2 + 55b) - (21b + 35)$
$\qquad = 11b(3b + 5) - 7(3b + 5)$
$\qquad = (3b + 5)(11b - 7)$

91. $18y^2 - 39y + 20$
$18 \cdot 20 = 360$
Factors of 360 whose sum is -39: -15 and -24
$18y^2 - 39y + 20 = 18y^2 - 15y - 24y + 20$
$\qquad = (18y^2 - 15y) - (24y - 20)$
$\qquad = 3y(6y - 5) - 4(6y - 5)$
$\qquad = (6y - 5)(3y - 4)$

93. $15x^2 - 82x + 24$
$15 \cdot 24 = 360$
There are no factors of 360 whose sum is -82.
$15x^2 - 82x + 24$ is nonfactorable over the integers.

95. $10z^2 - 29z + 10$
$10 \cdot 10 = 100$
Factors of 100 whose sum is -29: -4 and -25
$10z^2 - 29z + 10 = 10z^2 - 4z - 25z + 10$
$\qquad = (10z^2 - 4z) - (25z - 10)$
$\qquad = 2z(5z - 2) - 5(5z - 2)$
$\qquad = (5z - 2)(2z - 5)$

97. $36z^2 + 72z + 35$
$36 \cdot 35 = 1260$
Factors of 1260 whose sum is 72: 30 and 42
$36z^2 + 72z + 35 = 36z^2 + 30z + 42z + 35$
$\qquad = (36z^2 + 30z) + (42z + 35)$
$\qquad = 6z(6z + 5) + 7(6z + 5)$
$\qquad = (6z + 5)(6z + 7)$

99. $14y^2 - 29y + 12$
$14 \cdot 12 = 168$
Factors of 168 whose sum is -29: -21 and -8
$14y^2 - 29y + 12 = 14y^2 - 21y - 8y + 12$
$$= (14y^2 - 21y) - (8y - 12)$$
$$= 7y(2y - 3) - 4(2y - 3)$$
$$= (2y - 3)(7y - 4)$$

101. $6x^2 + 35x - 6$
$6(-6) = -36$
Factors of -36 whose sum is 35: 36 and -1
$6x^2 + 35x - 6 = 6x^2 + 36x - x - 6$
$$= (6x^2 + 36x) - (x + 6)$$
$$= 6x(x + 6) - 1(x + 6)$$
$$= (x + 6)(6x - 1)$$

103. $12x^2 + 33x - 9$
The GCF is 3.
$12x^2 + 33x - 9 = 3(4x^2 + 11x - 3)$
$4(-3) = -12$
Factors of -12 whose sum is 11: 12 and -1
$12x^2 + 33x - 9 = 3(4x^2 + 11x - 3)$
$$= 3(4x^2 + 12x - x - 3)$$
$$= 3[(4x^2 + 12x) - (x + 3)]$$
$$= 3[4x(x + 3) - 1(x + 3)]$$
$$= 3(x + 3)(4x - 1)$$

105. $30y^2 + 10y - 20$
The GCF is 10.
$30y^2 + 10y - 20 = 10(3y^2 + y - 2)$
$3(-2) = -6$
Factors of -6 whose sum is 1: -2 and 3
$30y^2 + 10y - 20 = 10(3y^2 + y - 2)$
$$= 10(3y^2 - 2y + 3y - 2)$$
$$= 10[(3y^2 - 2y) + (3y - 2)]$$
$$= 10[y(3y - 2) + 1(3y - 2)]$$
$$= 10(3y - 2)(y + 1)$$

107. $2x^3 - 3x^2 - 5x$
The GCF is x.
$2x^3 - 3x^2 - 5x = x(2x^2 - 3x - 5)$
$2(-5) = -10$
Factors of -10 whose sum is -3: -5 and 2
$2x^3 - 3x^2 - 5x = x(2x^2 - 3x - 5)$
$$= x(2x^2 - 5x + 2x - 5)$$
$$= x[(2x^2 - 5x) + (2x - 5)]$$
$$= x[x(2x - 5) + 1(2x - 5)]$$
$$= x(2x - 5)(x + 1)$$

109. $2a^2 - 9ab + 9b^2$
$2 \cdot 9 = 18$
Factors of 18 whose sum is -9: -3 and -6
$2a^2 - 9ab + 9b^2 = 2a^2 - 3ab - 6ab + 9b^2$
$$= (2a^2 - 3ab) - (6ab - 9b^2)$$
$$= a(2a - 3b) - 3b(2a - 3b)$$
$$= (2a - 3b)(a - 3b)$$

111. $2y^2 + 7yz + 5z^2$
$2 \cdot 5 = 10$
Factors of 10 whose sum is 7: 2 and 5
$2y^2 + 7yz + 5z^2 = 2y^2 + 2yz + 5yz + 5z^2$
$$= (2y^2 + 2yz) + (5yz + 5z^2)$$
$$= 2y(y + z) + 5z(y + z)$$
$$= (y + z)(2y + 5z)$$

113. $18 + 17x - x^2$
$18(-1) = -18$
Factors of -18 whose sum is 17: 18 and -1
$18 + 17x - x^2 = 18 + 18x - x - x^2$
$$= (18 + 18x) - (x + x^2)$$
$$= 18(1 + x) - x(1 + x)$$
$$= (1 + x)(18 - x)$$

115. $360y^2 + 4y - 4$
The GCF is 4.
$360y^2 + 4y - 4 = 4(90y^2 + y - 1)$
$90(-1) = -90$
Factors of -90 whose sum is 1: -9 and 10
$360y^2 + 4y - 4 = 4(90y^2 + y - 1)$
$$= 4(90y^2 - 9y + 10y - 1)$$
$$= 4[(90y^2 - 9y) + (10y - 1)]$$
$$= 4[9y(10y - 1) + 1(10y - 1)]$$
$$= 4(10y - 1)(9y + 1)$$

117. $16t^2 + 40t - 96$
The GCF is 8.
$16t^2 + 40t - 96 = 8(2t^2 + 5t - 12)$
$2(-12) = -24$
Factors of -24 whose sum is 5: 8 and -3
$16t^2 + 40t - 96 = 8(2t^2 + 5t - 12)$
$$= 8(2t^2 + 8t - 3t - 12)$$
$$= 8[(2t^2 + 8t) - (3t + 12)]$$
$$= 8[2t(t + 4) - 3(t + 4)]$$
$$= 8(t + 4)(2t - 3)$$

119. $6p^3 + 5p^2 + p$
The GCF is p.
$6p^3 + 5p^2 + p = p(6p^2 + 5p + 1)$
$6 \cdot 1 = 6$
Factors of 6 whose sum is 5: 2 and 3
$6p^3 + 5p^2 + p = p(6p^2 + 5p + 1)$
$$= p(6p^2 + 2p + 3p + 1)$$
$$= p[(6p^2 + 2p) + (3p + 1)]$$
$$= p[2p(3p + 1) + 1(3p + 1)]$$
$$= p(3p + 1)(2p + 1)$$

121. $30z^2 - 87z + 30$

The GCF is 3.

$30z^2 - 87z + 30 = 3(10z^2 - 29z + 10)$

$10 \cdot 10 = 100$

Factors of 100 whose sum is -29: -25 and -4

$30z^2 - 87z + 30 = 3(10z^2 - 29z + 10)$

$\qquad\qquad\qquad = 3(10z^2 - 25z - 4z + 10)$

$\qquad\qquad\qquad = 3[(10z^2 - 25z) - (4z - 10)]$

$\qquad\qquad\qquad = 3[5z(2z - 5) - 2(2z - 5)]$

$\qquad\qquad\qquad = 3(2z - 5)(5z - 2)$

123. $42a^3 + 45a^2 - 27a$

The GCF is $3a$.

$42a^3 + 45a^2 - 27a = 3a(14a^2 + 15a - 9)$

$14(-9) = -126$

Factors of -126 whose sum is 15: 21 and -6

$42a^3 + 45a^2 - 27a = 3a(14a^2 + 15a - 9)$

$\qquad\qquad\qquad = 3a(14a^2 + 21a - 6a - 9)$

$\qquad\qquad\qquad = 3a[(14a^2 + 21a) - (6a + 9)]$

$\qquad\qquad\qquad = 3a[7a(2a + 3) - 3(2a + 3)]$

$\qquad\qquad\qquad = 3a(2a + 3)(7a - 3)$

125. $9x^2y - 30xy^2 + 25y^3$

The GCF is y.

$9x^2y - 30xy^2 + 25y^3 = y(9x^2 - 30xy + 25y^2)$

$9 \cdot 25 = 225$

Factors of 225 whose sum is -30: -15 and -15

$9x^2y - 30xy^2 + 25y^3 = y(9x^2 - 30xy + 25y^2)$

$\qquad\qquad\qquad = y(9x^2 - 15xy - 15xy + 25y^2)$

$\qquad\qquad\qquad = y[(9x^2 - 15xy) - (15xy - 25y^2)]$

$\qquad\qquad\qquad = y[3x(3x - 5y) - 5y(3x - 5y)]$

$\qquad\qquad\qquad = y(3x - 5y)(3x - 5y)$

127. $9x^3y - 24x^2y^2 + 16xy^3$

The GCF is xy.

$9x^3y - 24x^2y^2 + 16xy^3 = xy(9x^2 - 24xy + 16y^2)$

$9 \cdot 16 = 144$

Factors of 144 whose sum is -24: -12 and -12

$9x^3y - 24x^2y^2 + 16xy^3 = xy(9x^2 - 24xy + 16y^2)$

$\qquad\qquad\qquad = xy(9x^2 - 12xy - 12xy + 16y^2)$

$\qquad\qquad\qquad = xy[(9x^2 - 12xy) - (12xy - 16y^2)]$

$\qquad\qquad\qquad = xy[3x(3x - 4y) - 4y(3x - 4y)]$

$\qquad\qquad\qquad = xy(3x - 4y)(3x - 4y)$

Applying Concepts 8.3

129. The GCF is $2y$.

$6y + 8y^3 - 26y^2 = 2y(3 + 4y^2 - 13y)$

Factor the trinomial.

$3 + 4y^2 - 13y = 4y^2 - 13y + 3$

Factors of 4: 1, 4   Factors of 3: -1, -3

$\qquad\qquad$ 2, 2

Trial Factors	Middle Term
$(y - 1)(4y - 3)$	$-3y - 4y = -7y$
$(y - 3)(4y - 1)$	$-y - 12y = -13y$
$(2y - 1)(2y - 3)$	$-6y - 2y = -8y$

$6y + 8y^3 - 26y^2 = 2y(y - 3)(4y - 1)$

131. The GCF is ab.

$a^3b - 24ab - 2a^2b = ab(a^2 - 24 - 2a)$

Factor the trinomial.

$a^2 - 24 - 2a = a^2 - 2a - 24$

Factors of 1: 1, 1 Factors of −24: +1, −24

−1, +24

+2, −12

−2, +12

+3, −8

−3, +8

+4, −6

−4, +6

Trial Factors	Middle Term
$(a+1)(a-24)$	$-24a + a = -23a$
$(a-1)(a+24)$	$24a - a = 23a$
$(a+2)(a-12)$	$-12a + 2a = -10a$
$(a-2)(a+12)$	$12a - 2a = 10a$
$(a+3)(a-8)$	$-8a + 3a = -5a$
$(a-3)(a+8)$	$8a - 3a = 5a$
$(a+4)(a-6)$	$-6a + 4a = -2a$
$(a-4)(a+6)$	$6a - 4a = 2a$

$a^3b - 24ab - 2a^2b = ab(a+4)(a-6)$

133. The GCF is $5t$.

$25t^2 + 60t - 10t^3 = 5t(5t + 12 - 2t^2)$

Factor the trinomial.

$5t + 12 - 2t^2 = 12 + 5t - 2t^2$

Factors of 12: 1, 12 Factors of −2: +1, −2

2, 6 −1, +2

3, 4

Trial Factors	Middle Term
$(1+t)(12-2t)$	Common factor
$(1-t)(12+2t)$	Common factor
$(1+2t)(12-t)$	$-t + 24t = 23t$
$(1-2t)(12+t)$	$t - 24t = -23t$
$(2+t)(6-2t)$	Common factor
$(2-t)(6+2t)$	Common factor
$(2+2t)(6-t)$	Common factor
$(2-2t)(6+t)$	Common factor
$(3+t)(4-2t)$	Common factor
$(3-t)(4+2t)$	Common factor
$(3+2t)(4-t)$	$-3t + 8t = 5t$
$(3-2t)(4+t)$	$3t - 8t = -5t$

$25t^2 + 60t - 10t^3 = 5t(3+2t)(4-t)$

135. $2(y+2)^2 - (y+2) - 3$

$2(y^2 + 4y + 4) - (y+2) - 3$

$2y^2 + 8y + 8 - y - 2 - 3$

$2y^2 + 7y + 3$

Factors of 2: 1, 2 Factors of 3: 1, 3

Trial Factors	Middle Term
$(2y+1)(y+3)$	$6y + y = 7y$
$(2y+3)(y+1)$	$2y + 3y = 5y$

$2(y+2)^2 - (y+2) - 3 = (2y+1)(y+3)$

137. $10(x+1)^2 - 11(x+1) - 6$

$10(x^2 + 2x + 1) - 11(x+1) - 6$

$10x^2 + 20x + 10 - 11x - 11 - 6$

$10x^2 + 9x - 7$

Factors of 10: 1, 10 Factors of 7: −1, 7

2, 5 1, −7

Trial Factors	Middle Term
$(x-1)(10x+7)$	$7x - 10x = -3x$
$(x+1)(10x-7)$	$-7x + 10x = 3x$
$(x+7)(10x-1)$	$-x + 70x = 69x$
$(x-7)(10x+1)$	$x - 70x = -69x$
$(2x-1)(5x+7)$	$14x - 5x = 9x$
$(2x+1)(5x-7)$	$-14x + 5x = -9x$
$(2x-7)(5x+1)$	$2x - 35x = -33x$
$(2x+7)(5x-1)$	$-2x + 35x = 33x$

$10(x+1)^2 - 11(x+1) - 6 = (2x-1)(5x+7)$

139. Factors of 2: 1, 2 Factors of 3: +1, +3

−1, −3

Trial Factors	Middle Term
$(x+1)(2x+3)$	$3x + 2x = 5x$
$(x-1)(2x-3)$	$-3x - 2x = -5x$
$(x+3)(2x+1)$	$x + 6x = 7x$
$(x-3)(2x-1)$	$-x - 6x = -7x$

$k = 5, -5, 7,$ or -7

141. Factors of 3: 1, 3 Factors of 2: +1, +2

−1, −2

Trial Factors	Middle Term
$(x+1)(3x+2)$	$2x + 3x = 5x$
$(x-1)(3x-2)$	$-2x - 3x = -5x$
$(x+2)(3x+1)$	$x + 6x = 7x$
$(x-2)(3x-1)$	$-x - 6x = -7x$

$k = 5, -5, 7,$ or -7

143. Factors of 2: 1, 2 Factors of 5: +1, +5

−1, −5

Trial Factors	Middle Term
$(x+1)(2x+5)$	$5x + 2x = 7x$
$(x-1)(2x-5)$	$-5x - 2x = -7x$
$(x+5)(2x+1)$	$x + 10x = 11x$
$(x-5)(2x-1)$	$-x - 10x = -11x$

$k = 7, -7, 11,$ or -11

145. If $x + 2$ is a factor of $x^3 - 2x^2 - 5x + 6$.

$$
\begin{array}{r}
x^2 - 4x + 3 \\
x + 2 \overline{\smash{)}x^3 - 2x^2 - 5x + 6} \\
\underline{x^3 + 2x^2} \\
-4x^2 - 5x \\
\underline{-4x^2 - 8x} \\
3x + 6 \\
\underline{3x + 6} \\
0
\end{array}
$$

$x^2 - 4x + 3 = (x - 3)(x - 1)$
$x^3 - 2x^2 - 5x + 6 = (x + 2)(x - 3)(x - 1)$

Section 8.4

Concept Review 8.4

1. Never true
$x^2 - 12$ is not the difference of two squares because 12 is not a perfect square.

3. Sometimes true
$x^2 + 4$ is nonfactorable over the integers.

5. Always true

7. Never true
A perfect-square trinomial contains three terms, not four. If a polynomial consists of four terms, try to factor it by grouping.

Objective 8.4.1 Exercises

1. Answers will vary.

   a. $x^2 - 16$

   b. $(x + 7)(x - 7)$

   c. $x^2 + 10x + 25$

   d. $(x - 3)^2$

   e. $x^2 + 36$

3. $4; 25x^6; 100x^4y^4$

5. $4z^4$

7. $9a^2y^3$

9. $x^2 - 4 = x^2 - 2^2 = (x + 2)(x - 2)$

11. $a^2 - 81 = a^2 - 9^2 = (a + 9)(a - 9)$

13. $4x^2 - 1 = (2x)^2 - 1^2 = (2x + 1)(2x - 1)$

15. $y^2 + 2y + 1$
$y^2 = (y)^2, 1 = 1^2$
$(y + 1)^2 = y^2 + 2(y)(1) + 1^2 = y^2 + 2y + 1$
The factorization checks.
$y^2 + 2y + 1 = (y + 1)^2$

17. $a^2 - 2a + 1$
$a^2 = (a)^2, 1 = 1^2$
$(a - 1)^2 = a^2 + 2(a)(-1) + (-1)^2 = a^2 - 2a + 1$
The factorization checks.
$a^2 - 2a + 1 = (a - 1)^2$

19. $z^2 - 18z - 81$
$z = (z)^2$
-81 is not a square because $\sqrt{-81}$ is not a real number.
$z^2 - 18z - 81$ is nonfactorable over the integers.

21. $x^6 - 9 = (x^3)^2 - 3^2 = (x^3 + 3)(x^3 - 3)$

23. $25x^2 - 1 = (5x)^2 - 1^2 = (5x + 1)(5x - 1)$

25. $1 - 49x^2 = 1^2 - (7x)^2 = (1 + 7x)(1 - 7x)$

27. $x^2 + 2xy + y^2$
$x^2 = (x)^2, y^2 = (y)^2$
$(x + y)^2 = x^2 + 2(x)(y) + y^2 = x^2 + 2xy + y^2$
The factorization checks.
$x^2 + 2xy + y^2 = (x + y)^2$

29. $4a^2 + 4a + 1$
$4a^2 = (2a)^2, 1 = 1^2$
$(2a + 1)^2 = (2a)^2 + 2(2a)(1) + 1^2 = 4a^2 + 4a + 1$
The factorization checks.
$4a^2 + 4a + 1 = (2a + 1)^2$

31. $64a^2 - 16a + 1$
$64a^2 = (8a)^2, 1 = 1^2$
$(8a - 1)^2 = (8a)^2 + 2(8a)(-1) + (-1)^2$
$ = 64a^2 - 16a + 1$
The factorization checks.
$64a^2 - 16a + 1 = (8a - 1)^2$

33. $t^2 + 36$ is the sum, not the difference, of two squares. $t^2 + 36$ is nonfactorable over the integers.

35. $x^4 - y^2 = (x^2)^2 - (y)^2 = (x^2 + y)(x^2 - y)$

37. $9x^2 - 16y^2 = (3x)^2 - (4y)^2 = (3x + 4y)(3x - 4y)$

39. $16b^2 + 8b + 1$
$16b^2 = (4b)^2, 1 = 1^2$
$(4b + 1)^2 = (4b)^2 + 2(4b)(1) + 1^2 = 16b^2 + 8b + 1$
The factorization checks.
$16b^2 + 8h + 1 = (4b + 1)^2$

41. $4b^2 + 28b + 49$
$4b^2 = (2b)^2, 49 = 7^2$
$(2b + 7)^2 = (2b)^2 + 2(2b)(7) + 7^2$
$\qquad = 4b^2 + 28b + 49$
The factorization checks.
$4b^2 + 28b + 49 = (2b + 7)^2$

43. $25a^2 + 30ab + 9b^2$
$25a^2 = (5a)^2, 9b^2 = (3b)^2$
$(5a + 3b)^2 = (5a)^2 + 2(5a)(3b) + (3b)^2$
$\qquad = 25a^2 + 30ab + 9b^2$
The factorization checks.
$25a^2 + 30ab + 9b^2 = (5a + 3b)^2$

45. $x^2y^2 - 4 = (xy)^2 - 2^2 = (xy + 2)(xy - 2)$

47. $16 - x^2y^2 = 4^2 - (xy)^2 = (4 + xy)(4 - xy)$

49. $4y^2 - 36yz + 81z^2$
$4y^2 = (2y)^2, 81z^2 = (9z)^2$
$(2y - 9z)^2 = (2y)^2 + 2(2y)(-9z) + (-9z)^2$
$\qquad = 4y^2 - 36yz + 81z^2$
The factorization checks.
$4y^2 - 36yz + 81z^2 = (2y - 9z)^2$

51. $9a^2b^2 - 6ab + 1$
$9a^2b^2 = (3ab)^2, 1 = 1^2$
$(3ab - 1)^2 = (3ab)^2 + 2(3ab)(-1) + (-1)^2$
$\qquad = 9a^2b^2 - 6ab + 1$
The factorization checks.
$9a^2b^2 - 6ab + 1 = (3ab - 1)^2$

53. $m^4 - 256 = (m^2)^2 - 16^2 = (m^2 + 16)(m^2 - 16)$
$\qquad = (m^2 + 16)(m^2 - 4^2)$
$\qquad = (m^2 + 16)(m + 4)(m - 4)$

55. $9x^2 + 13x + 4$
$9x^2 = (3x)^2, 4 = 2^2$
$(3x + 2)^2 = (3x)^2 + 2(3x)(2) + 2^2 = 9x^2 + 12x + 4$
The factorization does not check.
Factors of 36 whose sum is 13: 4 and 9
$9x^2 + 13x + 4 = 9x^2 + 4x + 9x + 4$
$\qquad = (9x^2 + 4x)(9x + 4)$
$\qquad = x(9x + 4) + 1(9x + 4)$
$\qquad = (9x + 4)(x + 1)$

57. $y^8 - 81 = (y^4)^2 - 9^2 = (y^4 + 9)(y^4 - 9)$
$\qquad = (y^4 + 9)[(y^2)^2 - 3^2]$
$\qquad = (y^4 + 9)(y^2 + 3)(y^2 - 3)$

Objective 8.4.2 Exercises

59. $2x^2 - 18 = 2(x^2 - 9) = 2(x + 3)(x - 3)$

61. $x^4 + 2x^3 - 35x^2 = x^2(x^2 + 2x - 35)$
$\qquad = x^2(x + 7)(x - 5)$

63. $5b^2 + 75b + 180 = 5(b^2 + 15b + 36)$
$\qquad = 5(b + 3)(b + 12)$

65. $3a^2 + 36a + 10$ is nonfactorable over the integers.

67. $2x^2y + 16xy - 66y = 2y(x^2 + 8x - 33)$
$\qquad = 2y(x - 3)(x + 11)$

69. $x^3 - 6x^2 - 5x = x(x^2 - 6x - 5)$

71. $3y^2 - 36 = 3(y^2 - 12)$

73. $20a^2 + 12a + 1 = 20a^2 + 2a + 10a + 1$
$\qquad = (20a^2 + 2a) + (10a + 1)$
$\qquad = 2a(10a + 1) + 1(10a + 1)$
$\qquad = (10a + 1)(2a + 1)$

75. $x^2y^2 - 7xy^2 - 8y^2 = y^2(x^2 - 7x - 8)$
$\qquad = y^2(x - 8)(x + 1)$

77. $10a^2 - 5ab - 15b^2 = 5(2a^2 - ab - 3b^2)$
$\qquad = 5(2a^2 - 3ab + 2ab - 3b^2)$
$\qquad = 5[(2a^2 - 3ab) + (2ab - 3b^2)]$
$\qquad = 5[a(2a - 3b) + b(2a - 3b)]$
$\qquad = 5(2a - 3b)(a + b)$

79. $50 - 2x^2 = 2(25 - x^2) = 2(5 + x)(5 - x)$

81. $12a^3b - a^2b^2 - ab^3$
$= ab(12a^2 - ab - b^2)$
$= ab(12a^2 - 4ab + 3ab - b^2)$
$= ab[(12a^2 - 4ab) + (3ab - b^2)]$
$= ab[4a(3a - b) + b(3a - b)]$
$= ab(3a - b)(4a + b)$

83. $2ax - 2a + 2bx - 2b = 2(ax - a + bx - b)$
$\qquad = 2[(ax - a) + (bx - b)]$
$\qquad = 2[a(x - 1) + b(x - 1)]$
$\qquad = 2(x - 1)(a + b)$

85. $12a^3 - 12a^2 + 3a = 3a(4a^2 - 4a + 1)$
$\qquad = 3a(2a - 1)^2$

87. $243 + 3a^2 = 3(81 + a^2)$

89. $12a^3 - 46a^2 + 40a = 2a(6a^2 - 23a + 20)$
$\qquad = 2a(6a^2 - 15a - 8a + 20)$
$\qquad = 2a[(6a^2 - 15a) - (8a - 20)]$
$\qquad = 2a[3a(2a - 5) - 4(2a - 5)]$
$\qquad = 2a(2a - 5)(3a - 4)$

91. $x^3 - 2x^2 - x + 2 = (x^3 - 2x^2) - (x - 2)$
$\qquad = x^2(x - 2) - 1(x - 2)$
$\qquad = (x - 2)(x^2 - 1)$
$\qquad = (x - 2)(x + 1)(x - 1)$

93. $4a^3 + 20a^2 + 25a = a(4a^2 + 20a + 25)$
$\qquad = a(2a + 5)^2$

95. $27a^2b - 18ab + 3b = 3b(9a^2 - 6a + 1)$
$\qquad\qquad\qquad\quad = 3b(3a - 1)^2$

97. $48 - 12x - 6x^2 = 6(8 - 2x - x^2)$
$\qquad\qquad\quad = 6(8 + 2x - 4x - x^2)$
$\qquad\qquad\quad = 6[(8 + 2x) - (4x + x^2)]$
$\qquad\qquad\quad = 6[2(4 + x) - x(4 + x)]$
$\qquad\qquad\quad = 6(4 + x)(2 - x)$

99. $ax^2 - 4a + bx^2 - 4b = (ax^2 - 4a) + (bx^2 - 4b)$
$\qquad\qquad\qquad\quad = a(x^2 - 4) + b(x^2 - 4)$
$\qquad\qquad\qquad\quad = (x^2 - 4)(a + b)$
$\qquad\qquad\qquad\quad = (x + 2)(x - 2)(a + b)$

101. $x^4 - x^2y^2 = x^2(x^2 - y^2) = x^2(x + y)(x - y)$

103. $18a^3 + 24a^2 + 8a = 2a(9a^2 + 12a + 4)$
$\qquad\qquad\qquad\quad = 2a(3a + 2)^2$

105. $2b + ab - 6a^2b = b(2 + a - 6a^2)$
$\qquad\qquad\qquad = b(2 - 3a + 4a - 6a^2)$
$\qquad\qquad\qquad = b[(2 - 3a) + (4a - 6a^2)]$
$\qquad\qquad\qquad = b[1(2 - 3a) + 2a(2 - 3a)]$
$\qquad\qquad\qquad = b(2 - 3a)(1 + 2a)$

107. $4x - 20 - x^3 + 5x^2 = (4x - 20) - (x^3 - 5x^2)$
$\qquad\qquad\qquad = 4(x - 5) - x^2(x - 5)$
$\qquad\qquad\qquad = (x - 5)(4 - x^2)$
$\qquad\qquad\qquad = (x - 5)(2 + x)(2 - x)$

109. $72xy^2 + 48xy + 8x = 8x(9y^2 + 6y + 1)$
$\qquad\qquad\qquad = 8x(3y + 1)^2$

111. $15y^2 - 2xy^2 - x^2y^2 = y^2(15 - 2x - x^2)$
$\qquad\qquad\qquad = y^2(15 + 3x - 5x - x^2)$
$\qquad\qquad\qquad = y^2[(15 + 3x) - (5x + x^2)]$
$\qquad\qquad\qquad = y^2[3(5 + x) - x(5 + x)]$
$\qquad\qquad\qquad = y^2(5 + x)(3 - x)$

113. $y^3 - 9y = y(y^2 - 9) = y(y + 3)(y - 3)$

115. $2x^4y^2 - 2x^2y^2 = 2x^2y^2(x^2 - 1)$
$\qquad\qquad\qquad = 2x^2y^2(x + 1)(x - 1)$

117. $x^9 - x^5 = x^5(x^4 - 1) = x^5(x^2 + 1)(x^2 - 1)$
$\qquad\qquad\quad = x^5(x^2 + 1)(x + 1)(x - 1)$

119. $24x^3y + 14x^2y - 20xy = 2xy(12x^2 + 7x - 10)$
$\qquad\qquad\qquad = 2xy(12x^2 + 15x - 8x - 10)$
$\qquad\qquad\qquad = 2xy[(12x^2 + 15x) - (8x + 10)]$
$\qquad\qquad\qquad = 2xy[3x(4x + 5) - 2(4x + 5)]$
$\qquad\qquad\qquad = 2xy(4x + 5)(3x - 2)$

121. $4x^4y^2 - 20x^3y^2 + 25x^2y^2 = x^2y^2(4x^2 - 20x + 25)$
$\qquad\qquad\qquad = x^2y^2(2x - 5)^2$

123. $x^3 - 2x^2 - 4x + 8 = (x^3 - 2x^2) - (4x - 8)$
$\qquad\qquad\qquad = x^2(x - 2) - 4(x - 2)$
$\qquad\qquad\qquad = (x - 2)(x^2 - 4)$
$\qquad\qquad\qquad = (x - 2)(x + 2)(x - 2)$
$\qquad\qquad\qquad = (x - 2)^2(x + 2)$

125. $8xy^2 - 20x^2y^2 + 12x^3y^2 = 4xy^2(2 - 5x + 3x^2)$
$\qquad\qquad\qquad = 4xy^2(2 - 2x - 3x + 3x^2)$
$\qquad\qquad\qquad = 4xy^2[(2 - 2x) - (3x - 3x^2)]$
$\qquad\qquad\qquad = 4xy^2[2(1 - x) - 3x(1 - x)]$
$\qquad\qquad\qquad = 4xy^2(1 - x)(2 - 3x)$

127. $36a^3b - 62a^2b^2 + 12ab^3 = 2ab(18a^2 - 31ab + 6b^2)$
$\qquad\qquad\qquad = 2ab(18a^2 - 27ab - 4ab + 6b^2)$
$\qquad\qquad\qquad = 2ab[(18a^2 - 27ab) - (4ab - 6b^2)]$
$\qquad\qquad\qquad = 2ab[9a(2a - 3b) - 2b(2a - 3b)]$
$\qquad\qquad\qquad = 2ab(2a - 3b)(9a - 2b)$

129. $5x^2y^2 - 11x^3y^2 - 12x^4y^2 = x^2y^2(5 - 11x - 12x^2)$
$$= x^2y^2(5 + 4x - 15x - 12x^2)$$
$$= x^2y^2[(5 + 4x) - (15x + 12x^2)]$$
$$= x^2y^2[1(5 + 4x) - 3x(5 + 4x)]$$
$$= x^2y^2(5 + 4x)(1 - 3x)$$

131. $(4x - 3)^2 - y^2 = (4x - 3 - y)(4x - 3 + y)$

133. $(x^2 - 4x + 4) - y^2 = (x - 2)^2 - y^2$
$$= (x - 2 - y)(x - 2 + y)$$

Applying Concepts 8.4

135. $4x^2 - kx + 9$
$4x^2 = (2x)^2,\ 9 = 3^2$
$(2x + 3)^2 = (2x)^2 + 2(2x)(3) + 3^2 = 4x^2 + 12x + 9$
$(2x - 3)^2 = (2x)^2 + 2(2x)(-3) + (-3)^2$
$$= 4x^2 - 12x + 9$$
$-kx = 12x \quad -kx = -12x$
$kx = -12x \quad kx = 12x$
$k = -12 \qquad k = 12$
$k = 12$ or -12

137. $36x^2 + kxy + y^2$
$36x^2 = (6x)^2,\ y^2 = (y)^2$
$(6x + y)^2 = (6x)^2 + 2(6x)(y) + y^2$
$$= 36x^2 + 12xy + y^2$$
$(6x - y)^2 = (6x)^2 + 2(6x)(-y) + (-y)^2$
$$= 36x^2 - 12xy + y^2$$
$kxy = 12xy \quad kxy = -12xy$
$k = 12 \qquad k = -12$
$k = 12$ or -12

139. $x^2 + 6x + k$
The trinomial is a perfect-square trinomial. Find a positive number that, when doubled, equals 6. Since $2(3) = 6$, the number is 3.
$(x + 3)^2 = x^2 + 6x + 9$
$k = 9$

141. $x^2 - 2x + k$
The trinomial is a perfect-square trinomial. Find a negative number that, when doubled, equals -2. Since $2(-1) = -2$, the number is -1.
$(x - 1)^2 = x^2 - 2x + 1$
$k = 1$

143. $2^3 \cdot 3^2 = 8 \cdot 9 = 72$
1, 4, 9, 36 are perfect-square factors of 72.

145. $300 \cdot 1 = 300$ not a perfect square
$300 \cdot 2 = 600$ not a perfect square
$300 \cdot 3 = 900$ a perfect square
3 is the smallest number by which 300 can be multiplied so that the product is a perfect square.

147. $x^3 + 8 = x^3 + 2^3$
$$= (x + 2)(x^2 - 2x + 4)$$

149. $y^3 - 27 = y^3 - 3^3$
$$= (y - 3)(y^2 + 3y + 9)$$

151. $y^3 + 64 = y^3 + 4^3$
$$= (y + 4)(y^2 - 4y + 16)$$

153. $8x^3 - 1 = (2x)^3 - 1^3$
$$= (2x - 1)(4x^2 + 2x + 1)$$

Section 8.5

Concept Review 8.5

1. Always true

3. Always true

5. Never true
   Set each factor to zero and find the solutions.
   $x = 8$ or $x = -6$

Objective 8.5.1 Exercises

3. $(y + 3)(y + 2) = 0$
$y + 3 = 0 \quad y + 2 = 0$
$\quad y = -3 \qquad y = -2$
The solutions are -2 and -3.

5. $(z - 7)(z - 3) = 0$
$z - 7 = 0 \quad z - 3 = 0$
$\quad z = 7 \qquad z = 3$
The solutions are 3 and 7.

7. $x(x - 5) = 0$
$x = 0 \quad x - 5 = 0$
$\qquad\quad\ x = 5$
The solutions are 0 and 5.

9. $a(a - 9) = 0$
$a = 0 \quad a - 9 = 0$
$\qquad\quad\ a = 9$
The solutions are 0 and 9.

11. $y(2y + 3) = 0$
$y = 0 \quad 2y + 3 = 0$
$\qquad\qquad\ 2y = -3$
$\qquad\qquad\ \ y = -\dfrac{3}{2}$
The solutions are $-\dfrac{3}{2}$ and 0.

13. $2a(3a-2)=0$
$2a=0 \quad 3a-2=0$
$a=0 \qquad 3a=2$
$\qquad\qquad a=\dfrac{2}{3}$
The solutions are $\dfrac{2}{3}$ and 0.

15. $(b+2)(b-5)=0$
$b+2=0 \quad b-5=0$
$\quad b=-2 \qquad b=5$
The solutions are -2 and 5.

17. $\qquad x^2-81=0$
$(x+9)(x-9)=0$
$x+9=0 \quad x-9=0$
$\quad x=-9 \qquad x=9$
The solutions are -9 and 9.

19. $\qquad 16x^2-49=0$
$(4x+7)(4x-7)=0$
$4x+7=0 \qquad 4x-7=0$
$\quad 4x=-7 \qquad\quad 4x=7$
$\qquad x=-\dfrac{7}{4} \qquad\quad x=\dfrac{7}{4}$
The solutions are $-\dfrac{7}{4}$ and $\dfrac{7}{4}$.

21. $x^2-8x+15=0$
$(x-3)(x-5)=0$
$x-3=0 \quad x-5=0$
$\quad x=3 \qquad x=5$
The solutions are 3 and 5.

23. $z^2+z-72=0$
$(z+9)(z-8)=0$
$z+9=0 \quad z-8=0$
$\quad z=-9 \qquad z=8$
The solutions are -9 and 8.

25. $2y^2-y-1=0$
$(2y+1)(y-1)=0$
$2y+1=0 \qquad y-1=0$
$\quad 2y=-1 \qquad\quad y=1$
$\quad y=-\dfrac{1}{2}$
The solutions are $-\dfrac{1}{2}$ and 1.

27. $3a^2+14a+8=0$
$(3a+2)(a+4)=0$
$3a+2=0 \qquad a+4=0$
$\quad 3a=-2 \qquad\quad a=-4$
$\quad a=-\dfrac{2}{3}$
The solutions are $-\dfrac{2}{3}$ and -4.

29. $3y^2+12y-63=0$
$3(y^2+4y-21)=0$
$\quad y^2+4y-21=0$
$\quad (y+7)(y-3)=0$
$y+7=0 \quad y-3=0$
$\quad y=-7 \qquad y=3$
The solutions are -7 and 3.

31. $x^2-7x=0$
$x(x-7)=0$
$x=0 \quad x-7=0$
$\qquad\quad x=7$
The solutions are 0 and 7.

33. $\qquad a^2+5a=-4$
$\quad a^2+5a+4=0$
$(a+1)(a+4)=0$
$a+1=0 \quad a+4=0$
$\quad a=-1 \qquad a=-4$
The solutions are -1 and -4.

35. $\qquad y^2-5y=-6$
$\quad y^2-5y+6=0$
$(y-2)(y-3)=0$
$y-2=0 \quad y-3=0$
$\quad y=2 \qquad y=3$
The solutions are 2 and 3.

37. $\qquad 2t^2+7t=4$
$\quad 2t^2+7t-4=0$
$(2t-1)(t+4)=0$
$2t-1=0 \quad t+4=0$
$\quad 2t=1 \qquad\quad t=-4$
$\quad t=\dfrac{1}{2}$
The solutions are $\dfrac{1}{2}$ and -4.

39. $\qquad 3t^2-13t=-4$
$\quad 3t^2-13t+4=0$
$(3t-1)(t-4)=0$
$3t-1=0 \quad t-4=0$
$\quad 3t=1 \qquad\quad t=4$
$\quad t=\dfrac{1}{3}$
The solutions are $\dfrac{1}{3}$ and 4.

41. $\qquad x(x-12)=-27$
$\qquad x^2-12x=-27$
$\quad x^2-12x+27=0$
$(x-3)(x-9)=0$
$x-3=0 \quad x-9=0$
$\quad x=3 \qquad x=9$
The solutions are 3 and 9.

43.
$$y(y-7)=18$$
$$y^2-7y=18$$
$$y^2-7y-18=0$$
$$(y+2)(y-9)=0$$
$$y+2=0 \quad y-9=0$$
$$y=-2 \quad y=9$$
The solutions are -2 and 9.

45.
$$p(p+3)=-2$$
$$p^2+3p=-2$$
$$p^2+3p+2=0$$
$$(p+1)(p+2)=0$$
$$p+1=0 \quad p+2=0$$
$$p=-1 \quad p=-2$$
The solutions are -1 and -2.

47.
$$y(y+4)=45$$
$$y^2+4y=45$$
$$y^2+4y-45=0$$
$$(y+9)(y-5)=0$$
$$y+9=0 \quad y-5=0$$
$$y=-9 \quad y=5$$
The solutions are -9 and 5.

49.
$$x(x+3)=28$$
$$x^2+3x=28$$
$$x^2+3x-28=0$$
$$(x+7)(x-4)=0$$
$$x+7=0 \quad x-4=0$$
$$x=-7 \quad x=4$$
The solutions are -7 and 4.

51.
$$(x+8)(x-3)=-30$$
$$x^2+5x-24=-30$$
$$x^2+5x+6=0$$
$$(x+2)(x+3)=0$$
$$x+2=0 \quad x+3=0$$
$$x=-2 \quad x=-3$$
The solutions are -2 and -3.

53.
$$(y+3)(y+10)=-10$$
$$y^2+13y+30=-10$$
$$y^2+13y+40=0$$
$$(y+5)(y+8)=0$$
$$y+5=0 \quad y+8=0$$
$$y=-5 \quad y=-8$$
The solutions are -5 and -8.

55.
$$(z-8)(z+4)=-35$$
$$z^2-4z-32=-35$$
$$z^2-4z+3=0$$
$$(z-1)(z-3)=0$$
$$z-1=0 \quad z-3=0$$
$$z=1 \quad z=3$$
The solutions are 1 and 3.

57.
$$(a+3)(a+4)=72$$
$$a^2+7a+12=72$$
$$a^2+7a-60=0$$
$$(a+12)(a-5)=0$$
$$a+12=0 \quad a-5=0$$
$$a=-12 \quad a=5$$
The solutions are -12 and 5.

59.
$$(z+3)(z-10)=-42$$
$$z^2-7z-30=-42$$
$$z^2-7z+12=0$$
$$(z-3)(z-4)=0$$
$$z-3=0 \quad z-4=0$$
$$z=3 \quad z=4$$
The solutions are 3 and 4.

61.
$$(y+3)(2y+3)=5$$
$$2y^2+9y+9=5$$
$$2y^2+9y+4=0$$
$$(2y+1)(y+4)=0$$
$$2y+1=0 \quad y+4=0$$
$$2y=-1 \quad y=-4$$
$$y=-\frac{1}{2}$$
The solutions are $-\frac{1}{2}$ and -4.

63. Strategy
The positive number: x
The square of the positive number is six more than five times the positive number.

Solution
$$x^2=5x+6$$
$$x^2-5x-6=0$$
$$(x-6)(x+1)=0$$
$$x-6=0 \quad x+1=0$$
$$x=6 \quad x=-1$$
Because -1 is not a positive number, it is not a solution.
The number is 6.

65. Strategy
The first number: x
The second number: $6-x$
The sum of the squares of the two numbers is twenty.

Solution
$$x^2+(6-x)^2=20$$
$$x^2+36-12x+x^2=20$$
$$2x^2-12x+16=0$$
$$2(x^2-6x+8)=0$$
$$(x-4)(x-2)=0$$
$$x-4=0 \quad x-2=0$$
$$x=4 \quad x=2$$
$$6-x=2 \quad 6-x=4$$
The two numbers are 2 and 4.

67. Strategy

The first positive integer: x

The second positive integer: $x + 1$

The sum of the squares of the two integers is forty-one.

Solution

$$x^2 + (x+1)^2 = 41$$
$$x^2 + x^2 + 2x + 1 = 41$$
$$2x^2 + 2x - 40 = 0$$
$$2(x^2 + x - 20) = 0$$
$$(x+5)(x-4) = 0$$
$$x + 5 = 0 \quad x - 4 = 0$$
$$x = -5 \quad x = 4$$

Because -5 is not a positive integer, it is not a solution.

$x = 4$

$x + 1 = 5$

The two numbers are 4 and 5.

69. Strategy

First positive integer: x

Second positive integer: $x + 1$

The product of the two consecutive integers is 240.

Solution

$$x(x+1) = 240$$
$$x^2 + x = 240$$
$$x^2 + x - 240 = 0$$
$$(x+16)(x-15) = 0$$
$$x + 16 = 0 \quad x - 15 = 0$$
$$x = -16 \quad x = 15$$

Since -16 is not a positive number, it is not a solution.

$x = 15$

$x + 1 = 15 + 1 = 16$

The numbers are 15 and 16.

71. Strategy

Height of the triangle: h

Base of the triangle: $3h$

The area of the triangle is 54 ft^2.

The equation for the area of the triangle is $A = \dfrac{1}{2}bh$. Substitute in the equation and solve for h.

Solution

$$A = \frac{1}{2}bh$$
$$54 = \frac{1}{2}(3h)(h)$$
$$54 = \frac{3h^2}{2}$$
$$108 = 3h^2$$
$$36 = h^2$$
$$0 = h^2 - 36$$
$$0 = (h+6)(h-6)$$
$$h + 6 = 0 \quad h - 6 = 0$$
$$h = -6 \quad h = 6$$

Since the height cannot be a negative number, -6 is not a solution.

$h = 6$

$b = 3h = 3 \cdot 6 = 18$

The height is 6 ft.

The base is 18 ft.

73. Strategy

Width of the rectangle: W

Length of the rectangle: $2W + 2$

The area of the rectangle is 144 ft^2.

The equation for the area of a rectangle is $A = L \cdot W$. Substitute in the equation and solve for W.

Solution

$$A = L \cdot W$$
$$144 = (2W + 2)W$$
$$144 = 2W^2 + 2W$$
$$0 = 2W^2 + 2W - 144$$
$$0 = 2(W^2 + W - 72)$$
$$0 = 2(W+9)(W-8)$$
$$W + 9 = 0 \quad W - 8 = 0$$
$$W = -9 \quad W = 8$$

Since the width cannot be a negative number, -9 is not a solution.

$W = 8$

$2W + 2 = 2(8) + 2 = 16 + 2 = 18$

The width is 8 ft.

The length is 18 ft.

75. Strategy

Side of the original square: x

Side of the larger square: $x + 4$

The area of the larger square is $64\,\text{m}^2$.

The equation for the area of a square is $A = s^2$.

Substitute in the equation and solve for x.

Solution

$$A = s^2$$
$$64 = (x+4)^2$$
$$64 = x^2 + 8x + 16$$
$$0 = x^2 + 8x - 48$$
$$0 = (x+12)(x-4)$$
$$x + 12 = 0 \qquad x - 4 = 0$$
$$x = -12 \qquad x = 4$$

Since the length of a side of a square cannot be a negative number, -12 is not a solution.

The length of a side of the original square is 4 m.

77. Strategy

Radius of the original circle: r

Radius of the larger circle: $r + 3$

The area of the largest circle is $100\,\text{in}^2$ more than the area of the original circle.

The equation for the area of a circle is $A = \pi r^2$.

Substitute in the equation and solve for r.

Solution

$$\pi r^2 + 100 = \pi(r+3)^2$$
$$\pi r^2 + 100 = \pi(r^2 + 6r + 9)$$
$$\pi r^2 + 100 = \pi r^2 + 6\pi r + 9\pi$$
$$100 - 9\pi = 6\pi r$$
$$\frac{100 - 9\pi}{6\pi} = r$$
$$3.81 \approx r$$

The radius of the original circle is approximately 3.81 in.

79. Strategy

The width of the uniform border: x

Type area width: $6 - 2x$

Type area length: $9 - 2x$

The required area is $28\,\text{in}^2$. The equation for the area of a rectangle is $A = LW$.

Solution

$$(6 - 2x)(9 - 2x) = 28$$
$$54 - 30x + 4x^2 = 28$$
$$4x^2 - 30x + 26 = 0$$
$$2(2x^2 - 15x + 13) = 0$$
$$(2x - 13)(x - 1) = 0$$
$$2x - 13 = 0 \qquad x - 1 = 0$$
$$2x = 13 \qquad x = 1$$
$$x = \frac{13}{2}$$

The solution $\frac{13}{2}$ is not possible because it is wider than the page.

$$x = 1$$
$$6 - 2x = 6 - 2(1) = 4$$
$$9 - 2x = 9 - 2(1) = 7$$

The dimensions of the type area are 4 in. by 7 in.

81. Strategy

Replace d by 1600 and v by 0 in the given equation and solve for t.

Solution

$$d = vt + 16t^2$$
$$1600 = 0t + 16t^2$$
$$1600 = 16t^2$$
$$100 = t^2$$
$$0 = t^2 - 100$$
$$0 = (t + 10)(t - 10)$$
$$t + 10 = 0 \qquad t - 10 = 0$$
$$t = -10 \qquad t = 10$$

The solution -10 is not possible.

The object will hit the ground 10 s later.

83. Strategy

Replace S by 78 in the given equation and solve for n.

Solution

$$S = \frac{n^2 + n}{2}$$
$$78 = \frac{n^2 + n}{2}$$
$$156 = n^2 + n$$
$$0 = n^2 + n - 156$$
$$0 = (n + 13)(n - 12)$$
$$n + 13 = 0 \qquad n - 12 = 0$$
$$n = -13 \qquad n = 12$$

Since -13 is not a natural number, the solution -13 is not possible.

Twelve consecutive numbers beginning with 1 give a sum of 78.

85. Strategy
    Replace N by 28 in the given equation and solve for t.

    Solution
    $$N = \frac{t^2 - t}{2}$$
    $$28 = \frac{t^2 - t}{2}$$
    $$56 = t^2 - t$$
    $$0 = t^2 - t - 56$$
    $$0 = (t - 8)(t + 7)$$
    $$t - 8 = 0 \quad t + 7 = 0$$
    $$t = 8 \qquad t = -7$$
    The solution -7 is not possible.
    There are 8 teams in the league.

87. Strategy
    To find the time for the ball to reach a height of 64 ft, replace the variables h and v by their given values and solve for t.

    Solution
    $$h = vt - 16t^2$$
    $$64 = 64t - 16t^2$$
    $$16t^2 - 64t + 64 = 0$$
    $$16(t^2 - 4t + 4) = 0$$
    $$(t - 2)^2 = 0$$
    $$t - 2 = 0 \quad t - 2 = 0$$
    $$t = 2 \qquad t = 2$$
    2 is a double root. The time is 2 s.

Applying Concepts 8.5

89. $$2y(y + 4) = -5(y + 3)$$
    $$2y^2 + 8y = -5y - 15$$
    $$2y^2 + 13y + 15 = 0$$
    $$(2y + 3)(y + 5) = 0$$
    $$2y + 3 = 0 \qquad y + 5 = 0$$
    $$2y = -3 \qquad y = -5$$
    $$y = -\frac{3}{2}$$

    The solutions are $-\frac{3}{2}$ and -5.

91. $$(a - 3)^2 = 36$$
    $$a^2 - 6a + 9 = 36$$
    $$a^2 - 6a - 27 = 0$$
    $$(a - 9)(a + 3) = 0$$
    $$a - 9 = 0 \quad a + 3 = 0$$
    $$a = 9 \qquad a = -3$$

93. $$p^3 = 9p^2$$
    $$p^3 - 9p^2 = 0$$
    $$p^2(p - 9) = 0$$
    $$p^2 = 0 \quad p - 9 = 0$$
    $$p = 0 \qquad p = 9$$
    The solutions are 0 and 9.

95. $$(2z - 3)(z + 5) = (z + 1)(z + 3)$$
    $$2z^2 + 7z - 15 = z^2 + 4z + 3$$
    $$z^2 + 3z - 18 = 0$$
    $$(z + 6)(z - 3) = 0$$
    $$z + 6 = 0 \quad z - 3 = 0$$
    $$z = -6 \qquad z = 3$$
    The solutions are -6 and 3.

97. $$n(n + 5) = -4$$
    $$n^2 + 5n = -4$$
    $$n^2 + 5n + 4 = 0$$
    $$(n + 4)(n + 1) = 0$$
    $$n + 4 = 0 \quad n + 1 = 0$$
    $$n = -4 \qquad n = -1$$
    $$3n^2 = 3(-4)^2 = 3(16) = 48$$
    or
    $$3n^2 = 3(-1)^2 = 3(1) = 3$$

99. Strategy
    The amount by which the length and width are increased: x
    Length of the larger rectangle: $x + 7$
    Width of the larger rectangle: $x + 4$
    Area of the larger rectangle
    = Area of smaller rectangle + increase in area
    = $7(4) + 42 = 28 + 42 = 70$

    Solution
    $$A = L \cdot W$$
    $$70 = (x + 7)(x + 4)$$
    $$70 = x^2 + 11x + 28$$
    $$0 = x^2 + 11x - 42$$
    $$0 = (x + 14)(x - 3)$$
    $$x + 14 = 0 \qquad x - 3 = 0$$
    $$x = -14 \qquad x = 3$$
    Since the length cannot be a negative number, the solution -14 is not possible.
    $$x + 7 = 3 + 7 = 10$$
    $$x + 4 = 3 + 4 = 7$$
    The length of the larger rectangle is 10 cm.
    The width of the larger rectangle is 7 cm.

Chapter Review Exercises

1. $14y^9 - 49y^6 + 7y^3 = 7y^3(2y^6 - 7y^3 + 1)$

2. $3a^2 - 12a + ab - 4b = (3a^2 - 12a) + (ab - 4b)$
   $= 3a(a-4) + b(a-4)$
   $= (a-4)(3a+b)$

3. $c^2 + 8c + 12 = (c+2)(c+6)$

4. $a^3 - 5a^2 + 6a = a(a^2 - 5a + 6)$
   $= a(a-2)(a-3)$

5. $6x^2 - 29x + 28 = 6x^2 - 8x - 21x + 28$
   $= (6x^2 - 8x) - (21x - 28)$
   $= 2x(3x-4) - 7(3x-4)$
   $= (3x-4)(2x-7)$

6. $3y^2 + 16y - 12 = 3y^2 + 18y - 2y - 12$
   $= (3y^2 + 18y) - (2y + 12)$
   $= 3y(y+6) - 2(y+6)$
   $= (y+6)(3y-2)$

7. $18a^2 - 3a - 10 = 18a^2 - 15a + 12a - 10$
   $= (18a^2 - 15a) + (12a - 10)$
   $= 3a(6a-5) + 2(6a-5)$
   $= (6a-5)(3a+2)$

8. $a^2b^2 - 1 = (ab)^2 - 1^2$
   $= (ab+1)(ab-1)$

9. $4y^2 - 16y + 16 = 4(y^2 - 4y + 4)$
   $= 4(y-2)^2$

10. $a(5a+1) = 0$
    $a = 0 \quad 5a + 1 = 0$
    $\qquad\qquad 5a = -1$
    $\qquad\qquad a = -\dfrac{1}{5}$

    The solutions are 0 and $-\dfrac{1}{5}$.

11. $12a^2b + 3ab^2 = 3ab(4a + b)$

12. $b^2 - 13b + 30 = (b-3)(b-10)$

13. $10x^2 + 25x + 4xy + 10y$
    $= (10x^2 + 25x) + (4xy + 10y)$
    $= 5x(2x+5) + 2y(2x+5)$
    $= (2x+5)(5x+2y)$

14. $3a^2 - 15a - 42 = 3(a^2 - 5a - 14)$
    $= 3(a+2)(a-7)$

15. $n^4 - 2n^3 - 3n^2 = n^2(n^2 - 2n - 3)$
    $= n^2(n+1)(n-3)$

16. $2x^2 - 5x + 6$ is nonfactorable over the integers.

17. $6x^2 - 7x + 2 = 6x^2 - 3x - 4x + 2$
    $= (6x^2 - 3x) - (4x - 2)$
    $= 3x(2x-1) - 2(2x-1)$
    $= (2x-1)(3x-2)$

18. $16x^2 + 49$ is the sum, not the difference, of two squares. It is nonfactorable over the integers.

19. $(x-2)(2x-3) = 0$
    $x - 2 = 0 \quad 2x - 3 = 0$
    $\quad x = 2 \qquad 2x = 3$
    $\qquad\qquad\qquad x = \dfrac{3}{2}$

    The solutions are 2 and $\dfrac{3}{2}$.

20. $7x^2 - 7 = 7(x^2 - 1)$
    $= 7(x+1)(x-1)$

21. $3x^5 - 9x^4 - 4x^3 = x^3(3x^2 - 9x - 4)$

22. $4x(x-3) - 5(3-x) = 4x(x-3) + 5(x-3)$
    $= (x-3)(4x+5)$

23. $a^2 + 5a - 14 = (a+7)(a-2)$

24. $y^2 + 5y - 36 = (y+9)(y-4)$

25. $5x^2 - 50x - 120 = 5(x^2 - 10x - 24)$
    $= 5(x-12)(x+2)$

26. $(x+1)(x-5) = 16$
    $x^2 - 4x - 5 = 16$
    $x^2 - 4x - 21 = 0$
    $(x-7)(x+3) = 0$
    $x - 7 = 0 \quad x + 3 = 0$
    $\quad x = 7 \qquad x = -3$

    The solutions are 7 and -3.

27. $7a^2 + 17a + 6 = 7a^2 + 3a + 14a + 6$
    $= (7a^2 + 3a) + (14a + 6)$
    $= a(7a+3) + 2(7a+3)$
    $= (7a+3)(a+2)$

28. $4x^2 + 83x + 60 = 4x^2 + 80x + 3x + 60$
    $= (4x^2 + 80x) + (3x + 60)$
    $= 4x(x+20) + 3(x+20)$
    $= (x+20)(4x+3)$

29. $9y^4 - 25z^2 = (3y^2)^2 - (5z)^2$
    $= (3y^2 + 5z)(3y^2 - 5z)$

30. $5x^2 - 5x - 30 = 5(x^2 - x - 6)$
    $= 5(x+2)(x-3)$

31. $6 - 6y^2 = 5y$

$$0 = 6y^2 + 5y - 6$$
$$0 = (3y - 2)(2y + 3)$$
$$3y - 2 = 0 \quad 2y + 3 = 0$$
$$3y = 2 \qquad 2y = -3$$
$$y = \frac{2}{3} \qquad y = -\frac{3}{2}$$

The solutions are $\dfrac{2}{3}$ and $-\dfrac{3}{2}$.

32. $12b^3 - 58b^2 + 56b = 2b(6b^2 - 29b + 28)$
$$= 2b(6b^2 - 8b - 21b + 28)$$
$$= 2b[(6b^2 - 8b) - (21b - 28)]$$
$$= 2b[2b(3b - 4) - 7(3b - 4)]$$
$$= 2b(3b - 4)(2b - 7)$$

33. $5x^3 + 10x^2 + 35x = 5x(x^2 + 2x + 7)$

34. $x^2 - 23x + 42 = (x - 2)(x - 21)$

35. $a(3a + 2) - 7(3a + 2) = (3a + 2)(a - 7)$

36. $8x^2 - 38x + 45 = 8x^2 - 18x - 20x + 45$
$$= (8x^2 - 18x) - (20x - 45)$$
$$= 2x(4x - 9) - 5(4x - 9)$$
$$= (4x - 9)(2x - 5)$$

37. $10a^2x - 130ax + 360x = 10x(a^2 - 13a + 36)$
$$= 10x(a - 4)(a - 9)$$

38. $2a^2 - 19a - 60 = 2a^2 + 5a - 24a - 60$
$$= (2a^2 + 5a) - (24a + 60)$$
$$= a(2a + 5) - 12(2a + 5)$$
$$= (2a + 5)(a - 12)$$

39. $21ax - 35bx - 10by + 6ay$
$$= (21ax - 35bx) - (10by - 6ay)$$
$$= 7x(3a - 5b) + 2y(3a - 5b)$$
$$= (3a - 5b)(7x + 2y)$$

40. $a^6 - 100 = (a^3)^2 - 10^2$
$$= (a^3 + 10)(a^3 - 10)$$

41. $16a^2 = (4a)^2, \; 1 = 1^2, \; 2(4a)(1) = 8a$
$$16a^2 + 8a + 1 = (4a + 1)^2$$

42. $4x^2 + 27x = 7$
$$4x^2 + 27x - 7 = 0$$
$$(4x - 1)(x + 7) = 0$$
$$4x - 1 = 0 \quad x + 7 = 0$$
$$4x = 1 \qquad x = -7$$
$$x = \frac{1}{4}$$

The solutions are $\dfrac{1}{4}$ and -7.

43. $20a^2 + 10a - 280 = 10(2a^2 + a - 28)$
$$= 10(2a^2 + 8a - 7a - 28)$$
$$= 10[(2a^2 + 8a) - (7a + 28)]$$
$$= 10[2a(a + 4) - 7(a + 4)]$$
$$= 10(a + 4)(2a - 7)$$

44. $6x - 18 = 6(x - 3)$

45. $3x^4y + 2x^3y + 6x^2y = x^2y(3x^2 + 2x + 6)$

46. $d^2 + 3d - 40 = (d - 5)(d + 8)$

47. $24x^2 - 12xy + 10y - 20x$
$$= 2(12x^2 - 6xy + 5y - 10x)$$
$$= 2[(12x^2 - 6xy) + (5y - 10x)]$$
$$= 2[6x(2x - y) + 5(y - 2x)]$$
$$= 2[6x(2x - y) - 5(2x - y)]$$
$$= 2(2x - y)(6x - 5)$$

48. $4x^3 - 20x^2 - 24x = 4x(x^2 - 5x - 6)$
$$= 4x(x + 1)(x - 6)$$

49. $x^2 - 8x - 20 = 0$
$$(x - 10)(x + 2) = 0$$
$$x - 10 = 0 \quad x + 2 = 0$$
$$x = 10 \qquad x = -2$$

The solutions are 10 and -2.

50. $3x^2 - 17x + 10 = 3x^2 - 2x - 15x + 10$
$$= (3x^2 - 2x) - (15x - 10)$$
$$= x(3x - 2) - 5(3x - 2)$$
$$= (3x - 2)(x - 5)$$

51. $16x^2 - 94x + 33 = 16x^2 - 6x - 88x + 33$
$$= (16x^2 - 6x) - (88x - 33)$$
$$= 2x(8x - 3) - 11(8x - 3)$$
$$= (8x - 3)(2x - 11)$$

52. $9x^2 = (3x)^2, \; 25 = 5^2, \; 2(3x)(-5) = -30x$
$$9x^2 - 30x + 25 = (3x - 5)^2$$

53. $12y^2 + 16y - 3 = 12y^2 + 18y - 2y - 3$
$$= (12y^2 + 18y) - (2y + 3)$$
$$= 6y(2y + 3) - 1(2y + 3)$$
$$= (2y + 3)(6y - 1)$$

54. $3x^2 + 36x + 108 = 3(x^2 + 12x + 36)$
$$= 3(x + 6)^2$$

55. Strategy
Width of the field: W
Length of the field: $2W$
The area is 5000 yd^2.
The equation for the area of a rectangle is
$A = L \cdot W$.

Solution
$A = L \cdot W$
$5000 = 2W(W)$
$5000 = 2W^2$
$2500 = W^2$
$0 = W^2 - 2500$
$0 = (W + 50)(W - 50)$
$W + 50 = 0 \qquad W - 50 = 0$
$\qquad W = -50 \qquad\quad W = 50$
Since the width cannot be a negative number, the solution -50 is not possible
$W = 50$
$2W = 2(50) = 100$
The width is 50 yd.
The length is 100 yd.

56. Strategy
Width of the field: W
Length of the field: $2W - 20$
The area is 6000 yd^2.
The equation for the area of a rectangle is
$A = L \cdot W$.

Solution
$A = L \cdot W$
$6000 = (2W - 20)W$
$6000 = 2W^2 - 20W$
$0 = 2W^2 - 20W - 6000$
$0 = 2(W^2 - 10W - 3000)$
$0 = 2(W + 50)(W - 60)$
$W + 50 = 0 \qquad W - 60 = 0$
$\qquad W = -50 \qquad\quad W = 60$
Since the width cannot be a negative number, the solution -50 is not possible.
$W = 60$
$2W - 20 = 2(60) - 20 = 120 - 20 = 100$
The width is 60 yd.
The length is 100 yd.

57. Strategy
First integer: n
Second integer: $n + 1$
The sum of the squares of the two integers is 41.

Solution
$n^2 + (n+1)^2 = 41$
$n^2 + n^2 + 2n + 1 = 41$
$2n^2 + 2n + 1 = 41$
$2n^2 + 2n - 40 = 0$
$2(n^2 + n - 20) = 0$
$2(n + 5)(n - 4) = 0$
$n + 5 = 0 \qquad n - 4 = 0$
$\quad n = -5 \qquad\quad n = 4$
Since -5 is not a positive integer, the solution -5 is not possible.
$n = 4$
$n + 1 = 4 + 1 = 5$
The two integers are 4 and 5.

58. Strategy
Replace s by 400 in the given equation and solve for d.

Solution
$s = d^2$
$400 = d^2$
$0 = d^2 - 400$
$0 = (d + 20)(d - 20)$
$d + 20 = 0 \qquad d - 20 = 0$
$\quad d = -20 \qquad\quad d = 20$
The solution -20 is not possible.
The distance is 20 ft.

59. Strategy
Width of the uniform border: x
New width: $12 + 2x$
New length: $15 + 2x$
The area is 270 ft^2.
The equation for the area of a rectangle is
$A = L \cdot W$.

Solution
$A = L \cdot W$
$270 = (15 + 2x)(12 + 2x)$
$270 = 180 + 54x + 4x^2$
$0 = 4x^2 + 54x - 90$
$0 = (2x - 3)(2x + 30)$
$2x - 3 = 0 \qquad 2x + 30 = 0$
$\quad 2x = 3 \qquad\qquad 2x = -30$
$\quad\; x = \dfrac{3}{2} \qquad\qquad x = -15$
The solution -15 is not possible.
$x = \dfrac{3}{2}$

$12 + 2x = 12 + 2\left(\dfrac{3}{2}\right) = 12 + 3 = 15$

The width is 15 ft.

60. Strategy
Length of a side of the original square: x
Length of a side of the larger square: $x + 4$
The area of the larger square is 576 ft^2.
The equation for the area of a square is $A = s^2$.

Solution
$A = s^2$
$576 = (x+4)^2$
$576 = x^2 + 8x + 16$
$0 = x^2 + 8x - 560$
$0 = (x+28)(x-20)$
The solution -28 is not possible.
The length of a side of the original square is 20 ft.

Chapter Test

1. $6x^2y^2 + 9xy^2 + 12y^2 = 3y^2(2x^2 + 3x + 4)$

2. $6x^3 - 8x^2 + 10x = 2x(3x^2 - 4x + 5)$

3. $p^2 + 5p + 6 = (p+2)(p+3)$

4. $a(x-2) + b(2-x) = a(x-2) - b(x-2)$
$= (x-2)(a-b)$

5. $(2a-3)(a+7) = 0$
$2a - 3 = 0 \quad a + 7 = 0$
$2a = 3 \quad\quad a = -7$
$a = \dfrac{3}{2}$

The solutions are $\dfrac{3}{2}$ and -7.

6. $a^2 - 19a + 48 = (a-3)(a-16)$

7. $x^3 + 2x^2 - 15x = x(x^2 + 2x - 15)$
$= x(x+5)(x-3)$

8. $8x^2 + 20x - 48 = 4(2x^2 + 5x - 12)$
$= 4(2x^2 + 8x - 3x - 12)$
$= 4[(2x^2 + 8x) - (3x + 12)]$
$= 4[2x(x+4) - 3(x+4)]$
$= 4(x+4)(2x-3)$

9. $ab + 6a - 3b - 18 = (ab + 6a) - (3b + 18)$
$= a(b+6) - 3(b+6)$
$= (b+6)(a-3)$

10. $4x^2 - 1 = 0$
$(2x+1)(2x-1) = 0$
$2x + 1 = 0 \quad 2x - 1 = 0$
$2x = -1 \quad\quad 2x = 1$
$x = -\dfrac{1}{2} \quad\quad x = \dfrac{1}{2}$

The solutions are $-\dfrac{1}{2}$ and $\dfrac{1}{2}$.

11. $6x^2 + 19x + 8 = 6x^2 + 16x + 3x + 8$
$= (6x^2 + 16x) + (3x + 8)$
$= 2x(3x+8) + 1(3x+8)$
$= (3x+8)(2x+1)$

12. $x^2 - 9x - 36 = (x+3)(x-12)$

13. $2b^2 - 32 = 2(b^2 - 16)$
$= 2(b^2 - 4^2)$
$= 2(b+4)(b-4)$

14. $4a^2 = (2a)^2, 9b^2 = (3b)^2, 2(2a)(-3b) = -12ab$
$4a^2 - 12ab + 9b^2 = (2a-3b)^2$

15. $px + x - p - 1 = (px + x) - (p+1)$
$= x(p+1) - 1(p+1)$
$= (p+1)(x-1)$

16. $5x^2 - 45x - 15 = 5(x^2 - 9x - 3)$

17. $2x^2 + 4x - 5$ is nonfactorable over the integers.

18. $4x^2 - 49y^2 = (2x)^2 - (7y)^2$
$= (2x + 7y)(2x - 7y)$

19. $x(x-8) = -15$
$x^2 - 8x = -15$
$x^2 - 8x + 15 = 0$
$(x-3)(x-5) = 0$
$x - 3 = 0 \quad x - 5 = 0$
$x = 3 \quad\quad x = 5$
The solutions are 3 and 5.

20. $p^2 = (p)^2, 36 = 6^2, 2(p)(6) = 12p$
$p^2 + 12p + 36 = (p+6)^2$

21. $18x^2 - 48xy + 32y^2 = 2(9x^2 - 24xy + 16y^2)$
$= 2(3x - 4y)^2$

22. $2y^4 - 14y^3 - 16y^2 = 2y^2(y^2 - 7y - 8)$
$= 2y^2(y+1)(y-8)$

23. Strategy

Width: x

Length: $3 + 2x$

The area of the rectangle is 90 cm^2.

The equation for the area of a rectangle is $A = L \cdot W$. Substitute in the equation and solve for x.

Solution

$A = L \cdot W$

$90 = (3 + 2x)x$

$90 = 3x + 2x^2$

$0 = 2x^2 + 3x - 90$

$0 = (2x + 15)(x - 6)$

$2x + 15 = 0 \qquad x - 6 = 0$

$\qquad 2x = -15 \qquad\quad x = 6$

$\qquad\quad x = -\dfrac{15}{2}$

Since the width cannot be a negative number, $-\dfrac{15}{2}$ is not a solution.

$x = 6$

$3 + 2x = 3 + 2 \cdot 6 = 3 + 12 = 15$

The width is 6 cm.

The length is 15 cm.

24. Strategy

Height of the triangle: x

Base of the triangle: $3x$

The area of the triangle is 24 in^2.

The equation for the area of a triangle is $A = \dfrac{1}{2}bh$. Substitute in the equation and solve for x.

Solution

$A = \dfrac{1}{2}bh$

$24 = \dfrac{1}{2}(3x)(x)$

$24 = \dfrac{1}{2}(3x^2)$

$48 = 3x^2$

$0 = 3x^2 - 48$

$0 = 3(x^2 - 16)$

$0 = 3(x + 4)(x - 4)$

$x + 4 = 0 \qquad x - 4 = 0$

$\quad x = -4 \qquad\quad x = 4$

Since the height cannot be a negative number, -4 is not a solution.

$x = 4$

$3x = 3(4) = 12$

The length of the base is 12 in.

25. Strategy

First negative integer: x

Second negative integer: $x + 1$

The product of the integers is 156.

Solution

$x(x + 1) = 156$

$x^2 + x = 156$

$x^2 + x - 156 = 0$

$(x + 13)(x - 12) = 0$

$x + 13 = 0 \qquad x - 12 = 0$

$\quad x = -13 \qquad\quad x = 12$

Since 12 is not a negative number, it is not a solution. The integers are -13 and -12.

Cumulative Review Exercises

1. $4 - (-5) - 6 - 11 = 4 + 5 + (-6) + (-11)$

$\qquad\qquad\qquad\quad = 9 + (-6) + (-11)$

$\qquad\qquad\qquad\quad = 3 + (-11)$

$\qquad\qquad\qquad\quad = -8$

2.

$$0.046.\overline{)0.372.00}$$

$$\begin{array}{r} 8.08 \\ \hline 368 \\ \hline 40 \\ 0 \\ \hline 400 \\ 368 \\ \hline 32 \end{array}$$

$0.372 \div (-0.046) \approx -8.1$

3. $(3 - 7)^2 \div (-2) - 3 \cdot (-4) = (-4)^2 \div (-2) - 3 \cdot (-4)$

$\qquad\qquad\qquad\qquad\qquad = 16 \div (-2) - 3 \cdot (-4)$

$\qquad\qquad\qquad\qquad\qquad = 8 - 3 \cdot (-4)$

$\qquad\qquad\qquad\qquad\qquad = -8 - (-12)$

$\qquad\qquad\qquad\qquad\qquad = -8 + 12$

$\qquad\qquad\qquad\qquad\qquad = 4$

4. $-2a^2 \div (2b) - c$

$-2(-4)^2 \div (2 \cdot 2) - (-1) = -2(16) \div 4 + 1$

$\qquad\qquad\qquad\qquad\qquad = -32 \div 4 + 1$

$\qquad\qquad\qquad\qquad\qquad = -8 + 1$

$\qquad\qquad\qquad\qquad\qquad = -7$

5. $(3 + 8) + 7 = 3 + (8 + 7)$

The Associative Property of Addition

6. $-\dfrac{3}{4}(-24x^2) = 18x^2$

7. $-2[3x - 4(3 - 2x) - 8x] = 2[3x - 12 + 8x - 8x]$

$\qquad\qquad\qquad\qquad\qquad = -2[3x - 12]$

$\qquad\qquad\qquad\qquad\qquad = -6x + 24$

8.
$$-\frac{5}{7}x = -\frac{10}{21}$$
$$-\frac{7}{5}\left(-\frac{5}{7}x\right) = -\frac{7}{5}\left(-\frac{10}{21}\right)$$
$$x = \frac{2}{3}$$
The solution is $\frac{2}{3}$.

9.
$$4 + 3(x-2) = 13$$
$$4 + 3x - 6 = 13$$
$$3x - 2 = 13$$
$$3x - 2 + 2 = 13 + 2$$
$$3x = 15$$
$$\frac{3x}{3} = \frac{15}{3}$$
$$x = 5$$
The solution is 5.

10.
$$3x - 2 = 12 - 5x$$
$$3x + 5x - 2 = 12 - 5x + 5x$$
$$8x - 2 = 12$$
$$8x - 2 + 2 = 12 + 2$$
$$8x = 14$$
$$\frac{8x}{8} = \frac{14}{8}$$
$$x = \frac{7}{4}$$
The solution is $\frac{7}{4}$.

11.
$$-2 + 4[3x - 2(4-x) - 3] = 4x + 2$$
$$-2 + 4[3x - 8 + 2x - 3] = 4x + 2$$
$$-2 + 4[5x - 11] = 4x + 2$$
$$-2 + 20x - 44 = 4x + 2$$
$$20x - 46 = 4x + 2$$
$$20x - 4x - 46 = 4x - 4x + 2$$
$$16x - 46 = 2$$
$$16x - 46 + 46 = 2 + 46$$
$$16x = 48$$
$$\frac{16x}{16} = \frac{48}{16}$$
$$x = 3$$
The solution is 3.

12.
$$PB = A$$
$$1.20B = 42$$
$$\frac{1.20B}{1.20} = \frac{42}{1.20}$$
$$B = 35$$
120% of 35 is 42.

13.
$$-4x - 2 \geq 10$$
$$-4x - 2 + 2 \geq 10 + 2$$
$$-4x \geq 12$$
$$\frac{-4x}{-4} \leq \frac{12}{-4}$$
$$x \leq -3$$

14.
$$9 - 2(4x - 5) < 3(7 - 6x)$$
$$9 - 8x + 10 < 21 - 18x$$
$$19 - 8x < 21 - 18x$$
$$19 - 8x + 18x < 21 - 18x + 18x$$
$$19 + 10x < 21$$
$$-19 + 19 + 10x < -19 + 21$$
$$10x < 2$$
$$\frac{10x}{10} < \frac{2}{10}$$
$$x < \frac{1}{5}$$

15.

16.

17. The domain is $\{-5, -3, -1, 1, 3\}$
The range is $\{-4, -2, 0, 2, 4\}$.
No two ordered pairs have the same first coordinate.
The relation is a function.

18.
$$f(x) = 6x - 5$$
$$f(11) = 6(11) - 5 = 66 - 5 = 61$$
The value of the function at $x = 11$ is 61.

19.
$$x + 3y > 2$$
$$3y > -x + 2$$
$$y > -\frac{1}{3}x + \frac{2}{3}$$

20.
$$6x + y = 7$$
$$x - 3y = 17$$
$$x = 3y + 17$$
$$6(3y + 17) + y = 7$$
$$18y + 102 + y = 7$$
$$19y + 102 = 7$$
$$19y = -95$$
$$y = -5$$
$$x = 3(-5) + 17$$
$$x = -15 + 17$$
$$x = 2$$
The solution is $(2, -5)$.

21. $2x - 3y = -4$
$5x + y = 7$

$2x - 3y = -4$
$3(5x + y) = 3(7)$

$2x - 3y = -4$
$15x + 3y = 21$
$\quad\quad 17x = 17$
$\quad\quad\quad x = 1$

$5(1) + y = 7$
$\quad 5 + y = 7$
$\quad\quad\; y = 2$

The solution is (1, 2).

22. $(3y^3 - 5y^2 - 6) + (2y^2 - 8y + 1)$
$= 3y^3 + (-5y^2 + 2y^2) - 8y + (-6 + 1)$
$= 3y^3 - 3y^2 - 8y - 5$

23. $(-3a^4 b^2)^3 = (-3)^3 a^{12} b^6 = -27a^{12} b^6$

24.
$$\begin{array}{r} x^2 - 5x + 4 \\ \underline{x + 2} \\ 2x^2 - 10x + 8 \\ \underline{x^3 - 5x^2 + 4x \quad\;\;} \\ x^3 - 3x^2 - 6x + 8 \end{array}$$

25.
$$\begin{array}{r} 4x + 8 \\ 2x - 3\overline{\smash{\big)}\,8x^2 + 4x - 3} \\ \underline{8x^2 - 12x \quad\;\;} \\ 16x - 3 \\ \underline{16x - 24} \\ 21 \end{array}$$

$(8x^2 + 4x - 3) \div (2x - 3) = 4x + 8 + \dfrac{21}{2x - 3}$

26. $(x^{-4} y^2)^3 = x^{-12} y^6 = \dfrac{y^6}{x^{12}}$

27. $3a - 3b - ax + bx = (3a - 3b) - (ax - bx)$
$\quad\quad\quad\quad\quad\quad\quad\quad = 3(a - b) - x(a - b)$
$\quad\quad\quad\quad\quad\quad\quad\quad = (a - b)(3 - x)$

28. $x^2 + 3xy - 10y^2 = (x + 5y)(x - 2y)$

29. $6a^4 + 22a^3 + 12a^2 = 2a^2(3a^2 + 11a + 6)$
$\quad\quad\quad\quad\quad\quad\quad = 2a^2(3a^2 + 2a + 9a + 6)$
$\quad\quad\quad\quad\quad\quad\quad = 2a^2[(3a^2 + 2a) + (9a + 6)]$
$\quad\quad\quad\quad\quad\quad\quad = 2a^2[a(3a + 2) + 3(3a + 2)]$
$\quad\quad\quad\quad\quad\quad\quad = 2a^2(3a + 2)(a + 3)$

30. $25a^2 - 36b^2 = (5a)^2 - (6b)^2$
$\quad\quad\quad\quad\quad = (5a + 6b)(5a - 6b)$

31. $12x^2 - 36xy + 27y^2 = 3(4x^2 - 12xy + 9y^2)$
$\quad\quad\quad\quad\quad\quad\quad = 3(2x - 3y)^2$

32. $3x^2 + 11x - 20 = 0$
$(3x - 4)(x + 5) = 0$
$3x - 4 = 0 \quad\quad x + 5 = 0$
$\quad 3x = 4 \quad\quad\quad\; x = -5$
$\quad\;\; x = \dfrac{4}{3}$

The solutions are $\dfrac{4}{3}$ and -5.

33. $\quad f(x) = \dfrac{4}{5}x - 3$

$f(-10) = \dfrac{4}{5}(-10) - 3 = -8 - 3 = -11$

$f(-5) = \dfrac{4}{5}(-5) - 3 = -4 - 3 = -7$

$f(0) = \dfrac{4}{5}(0) - 3 = 0 - 3 = -3$

$f(5) = \dfrac{4}{5}(5) - 3 = 4 - 3 = 1$

$f(10) = \dfrac{4}{5}(10) - 3 = 8 - 3 = 5$

The range is $\{-11, -7, -3, 1, 5\}$.

34. Strategy
To find the average:
• Add the seven temperature readings.
• Divide the sum by 7.

Solution
$-4 + (-7) + 2 + 0 + (-1) + (-6) + (-5)$
$= -11 + 2 + 0 + (-1) + (-6) + (-5)$
$= -9 + 0 + (-1) + (-6) + (-5)$
$= -9 + (-1) + (-6) + (-5)$
$= -10 + (-6) + (-5)$
$= -16 + (-5)$
$= -21$
$-21 \div 7 = -3$
The average daily high temperature was $-3°C$.

35. Strategy
• Length: L
Width: $0.40L$
• Use the formula for the perimeter of a
rectangle.

Solution
$\quad\quad 2L + 2W = P$
$2L + 2(0.40L) = 42$
$\quad 2L + 0.8L = 42$
$\quad\quad\quad 2.8L = 42$
$\quad\quad\quad\quad\; L = 15$
$0.40L = 0.40(15) = 6$
The length is 15 cm.
The width is 6 cm.

36. Strategy
To find the length of each piece, write and solve an equation using x to represent the length of the shorter piece and $10 - x$ to represent the length of the longer piece.

Solution

Four times the length of the shorter piece	is	2 ft less than three times the length of the longer piece

$4x = 3(10 - x) - 2$
$4x = 30 - 3x - 2$
$4x = 28 - 3x$
$7x = 28$
$x = 4$
$10 - x = 10 - 4 = 6$
The pieces measure 4 ft and 6 ft.

37. Strategy
To find the maximum number of miles:
• Write an expression for the cost of each car using x to represent the number of miles driven per day.
• Write and solve an inequality.

Solution

Cost of a Company A car	is less than	Cost of a Company B car

$6 + 0.25x < 15 + 0.10x$
$6 + 0.25x - 0.10x < 15 + 0.10x - 0.10x$
$6 + 0.15x < 15$
$-6 + 6 + 0.15x < -6 + 15$
$0.15x < 9$
$\dfrac{0.15x}{0.15} < \dfrac{9}{0.15}$
$x < 60$

$6x < 360$
The maximum number of miles you can drive a Company A car is 359 mi.

38. Strategy
• Amount invested at 11%: x

	Principal	Rate	Interest
Amount at 8%	4000	0.08	0.08(400)
Amount at 11%	x	0.11	0.11x

• The total annual interest earned is $1035.

Solution
$0.08(4000) + 0.11x = 1035$
$320 + 0.11x = 1035$
$0.11x = 715$
$x = 6500$
$6500 should be invested at 11%.

39. Strategy

Given: $R = \$165$

  $S = \$99$

Unknown: r

Use the equation $S = R - rR$.

Solution

$S = R - rR$

$99 = 165 - r(165)$

$-66 = -165r$

$0.40 = r$

The discount rate is 40%.

40. Strategy

• First integer: n

Middle integer: $n + 2$

Third integer: $n + 4$

• Five times the middle integer is twelve more than twice the sum of the first and third.

Solution

$5(n + 2) = 2(n + n + 4) + 12$

$5n + 10 = 2(2n + 4) + 12$

$5n + 10 = 4n + 8 + 12$

$5n + 10 = 4n + 20$

$n + 10 = 20$

$n = 10$

$n + 2 = 10 + 2 = 12$

$n + 4 = 10 + 4 = 14$

The integers are 10, 12, and 14.

Chapter 9: Rational Expressions

Prep Test

1. 36   [1.3.2]

2. $\dfrac{3x}{9y^3}$ [7.4.1]

3. $\dfrac{3}{4}-\dfrac{8}{9}=\dfrac{27}{36}-\dfrac{32}{36}=-\dfrac{5}{36}$ [1.3.2]

4. $\left(-\dfrac{8}{11}\right)\div\dfrac{4}{5}=-\dfrac{8}{11}\cdot\dfrac{5}{4}=-\dfrac{10}{11}$ [1.3.3]

5. No. $\dfrac{0}{a}=0$ but $\dfrac{a}{0}$ is undefined.   [1.2.4]

6. $\dfrac{2}{3}x-\dfrac{3}{4}=\dfrac{5}{6}$ [3.2.1]

$\dfrac{2}{3}x=\dfrac{5}{6}+\dfrac{3}{4}$

$\dfrac{2}{3}x=\dfrac{19}{12}$

$x=\dfrac{19}{8}$

7.

$y=a$ because corresponding angles have the same measure. $y+50°=180°$ because adjacent angles of intersecting lines are supplementary angles. Substitute a for y and solve for a.

$a+50=180$

$a=130°$

8. $(x-6)(x+2)$ [8.2.1]

9. $(2x-3)(x+1)$ [8.3.1/8.3.2]

10. Jean's time: t [4.7.1]
Anthony's time: $t+10$

Anthony and Jean jog the same distance.
$9(t+10)=12t$
$9t+90=12t$
$90=3t$
$30=t$
9:10AM + 30 min = 9:40AM
Jean will catch up to Anthony at 9:40AM.

Go Figure

The first mouse proceeds at 2 ft/s for
(6s + 18s = 24s)
Using D = RT, the first mouse travels
  (2)(24) = 48 ft
The second mouse travels a distance of
  (3)(18) = 54 ft.
The distance between the two mice is
54 − 48 = 6 ft.

Section 9.1

Concept Review 9.1

1. Always true

3. Never true
To multiply rational expressions, multiply the numerators and multiply the denominators.

5. Always true

7. Never true
To divide two rational expressions, multiply the first expression by the reciprocal of the second expression.

Objective 9.1.1 Exercises

5. $\dfrac{9x^3}{12x^4}=\dfrac{\overset{1}{\cancel{3}}\cdot3x^3}{2\cdot2\cdot\cancel{3}x^4_{1}}=\dfrac{3}{4x}$

7. $\dfrac{(x+3)^2}{(x+3)^3}=(x+3)^{2-3}=(x+3)^{-1}=\dfrac{1}{x+3}$

9. $\dfrac{3n-4}{4-3n}=\dfrac{-(4-3n)}{4-3n}=-1$

11. $\dfrac{6y(y+2)}{9y^2(y+2)}=\dfrac{2\cdot3y(y+2)}{3\cdot3y^2(y+2)}=\dfrac{2}{3y}$

13. $\dfrac{6x(x-5)}{8x^2(5-x)}=\dfrac{2\cdot3x(x-5)}{2\cdot2\cdot2x^2(5-x)}=-\dfrac{3}{4x}$

15. $\dfrac{a^2+4a}{ab+4b}=\dfrac{a(a+4)}{b(a+4)}=\dfrac{a}{b}$

17. $\dfrac{4-6x}{3x^2-2x}=\dfrac{2(2-3x)}{x(3x-2)}=-\dfrac{2}{x}$

19. $\dfrac{y^2-3y+2}{y^2-4y+3}=\dfrac{(y-1)(y-2)}{(y-1)(y-3)}=\dfrac{y-2}{y-3}$

21. $\dfrac{x^2+3x-10}{x^2+2x-8}=\dfrac{(x+5)(x-2)}{(x+4)(x-2)}=\dfrac{x+5}{x+4}$

23. $\dfrac{x^2+x-12}{x^2-6x+9}=\dfrac{(x+4)(x-3)}{(x-3)(x-3)}=\dfrac{x+4}{x-3}$

25. $\dfrac{x^2-3x-10}{25-x^2}=\dfrac{(x+2)(x-5)}{(5+x)(5-x)}=-\dfrac{x+2}{x+5}$

27. $\dfrac{2x^3+2x^2-4x}{x^3+2x^2-3x}=\dfrac{2x(x^2+x-2)}{x(x^2+2x-3)}=\dfrac{2x(x+2)(x-1)}{x(x+3)(x-1)}=\dfrac{2(x+2)}{x+3}$

29. $\dfrac{6x^2-7x+2}{6x^2+5x-6}=\dfrac{(2x-1)(3x-2)}{(2x+3)(3x-2)}=\dfrac{2x-1}{2x+3}$

Objective 9.1.2 Exercises

31. $\dfrac{8x^2}{9y^3}\cdot\dfrac{3y^2}{4x^3}=\dfrac{\overset{1}{2}\cdot\overset{1}{2}\cdot 2x^2\cdot\overset{1}{3}y^2}{\underset{1}{3}\cdot 3y^3\cdot\underset{1}{2\cdot 2}x^3}=\dfrac{2}{3xy}$

33. $\dfrac{12x^3y^4}{7a^2b^3}\cdot\dfrac{14a^3b^4}{9x^2y^2}=\dfrac{2\cdot 2\cdot\overset{1}{3}x^3y^4\cdot 2\cdot\overset{1}{7}a^3b^4}{\underset{1}{7}a^2b^3\cdot\underset{1}{3}\cdot 3x^2y^2}=\dfrac{8abxy^2}{3}$

35. $\dfrac{3x-6}{5x-20}\cdot\dfrac{10x-40}{27x-54}=\dfrac{3(x-2)}{5(x-4)}\cdot\dfrac{10(x-4)}{27(x-2)}=\dfrac{3(x-2)\cdot 2\cdot 5(x-4)}{5(x-4)\cdot 3\cdot 3\cdot 3(x-2)}=\dfrac{2}{9}$

37. $\dfrac{3x^2+2x}{3xy-3y}\cdot\dfrac{3xy^3-3y^3}{3x^3+2x^2}=\dfrac{x(3x+2)}{y(3x-3)}\cdot\dfrac{y^3(3x-3)}{x^2(3x+2)}=\dfrac{x(3x+2)y^3(3x-3)}{y(3x-3)x^2(3x+2)}=\dfrac{y^2}{x}$

39. $\dfrac{x^2+5x+4}{x^3y^2}\cdot\dfrac{x^2y^3}{x^2+2x+1}=\dfrac{(x+1)(x+4)}{x^3y^2}\cdot\dfrac{x^2y^3}{(x+1)(x+1)}=\dfrac{(x+1)(x+4)x^2y^3}{x^3y^2(x+1)(x+1)}=\dfrac{y(x+4)}{x(x+1)}$

41. $\dfrac{x^4y^2}{x^2+3x-28}\cdot\dfrac{x^2-49}{xy^4}=\dfrac{x^4y^2}{(x+7)(x-4)}\cdot\dfrac{(x+7)(x-7)}{xy^4}=\dfrac{x^4y^2(x+7)(x-7)}{(x+7)(x-4)xy^4}=\dfrac{x^3(x-7)}{y^2(x-4)}$

43. $\dfrac{2x^2-5x}{2xy+y}\cdot\dfrac{2xy^2+y^2}{5x^2-2x^3}=\dfrac{x(2x-5)}{y(2x+1)}\cdot\dfrac{y^2(2x+1)}{x^2(5-2x)}=\dfrac{x(2x-5)y^2(2x+1)}{y(2x+1)x^2(5-2x)}=-\dfrac{y}{x}$

45. $\dfrac{x^2-2x-24}{x^2-5x-6}\cdot\dfrac{x^2+5x+6}{x^2+6x+8}=\dfrac{(x+4)(x-6)}{(x+1)(x-6)}\cdot\dfrac{(x+2)(x+3)}{(x+2)(x+4)}=\dfrac{(x+4)(x-6)(x+2)(x+3)}{(x+1)(x-6)(x+2)(x+4)}=\dfrac{x+3}{x+1}$

47. $\dfrac{x^2+2x-35}{x^2+4x-21}\cdot\dfrac{x^2+3x-18}{x^2+9x+18}=\dfrac{(x+7)(x-5)}{(x+7)(x-3)}\cdot\dfrac{(x+6)(x-3)}{(x+3)(x+6)}=\dfrac{(x+7)(x-5)(x+6)(x-3)}{(x+7)(x-3)(x+3)(x+6)}=\dfrac{x-5}{x+3}$

49. $\dfrac{x^2-3x-4}{x^2+6x+5}\cdot\dfrac{x^2+5x+6}{8+2x-x^2}=\dfrac{(x+1)(x-4)}{(x+1)(x+5)}\cdot\dfrac{(x+2)(x+3)}{(4-x)(2+x)}=\dfrac{(x+1)(x-4)(x+2)(x+3)}{(x+1)(x+5)(4-x)(2+x)}=-\dfrac{x+3}{x+5}$

51. $\dfrac{12x^2-6x}{x^2+6x+5}\cdot\dfrac{2x^4+10x^3}{4x^2-1}=\dfrac{6x(2x-1)}{(x+1)(x+5)}\cdot\dfrac{2x^3(x+5)}{(2x+1)(2x-1)}$

$=\dfrac{2\cdot 3x(2x-1)\cdot 2x^3(x+5)}{(x+1)(x+5)(2x+1)(2x-1)}$

$=\dfrac{12x^4}{(x+1)(2x+1)}$

53. $\dfrac{16+6x-x^2}{x^2-10x-24}\cdot\dfrac{x^2-6x-27}{x^2-17x+72}=\dfrac{(8-x)(2+x)}{(x+2)(x-12)}\cdot\dfrac{(x+3)(x-9)}{(x-8)(x-9)}=\dfrac{(8-x)(2+x)(x+3)(x-9)}{(x+2)(x-12)(x-8)(x-9)}=-\dfrac{x+3}{x-12}$

55. $\dfrac{2x^2+5x+2}{2x^2+7x+3}\cdot\dfrac{x^2-7x-30}{x^2-6x-40}=\dfrac{(2x+1)(x+2)}{(2x+1)(x+3)}\cdot\dfrac{(x+3)(x-10)}{(x+4)(x-10)}=\dfrac{(2x+1)(x+2)(x+3)(x-10)}{(2x+1)(x+3)(x+4)(x-10)}=\dfrac{x+2}{x+4}$

57. $\dfrac{2x^2+x-3}{2x^2-x-6}\cdot\dfrac{2x^2-9x+10}{2x^2-3x+1}=\dfrac{(2x+3)(x-1)}{(2x+3)(x-2)}\cdot\dfrac{(2x-5)(x-2)}{(2x-1)(x-1)}=\dfrac{(2x+3)(x-1)(2x-5)(x-2)}{(2x+3)(x-2)(2x-1)(x-1)}=\dfrac{2x-5}{2x-1}$

59. $\dfrac{6x^2-11x+4}{6x^2+x-2}\cdot\dfrac{12x^2+11x+2}{8x^2+14x+3}=\dfrac{(2x-1)(3x-4)}{(3x+2)(2x-1)}\cdot\dfrac{(4x+1)(3x+2)}{(2x+3)(4x+1)}$

$$=\dfrac{(2x-1)(3x-4)(4x+1)(3x+2)}{(3x+2)(2x-1)(2x+3)(4x+1)}$$

$$=\dfrac{3x-4}{2x+3}$$

Objective 9.1.3 Exercises

63. $\dfrac{4x^2y^3}{15a^2b^3}\div\dfrac{6xy}{5a^3b^5}=\dfrac{4x^2y^3}{15a^2b^3}\cdot\dfrac{5a^3b^5}{6xy}=\dfrac{\overset{1}{2}\cdot2x^2y^3\cdot\overset{1}{5}a^3b^5}{3\cdot\underset{1}{5}a^2b^3\cdot\underset{1}{2}3xy}=\dfrac{2ab^2xy^2}{9}$

65. $\dfrac{6x-12}{8x+32}\div\dfrac{18x-36}{10x+40}=\dfrac{6x-12}{8x+32}\cdot\dfrac{10x+40}{18x-36}=\dfrac{6(x-2)}{8(x+4)}\cdot\dfrac{10(x+4)}{18(x-2)}=\dfrac{\overset{1}{2}\cdot\overset{1}{3}(x-2)\cdot\overset{1}{2}\cdot5(x+4)}{\underset{1}{2}\cdot\underset{1}{2}\cdot2(x+4)\cdot2\cdot\underset{1}{3}\cdot3(x-2)}=\dfrac{5}{12}$

67. $\dfrac{6x^3+7x^2}{12x-3}\div\dfrac{6x^2+7x}{36x-9}=\dfrac{6x^3+7x^2}{12x-3}\cdot\dfrac{36x-9}{6x^2+7x}=\dfrac{x^2(6x+7)}{3(4x-1)}\cdot\dfrac{9(4x-1)}{x(6x+7)}=\dfrac{x^2(6x+7)\cdot3\cdot3(4x-1)}{3(4x-1)x(6x+7)}=3x$

69. $\dfrac{x^2+4x+3}{x^2y}\div\dfrac{x^2+2x+1}{xy^2}=\dfrac{x^2+4x+3}{x^2y}\cdot\dfrac{xy^2}{x^2+2x+1}=\dfrac{(x+1)(x+3)}{x^2y}\cdot\dfrac{xy^2}{(x+1)(x+1)}=\dfrac{y(x+3)}{x(x+1)}$

71. $\dfrac{x^2-49}{x^4y^3}\div\dfrac{x^2-14x+49}{x^4y^3}=\dfrac{x^2-49}{x^4y^3}\cdot\dfrac{x^4y^3}{x^2-14x+49}=\dfrac{(x+7)(x-7)}{x^4y^3}\cdot\dfrac{x^4y^3}{(x-7)(x-7)}=\dfrac{x+7}{x-7}$

73. $\dfrac{4ax-8a}{c^2}\div\dfrac{2y-xy}{c^3}=\dfrac{4ax-8a}{c^2}\cdot\dfrac{c^3}{2y-xy}=\dfrac{4a(x-2)}{c^2}\cdot\dfrac{c^3}{y(2-x)}=-\dfrac{4ac}{y}$

75. $\dfrac{x^2-5x+6}{x^2-9x+18}\div\dfrac{x^2-6x+8}{x^2-9x+20}=\dfrac{x^2-5x+6}{x^2-9x+18}\cdot\dfrac{x^2-9x+20}{x^2-6x+8}=\dfrac{(x-2)(x-3)}{(x-3)(x-6)}\cdot\dfrac{(x-4)(x-5)}{(x-2)(x-4)}=\dfrac{x-5}{x-6}$

77. $\dfrac{x^2+2x-15}{x^2-4x-45}\div\dfrac{x^2+x-12}{x^2-5x-36}=\dfrac{x^2+2x-15}{x^2-4x-45}\cdot\dfrac{x^2-5x-36}{x^2+x-12}=\dfrac{(x+5)(x-3)}{(x+5)(x-9)}\cdot\dfrac{(x+4)(x-9)}{(x+4)(x-3)}=1$

79. $\dfrac{8+2x-x^2}{x^2+7x+10}\div\dfrac{x^2-11x+28}{x^2-x-42}=\dfrac{8+2x-x^2}{x^2+7x+10}\cdot\dfrac{x^2-x-42}{x^2-11x+28}=\dfrac{(2+x)(4-x)}{(x+2)(x+5)}\cdot\dfrac{(x+6)(x-7)}{(x-4)(x-7)}=-\dfrac{x+6}{x+5}$

81. $\dfrac{2x^2-3x-20}{2x^2-7x-30}\div\dfrac{2x^2-5x-12}{4x^2+12x+9}=\dfrac{2x^2-3x-20}{2x^2-7x-30}\cdot\dfrac{4x^2+12x+9}{2x^2-5x-12}=\dfrac{(2x+5)(x-4)}{(2x+5)(x-6)}\cdot\dfrac{(2x+3)(2x+3)}{(2x+3)(x-4)}=\dfrac{2x+3}{x-6}$

83. $\dfrac{9x^2-16}{6x^2-11x+4}\div\dfrac{6x^2+11x+4}{8x^2+10x+3}=\dfrac{9x^2-16}{6x^2-11x+4}\cdot\dfrac{8x^2+10x+3}{6x^2+11x+4}=\dfrac{(3x+4)(3x-4)}{(3x-4)(2x-1)}\cdot\dfrac{(4x+3)(2x+1)}{(3x+4)(2x+1)}=\dfrac{4x+3}{2x-1}$

85. $\dfrac{8x^2+18x-5}{10x^2-9x+2}\div\dfrac{8x^2+22x+15}{10x^2+11x-6}=\dfrac{8x^2+18x-5}{10x^2-9x+2}\cdot\dfrac{10x^2+11x-6}{8x^2+22x+15}$

$$=\dfrac{(2x+5)(4x-1)}{(2x-1)(5x-2)}\cdot\dfrac{(2x+3)(5x-2)}{(2x+3)(4x+5)}$$

$$=\dfrac{(2x+5)(4x-1)}{(2x-1)(4x+5)}$$

Applying Concepts 9.1

87. $(x+6)(x-1)=0$

$x+6=0\quad x-1=0$

$x=-6\quadx=1$

The fraction is undefined if $x=-6$ or $x=1$.

89. $x^2-1=0$

$(x+1)(x-1)=0$

$x+1=0\quad x-1=0$

$x=-1\quadx=1$

The fraction is undefined if $x=-1$ or $x=1$.

91. $x^2 - x - 6 = 0$
$(x-3)(x+2) = 0$
$x - 3 = 0 \quad x + 2 = 0$
$\quad x = 3 \qquad x = -2$
The fraction is undefined if $x = 3$ or $x = -2$.

93. $x^2 + 6x + 9 = 0$
$\quad (x+3)^2 = 0$
$x + 3 = 0$
$\quad x = -3$
The fraction is undefined if $x = -3$.

95. $6x^2 - 5x - 4 = 0$
$(3x-4)(2x+1) = 0$
$3x - 4 = 0 \quad 2x + 1 = 0$
$\quad 3x = 4 \qquad 2x = -1$
$\qquad x = \dfrac{4}{3} \qquad x = -\dfrac{1}{2}$
The fraction is undefined if $x = \dfrac{4}{3}$ or $x = -\dfrac{1}{2}$.

97. $\dfrac{ab}{3} \cdot \dfrac{a}{b^2} \div \dfrac{a}{4} = \dfrac{ab}{3} \cdot \dfrac{a}{b^2} \cdot \dfrac{4}{a} = \dfrac{4a^2 b}{3ab^2} = \dfrac{4a}{3b}$

99. $\left(\dfrac{c}{3}\right)^2 \div \left(\dfrac{c}{2} \cdot \dfrac{c}{4}\right) = \dfrac{c^2}{9} \div \dfrac{c^2}{8} = \dfrac{c^2}{9} \cdot \dfrac{8}{c^2} = \dfrac{8}{9}$

101. $\left(\dfrac{x-4}{y^2}\right)^3 \cdot \left(\dfrac{y}{4-x}\right)^2 = \dfrac{(x-4)^3}{y^6} \cdot \dfrac{y^2}{(4-x)^2} = \dfrac{(x-4)(x-4)(x-4)y^2}{y^6(4-x)(4-x)} = \dfrac{x-4}{y^4}$

103. $\dfrac{x^2+x-6}{x^2+7x+12} \cdot \dfrac{x^2+3x-4}{x^2+x-2} \div \dfrac{x^2-16}{x^2-4} = \dfrac{x^2+x-6}{x^2+7x+12} \cdot \dfrac{x^2+3x-4}{x^2+x-2} \cdot \dfrac{x^2-4}{x^2-16}$

$\qquad = \dfrac{(x+3)(x-2)}{(x+3)(x+4)} \cdot \dfrac{(x+4)(x-1)}{(x+2)(x-1)} \cdot \dfrac{(x+2)(x-2)}{(x+4)(x-4)}$

$\qquad = \dfrac{(x-2)(x-2)}{(x+4)(x-4)}$

105. Use the formula for the area of a triangle, $A = \dfrac{1}{2}bh$

and the formula for the area of a rectangle, $A = LW$.

shaded area $= \dfrac{1}{2}(x+4)(2x) = x(x+4)$

total area $= (x+8)(x+4)$

ratio $= \dfrac{x(x+4)}{(x+8)(x+4)} = \dfrac{x}{x+8}$

107. Students will choose different values of y. Two examples are shown below.

For $y = 2$, $\dfrac{1}{y-3} = \dfrac{1}{2-3} = \dfrac{1}{-1} = -1$.

For $y = 0$, $\dfrac{1}{y-3} = \dfrac{1}{0-3} = \dfrac{1}{-3} = -\dfrac{1}{3}$

Yes, it is possible for $\dfrac{1}{y-3}$ to be greater than 10,000,000. Choose a value of y that is close to, but greater than, 3.

For example, let $y = 3.00000001$

$\dfrac{1}{y-3} = \dfrac{1}{3.00000001 - 3} = \dfrac{1}{0.00000001} = 100,000,000$, which is greater than 10,000,000.

Section 9.2

Concept Review 9.2

1. Always true

3. Never true
The LCM of x^2, x^5, and x^8 is x^8.

5. Always true

Objective 9.2.1 Exercises

1. $8x^3 y = 2 \cdot 2 \cdot 2 \cdot x \cdot x \cdot x \cdot y$
$12xy^2 = 2 \cdot 2 \cdot 3 \cdot x \cdot y \cdot y$
LCM $= 2 \cdot 2 \cdot 2 \cdot 3 \cdot x \cdot x \cdot x \cdot y \cdot y = 24x^3 y^2$

3. $10x^4 y^2 = 2 \cdot 5 \cdot x \cdot x \cdot x \cdot x \cdot y \cdot y$
$15x^3 y = 3 \cdot 5 \cdot x \cdot x \cdot x \cdot y$
LCM $= 2 \cdot 3 \cdot 5 \cdot x \cdot x \cdot x \cdot x \cdot y \cdot y = 30x^4 y^2$

5. $8x^2 = 2 \cdot 2 \cdot 2 \cdot x \cdot x$
 $4x^2 + 8x = 4x(x+2) = 2 \cdot 2 \cdot x(x+2)$
 $\text{LCM} = 2 \cdot 2 \cdot 2 \cdot x \cdot x(x+2) = 8x^2(x+2)$

7. $2x^2 y = 2 \cdot x \cdot x \cdot y$
 $3x^2 + 12x = 3x(x+4)$
 $\text{LCM} = 2 \cdot 3 \cdot x \cdot x \cdot y(x+4) = 6x^2 y(x+4)$

9. $8x^2(x-1)^2 = 2 \cdot 2 \cdot 2 \cdot x \cdot x(x-1)(x-1)$
 $10x^3(x-1) = 2 \cdot 5 \cdot x \cdot x \cdot x(x-1)$
 $\text{LCM} = 2 \cdot 2 \cdot 2 \cdot 5 \cdot x \cdot x \cdot x(x-1)(x-1) = 40x^3(x-1)^2$

11. $4x - 12 = 4(x-3) = 2 \cdot 2(x-3)$
 $2x^2 - 12x + 18 = 2(x^2 - 6x + 9) = 2(x-3)(x-3)$
 $\text{LCM} = 2 \cdot 2(x-3)(x-3) = 4(x-3)^2$

13. $(2x-1)(x+4)$
 $(2x+1)(x+4)$
 $\text{LCM} = (2x-1)(x+4)(2x+1)$

15. $(x-7)(x+2)$
 $(x-7)^2 = (x-7)(x-7)$
 $\text{LCM} = (x+2)(x-7)(x-7)$
 $\quad\quad = (x+2)(x-7)^2$

17. $(x+4)(x-3)$
 $x+4$
 $x-3$
 $\text{LCM} = (x+4)(x-3)$

19. $x^2 + 3x - 10 = (x+5)(x-2)$
 $x^2 + 5x - 14 = (x+7)(x-2)$
 $\text{LCM} = (x+5)(x-2)(x+7)$

21. $x^2 - 10x + 21 = (x-3)(x-7)$
 $x^2 - 8x + 15 = (x-3)(x-5)$
 $\text{LCM} = (x-3)(x-7)(x-5)$

23. $x^2 + 7x + 10 = (x+2)(x+5)$
 $x^2 - 25 = (x+5)(x-5)$
 $\text{LCM} = (x+2)(x+5)(x-5)$

25. $2x^2 - 7x + 3 = (x-3)(2x-1)$
 $2x^2 + x - 1 = (x+1)(2x-1)$
 $\text{LCM} = (x-3)(2x-1)(x+1)$

27. $2x^2 - 9x + 10 = (x-2)(2x-5)$
 $2x^2 + x - 15 = (x+3)(2x-5)$
 $\text{LCM} = (x-2)(2x-5)(x+3)$

29. $15 + 2x - x^2 = (3+x)(5-x)$
 $x-5$
 $x+3$
 $\text{LCM} = (x+3)(x-5)$

31. $x^2 + 3x - 18 = (x+6)(x-3)$
 $3-x$
 $x+6$
 $\text{LCM} = (x+6)(x-3)$

Objective 9.2.2 Exercises

33. The LCM is x^2.
$$\frac{4}{x} = \frac{4}{x} \cdot \frac{x}{x} = \frac{4x}{x^2}$$
$$\frac{3}{x^2} = \frac{3}{x^2}$$

35. The LCM is $12y^2$.
$$\frac{x}{3y^2} = \frac{x}{3y^2} \cdot \frac{4}{4} = \frac{4x}{12y^2}$$
$$\frac{z}{4y} = \frac{z}{4y} \cdot \frac{3y}{3y} = -\frac{3yz}{12y^2}$$

37. The LCM is $x^2(x-3)$.
$$\frac{y}{x(x-3)} = \frac{y}{x(x-3)} \cdot \frac{x}{x} = \frac{xy}{x^2(x-3)}$$
$$\frac{6}{x^2} = \frac{6}{x^2} \cdot \frac{x-3}{x-3} = \frac{6x-18}{x^2(x-3)}$$

39. The LCM is $x(x-1)^2$.
$$\frac{9}{(x-1)^2} = \frac{9}{(x-1)^2} \cdot \frac{x}{x} = \frac{9x}{x(x-1)^2}$$
$$\frac{6}{x(x-1)} = \frac{6}{x(x-1)} \cdot \frac{x-1}{x-1} = \frac{6x-6}{x(x-1)^2}$$

41. $-\dfrac{5}{x(3-x)} = -\dfrac{5}{-x(x-3)} = \dfrac{5}{x(x-3)}$
 The LCM is $x(x-3)$.
$$\frac{3}{x-3} = \frac{3}{x-3} \cdot \frac{x}{x} = \frac{3x}{x(x-3)}$$
$$-\frac{5}{x(3-x)} = \frac{5}{x(x-3)}$$

43. $\dfrac{2}{5-x} = \dfrac{2}{-(x-5)} = -\dfrac{2}{x-5}$
 The LCM is $(x-5)^2$.
$$\frac{3}{(x-5)^2} = \frac{3}{(x-5)^2}$$
$$\frac{2}{5-x} = -\frac{2}{x-5} \cdot \frac{x-5}{x-5} = -\frac{2x-10}{(x-5)^2}$$

45. The LCM is $x^2(x+2)$.

$$\frac{3}{x^2+2x} = \frac{3}{x(x+2)} \cdot \frac{x}{x} = \frac{3x}{x^2(x+2)}$$

$$\frac{4}{x^2} = \frac{4}{x^2} \cdot \frac{x+2}{x+2} = \frac{4x+8}{x^2(x+2)}$$

47. The LCM is $(x+3)(x-4)$.

$$\frac{x-2}{x+3} = \frac{x-2}{x+3} \cdot \frac{x-4}{x-4} = \frac{x^2-6x+8}{(x+3)(x-4)}$$

$$\frac{x}{x-4} = \frac{x}{x-4} \cdot \frac{x+3}{x+3} = \frac{x^2+3x}{(x+3)(x-4)}$$

49. The LCM is $(x+2)(x-1)$.

$$\frac{3}{x^2+x-2} = \frac{3}{(x+2)(x-1)}$$

$$\frac{x}{x+2} = \frac{x}{x+2} \cdot \frac{x-1}{x-1} = \frac{x^2-x}{(x+2)(x-1)}$$

51. The LCM is $(2x-5)(x-2)$.

$$\frac{5}{2x^2-9x+10} = \frac{5}{(2x-5)(x-2)}$$

$$\frac{x-1}{2x-5} = \frac{x-1}{2x-5} \cdot \frac{x-2}{x-2} = \frac{x^2-3x+2}{(2x-5)(x-2)}$$

53. The LCM is $(x+3)(x-2)(x-3)$.

$$\frac{x}{x^2+x-6} = \frac{x}{(x+3)(x-2)} \cdot \frac{x-3}{x-3} = \frac{x^2-3x}{(x+3)(x-2)(x-3)}$$

$$\frac{2x}{x^2-9} = \frac{2x}{(x+3)(x-3)} \cdot \frac{x-2}{x-2} = \frac{2x^2-4x}{(x+3)(x-2)(x-3)}$$

55. $\dfrac{x}{9-x^2} = \dfrac{x}{-(x^2-9)} = -\dfrac{x}{x^2-9} = -\dfrac{x}{(x+3)(x-3)}$

The LCM is $(x+3)(x-3)^2$.

$$\frac{x}{9-x^2} = -\frac{x}{(x+3)(x-3)} \cdot \frac{x-3}{x-3} = -\frac{x^2-3x}{(x+3)(x-3)^2}$$

$$\frac{x-1}{x^2-6x+9} = \frac{x-1}{(x-3)(x-3)} \cdot \frac{x+3}{x+3} = \frac{x^2+2x-3}{(x+3)(x-3)^2}$$

57. $\dfrac{3}{20+x-x^2} = \dfrac{3}{-(x^2-x-20)} = -\dfrac{3}{x^2-x-20} = -\dfrac{3}{(x+4)(x-5)}$

The LCM is $(x+4)(x-5)$.

$$\frac{3x}{x-5} = \frac{3x}{x-5} \cdot \frac{x+4}{x+4} = \frac{3x^2+12x}{(x+4)(x-5)}$$

$$\frac{x}{x+4} = \frac{x}{x+4} \cdot \frac{x-5}{x-5} = \frac{x^2-5x}{(x+4)(x-5)}$$

$$\frac{3}{20+x-x^2} = -\frac{3}{(x+4)(x-5)}$$

Applying Concepts 9.2

59. The LCM is 10^4.

$$\frac{6}{x} = \frac{2}{3}$$

$$3x \cdot \frac{6}{x} = 3x \cdot \frac{2}{3}$$

$$18 = 2x$$

$$9 = x$$

61. The LCM is b.

$$b = \frac{b}{1} = \frac{b}{1} \cdot \frac{b}{b} = \frac{b^2}{b}$$

$$\frac{5}{b} = \frac{5}{b}$$

63. The LCM is $y-1$.

$$1 = \frac{1}{1} = \frac{1}{1} \cdot \frac{y-1}{y-1} = \frac{y-1}{y-1}$$

$$\frac{y}{y-1} = \frac{y}{y-1}$$

65. The LCM is $(x-1)^3$.

$$\frac{x^2+1}{(x-1)^3}=\frac{x^2+1}{(x-1)^3}$$

$$\frac{x+1}{(x-1)^2}=\frac{x+1}{(x-1)^2}\cdot\frac{x-1}{x-1}=\frac{x^2-1}{(x-1)^3}$$

$$\frac{1}{x-1}=\frac{1}{x-1}\cdot\frac{(x-1)^2}{(x-1)^2}=\frac{x^2-2x+1}{(x-1)^3}$$

67. $4a^2-4b^2=4(a^2-b^2)=4(a+b)(a-b)$
$8a-8b=8(a-b)$
The LCM is $8(a+b)(a-b)$.

$$\frac{b}{4a^2-4b^2}=\frac{b}{4(a+b)(a-b)}\cdot\frac{2}{2}=\frac{2b}{8(a+b)(a-b)}$$

$$\frac{a}{8a-8b}=\frac{a}{8(a-b)}\cdot\frac{a+b}{a+b}=\frac{a^2+ab}{8(a+b)(a-b)}$$

69. $x^2+2x+xy+2y=(x^2+2x)+(xy+2y)$
$\qquad\qquad\qquad\quad=x(x+2)+y(x+2)$
$\qquad\qquad\qquad\quad=(x+2)(x+y)$
$x^2+xy-2x-2y=(x^2+xy)-(2x+2y)$
$\qquad\qquad\qquad\quad=x(x+y)-2(x+y)$
$\qquad\qquad\qquad\quad=(x+y)(x-2)$
The LCM is $(x+y)(x+2)(x-2)$.

$$\frac{1}{x^2+2x+xy+2y}=\frac{1}{(x+2)(x+y)}\cdot\frac{x-2}{x-2}=\frac{x-2}{(x+y)(x+2)(x-2)}$$

$$\frac{1}{x^2+xy-2x-2y}=\frac{1}{(x+y)(x-2)}\cdot\frac{x+2}{x+2}=\frac{x+2}{(x+y)(x+2)(x-2)}$$

Section 9.3

Concept Review 9.3

1. Never true
To add two fractions, build the fractions so they have a common denominator. Then add the numerators and place their sum over the common denominator.

3. Never true
To add two rational expressions, first build both fractions so that they have a common denominator. The common denominator is usually the LCM of the denominators.

5. Always true

Objective 9.3.1 Exercises

1. $\dfrac{3}{y^2}+\dfrac{8}{y^2}=\dfrac{3+8}{y^2}=\dfrac{11}{y^2}$

3. $\dfrac{3}{x+4}-\dfrac{10}{x+4}=\dfrac{3-10}{x+4}=-\dfrac{7}{x+4}$

5. $\dfrac{3x}{2x+3}+\dfrac{5x}{2x+3}=\dfrac{3x+5x}{2x+3}=\dfrac{8x}{2x+3}$

7. $\dfrac{2x+1}{x-3}+\dfrac{3x+6}{x-3}=\dfrac{2x+1+(3x+6)}{x-3}=\dfrac{5x+7}{x-3}$

9. $\dfrac{5x-1}{x+9}-\dfrac{3x+4}{x+9}=\dfrac{5x-1-(3x+4)}{x+9}=\dfrac{5x-1-3x-4}{x+9}=\dfrac{2x-5}{x+9}$

11. $\dfrac{x-7}{2x+7}-\dfrac{4x-3}{2x+7}=\dfrac{x-7-(4x-3)}{2x+7}=\dfrac{x-7-4x+3}{2x+7}=\dfrac{-3x-4}{2x+7}$

13. $\dfrac{x}{x^2+2x-15}-\dfrac{3}{x^2+2x-15}=\dfrac{x-3}{x^2+2x-15}=\dfrac{x-3}{(x+5)(x-3)}=\dfrac{1}{x+5}$

15. $\dfrac{2x+3}{x^2-x-30}-\dfrac{x-2}{x^2-x-30}=\dfrac{2x+3-(x-2)}{x^2-x-30}=\dfrac{2x+3-x+2}{(x+5)(x-6)}=\dfrac{x+5}{(x+5)(x-6)}=\dfrac{1}{x-6}$

17. $\dfrac{4y+7}{2y^2+7y-4} - \dfrac{y-5}{2y^2+7y-4} = \dfrac{4y+7-(y-5)}{2y^2+7y-4} = \dfrac{4y+7-y+5}{2y^2+7y-4} = \dfrac{3y+12}{(y+4)(2y-1)} = \dfrac{3(y+4)}{(y+4)(2y-1)} = \dfrac{3}{2y-1}$

19. $\dfrac{2x^2+3x}{x^2-9x+20} + \dfrac{2x^2-3}{x^2-9x+20} - \dfrac{4x^2+2x+1}{x^2-9x+20}$

$= \dfrac{2x^2+3x+(2x^2-3)-(4x^2+2x+1)}{x^2-9x+20}$

$= \dfrac{2x^2+3x+2x^2-3-4x^2-2x-1}{x^2-9x+20}$

$= \dfrac{x-4}{(x-4)(x-5)}$

$= \dfrac{1}{x-5}$

Objective 9.3.2 Exercises

23. The LCM of the denominators is xy.

$\dfrac{4}{x} + \dfrac{5}{y} = \dfrac{4}{x} \cdot \dfrac{y}{y} + \dfrac{5}{y} \cdot \dfrac{x}{x} = \dfrac{4y}{xy} + \dfrac{5x}{xy} = \dfrac{4y+5x}{xy}$

25. The LCM of the denominators is $2x$.

$\dfrac{12}{x} - \dfrac{5}{2x} = \dfrac{12}{x} \cdot \dfrac{2}{2} - \dfrac{5}{2x}$

$= \dfrac{24}{2x} - \dfrac{5}{2x}$

$= \dfrac{24-5}{2x} = \dfrac{19}{2x}$

27. The LCM of the denominators is $12x$.

$\dfrac{1}{2x} - \dfrac{5}{4x} + \dfrac{7}{6x} = \dfrac{1}{2x} \cdot \dfrac{6}{6} - \dfrac{5}{4x} \cdot \dfrac{3}{3} + \dfrac{7}{6x} \cdot \dfrac{2}{2}$

$= \dfrac{6}{12x} - \dfrac{15}{12x} + \dfrac{14}{12x}$

$= \dfrac{6-15+14}{12x} = \dfrac{5}{12x}$

29. The LCM of the denominators is $6x^2$.

$\dfrac{5}{3x} - \dfrac{2}{x^2} + \dfrac{3}{2x} = \dfrac{5}{3x} \cdot \dfrac{2x}{2x} - \dfrac{2}{x^2} \cdot \dfrac{6}{6} + \dfrac{3}{2x} \cdot \dfrac{3x}{3x}$

$= \dfrac{10x}{6x^2} - \dfrac{12}{6x^2} + \dfrac{9x}{6x^2}$

$= \dfrac{10x-12+9x}{6x^2}$

$= \dfrac{19x-12}{6x^2}$

31. The LCM of the denominators is $20xy$.

$\dfrac{2}{x} - \dfrac{3}{2y} + \dfrac{3}{5x} - \dfrac{1}{4y} = \dfrac{2}{x} \cdot \dfrac{20y}{20y} - \dfrac{3}{2y} \cdot \dfrac{10x}{10x} + \dfrac{3}{5x} \cdot \dfrac{4y}{4y} - \dfrac{1}{4y} \cdot \dfrac{5x}{5x}$

$= \dfrac{40y}{20xy} - \dfrac{30x}{20xy} + \dfrac{12y}{20xy} - \dfrac{5x}{20xy}$

$= \dfrac{40y-30x+12y-5x}{20xy}$

$= \dfrac{52y-35x}{20xy}$

33. The LCM of the denominators is $15x$.

$$\frac{2x+1}{3x}+\frac{x-1}{5x}=\frac{2x+1}{3x}\cdot\frac{5}{5}+\frac{x-1}{5x}\cdot\frac{3}{3}$$

$$=\frac{10x+5}{15x}+\frac{3x-3}{15x}$$

$$=\frac{10x+5+(3x-3)}{15x}$$

$$=\frac{13x+2}{15x}$$

35. The LCM of the denominators is $24x$.

$$\frac{x-3}{6x}+\frac{x+4}{8x}=\frac{x-3}{6x}\cdot\frac{4}{4}+\frac{x+4}{8x}\cdot\frac{3}{3}$$

$$=\frac{4x-12}{24x}+\frac{3x+12}{24x}$$

$$=\frac{4x-12+(3x+12)}{24x}$$

$$=\frac{7x}{24x}=\frac{7}{24}$$

37. The LCM of the denominators is $45x$.

$$\frac{2x+9}{9x}-\frac{x-5}{5x}=\frac{2x+9}{9x}\cdot\frac{5}{5}-\frac{x-5}{5x}\cdot\frac{9}{9}$$

$$=\frac{10x+45}{45x}-\frac{9x-45}{45x}$$

$$=\frac{10x+45-(9x-45)}{45x}$$

$$=\frac{10x+45-9x+45}{45x}$$

$$=\frac{x+90}{45x}$$

39. The LCM of the denominators is $2x^2$.

$$\frac{x+4}{2x}-\frac{x-1}{x^2}=\frac{x+4}{2x}\cdot\frac{x}{x}-\frac{x-1}{x^2}\cdot\frac{2}{2}$$

$$=\frac{x^2+4x}{2x^2}-\frac{2x-2}{2x^2}$$

$$=\frac{x^2+4x-(2x-2)}{2x^2}$$

$$=\frac{x^2+4x-2x+2}{2x^2}$$

$$=\frac{x^2+2x+2}{2x^2}$$

41. The LCM of the denominators is $4x^2$.

$$\frac{x-10}{4x^2}+\frac{x+1}{2x}=\frac{x-10}{4x^2}+\frac{x+1}{2x}\cdot\frac{2x}{2x}$$

$$=\frac{x-10}{4x^2}+\frac{2x^2+2x}{4x^2}$$

$$=\frac{x-10+(2x^2+2x)}{4x^2}$$

$$=\frac{2x^2+3x-10}{4x^2}$$

43. The LCM of the denominators is $3y$.

$$y+\frac{8}{3y}=\frac{y}{1}\cdot\frac{3y}{3y}+\frac{8}{3y}=\frac{3y^2}{3y}+\frac{8}{3y}$$

$$=\frac{3y^2+8}{3y}$$

45. The LCM of the denominators is $x+4$.

$$\frac{4}{x+4}+x=\frac{4}{x+4}+\frac{x}{1}\cdot\frac{x+4}{x+4}$$

$$=\frac{4}{x+4}+\frac{x^2+4x}{x+4}$$

$$=\frac{4+(x^2+4x)}{x+4}$$

$$=\frac{x^2+4x+4}{x+4}$$

47. The LCM of the denominators is $x+1$.

$$5-\frac{x-2}{x+1}=\frac{5}{1}\cdot\frac{x+1}{x+1}-\frac{x-2}{x+1}$$

$$=\frac{5x+5}{x+1}-\frac{x-2}{x+1}$$

$$=\frac{5x+5-(x-2)}{x+1}$$

$$=\frac{5x+5-x+2}{x+1}$$

$$=\frac{4x+7}{x+1}$$

49. The LCM of the denominators is $12x^2$.

$$\frac{2x+1}{6x^2}-\frac{x-4}{4x}=\frac{2x+1}{6x^2}\cdot\frac{2}{2}-\frac{x-4}{4x}\cdot\frac{3x}{3x}$$

$$=\frac{4x+2}{12x^2}-\frac{3x^2-12x}{12x^2}$$

$$=\frac{4x+2-(3x^2-12x)}{12x^2}$$

$$=\frac{4x+2-3x^2+12x}{12x^2}$$

$$=\frac{-3x^2+16x+2}{12x^2}$$

51. The LCM of the denominators is x^2y.

$$\frac{x+2}{xy}-\frac{3x-2}{x^2y}=\frac{x+2}{xy}\cdot\frac{x}{x}-\frac{3x-2}{x^2y}$$

$$=\frac{x^2+2x}{x^2y}-\frac{3x-2}{x^2y}$$

$$=\frac{x^2+2x-(3x-2)}{x^2y}$$

$$=\frac{x^2+2x-3x+2}{x^2y}=\frac{x^2-x+2}{x^2y}$$

53. The LCM of the denominators is $12x^2y^2$.

$$\frac{4x-3}{3x^2y}+\frac{2x+1}{4xy^2}=\frac{4x-3}{3x^2y}\cdot\frac{4y}{4y}+\frac{2x+1}{4xy^2}\cdot\frac{3x}{3x}$$

$$=\frac{16xy-12y}{12x^2y^2}+\frac{6x^2+3x}{12x^2y^2}$$

$$=\frac{16xy-12y+(6x^2+3x)}{12x^2y^2}$$

$$=\frac{16xy-12y+6x^2+3x}{12x^2y^2}$$

55. The LCM of the denominators is $24x^2y$.

$$\frac{x-2}{8x^2}-\frac{x+7}{12xy}=\frac{x-2}{8x^2}\cdot\frac{3y}{3y}-\frac{x+7}{12xy}\cdot\frac{2x}{2x}$$

$$=\frac{3xy-6y}{24x^2y}-\frac{2x^2+14x}{24x^2y}$$

$$=\frac{3xy-6y-(2x^2+14x)}{24x^2y}$$

$$=\frac{3xy-6y-2x^2-14x}{24x^2y}$$

57. The LCM of the denominators is $(x-2)(x+3)$.

$$\frac{4}{x-2}+\frac{5}{x+3}=\frac{4}{x-2}\cdot\frac{x+3}{x+3}+\frac{5}{x+3}\cdot\frac{x-2}{x-2}$$

$$=\frac{4x+12}{(x-2)(x+3)}+\frac{5x-10}{(x-2)(x+3)}$$

$$=\frac{4x+12+(5x-10)}{(x-2)(x+3)}$$

$$=\frac{9x+2}{(x-2)(x+3)}$$

63. The LCM of the denominators is $(2x-1)(x-6)$.

$$\frac{4x}{2x-1}-\frac{5}{x-6}=\frac{4x}{2x-1}\cdot\frac{x-6}{x-6}-\frac{5}{x-6}\cdot\frac{2x-1}{2x-1}$$

$$=\frac{4x^2-24x}{(2x-1)(x-6)}-\frac{10x-5}{(2x-1)(x-6)}$$

$$=\frac{4x^2-24x-(10x-5)}{(2x-1)(x-6)}$$

$$=\frac{4x^2-24x-10x+5}{(2x-1)(x-6)}$$

$$=\frac{4x^2-34x+5}{(2x-1)(x-6)}$$

65. The LCM of the denominators is $a-7$.

$$\frac{2a}{a-7}+\frac{5}{7-a}=\frac{2a}{a-7}+\frac{5}{-(a-7)}\cdot\frac{-1}{-1}$$

$$=\frac{2a}{a-7}+\frac{-5}{a-7}$$

$$=\frac{2a+(-5)}{a-7}$$

$$=\frac{2a-5}{a-7}$$

59. The LCM of the denominators is $(x-7)(x+3)$.

$$\frac{6}{x-7}-\frac{4}{x+3}=\frac{6}{x-7}\cdot\frac{x+3}{x+3}-\frac{4}{x+3}\cdot\frac{x-7}{x-7}$$

$$=\frac{6x+18}{(x-7)(x+3)}-\frac{4x-28}{(x-7)(x+3)}$$

$$=\frac{6x+18-(4x-28)}{(x-7)(x+3)}$$

$$=\frac{6x+18-4x+28}{(x-7)(x+3)}$$

$$=\frac{2x+46}{(x-7)(x+3)}$$

$$=\frac{2(x+23)}{(x-7)(x+3)}$$

61. The LCM of the denominators is $(x+1)(x-3)$.

$$\frac{2x}{x+1}+\frac{1}{x-3}=\frac{2x}{x+1}\cdot\frac{x-3}{x-3}+\frac{1}{x-3}\cdot\frac{x+1}{x+1}$$

$$=\frac{2x^2-6x}{(x+1)(x-3)}+\frac{x+1}{(x+1)(x-3)}$$

$$=\frac{2x^2-6x+(x+1)}{(x+1)(x-3)}$$

$$=\frac{2x^2-5x+1}{(x+1)(x-3)}$$

67. The LCM of the denominators is $(x + 3)(x - 3)$.

$$\frac{x}{x^2 - 9} + \frac{3}{x - 3} = \frac{x}{(x+3)(x-3)} + \frac{3}{x-3} \cdot \frac{x+3}{x+3}$$

$$= \frac{x}{(x+3)(x-3)} + \frac{3x+9}{(x+3)(x-3)}$$

$$= \frac{x + (3x + 9)}{(x+3)(x-3)}$$

$$= \frac{4x + 9}{(x+3)(x-3)}$$

69. The LCM of the denominators is $(x + 2)(x - 3)$.

$$\frac{2x}{x^2 - x - 6} - \frac{3}{x + 2} = \frac{2x}{(x+2)(x-3)} - \frac{3}{x+2} \cdot \frac{x-3}{x-3}$$

$$= \frac{2x}{(x+2)(x-3)} - \frac{3x-9}{(x+2)(x-3)}$$

$$= \frac{2x - (3x - 9)}{(x+2)(x-3)}$$

$$= \frac{2x - 3x + 9}{(x+2)(x-3)}$$

$$= \frac{-x + 9}{(x+2)(x-3)}$$

71. The LCM of the denominators is $(x - 5)(x - 5)$.

$$\frac{3x - 1}{x^2 - 10x + 25} - \frac{3}{x - 5} = \frac{3x-1}{(x-5)(x-5)} - \frac{3}{x-5} \cdot \frac{x-5}{x-5}$$

$$= \frac{3x-1}{(x-5)(x-5)} - \frac{3x-15}{(x-5)(x-5)}$$

$$= \frac{3x - 1 - (3x - 15)}{(x-5)(x-5)}$$

$$= \frac{3x - 1 - 3x + 15}{(x-5)(x-5)}$$

$$= \frac{14}{(x-5)(x-5)}$$

73. The LCM of the denominators is $(x + 6)(x - 7)$.

$$\frac{x + 4}{x^2 - x - 42} + \frac{3}{7 - x} = \frac{x+4}{(x+6)(x-7)} + \frac{3}{-(x-7)} \cdot \frac{x+6}{x+6}$$

$$= \frac{x+4}{(x+6)(x-7)} + \frac{-3(x+6)}{(x+6)(x-7)}$$

$$= \frac{x+4}{(x+6)(x-7)} + \frac{-3x-18}{(x+6)(x-7)}$$

$$= \frac{x + 4 + (-3x - 18)}{(x+6)(x-7)}$$

$$= \frac{-2x - 14}{(x+6)(x-7)}$$

$$= \frac{-2(x+7)}{(x+6)(x-7)}$$

75. The LCM of the denominators is $2x(x+2)$.

$$\frac{x}{2x+4} - \frac{2}{x^2+2x} = \frac{x}{2(x+2)} \cdot \frac{x}{x} - \frac{2}{x(x+2)} \cdot \frac{2}{2}$$

$$= \frac{x^2}{2x(x+2)} - \frac{4}{2x(x+4)}$$

$$= \frac{x^2-4}{2x(x+2)}$$

$$= \frac{(x-2)(x+2)}{2x(x+2)}$$

$$= \frac{x-2}{2x}$$

77. The LCM of the denominators is $(x-2)(x+1)(x-1)$.

$$\frac{x-1}{x^2-x-2} + \frac{3}{x^2-3x+2} = \frac{x-1}{(x-2)(x+1)} \cdot \frac{x-1}{x-1} + \frac{3}{(x-2)(x-1)} \cdot \frac{x+1}{x+1}$$

$$= \frac{x^2-2x+1}{(x-2)(x+1)(x-1)} + \frac{3x+3}{(x-2)(x+1)(x-1)}$$

$$= \frac{x^2-2x+1+3x+3}{(x-2)(x+1)(x-1)}$$

$$= \frac{x^2+x+4}{(x-2)(x+1)(x-1)}$$

79. The LCM of the denominators is $(x+1)(x-6)$.

$$\frac{1}{x+1} + \frac{x}{x-6} - \frac{5x-2}{x^2-5x-6} = \frac{1}{x+1} \cdot \frac{x-6}{x-6} + \frac{x}{x-6} \cdot \frac{x+1}{x+1} - \frac{5x-2}{(x+1)(x-6)}$$

$$= \frac{x-6}{(x+1)(x-6)} + \frac{x^2+x}{(x+1)(x-6)} - \frac{5x-2}{(x+1)(x-6)}$$

$$= \frac{x-6+(x^2+x)-(5x-2)}{(x+1)(x-6)}$$

$$= \frac{x-6+x^2+x-5x+2}{(x+1)(x-6)}$$

$$= \frac{x^2-3x-4}{(x+1)(x-6)}$$

$$= \frac{(x+1)(x-4)}{(x+1)(x-6)}$$

$$= \frac{x-4}{x-6}$$

81. The LCM of the denominators is $(x-1)(x-3)$.

$$\frac{3x+1}{x-1} - \frac{x-1}{x-3} + \frac{x+1}{x^2-4x+3} = \frac{3x+1}{x-1} \cdot \frac{x-3}{x-3} - \frac{x-1}{x-3} \cdot \frac{x-1}{x-1} + \frac{x+1}{(x-1)(x-3)}$$

$$= \frac{3x^2-8x-3}{(x-1)(x-3)} - \frac{x^2-2x+1}{(x-1)(x-3)} + \frac{x+1}{(x-1)(x-3)}$$

$$= \frac{3x^2-8x-3-(x^2-2x+1)+(x+1)}{(x-1)(x-3)}$$

$$= \frac{3x^2-8x-3-x^2+2x-1+x+1}{(x-1)(x-3)}$$

$$= \frac{2x^2-5x-3}{(x-1)(x-3)}$$

$$= \frac{(x-3)(2x+1)}{(x-1)(x-3)}$$

$$= \frac{2x+1}{x-1}$$

Applying Concepts 9.3

83. The LCM of the denominators is $a - b$.

$$\frac{a}{a-b} + \frac{b}{b-a} + 1 = \frac{a}{a-b} + \frac{b}{(-1)(a-b)} + \frac{1}{1} \cdot \frac{a-b}{a-b}$$

$$= \frac{a}{a-b} + \frac{-b}{a-b} + \frac{a-b}{a-b}$$

$$= \frac{a-b+(a-b)}{a-b}$$

$$= \frac{2(a-b)}{(a-b)} = 2$$

85. The LCM of the denominators is $b + 4$.

$$b - 3 + \frac{5}{b+4} = \frac{b}{1} \cdot \frac{b+4}{b+4} - \frac{3}{1} \cdot \frac{b+4}{b+4} + \frac{5}{b+4}$$

$$= \frac{b^2 + 4b}{b+4} - \frac{3b+12}{b+4} + \frac{5}{b+4}$$

$$= \frac{b^2 + 4b - (3b+12) + 5}{b+4}$$

$$= \frac{b^2 + 4b - 3b - 12 + 5}{b+4}$$

$$= \frac{b^2 + b - 7}{b+4}$$

87. The LCM of the denominators is $(n-1)^2$.

$$\frac{(n+1)^2}{(n-1)^2} - 1 = \frac{(n+1)^2}{(n-1)^2} - \frac{1}{1} \cdot \frac{(n-1)^2}{(n-1)^2}$$

$$= \frac{n^2 + 2n + 1}{(n-1)^2} - \frac{n^2 - 2n + 1}{(n-1)^2}$$

$$= \frac{n^2 + 2n + 1 - (n^2 - 2n + 1)}{(n-1)^2}$$

$$= \frac{n^2 + 2n + 1 - n^2 + 2n - 1}{(n-1)^2}$$

$$= \frac{4n}{(n-1)^2}$$

89. $\dfrac{2x+9}{3-x} = \dfrac{2x+9}{-(x-3)} = \dfrac{-(2x+9)}{x-3}$

The LCM of the denominators is $(x-3)(x+7)$.

$$\frac{2x+9}{3-x} + \frac{x+5}{x+7} - \frac{2x^2+3x-3}{x^2+4x-21} = \frac{-(2x+9)}{x-3} \cdot \frac{x+7}{x+7} + \frac{x+5}{x+7} \cdot \frac{x-3}{x-3} - \frac{2x^2+3x-3}{(x-3)(x+7)}$$

$$= \frac{-2x^2 - 23x - 63}{(x-3)(x+7)} + \frac{x^2 + 2x - 15}{(x-3)(x+7)} - \frac{2x^2 + 3x - 3}{(x-3)(x+7)}$$

$$= \frac{-2x^2 - 23x - 63 + x^2 + 2x - 15 - 2x^2 - 3x + 3}{(x-3)(x+7)}$$

$$= \frac{-3x^2 - 24x - 75}{(x-3)(x+7)}$$

$$= \frac{-3(x^2 + 8x + 25)}{(x-3)(x+7)}$$

91. $\dfrac{x^2+x-6}{x^2+2x-8} \cdot \dfrac{x^2+5x+4}{x^2+2x-3} - \dfrac{2}{x-1} = \dfrac{(x+3)(x-2)}{(x+4)(x-2)} \cdot \dfrac{(x+4)(x+1)}{(x+3)(x-1)} - \dfrac{2}{x-1}$

$$= \frac{(x+3)(x-2)(x+4)(x+1)}{(x+4)(x-2)(x+3)(x-1)} - \frac{2}{x-1}$$

$$= \frac{x+1}{x-1} - \frac{2}{x-1}$$

$$= \frac{x+1-2}{x-1}$$

$$= \frac{x-1}{x-1}$$

$$= 1$$

93. $\dfrac{x^2-9}{x^2+6x+9} \div \dfrac{x^2+x-20}{x^2-x-12} + \dfrac{1}{x+1} = \dfrac{x^2-9}{x^2+6x+9} \cdot \dfrac{x^2-x-12}{x^2+x-20} + \dfrac{1}{x+1}$

$$= \frac{(x+3)(x-3)}{(x+3)(x+3)} \cdot \frac{(x-4)(x+3)}{(x+5)(x-4)} + \frac{1}{x+1}$$

$$= \frac{(x+3)(x-3)(x-4)(x+3)}{(x+3)(x+3)(x+5)(x-4)} + \frac{1}{x+1}$$

$$= \frac{x-3}{x+5} + \frac{1}{x+1}$$

$$= \frac{x-3}{x+5} \cdot \frac{x+1}{x+1} + \frac{1}{x+1} \cdot \frac{x+5}{x+5}$$

$$= \frac{x^2-2x-3}{(x+5)(x+1)} + \frac{x+5}{(x+5)(x+1)}$$

$$= \frac{x^2-2x-3+(x+5)}{(x+5)(x+1)}$$

$$= \frac{x^2-x+2}{(x+5)(x+1)}$$

95. $\dfrac{1}{1\cdot2} + \dfrac{1}{2\cdot3} = \dfrac{1}{2} + \dfrac{1}{6} = \dfrac{3}{6} + \dfrac{1}{6} = \dfrac{4}{6} = \dfrac{2}{3}\left(\text{or } \dfrac{2}{2+1}\right)$

$\dfrac{1}{1\cdot2} + \dfrac{1}{2\cdot3} + \dfrac{1}{3\cdot4} = \dfrac{1}{2} + \dfrac{1}{6} + \dfrac{1}{12} = \dfrac{6}{12} + \dfrac{2}{12} + \dfrac{1}{12} = \dfrac{9}{12} = \dfrac{3}{4}\left(\text{or } \dfrac{3}{3+1}\right)$

$\dfrac{1}{1\cdot2} + \dfrac{1}{2\cdot3} + \dfrac{1}{3\cdot4} + \dfrac{1}{4\cdot5} = \dfrac{1}{2} + \dfrac{1}{6} + \dfrac{1}{12} + \dfrac{1}{20} = \dfrac{30}{60} + \dfrac{10}{60} + \dfrac{5}{60} + \dfrac{3}{60} = \dfrac{48}{60} = \dfrac{4}{5}\left(\text{or } \dfrac{4}{4+1}\right)$

Sum of 50 terms is $\dfrac{50}{51}$. Sum of 100 terms is $\dfrac{100}{101}$. Sum of 1000 terms is $\dfrac{1000}{1001}$.

97. $\dfrac{6x+7y}{xy} = \dfrac{6x}{xy} + \dfrac{7y}{xy} = \dfrac{6}{y} + \dfrac{7}{x}$

99. $\dfrac{2mn^2+8m^2n}{m^3n^3} = \dfrac{2mn^2}{m^3n^3} + \dfrac{8m^2n}{m^3n^3} = \dfrac{2}{m^2n} + \dfrac{8}{mn^2}$

Section 9.4

Concept Review 9.4

1. Sometimes true
A complex fraction is a fraction that has a fraction in the numerator, in the denominator, or in both the numerator and the denominator.

3. Never true
To simplify a complex fraction, multiply the numerator and denominator of the complex fraction by the LCM of the denominators of the fractions in the numerator and denominator.

5. Always true

Objective 9.4.1 Exercises

1. The LCM of x and x^2 is x^2.

$\dfrac{1+\frac{3}{x}}{1-\frac{9}{x^2}} = \dfrac{1+\frac{3}{x}}{1-\frac{9}{x^2}} \cdot \dfrac{x^2}{x^2} = \dfrac{1\cdot x^2 + \frac{3}{x}\cdot x^2}{1\cdot x^2 - \frac{9}{x^2}\cdot x^2}$

$= \dfrac{x^2+3x}{x^2-9} = \dfrac{x(x+3)}{(x+3)(x-3)} = \dfrac{x}{x-3}$

3. The LCM is $x+4$.

$\dfrac{2-\frac{8}{x+4}}{3-\frac{12}{x+4}} = \dfrac{2-\frac{8}{x+4}}{3-\frac{12}{x+4}} \cdot \dfrac{x+4}{x+4}$

$= \dfrac{2(x+4)-\frac{8}{x+4}(x+4)}{3(x+4)-\frac{12}{x+4}(x+4)}$

$= \dfrac{2(x+4)-8}{3(x+4)-12} = \dfrac{2x+8-8}{3x+12-12}$

$= \dfrac{2x}{3x} = \dfrac{2}{3}$

5. The LCM is $y-2$.

$\dfrac{1+\frac{5}{y-2}}{1-\frac{2}{y-2}} = \dfrac{1+\frac{5}{y-2}}{1-\frac{2}{y-2}} \cdot \dfrac{y-2}{y-2}$

$= \dfrac{1(y-2)+\frac{5}{y-2}(y-2)}{1(y-2)-\frac{2}{y-2}(y-2)}$

$= \dfrac{(y-2)+5}{(y-2)-2} = \dfrac{y-2+5}{y-2-2} = \dfrac{y+3}{y-4}$

7. The LCM is $x + 7$.

$$\frac{4 - \frac{2}{x+7}}{5 + \frac{1}{x+7}} = \frac{4 - \frac{2}{x+7}}{5 + \frac{1}{x+7}} \cdot \frac{x+7}{x+7}$$

$$= \frac{4(x+7) - \frac{2}{x+7}(x+7)}{5(x+7) + \frac{1}{x+7}(x+7)}$$

$$= \frac{4(x+7) - 2}{5(x+7) + 1} = \frac{4x + 28 - 2}{5x + 35 + 1}$$

$$= \frac{4x + 26}{5x + 36} = \frac{2(2x + 13)}{5x + 36}$$

9. The LCM is $x - 2$.

$$\frac{\frac{3}{x-2} + 3}{\frac{4}{x-2} + 4} = \frac{\frac{3}{x-2} + 3}{\frac{4}{x-2} + 4} \cdot \frac{x-2}{x-2}$$

$$= \frac{\frac{3}{x-2}(x-2) + 3(x-2)}{\frac{4}{x-2}(x-2) + 4(x-2)}$$

$$= \frac{3 + 3x - 6}{4 + 4x - 8}$$

$$= \frac{3x - 3}{4x - 4}$$

$$= \frac{3(x-1)}{4(x-1)}$$

$$= \frac{3}{4}$$

11. The LCM is x^2.

$$\frac{2 - \frac{3}{x} - \frac{2}{x^2}}{2 + \frac{5}{x} + \frac{2}{x^2}} = \frac{2 - \frac{3}{x} - \frac{2}{x^2}}{2 + \frac{5}{x} + \frac{2}{x^2}} \cdot \frac{x^2}{x^2}$$

$$= \frac{2 \cdot x^2 - \frac{3}{x} \cdot x^2 - \frac{2}{x^2} \cdot x^2}{2 \cdot x^2 + \frac{5}{x} \cdot x^2 + \frac{2}{x^2} \cdot x^2}$$

$$= \frac{2x^2 - 3x - 2}{2x^2 + 5x + 2}$$

$$= \frac{(2x+1)(x-2)}{(2x+1)(x+2)}$$

$$= \frac{x-2}{x+2}$$

13. The LCM of x and x^2 is x^2.

$$\frac{1 - \frac{1}{x} - \frac{6}{x^2}}{1 - \frac{9}{x^2}} = \frac{1 - \frac{1}{x} - \frac{6}{x^2}}{1 - \frac{9}{x^2}} \cdot \frac{x^2}{x^2}$$

$$= \frac{1 \cdot x^2 - \frac{1}{x} \cdot x^2 - \frac{6}{x^2} \cdot x^2}{1 \cdot x^2 - \frac{9}{x^2} \cdot x^2}$$

$$= \frac{x^2 - x - 6}{x^2 - 9} = \frac{(x+2)(x-3)}{(x+3)(x-3)}$$

$$= \frac{x+2}{x+3}$$

15. The LCM of x and x^2 is x^2.

$$\frac{1 - \frac{5}{x} - \frac{6}{x^2}}{1 + \frac{6}{x} + \frac{5}{x^2}} = \frac{1 - \frac{5}{x} - \frac{6}{x^2}}{1 + \frac{6}{x} + \frac{5}{x^2}} \cdot \frac{x^2}{x^2}$$

$$= \frac{1 \cdot x^2 - \frac{5}{x} \cdot x^2 - \frac{6}{x^2} \cdot x^2}{1 \cdot x^2 + \frac{6}{x} \cdot x^2 + \frac{5}{x^2} \cdot x^2}$$

$$= \frac{x^2 - 5x - 6}{x^2 + 6x + 5} = \frac{(x+1)(x-6)}{(x+1)(x+5)}$$

$$= \frac{x-6}{x+5}$$

17. The LCM of x and x^2 is x^2.

$$\frac{1 - \frac{6}{x} + \frac{8}{x^2}}{\frac{4}{x^2} + \frac{3}{x} - 1} = \frac{1 - \frac{6}{x} + \frac{8}{x^2}}{\frac{4}{x^2} + \frac{3}{x} - 1} \cdot \frac{x^2}{x^2}$$

$$= \frac{1 \cdot x^2 - \frac{6}{x} \cdot x^2 + \frac{8}{x^2} \cdot x^2}{\frac{4}{x^2} \cdot x^2 + \frac{3}{x} \cdot x^2 - 1 \cdot x^2}$$

$$= \frac{x^2 - 6x + 8}{4 + 3x - x^2} = \frac{(x-2)(x-4)}{(1+x)(4-x)}$$

$$= -\frac{x-2}{x+1}$$

19. The LCM is $x + 3$.

$$\frac{x - \frac{4}{x+3}}{1 + \frac{1}{x+3}} = \frac{x - \frac{4}{x+3}}{1 + \frac{1}{x+3}} \cdot \frac{x+3}{x+3}$$

$$= \frac{x(x+3) - \frac{4}{x+3}(x+3)}{1(x+3) + \frac{1}{x+3}(x+3)}$$

$$= \frac{x^2 + 3x - 4}{x + 3 + 1} = \frac{(x+4)(x-1)}{(x+4)} = x - 1$$

21. The LCM is $2x + 1$.

$$\frac{1 - \frac{x}{2x+1}}{x - \frac{1}{2x+1}} = \frac{1 - \frac{x}{2x+1}}{x - \frac{1}{2x+1}} \cdot \frac{2x+1}{2x+1}$$

$$= \frac{1(2x+1) - \frac{x}{2x+1}(2x+1)}{x(2x+1) - \frac{1}{2x+1}(2x+1)}$$

$$= \frac{2x + 1 - x}{2x^2 + x - 1} = \frac{x+1}{(x+1)(2x-1)}$$

$$= \frac{1}{2x - 1}$$

23. The LCM is $x + 4$.

$$\frac{x - 5 + \frac{14}{x+4}}{x + 3 - \frac{2}{x+4}} = \frac{x - 5 + \frac{14}{x+4}}{x + 3 - \frac{2}{x+4}} \cdot \frac{x+4}{x+4}$$

$$= \frac{(x-5)(x+4) + \frac{14}{x+4}(x+4)}{(x+3)(x+4) - \frac{2}{x+4}(x+4)}$$

$$= \frac{x^2 - x - 20 + 14}{x^2 + 7x + 12 - 2} = \frac{x^2 - x - 6}{x^2 + 7x + 10}$$

$$= \frac{(x+2)(x-3)}{(x+2)(x+5)} = \frac{x-3}{x+5}$$

25. The LCM is $x - 6$.

$$\frac{x+3-\frac{10}{x-6}}{x+2-\frac{20}{x-6}} = \frac{x+3-\frac{10}{x-6}}{x+2-\frac{20}{x-6}} \cdot \frac{x-6}{x-6} = \frac{(x+3)(x-6)-\frac{10}{x-6}(x-6)}{(x+2)(x-6)-\frac{20}{x-6}(x-6)}$$

$$= \frac{x^2-3x-18-10}{x^2-4x-12-20} = \frac{x^2-3x-28}{x^2-4x-32} = \frac{(x+4)(x-7)}{(x+4)(x-8)} = \frac{x-7}{x-8}$$

27. The LCM is $2y + 3$.

$$\frac{y-6+\frac{22}{2y+3}}{y-5+\frac{11}{2y+3}} = \frac{y-6+\frac{22}{2y+3}}{y-5+\frac{11}{2y+3}} \cdot \frac{2y+3}{2y+3} = \frac{(y-6)(2y+3)+\frac{22}{2y+3}(2y+3)}{(y-5)(2y+3)+\frac{11}{2y+3}(2y+3)}$$

$$= \frac{2y^2-9y-18+22}{2y^2-7y-15+11} = \frac{2y^2-9y+4}{2y^2-7y-4} = \frac{(2y-1)(y-4)}{(2y+1)(y-4)} = \frac{2y-1}{2y+1}$$

29. The LCM is $2x - 3$.

$$\frac{x-\frac{2}{2x-3}}{2x-1-\frac{8}{2x-3}} = \frac{x-\frac{2}{2x-3}}{2x-1-\frac{8}{2x-3}} \cdot \frac{2x-3}{2x-3} = \frac{x(2x-3)-\frac{2}{2x-3}(2x-3)}{(2x-1)(2x-3)-\frac{8}{2x-3}(2x-3)}$$

$$= \frac{2x^2-3x-2}{4x^2-8x+3-8} = \frac{2x^2-3x-2}{4x^2-8x-5} = \frac{(2x+1)(x-2)}{(2x+1)(2x-5)} = \frac{x-2}{2x-5}$$

31. The LCM of $x + 1$ and $x - 2$ is $(x + 1)(x - 2)$.

$$\frac{1-\frac{2}{x+1}}{1+\frac{1}{x-2}} = \frac{1-\frac{2}{x+1}}{1+\frac{1}{x-2}} \cdot \frac{(x+1)(x-2)}{(x+1)(x-2)}$$

$$= \frac{1(x+1)(x-2)-\frac{2}{x+1}(x+1)(x-2)}{1(x+1)(x-2)+\frac{1}{x-2}(x+1)(x-2)}$$

$$= \frac{x^2-x-2-2(x-2)}{x^2-x-2+x+1} = \frac{x^2-x-2-2x+4}{x^2-1}$$

$$= \frac{x^2-3x+2}{x^2-1} = \frac{(x-2)(x-1)}{(x+1)(x-1)} = \frac{x-2}{x+1}$$

33. The LCM of $x + 4$ and $x - 1$ is $(x + 4)(x - 1)$.

$$\frac{1-\frac{2}{x+4}}{1+\frac{3}{x-1}} = \frac{1-\frac{2}{x+4}}{1+\frac{3}{x-1}} \cdot \frac{(x+4)(x-1)}{(x+4)(x-1)} = \frac{1(x+4)(x-1)-\frac{2}{x+4}(x+4)(x-1)}{1(x+4)(x-1)+\frac{3}{x-1}(x+4)(x-1)}$$

$$= \frac{x^2+3x-4-2(x-1)}{x^2+3x-4+3(x+4)} = \frac{x^2+3x-4-2x+2}{x^2+3x-4+3x+12} = \frac{x^2+x-2}{x^2+6x+8}$$

$$= \frac{(x+2)(x-1)}{(x+2)(x+4)} = \frac{x-1}{x+4}$$

35. The LCM of x and $x - 1$ is $x(x - 1)$.

$$\frac{\frac{1}{x}-\frac{2}{x-1}}{\frac{3}{x}+\frac{1}{x-1}} = \frac{\frac{1}{x}-\frac{2}{x-1}}{\frac{3}{x}+\frac{1}{x-1}} \cdot \frac{x(x-1)}{x(x-1)} = \frac{\frac{1}{x}(x)(x-1)-\frac{2}{x-1}(x)(x-1)}{\frac{3}{x}(x)(x-1)+\frac{1}{x-1}(x)(x-1)}$$

$$= \frac{x-1-2x}{3(x-1)+x} = \frac{-x-1}{3x-3+x} = \frac{-x-1}{4x-3}$$

37. The LCM of x and $2x - 1$ is $x(2x - 1)$.

$$\frac{\frac{3}{2x-1}-\frac{1}{x}}{\frac{4}{x}+\frac{2}{2x-1}} = \frac{\frac{3}{2x-1}-\frac{1}{x}}{\frac{4}{x}+\frac{2}{2x-1}} \cdot \frac{x(2x-1)}{x(2x-1)} = \frac{\frac{3}{2x-1}(x)(2x-1)-\frac{1}{x}(x)(2x-1)}{\frac{4}{x}(x)(2x-1)+\frac{2}{2x-1}(x)(2x-1)}$$

$$= \frac{3x-(2x-1)}{4(2x-1)+2x} = \frac{3x-2x+1}{8x-4+2x} = \frac{x+1}{10x-4} = \frac{x+1}{2(5x-2)}$$

39. The LCM of $b - 4$ and $b + 1$ is $(b - 4)(b + 1)$.

$$\frac{\frac{3}{b-4}-\frac{2}{b+1}}{\frac{5}{b+1}-\frac{1}{b-4}} = \frac{\frac{3}{b-4}-\frac{2}{b+1}}{\frac{5}{b+1}-\frac{1}{b-4}} \cdot \frac{(b-4)(b+1)}{(b-4)(b+1)} = \frac{\frac{3}{b-4}(b-4)(b+1)-\frac{2}{b+1}(b-4)(b+1)}{\frac{5}{b+1}(b-4)(b+1)-\frac{1}{b-4}(b-4)(b+1)}$$

$$= \frac{3(b+1)-2(b-4)}{5(b-4)-(b+1)} = \frac{3b+3-2b+8}{5b-20-b-1} = \frac{b+11}{4b-21}$$

Applying Concepts 9.4

41. $1 + \dfrac{1}{1+\frac{1}{2}} = 1 + \dfrac{1}{1+\frac{1}{2}} \cdot \dfrac{2}{2} = 1 + \dfrac{1 \cdot 2}{1 \cdot 2 + \frac{1}{2} \cdot 2}$

$= 1 + \dfrac{2}{2+1} = 1 + \dfrac{2}{3} = \dfrac{3}{3} + \dfrac{2}{3} = \dfrac{5}{3}$

43. $1 - \dfrac{1}{1-\frac{1}{x}} = 1 - \dfrac{1}{1-\frac{1}{x}} \cdot \dfrac{x}{x} = 1 - \dfrac{1 \cdot x}{1 \cdot x - \frac{1}{x} \cdot x} = 1 - \dfrac{x}{x-1}$

$= \dfrac{x-1}{x-1} - \dfrac{x}{x-1} = \dfrac{x-1-x}{x-1} = \dfrac{-1}{x-1} = -\dfrac{1}{x-1}$

45. $\dfrac{a^{-1} - b^{-1}}{a^{-2} - b^{-2}} = \dfrac{\frac{1}{a} - \frac{1}{b}}{\frac{1}{a^2} - \frac{1}{b^2}} = \dfrac{\frac{1}{a} - \frac{1}{b}}{\frac{1}{a^2} - \frac{1}{b^2}} \cdot \dfrac{a^2 b^2}{a^2 b^2}$

$= \dfrac{\frac{1}{a} \cdot a^2 b^2 - \frac{1}{b} \cdot a^2 b^2}{\frac{1}{a^2} \cdot a^2 b^2 - \frac{1}{b^2} \cdot a^2 b^2} = \dfrac{ab^2 - a^2 b}{b^2 - a^2} = \dfrac{ab(b-a)}{(b+a)(b-a)} = \dfrac{ab}{b+a}$

47. $\left(\dfrac{y}{4} - \dfrac{4}{y}\right) \div \left(\dfrac{4}{y} - 3 + \dfrac{y}{2}\right) = \dfrac{\frac{y}{4} - \frac{4}{y}}{\frac{4}{y} - 3 + \frac{y}{2}} = \dfrac{\frac{y}{4} - \frac{4}{y}}{\frac{4}{y} - 3 + \frac{y}{2}} \cdot \dfrac{4y}{4y}$

$= \dfrac{\frac{y}{4} \cdot 4y - \frac{4}{y} \cdot 4y}{\frac{4}{y} \cdot 4y - 3 \cdot 4y + \frac{y}{2} \cdot 4y} = \dfrac{y^2 - 16}{16 - 12y + 2y^2}$

$= \dfrac{(y+4)(y-4)}{2(y^2 - 6y + 8)} = \dfrac{(y+4)(y-4)}{2(y-4)(y-2)} = \dfrac{y+4}{2(y-2)}$

Section 9.5

Concept Review 9.5

1. Always true

3. Always true

5. Sometimes true
A quadratic may have two solutions, it may have one solution, or it may not have a real number solution.

7. Always true

Objective 9.5.1 Exercises

3. $2 + \dfrac{5}{x} = 7$ The LCM is x.

$\dfrac{x}{1}\left(2 + \dfrac{5}{x}\right) = \dfrac{x}{1} \cdot 7$

$2x + \dfrac{x}{1} \cdot \dfrac{5}{x} = 7x$

$2x + 5 = 7x$

$5 = 5x$

$1 = x$

1 checks as a solution.
The solution is 1.

5. $1 - \dfrac{9}{x} = 4$ The LCM is x.

$\dfrac{x}{1}\left(1 - \dfrac{9}{x}\right) = \dfrac{x}{1} \cdot 4$

$x - \dfrac{x}{1} \cdot \dfrac{9}{x} = 4x$

$x - 9 = 4x$

$-9 = 3x$

$-3 = x$

-3 checks as a solution.
The solution is -3.

7. $\dfrac{2}{y} + 5 = 9$ The LCM is y.

$\dfrac{y}{1}\left(\dfrac{2}{y} + 5\right) = \dfrac{y}{1} \cdot 9$

$\dfrac{y}{1} \cdot \dfrac{2}{y} + \dfrac{y}{1} \cdot 5 = 9y$

$2 + 5y = 9y$

$2 = 4y$

$\dfrac{1}{2} = y$

$\dfrac{1}{2}$ checks as a solution.

The solution is $\dfrac{1}{2}$.

9.
$$\frac{4x}{x-4}+5=\frac{5x}{x-4} \quad \text{The LCM is } x-4.$$

$$\frac{x-4}{1}\left(\frac{4x}{x-4}+5\right)=\frac{x-4}{1}\cdot\frac{5x}{x-4}$$

$$\frac{x-4}{1}\cdot\frac{4x}{x-4}+\frac{x-4}{1}\cdot 5=5x$$

$$4x+(x-4)5=5x$$
$$4x+5x-20=5x$$
$$9x-20=5x$$
$$-20=-4x$$
$$5=x$$

5 checks as a solution.
The solution is 5.

11.
$$2+\frac{3}{a-3}=\frac{a}{a-3} \quad \text{The LCM is } a-3.$$

$$\frac{a-3}{1}\left(2+\frac{3}{a-3}\right)=\frac{a-3}{1}\cdot\frac{a}{a-3}$$

$$\frac{a-3}{1}\cdot 2+\frac{a-3}{1}\cdot\frac{3}{a-3}=a$$

$$(a-3)2+3=a$$
$$2a-6+3=a$$
$$2a-3=a$$
$$a-3=0$$
$$a=3$$

3 does not check as solution.
The equation has no solution.

13.
$$\frac{x}{x-1}=\frac{8}{x+2} \quad \text{The LCM is } (x-1)(x+2).$$

$$\frac{(x-1)(x+2)}{1}\cdot\frac{x}{x-1}=\frac{(x-1)(x+2)}{1}\cdot\frac{8}{x+2}$$

$$(x+2)x=(x-1)8$$
$$x^2+2x=8x-8$$
$$x^2-6x+8=0$$
$$(x-4)(x-2)=0$$

Both 4 and 2 check as solutions.
The solutions are 4 and 2.

15.
$$\frac{2x}{x+4}=\frac{3}{x-1} \quad \text{The LCM is } (x+4)(x-1).$$

$$\frac{(x+4)(x-1)}{1}\cdot\frac{2x}{x+4}=\frac{(x+4)(x-1)}{1}\cdot\frac{3}{x-1}$$

$$(x-1)2x=(x+4)3$$
$$2x^2-2x=3x+12$$
$$2x^2-5x-12=0$$
$$(2x+3)(x-4)=0$$
$$2x+3=0 \qquad x-4=0$$
$$2x=-3 \qquad x=4$$
$$x=-\frac{3}{2}$$

Both $-\frac{3}{2}$ and 4 check as solutions.

The solutions are $-\frac{3}{2}$ and 4.

17.
$$x+\frac{6}{x-2}=\frac{3x}{x-2} \quad \text{The LCM is } x-2.$$

$$\frac{x-2}{1}\left(x+\frac{6}{x-2}\right)=\frac{x-2}{1}\cdot\frac{3x}{x-2}$$

$$\frac{x-2}{1}\cdot x+\frac{x-2}{1}\cdot\frac{6}{x-2}=3x$$

$$(x-2)x+6=3x$$
$$x^2-2x+6=3x$$
$$x^2-5x+6=0$$
$$(x-3)(x-2)=0$$
$$x-3=0 \quad x-2=0$$
$$x=3 \qquad x=2$$

2 does not check as a solution.
3 does check as a solution.
The solution is 3.

19.
$$\frac{x}{x+2}+\frac{2}{x-2}=\frac{x+6}{x^2-4}$$ The LCM is $(x+2)(x-2)$.

$$\frac{(x+2)(x-2)}{1}\left[\frac{x}{x+2}+\frac{2}{x-2}\right]=\frac{(x+2)(x-2)}{1}\cdot\frac{x+6}{x^2-4}$$

$$\frac{(x+2)(x-2)}{1}\cdot\frac{x}{x+2}+\frac{(x+2)(x-2)}{1}\cdot\frac{2}{x-2}=\frac{(x+2)(x-2)}{1}\cdot\frac{x+6}{(x+2)(x-2)}$$

$$(x-2)x+(x+2)2=x+6$$
$$x^2-2x+2x+4=x+6$$
$$x^2+4=x+6$$
$$x^2-x-2=0$$
$$(x-2)(x+1)=0$$

$x-2=0 \quad x+1=0$
$\quad x=2 \qquad x=-1$

2 does not check as a solution.
−1 does check as a solution.
The solution is −1.

21.
$$\frac{8}{y}=\frac{2}{y-2}+1$$ The LCM is $y(y-2)$.

$$\frac{y(y-2)}{1}\cdot\frac{8}{y}=\frac{y(y-2)}{1}\left(\frac{2}{y-2}+1\right)$$

$$(y-2)8=\frac{y(y-2)}{1}\cdot\frac{2}{y-2}+\frac{y(y-2)}{1}\cdot1$$

$$8y-16=2y+y^2-2y$$
$$8y-16=y^2$$
$$0=y^2-8y+16$$
$$0=(y-4)(y-4)$$

$y-4=0 \quad y-4=0$
$\quad y=4 \qquad y=4$

4 checks as a solution.
The solution is 4.

23. 2 cannot be a solution of the equation because the denominator of $\frac{x}{x-2}$ is 0 when $x=0$.

Objective 9.5.2 Exercises

25.
$$\frac{6}{x}=\frac{2}{3}$$
$$3x\cdot\frac{6}{x}=3x\cdot\frac{2}{3}$$
$$18=2x$$
$$9=x$$
The solution is 9.

27.
$$\frac{16}{9}=\frac{64}{x}$$
$$9x\cdot\frac{16}{9}=9x\cdot\frac{64}{x}$$
$$16x=576$$
$$x=36$$
The solution is 36.

29.
$$\frac{3}{5} = \frac{x-4}{10}$$
$$10 \cdot \frac{3}{5} = 10 \cdot \frac{x-4}{10}$$
$$6 = x - 4$$
$$10 = x$$
The solution is 10.

31.
$$\frac{2}{11} = \frac{20}{x-3}$$
$$11(x-3) \cdot \frac{2}{11} = 11(x-3) \cdot \frac{20}{x-3}$$
$$(x-3)2 = 220$$
$$2x - 6 = 220$$
$$2x = 226$$
$$x = 113$$
The solution is 113.

33.
$$\frac{16}{x-2} = \frac{8}{x}$$
$$x(x-2) \cdot \frac{16}{x-2} = x(x-2) \cdot \frac{8}{x}$$
$$16x = (x-2)8$$
$$16x = 8x - 16$$
$$8x = -16$$
$$x = -2$$
The solution is –2.

35.
$$\frac{x-6}{3} = \frac{x}{5}$$
$$15 \cdot \frac{x-6}{3} = 15 \cdot \frac{x}{5}$$
$$5(x-6) = 3x$$
$$5x - 30 = 3x$$
$$2x - 30 = 0$$
$$2x = 30$$
$$x = 15$$
The solution is 15.

37.
$$\frac{9}{x+2} = \frac{3}{x-2}$$
$$(x+2)(x-2) \cdot \frac{9}{x+2} = (x+2)(x-2) \cdot \frac{3}{x-2}$$
$$(x-2)9 = (x+2)3$$
$$9x - 18 = 3x + 6$$
$$6x - 18 = 6$$
$$6x = 24$$
$$x = 4$$
The solution is 4.

Objective 9.5.3 Exercises

39. Strategy

To find the number of people, write and solve a proportion using x to represent the number of voters who favored the amendment.

Solution
$$\frac{4}{7} = \frac{x}{35,000}$$
$$35,000 \cdot \frac{4}{7} = \frac{x}{35,000} \cdot 35,000$$
$$20,000 = x$$
20,000 people voted in favor of the amendment.

41. Strategy

To find the number of air vents, write and solve a proportion using x to represent the required number of air vents.

Solution
$$\frac{2}{300} = \frac{x}{21,000}$$
$$21,000 \cdot \frac{2}{300} = \frac{x}{21,000} \cdot 21,000$$
$$140 = x$$

140 air vents are required for a 21,000-ft^2 building.

43. Strategy

To find the amount of sugar, write and solve a proportion using x to represent the number of cups of sugar.

Solution

$$\frac{8ab^2}{15x^3 y} \cdot \frac{5xy^4}{16a^2 b} = \frac{\overset{1}{2} \cdot \overset{1}{2} \cdot \overset{1}{2} \cdot \overset{1}{5} ab^2 xy^4}{3 \cdot 5 \cdot 2 \cdot 2 \cdot 2 \cdot 2 x^3 y a^2 b} = \frac{by^3}{6ax^2}$$

For 2 c of boiling water, 6 c of sugar are required.

45. Strategy
To find the number of fish in the lake, write and solve a proportion using x to represent the number of fish in the lake.

Solution

$$\frac{4}{80} = \frac{40}{x}$$

$$80x\left(\frac{4}{80}\right) = 80x\left(\frac{40}{x}\right)$$

$$4x = 3200$$

$$x = 800$$

There are approximately 800 fish in the lake.

47. Strategy
To determine if the shipment will be accepted, compare the ratios of defective bearings to total bearings.

Solution

$$\frac{400}{20,000} = \frac{2}{100}$$

$$\frac{2}{100} < \frac{3}{100}$$

Yes, the shipment will be accepted.

49. Strategy
To find the height of the person, write and solve a proportion using x to represent the height of the person.

Solution

$$\frac{1}{54} = \frac{1.25}{x}$$

$$54x\left(\frac{1}{54}\right) = 54x\left(\frac{1.25}{x}\right)$$

$$x = 67.5$$

The person is 67.5 in.

51. Strategy
To find the number of miles between the two cities, write and solve a proportion, using x to represent the distance between the two cities.

Solution

$$\frac{100}{\frac{3}{4}} = \frac{x}{5\frac{5}{8}}$$

$$\frac{100}{\frac{3}{4}} = \frac{x}{\frac{45}{8}}$$

$$\frac{1}{8} \cdot \frac{100}{\frac{3}{4}} = \frac{1}{8} \cdot \frac{x}{\frac{45}{8}}$$

$$\frac{100}{6} = \frac{x}{45}$$

$$90 \cdot \frac{100}{6} = 90 \cdot \frac{x}{45}$$

$$1500 = 2x$$

$$\frac{1500}{2} = x$$

$$x = 750$$

The distance between the two cities is 750 mi.

53. Strategy
To determine the recommended area, write and solve a proportion using x to represent the area of the window.

Solution

$$\frac{5}{12} = \frac{x}{8 \cdot 12}$$

$$\frac{5}{12} = \frac{x}{96}$$

$$96\left(\frac{5}{12}\right) = 96\left(\frac{x}{96}\right)$$

$$40 = x$$

The recommended area of the window is 40 ft^2.

55. Strategy
To find the number of milliliters of syrup, write and solve a proportion using x to represent the amount of syrup.

Solution

$$\frac{4}{7} = \frac{x}{280}$$

$$280 \cdot \frac{4}{7} = 280 \cdot \frac{x}{280}$$

$$160 = x$$

160 ml of syrup are in 280 ml of soft drink.

57. Strategy
To find the additional number of acres, write and solve a proportion using x to represent the additional number of acres. Then $50 + x$ is the total number of acres.

Solution

$$\frac{50}{1100} = \frac{50 + x}{1320}$$

$$\frac{1}{22} = \frac{50 + x}{1320}$$

$$1320\left(\frac{1}{22}\right) = 1320\left(\frac{50 + x}{1320}\right)$$

$$60 = 50 + x$$

$$10 = x$$

10 additional acres must be planted.

59. Strategy
To find the additional amount of coffee needed, write and solve a proportion using x to represent the additional number of gallons. Then $5 + x$ is the total number of gallons.

Solution

$$\frac{5}{50} = \frac{5 + x}{70}$$

$$\frac{1}{10} = \frac{5 + x}{70}$$

$$70\left(\frac{1}{10}\right) = 70\left(\frac{5 + x}{70}\right)$$

$$7 = 5 + x$$

$$2 = x$$

To serve 70 people, 2 additional gallons of coffee are needed.

Objective 9.5.4 Exercises

61. Strategy

Triangle *ABC* is similar to Triangle *DEF*. Write and solve a proportion, using x to represent the length of *DE*.

Solution

$$\frac{AC}{DF} = \frac{AB}{DE}$$
$$\frac{5}{8} = \frac{8}{x}$$
$$8x \cdot \frac{5}{8} = 8x \cdot \frac{8}{x}$$
$$5x = 64$$
$$x = 12.8$$

The length of *DE* is 12.8 in.

63. Strategy

Triangle *ABC* is similar to triangle *DEF*. Write and solve a proportion, using x to represent the height of triangle *DEF*.

Solution

$$\frac{CB}{FE} = \frac{\text{height of triangle } ABC}{\text{height of triangle } DEF}$$
$$\frac{9}{14} = \frac{3}{x}$$
$$14x \cdot \frac{9}{14} = 14x \cdot \frac{3}{x}$$
$$9x = 42$$
$$x \approx 4.67$$

The height of triangle *DEF* is approximately 4.7 ft.

65. Strategy

To find the perimeter of triangle *ABC*:
- Use a proportion to find the length of *AC*.
- Use a proportion to find the length of *BC*.
- Add the measures of the three sides of triangle *ABC*.

Solution

$$\frac{AB}{DE} = \frac{AC}{DF}$$
$$\frac{4}{5} = \frac{x}{7.5}$$
$$15 \cdot \frac{4}{5} = 15 \cdot \frac{x}{7.5}$$
$$12 = 2x$$
$$6 = x$$
$$AC = 6$$
$$\frac{AB}{DE} = \frac{BC}{EF}$$
$$\frac{4}{5} = \frac{x}{10}$$
$$10 \cdot \frac{4}{5} = 10 \cdot \frac{x}{10}$$
$$8 = x$$
$$BC = 8$$

Perimeter $= AB + AC + BC$
$$= 4 + 6 + 8 = 18$$

The perimeter of triangle *ABC* is 18 m.

67. Strategy

To find the area of triangle *ABC*:
- Use a proportion to find the height of triangle *ABC*. Let h represent the height of triangle *ABC*.
- Use the formula for the area of a triangle to find the area. The base of triangle *ABC* is 12 cm.

Solution

$$\frac{AB}{DE} = \frac{\text{height of triangle } ABC}{\text{height of triangle } DEF}$$
$$\frac{12}{22.5} = \frac{h}{15}$$
$$(22.5)(15) \cdot \frac{12}{22.5} = (22.5)(15) \cdot \frac{h}{15}$$
$$180 = 22.5h$$
$$8 = h$$

$$A = \frac{1}{2}bh$$
$$= \frac{1}{2}(12)(8)$$
$$= 48$$

The area of triangle *ABC* is 48 cm^2.

69. Strategy

$\angle BDE = \angle BAC$ and $\angle B = \angle B$, thus triangle *BDE* is similar to triangle *BAC* because two angles of triangle *BDE* are equal to two angles of triangle *BAC*. Write and solve a proportion to find the length of *BC*.
$AB = AD + DB = 12 + 8 = 20$

Solution

$$\frac{BD}{AB} = \frac{BE}{BC}$$
$$\frac{8}{20} = \frac{6}{x}$$
$$20x \cdot \frac{8}{20} = 20x \cdot \frac{6}{x}$$
$$8x = 120$$
$$x = 15$$

The measure of *BC* is 15 m.

71. Strategy

$\angle OMN = \angle OPQ$ and $\angle MON = \angle POQ$, thus triangle *NMO* is similar to triangle *QPO*. Solve a proportion to find the length of *QO*.

Solution

$$\frac{MO}{PO} = \frac{NO}{QO}$$
$$\frac{20}{8} = \frac{25}{x}$$
$$8x \cdot \frac{20}{8} = 8x \cdot \frac{25}{x}$$
$$20x = 200$$
$$x = 10$$

The measure of *QO* is 10 ft.

73. Strategy

Triangle *MNO* is similar to triangle *OPQ*. Solve one proportion to find the length of *OP*. Solve another proportion to find the length of *OQ*. Let *x* represent the length of *OQ* and $20 - x$ represent the length of *OM*. Let *y* represent the length of *OP*. The perimeter of Triangle *QPO* is the sum of the measures of the sides of the triangle.

Solution

$$\frac{MN}{PQ} = \frac{NO}{OP}$$

$$\frac{9}{3} = \frac{12}{y}$$

$$3y \cdot \frac{9}{3} = 3y \cdot \frac{12}{y}$$

$$9y = 36$$

$$y = 4$$

$$\frac{MN}{PQ} = \frac{OM}{OQ}$$

$$\frac{9}{3} = \frac{20 - x}{x}$$

$$3x \cdot \frac{9}{3} = 3x \cdot \frac{20 - x}{x}$$

$$9x = 60 - 3x$$

$$12x = 60$$

$$x = 5$$

Perimeter $= PQ + OP + OQ$

$$= 3 + 4 + 5 = 12$$

The perimeter of Triangle *OPQ* is 12 m.

75. Strategy

The triangle the person makes with his shadow is similar to the triangle that the flagpole makes with its shadow. Write and solve a proportion to find the height of the flagpole. 5 ft 9 in = 5.75 ft.

Solution

$$\frac{\text{height of person}}{\text{height of flagpole}} = \frac{\text{length of person's shadow}}{\text{length of flagpole's shadow}}$$

$$\frac{5.75}{h} = \frac{12}{30}$$

$$30h \cdot \frac{5.75}{h} = 30h \cdot \frac{12}{30}$$

$$172.5 = 12h$$

$$14.375 = h$$

The height of the flagpole is 14.375 ft.

Applying Concepts 9.5

77. $\frac{3}{4}a = \frac{1}{2}(3 - a) + \frac{a - 2}{4}$ The LCM is 4.

$$4 \cdot \frac{3}{4}a = 4\left(\frac{1}{2}(3 - a) + \frac{a - 2}{4}\right)$$

$$3a = 2(3 - a) + (a - 2)$$

$$3a = 6 - 2a + a - 2$$

$$3a = 4 - a$$

$$4a = 4$$

$$a = 1$$

1 checks as a solution.
The solution is 1.

79.

$$\frac{x + 1}{x^2 + x - 2} = \frac{x + 2}{x^2 - 1} + \frac{3}{x + 2}$$ The LCM is $(x + 2)(x - 1)(x + 1)$.

$$\frac{x + 1}{(x + 2)(x - 1)} = \frac{x + 2}{(x + 1)(x - 1)} + \frac{3}{x + 2}$$

$$\frac{(x + 2)(x - 1)(x + 1)}{1} \cdot \frac{x + 1}{(x + 2)(x - 1)} = \frac{(x + 2)(x - 1)(x + 1)}{1}\left(\frac{x + 2}{(x + 1)(x - 1)} + \frac{3}{x + 2}\right)$$

$$(x + 1)(x + 1) = \frac{(x + 2)(x - 1)(x + 1)}{1} \cdot \frac{x + 2}{(x + 1)(x - 1)} + \frac{(x + 2)(x - 1)(x + 1)}{1} \cdot \frac{3}{x + 2}$$

$$x^2 + 2x + 1 = (x + 2)(x + 2) + 3(x - 1)(x + 1)$$

$$x^2 + 2x + 1 = x^2 + 4x + 4 + 3(x^2 - 1)$$

$$x^2 + 2x + 1 = x^2 + 4x + 4 + 3x^2 - 3$$

$$x^2 + 2x + 1 = 4x^2 + 4x + 1$$

$$0 = 3x^2 + 2x$$

$$0 = x(3x + 2)$$

$$x = 0 \quad 3x + 2 = 0$$

$$3x = -2$$

$$x = -\frac{2}{3}$$

Both 0 and $-\frac{2}{3}$ check as solutions.

The solutions are 0 and $-\frac{2}{3}$.

81. The first integer: x

The next consecutive integer: $x + 1$

The multiplicative inverse of the first integer: $\dfrac{1}{x}$

The multiplicative inverse of the second integer: $\dfrac{1}{x+1}$

The sum of the multiplicative inverses of the two integers	is	$-\dfrac{7}{12}$

$$\frac{1}{x} + \frac{1}{x+1} = -\frac{7}{12}$$

$$12x(x+1)\cdot\frac{1}{x} + 12x(x+1)\cdot\frac{1}{x+1} = 12x(x+1)\left(\frac{-7}{12}\right)$$

$$12x + 12 + 12x = -7x^2 - 7x$$

$$7x^2 + 31x + 12 = 0$$

$$(7x+3)(x+4) = 0$$

$$7x + 3 = 0 \qquad x + 4 = 0$$

$$7x = -3 \qquad x = -4$$

$$x = -\frac{3}{7}$$

Since $-\dfrac{3}{7}$ is not an integer, it is not a solution.

$x = -4$

$x + 1 = -4 + 1 = -3$

The integers are -3 and -4.

83. Strategy

To find the person's share, write and solve a proportion using y to represent the first person's share.

$\$25 + \$30 + \$35 = \$90$, the total amount spent.

Solution

$$\frac{25}{90} = \frac{y}{4.5}$$

$$\frac{5}{18} = \frac{2y}{9}$$

$$18\cdot\frac{5}{18} = 18\cdot\frac{2y}{9}$$

$$5 = 4y$$

$$y = \frac{5}{4}$$

$$y = 1.25$$

The first person's share is $1.25 million.

85. Strategy

To find the total cost of 24 photos, write and solve a proportion using x to represent the cost of 24 photos minus the sitting fee. Then $x + 4$ is the total cost of 24 photos.

Solution.

$$\frac{10-4}{10} = \frac{x}{24}$$

$$\frac{6}{10} = \frac{x}{24}$$

$$\frac{3}{5} = \frac{x}{24}$$

$$120\cdot\frac{3}{5} = 120\cdot\frac{x}{24}$$

$$72 = 5x$$

$$14.40 = x$$

$$x + 4 = 14.40 + 4 = 18.40$$

The total cost of 24 photos is $18.40.

87. a. Strategy

To find Eratosthenes' calculation of Earth's circumference, write and solve a proportion using x to represent the circumference.

Solution

$$\frac{7.5}{360} = \frac{520}{x}$$

$$x = 520\cdot\frac{360}{7.5}$$

$$x = 24{,}960$$

The calculation was 24,960 miles.

87. b. The difference is

$24{,}960 - 24{,}800 = 160$ miles.

Section 9.6

Concept Review 9.6

1. Always true

3. Always true

5. Always true

Objective 9.6.1 Exercises

1. $A = \dfrac{1}{2}bh$

$$2\cdot A = 2\cdot\frac{1}{2}bh$$

$$2A = bh$$

$$\frac{2A}{b} = \frac{bh}{b}$$

$$\frac{2A}{b} = h$$

3. $d = rt$

$$\frac{d}{r} = \frac{rt}{t}$$

$$\frac{d}{r} = t$$

5. $PV = nRT$

$\dfrac{PV}{nR} = \dfrac{nRT}{nR}$

$\dfrac{PV}{nR} = T$

7. $P = 2L + 2W$

$P - 2W = 2L + 2W - 2W$

$P - 2W = 2L$

$\dfrac{P - 2W}{2} = \dfrac{2L}{2}$

$\dfrac{P - 2W}{2} = L$

9. $A = \dfrac{1}{2}h(b_1 + b_2)$

$2 \cdot A = 2 \cdot \dfrac{1}{2}h(b_1 + b_2)$

$2A = h(b_1 + b_2)$

$2A = hb_1 + hb_2$

$2A - hb_2 = hb_1 + hb_2 - hb_2$

$2A - hb_2 = hb_1$

$\dfrac{2A - hb_2}{h} = \dfrac{hb_1}{h}$

$\dfrac{2A - hb_2}{h} = b_1$

11. $V = \dfrac{1}{3}Ah$

$3 \cdot V = 3 \cdot \dfrac{1}{3}Ah$

$3V = Ah$

$\dfrac{3V}{A} = \dfrac{Ah}{A}$

$\dfrac{3V}{A} = h$

13. $R = \dfrac{C - S}{t}$

$t \cdot R = t \cdot \dfrac{C - S}{t}$

$Rt = C - S$

$Rt - C = C - C - S$

$Rt - C = -S$

$-1(Rt - C) = -1(-S)$

$C - Rt = S$

15. $A = P + Prt$

$A = P(1 + rt)$

$\dfrac{A}{1 + rt} = \dfrac{P(1 + rt)}{1 + rt}$

$\dfrac{A}{1 + rt} = P$

17. $A = Sw + w$

$A = w(S + 1)$

$\dfrac{A}{S + 1} = \dfrac{w(S + 1)}{S + 1}$

$\dfrac{A}{S + 1} = w$

Applying Concepts 9.6

19. a. $S = 2\pi rh + 2\pi r^2$

$S - 2\pi r^2 = 2\pi rh$

$\dfrac{S - 2\pi r^2}{2\pi r} = \dfrac{2\pi rh}{2\pi r}$

$\dfrac{S - 2\pi r^2}{2\pi r} = h$

b. $h = \dfrac{S - 2\pi r^2}{2\pi r}$

$h = \dfrac{12\pi - 2\pi(1)^2}{2\pi(1)}$

$h = \dfrac{10\pi}{2\pi}$

$h = 5$

The height is 5 in.

c. $h = \dfrac{S - 2\pi r^2}{2\pi r}$

$h = \dfrac{24\pi - 2\pi(2)^2}{2\pi(2)}$

$h = \dfrac{24\pi - 8\pi}{4\pi}$

$h = \dfrac{16\pi}{4\pi}$

$h = 4$

The height is 4 in.

21. a. $S = C + Cr$

$S - C = Cr$

$\dfrac{S - C}{C} = r$

b. $r = \dfrac{S - C}{C}$

$r = \dfrac{180 - 108}{108}$

$r = \dfrac{72}{108}$

$r = 0.6666$

The markup rate is $66.\overline{66}\%$.

c. $r = \dfrac{S - C}{C}$

$r = \dfrac{172 - 120.40}{120}$

$r = \dfrac{51.60}{120}$

$r = 0.43$

The markup rate is 43%.

238   *Chapter 9: Rational Expressions*

Section 9.7

Concept Review 9.7

1. Never true
   The work problems require use of the equation
   Rate of work · Time worked = Part of task
   completed

3. Always true

5. Never true
   Uniform motion problems are based on the
   equation
   Distance = Rate · Time

Objective 9.7.1 Exercises

3. Strategy
   • Time for the sprinklers working together: t

	Rate	Time	Part
First sprinkler	$\frac{1}{3}$	t	$\frac{t}{3}$
Second sprinkler	$\frac{1}{6}$	t	$\frac{t}{6}$

   • The sum of the parts done by each sprinkler
   must equal 1.

   Solution
   $$\frac{t}{3}+\frac{t}{6}=1$$
   $$6\left(\frac{t}{3}+\frac{t}{6}\right)=6\cdot1$$
   $$2t+t=6$$
   $$3t=6$$
   $$t=2$$
   It will take 2 h to fill the fountain with both
   sprinklers operating.

5. Strategy
   • Time to remove the earth with both skiploaders
   working together: t

	Rate	Time	Part
First skiploader	$\frac{1}{12}$	t	$\frac{t}{12}$
Larger skiploader	$\frac{1}{4}$	t	$\frac{t}{4}$

   • The sum of the parts of the task completed by
   each skiploader must equal 1.

   Solution
   $$\frac{t}{12}+\frac{t}{4}=1$$
   $$12\left(\frac{t}{12}+\frac{t}{4}\right)=12\cdot1$$
   $$t+3t=12$$
   $$4t=12$$
   $$t=3$$
   With both skiploaders working together, it would
   take 3 h to remove the earth.

7. Strategy
   • Time for the computers working together: t

	Rate	Time	Part
First computer	$\frac{1}{75}$	t	$\frac{t}{75}$
Second computer	$\frac{1}{50}$	t	$\frac{t}{50}$

   • The sum of the parts of the task completed by
   each computer must equal 1.

   Solution
   $$\frac{t}{75}+\frac{t}{50}=1$$
   $$150\left(\frac{t}{75}+\frac{t}{50}\right)=150\cdot1$$
   $$2t+3t=150$$
   $$5t=150$$
   $$t=30$$
   With both computers working, it would take 30 h
   to solve the problem.

9. Strategy
- Time using both air conditions: t

	Rate	Time	Part
Smaller air conditioner	$\dfrac{1}{60}$	t	$\dfrac{t}{60}$
Large air conditioner	$\dfrac{1}{40}$	t	$\dfrac{t}{40}$

- The sum of the parts done by each air conditioner must be 1.

Solution
$$\frac{t}{60}+\frac{t}{40}=1$$
$$120\left(\frac{t}{60}+\frac{t}{40}\right)=1\cdot 120$$
$$2t+3t=120$$
$$5t=120$$
$$t=24$$

Working together, the air conditioners can cool the room 5°F in 24 minutes.

11. Strategy
- Time for the second welder to complete the task: t

	Rate	Time	Part
First welder	$\dfrac{1}{10}$	6	$\dfrac{6}{10}$
Second welder	$\dfrac{1}{t}$	6	$\dfrac{6}{t}$

- The sum of the parts of the task completed by each welder must equal 1.

Solution
$$\frac{6}{10}+\frac{6}{t}=1$$
$$10t\left(\frac{6}{10}+\frac{6}{t}\right)=10t\cdot 1$$
$$6t+60=10t$$
$$60=4t$$
$$15=t$$

The second welder, working alone, would take 15 h.

13. Strategy
- Time for the second pipeline, alone: t

	Rate	Time	Part
First pipeline	$\dfrac{1}{45}$	30	$\dfrac{30}{45}$
Second pipeline	$\dfrac{1}{t}$	30	$\dfrac{30}{t}$

- The sum of the parts done by each pipeline must equal 1.

Solution
$$\frac{30}{45}+\frac{30}{t}=1$$
$$45t\left(\frac{30}{45}+\frac{30}{t}\right)=1\cdot 45t$$
$$30t+1350=45t$$
$$1350=15t$$
$$90=t$$

It would take the second pipeline 90 min, working alone.

15. Strategy
- Time for the older reaper to harvest the field: t

	Rate	Time	Part
New reaper	$\dfrac{1}{1.5}$	1	$\dfrac{1}{1.5}$
Older reaper	$\dfrac{1}{t}$	1	$\dfrac{1}{t}$

- The sum of the parts done by each reaper must equal 1.

Solution
It would take the older reaper 3 h to harvest the field.

17. Strategy
• Time for the second technician to complete the task.

	Rate	Time	Part
First technician	$\frac{1}{4}$	2	$\frac{2}{4}$
Second technician	$\frac{1}{6}$	t	$\frac{t}{6}$

• The sum of the parts of the task completed by each worker must equal 1.

Solution
$$\frac{2}{4}+\frac{t}{6}=1$$
$$12\left(\frac{2}{4}+\frac{t}{6}\right)=12\cdot1$$
$$6+2t=12$$
$$2t=6$$
$$t=3$$

It would take the second technician 3 h to complete the task.

19. Strategy
• Time for the smaller heating unit: t

	Rate	Time	Part
Large heating unit	$\frac{1}{8}$	2	$\frac{2}{8}$
Small heating unit	$\frac{1}{t}$	$2+9$	$\frac{11}{t}$

• The sum of the parts of the task completed by each heating unit must equal 1.

Solution
$$\frac{2}{8}+\frac{11}{t}=1$$
$$8t\left(\frac{2}{8}+\frac{11}{t}\right)=8t\cdot1$$
$$2t+88=8t$$
$$88=6t$$
$$14\frac{2}{3}=t$$

The small heating unit would take $14\frac{2}{3}$ h to heat the pool.

21. Strategy
• Time for one welder to complete the task: t

	Rate	Time	Part
First welder	$\frac{1}{t}$	10	$\frac{10}{t}$
Second welder	$\frac{1}{t}$	30	$\frac{30}{t}$

• The sum of the parts of the task completed by each welder must equal 1.

Solution
$$\frac{10}{t}+\frac{30}{t}=1$$
$$t\left(\frac{10}{t}+\frac{30}{t}\right)=t\cdot1$$
$$10+30=t$$
$$40=t$$

It would take one welder 40 h to complete the task.

Objective 9.7.2 Exercises

23. • Rate of hiking: r
Rate of car: $9r$

	Distance	Rate	Time
Hiking	5	r	$\frac{5}{r}$
By car	90	$9r$	$\frac{90}{9r}$

• The total time spent hiking and driving was 3 h.

Solution
$$\frac{5}{r}+\frac{90}{9r}=3$$
$$9r\left(\frac{5}{r}+\frac{90}{9r}\right)=9r\cdot3$$
$$45+90=27r$$
$$135=27r$$
$$5=r$$
The rate of hiking is 5 mph.

25. Strategy
- Rate of the helicopter: r
  Rate of the jet: $4r$

	Distance	Rate	Time
Helicopter	180	r	$\dfrac{180}{r}$
Jet	1080	$4r$	$\dfrac{1080}{4r}$

- The total time for the trip was 5 h.

Solution
$$\frac{180}{r}+\frac{1080}{4r}=5$$
$$4r\left(\frac{180}{r}+\frac{1080}{4r}\right)=4r\cdot 5$$
$$720+1080=20r$$
$$1800=20r$$
$$90=r$$
$$4r=90\cdot 4=360$$
The rate of the jet is 360 mph.

27. Strategy
- Original rate: r
  Increased rate: $r+2$

	Distance	Rate	Time
First leg of trip	15	r	$\dfrac{15}{r}$
Second leg of trip	19	$r+2$	$\dfrac{19}{r+2}$

- The total sailing time is 4 h.

Solution
$$\frac{15}{r}+\frac{19}{r+2}=4$$
$$r(r+2)\left(\frac{15}{r}+\frac{19}{r+2}\right)=r(r+2)\cdot 4$$
$$15(r+2)+19r=4r(r+2)$$
$$15r+30+19r=4r^2+8r$$
$$34r+30=4r^2+8r$$
$$0=4r^2-26r-30$$
$$0=2(2r^2-13r-15)$$
$$0=2(2r-15)(r+1)$$
$$2r-15=0 \qquad r+1=0$$
$$2r=15 \qquad\quad r=-1$$
$$r=\frac{15}{2}=7.5$$
The solution -1 is not possible because the rate cannot be negative. The rate for the first leg of the trip is 7.5 mph.

29. Strategy
- Distance the family can travel: d

	Distance	Rate	Time
Downstream	d	$15+3$	$\dfrac{d}{18}$
Upstream	d	$15-3$	$\dfrac{d}{12}$

- The sum of the times must equal 3.

Solution
$$\frac{d}{18}+\frac{d}{12}=3$$
$$36\left(\frac{d}{18}+\frac{d}{12}\right)=3\cdot 36$$
$$2d+3d=108$$
$$5d=108$$
$$d=21.6$$
The family can travel 21.6 mi down the river.

31. Strategy
- Rate in traffic: r
  Rate on expressway: $r+20$

	Distance	Rate	Time
In congested traffic	10	r	$\dfrac{10}{r}$
On expressway	20	$r+20$	$\dfrac{20}{r+20}$

- The total time for the trip is 1 h.

Solution
$$\frac{10}{r}+\frac{20}{r+20}=1$$
$$r(r+20)\left(\frac{10}{r}+\frac{20}{r+20}\right)=r(r+20)\cdot 1$$
$$10(r+20)+20r=r(r+20)$$
$$10r+200+20r=r^2+20r$$
$$30r+200=r^2+20r$$
$$0=r^2-10r-200$$
$$0=(r-20)(r+10)$$
The solution -10 is not possible because the rate cannot be negative.
The rate in congested traffic is 20 mph.

33. Strategy
• Rate of the current: r

	Distance	Rate	Time
Downstream	75	$20 + r$	$\dfrac{75}{20+r}$
Upstream	45	$20 - r$	$\dfrac{45}{20-r}$

• The times to travel in each direction are equal.

Solution

$$\frac{75}{20+r} = \frac{45}{20-r}$$
$$(20+r)(20-r) \cdot \frac{75}{20+r} = \frac{45}{20-r} \cdot (20+r)(20-r)$$
$$(20-r)75 = 45(20+r)$$
$$1500 - 75r = 900 + 45r$$
$$600 = 120r$$
$$5 = r$$

The rate of the current is 5 mph.

35. Strategy
• Rate of the freight train: r
  Rate of the express train: $r + 20$

	Distance	Rate	Time
Freight train	360	r	$\dfrac{360}{r}$
Express train	600	$r + 20$	$\dfrac{600}{r+20}$

• The time of the freight train equals the time of the express train.

Solution

$$\frac{360}{r} = \frac{600}{r+20}$$
$$r(r+20)\left(\frac{360}{r}\right) = r(r+20)\left(\frac{600}{r+20}\right)$$
$$(r+20)(360) = 600r$$
$$360r + 7200 = 600r$$
$$7200 = 240r$$
$$30 = r$$
$$r + 20 = 30 + 20 = 50$$

The rate of the freight train is 30 mph.
The rate of the express train is 50 mph.

37. Strategy
• Rate of the cyclist: r
  Rate of the car: $r + 36$

	Distance	Rate	Time
Cyclist	96	r	$\dfrac{96}{r}$
Car	384	$r + 36$	$\dfrac{384}{r+36}$

• The time of the cyclist equals the time of the car.

Solution

$$\frac{96}{r} = \frac{384}{r+36}$$
$$r(r+36)\left(\frac{96}{r}\right) = r(r+36)\left(\frac{384}{r+36}\right)$$
$$(r+36)(96) = 384r$$
$$96r + 3456 = 384r$$
$$3456 = 288r$$
$$12 = r$$
$$r + 36 = 12 + 36 = 48$$

The rate of the car is 48 mph.

39. Strategy
  • Rate of the jet stream: r

	Distance	Rate	Time
With jet stream	2400	$550 + r$	$\dfrac{2400}{550+r}$
Against jet stream	2000	$550 - r$	$\dfrac{2000}{550-r}$

  • The time with the jet stream equals the time against the jet stream.

Solution

$$\frac{2400}{550+r} = \frac{2000}{550-r}$$

$$(550+r)(550-r)\left(\frac{2400}{550+r}\right) = (550+r)(550-r)\left(\frac{2000}{550-r}\right)$$

$$2400(550-r) = 2000(550+r)$$

$$1,320,000 - 2400r = 1,100,000 + 2000r$$

$$220,000 = 4400r$$

$$50 = r$$

The rate of the jet stream is 50 mph.

41. Strategy
  • Rate of the current: r

	Distance	Rate	Time
With current	25	$20 + r$	$\dfrac{25}{20+r}$
Against current	15	$20 - r$	$\dfrac{15}{20-r}$

  • The time spent rowing with the current equals the time spent rowing against the current.

Solution

$$\frac{25}{20+r} = \frac{15}{20-r}$$

$$(20+r)(20-r)\frac{25}{20+r} = (20+r)(20-r)\frac{15}{20-r}$$

$$(20-r)25 = (20+r)15$$

$$500 - 25r = 300 + 15r$$

$$200 = 40r$$

$$5 = r$$

The rate of the current is 5 mph.

Applying Concepts 9.7

43. Unknown time to fill the tank working together: t

	Rate	Time	Part
First pipe	$\frac{1}{2}$	t	$\frac{t}{2}$
Second pipe	$\frac{1}{4}$	t	$\frac{t}{4}$
Third pipe	$\frac{1}{5}$	t	$\frac{t}{5}$

• The sum of the parts of the task completed by the three pipes is 1.

Solution

$$\frac{t}{2}+\frac{t}{4}+\frac{t}{5}=1$$

$$20\left(\frac{t}{2}+\frac{t}{4}+\frac{t}{5}\right)=20\cdot1$$

$$10t+5t+4t=20$$

$$19t=20$$

$$t=\frac{20}{19}$$

$$t=1\frac{1}{19}$$

It will take $1\frac{1}{19}$ h to fill the tank with all three pipes operating.

45. Strategy
• Rate on foot: r
  Rate by canoe: $4r$

	Distance	Rate	Time
Canoe	32	$4r$	$\frac{8}{r}$
Foot	4	r	$\frac{4}{r}$

• The time walking is 1 h less than the time canoeing.

Solution

$$\frac{4}{r}=\frac{8}{r}-1$$

$$r\cdot\frac{4}{r}=r\left(\frac{8}{r}-1\right)$$

$$4=8-r$$

$$-4=-r$$

$$r=4$$

$$\frac{8}{r}=\frac{8}{4}=2$$

The time spent traveling by canoe is 2 h.

47. Strategy
 • Unknown rate for the bus to travel 150 mi under normal conditions: r

	Distance	Rate	Time
Normal conditions	150	r	$\dfrac{150}{r}$
Reduced conditions	150	$r - 10$	$\dfrac{150}{r - 10}$

 • The time traveling at the reduced rate is 30 min, or $\dfrac{1}{2}$ h, greater than the time driving under normal conditions.

Solution

$$\frac{150}{r} = \frac{150}{r-10} - \frac{1}{2}$$

$$(r)(r-10)(2) \cdot \frac{150}{r} = (r)(r-10)(2)\left(\frac{150}{r-10} - \frac{1}{2}\right)$$

$$300(r-10) = 300r - r(r-10)$$

$$300r - 3000 = 300r - r^2 + 10r$$

$$-3000 = 10r - r^2$$

$$0 = r^2 - 10r - 3000$$

$$0 = (r+50)(r-60)$$

$$r + 50 = 0 \qquad r - 60 = 0$$

$$r = -50 \qquad r = 60$$

−50 does not check as a solution.
60 checks as a solution.
The usual rate for the bus is 60 mph.

Chapter Review Exercises

1. $\dfrac{8ab^2}{15x^3 y} \cdot \dfrac{5xy^4}{16a^2 b} = \dfrac{\overset{1}{2} \cdot \overset{1}{2} \cdot \overset{1}{2} \cdot \overset{1}{5} ab^2 xy^4}{3 \cdot \underset{1}{5} \cdot \underset{1}{2} \cdot \underset{1}{2} \cdot \underset{1}{2} \cdot 2 x^3 y a^2 b} = \dfrac{by^3}{6ax^2}$

2. $\dfrac{5}{3x-4} + \dfrac{4}{2x+3} = \dfrac{5}{3x-4} \cdot \dfrac{2x+3}{2x+3} + \dfrac{4}{2x+3} \cdot \dfrac{3x-4}{3x-4}$

$\qquad = \dfrac{10x+15}{(3x-4)(2x+3)} + \dfrac{12x-16}{(3x-4)(2x+3)}$

$\qquad = \dfrac{10x+15+12x-16}{(3x-4)(2x+3)} = \dfrac{22x-1}{(3x-4)(2x+3)}$

3.
$$4x + 3y = 12$$
$$4x + 3y - 3y = -3y + 12$$
$$4x = -3y + 12$$
$$\frac{4x}{4} = \frac{-3y+12}{4}$$
$$x = -\frac{3}{4}y + 3$$

4. $\dfrac{16x^5 y^3}{24xy^{10}} = \dfrac{\overset{1}{2} \cdot \overset{1}{2} \cdot \overset{1}{2} \cdot 2 x^5 y^3}{\underset{1}{2} \cdot \underset{1}{2} \cdot \underset{1}{2} \cdot 3 x y^{10}}$

$\qquad = \dfrac{2x^4}{3y^7}$

5. $\dfrac{20x^2 - 45x}{6x^3 + 4x^2} \div \dfrac{40x^3 - 90x^2}{12x^2 + 8x} = \dfrac{20x^2 - 45x}{6x^3 + 4x^2} \cdot \dfrac{12x^2 + 8x}{40x^3 - 90x^2}$

$\qquad = \dfrac{5x(4x - 9)}{2x^2(3x + 2)} \cdot \dfrac{4x(3x + 2)}{10x^2(4x - 9)}$

$\qquad = \dfrac{5x(4x - 9) \cdot 4x(3x + 2)}{2x^2(3x + 2) \cdot 10x^2(4x - 9)}$

$\qquad = \dfrac{1}{x^2}$

6. $\dfrac{x - \frac{16}{5x-2}}{3x - 4 - \frac{88}{5x-2}} = \dfrac{x - \frac{16}{5x-2}}{3x - 4 - \frac{88}{5x-2}} \cdot \dfrac{5x - 2}{5x - 2}$

$\qquad = \dfrac{x(5x - 2) - \frac{16}{5x-2}(5x - 2)}{(3x - 4)(5x - 2) - \frac{88}{5x-2}(5x - 2)}$

$\qquad = \dfrac{5x^2 - 2x - 16}{15x^2 - 26x + 8 - 88}$

$\qquad = \dfrac{5x^2 - 2x - 16}{15x^2 - 26x - 80}$

$\qquad = \dfrac{(5x + 8)(x - 2)}{(5x + 8)(3x - 10)}$

$\qquad = \dfrac{x - 2}{3x - 10}$

7. $24a^2 b^5 = 2 \cdot 2 \cdot 2 \cdot 3a^2 b^5$

$36a^3 b = 2 \cdot 2 \cdot 3 \cdot 3a^3 b$

LCM $= 2 \cdot 2 \cdot 2 \cdot 3 \cdot 3a^3 b^5$

LCM $= 72a^3 b^5$

8. $\dfrac{5x}{3x + 7} - \dfrac{x}{3x + 7} = \dfrac{5x - x}{3x + 7} = \dfrac{4x}{3x + 7}$

9. The LCM is $16x^2$.

$\dfrac{3}{16x} = \dfrac{3}{16x} \cdot \dfrac{x}{x} = \dfrac{3x}{16x^2}$

$\dfrac{5}{8x^2} = \dfrac{5}{8x^2} \cdot \dfrac{2}{2} = \dfrac{10}{16x^2}$

10. $\dfrac{2x^2 - 13x - 45}{2x^2 - x - 15} = \dfrac{(2x + 5)(x - 9)}{(2x + 5)(x - 3)} = \dfrac{x - 9}{x - 3}$

11. $\dfrac{x^2 - 5x - 14}{x^2 - 3x - 10} \div \dfrac{x^2 - 4x - 21}{x^2 - 9x + 20} = \dfrac{x^2 - 5x - 14}{x^2 - 3x - 10} \cdot \dfrac{x^2 - 9x + 20}{x^2 - 4x - 21}$

$\qquad = \dfrac{(x - 7)(x + 2)}{(x - 5)(x + 2)} \cdot \dfrac{(x - 5)(x - 4)}{(x - 7)(x + 3)}$

$\qquad = \dfrac{x - 4}{x + 3}$

12. $\dfrac{2y}{5y - 7} + \dfrac{3}{7 - 5y} = \dfrac{2y}{5y - 7} + \dfrac{3}{-(5y - 7)} = \dfrac{2y}{5y - 7} - \dfrac{3}{5y - 7} = \dfrac{2y - 3}{5y - 7}$

13. $\dfrac{3x^3 + 10x^2}{10x - 2} \cdot \dfrac{20x - 4}{6x^4 + 20x^3} = \dfrac{x^2(3x + 10)}{2(5x - 1)} \cdot \dfrac{4(5x - 1)}{2x^3(3x + 10)}$

$\qquad = \dfrac{x^2(3x + 10) \cdot 4(5x - 1)}{2(5x - 1) \cdot 2x^3(3x + 10)}$

$\qquad = \dfrac{1}{x}$

14. $\dfrac{5x+3}{2x^2+5x-3}-\dfrac{3x+4}{2x^2+5x-3}=\dfrac{5x+3-(3x+4)}{2x^2+5x-3}$

$\qquad\qquad=\dfrac{5x+3-3x-4}{2x^2+5x-3}$

$\qquad\qquad=\dfrac{2x-1}{2x^2+5x-3}$

$\qquad\qquad=\dfrac{2x-1}{(2x-1)(x+3)}$

$\qquad\qquad=\dfrac{1}{x+3}$

15. $5x^4(x-7)^2=5x^4(x-7)^2$
$\quad\ 15x(x-7)=3\cdot5x(x-7)$
The LCM is $15x^4(x-7)^2$.

16. $\qquad\qquad\dfrac{6}{x-7}=\dfrac{8}{x-6}$

$(x-7)(x-6)\left(\dfrac{6}{x-7}\right)=(x-7)(x-6)\left(\dfrac{8}{x-6}\right)$

$\qquad\qquad(x-6)6=(x-7)8$

$\qquad\qquad 6x-36=8x-56$

$\qquad\qquad\ \ -36=2x-56$

$\qquad\qquad\ \ \ \ 20=2x$

$\qquad\qquad\ \ \ \ 10=x$

The solution is 10.

17. $\qquad\qquad\dfrac{x+8}{x+4}=1+\dfrac{5}{x+4}$

$(x+4)\left(\dfrac{x+8}{x+4}\right)=(x+4)\left(1+\dfrac{5}{x+4}\right)$

$\qquad\qquad x+8=x+4+5$

$\qquad\qquad x+8=x+9$

$\qquad\qquad\ \ \ \ 8\neq9$

The equation has no solution.

18. $\dfrac{12a^2b(4x-7)}{15ab^2(7-4x)}=\dfrac{2\cdot2\cdot3a^2b(4x-7)}{3\cdot5ab^2(7-4x)}=-\dfrac{4a}{5b}$

19. $\dfrac{5x-1}{x^2-9}+\dfrac{4x-3}{x^2-9}-\dfrac{8x-1}{x^2-9}=\dfrac{5x-1+(4x-3)-(8x-1)}{x^2-9}$

$\qquad\qquad=\dfrac{5x-1+4x-3-8x+1}{x^2-9}$

$\qquad\qquad=\dfrac{x-3}{x^2-9}$

$\qquad\qquad=\dfrac{x-3}{(x+3)(x-3)}$

$\qquad\qquad=\dfrac{1}{x+3}$

20. Strategy

To find the perimeter of triangle ABC:

• Use a proportion to find the length of AC.

• Add the measures of the three sides of triangle ABC.

Solution

$$\frac{BC}{EF} = \frac{AC}{DF}$$

$$\frac{6}{9} = \frac{x}{12}$$

$$36 \cdot \frac{6}{9} = 36 \cdot \frac{x}{12}$$

$$24 = 3x$$

$$8 = x$$

$$\text{Perimeter} = AB + BC + AC$$

$$= 10 + 6 + 8$$

$$= 24$$

The perimeter of triangle ABC is 24 in.

21.

$$\frac{20}{2x+3} = \frac{17x}{2x+3} - 5$$

$$(2x+3)\left(\frac{20}{2x+3}\right) = (2x+3)\left(\frac{17x}{2x+3} - 5\right)$$

$$20 = 17x - 5(2x+3)$$

$$20 = 17x - 10x - 15$$

$$20 = 7x - 15$$

$$35 = 7x$$

$$5 = x$$

The solution is 5.

22. $\dfrac{\frac{5}{x-1} - \frac{3}{x+3}}{\frac{6}{x+3} + \frac{2}{x-1}} = \dfrac{\frac{5}{x-1} - \frac{3}{x+3}}{\frac{6}{x+3} + \frac{2}{x-1}} \cdot \dfrac{(x-1)(x+3)}{(x-1)(x+3)}$

$$= \frac{\frac{5}{x-1}(x-1)(x+3) - \frac{3}{x+3}(x-1)(x+3)}{\frac{6}{x+3}(x-1)(x+3) + \frac{2}{x-1}(x-1)(x+3)}$$

$$= \frac{5(x+3) - 3(x-1)}{6(x-1) + 2(x+3)}$$

$$= \frac{5x+15 - 3x+3}{6x-6 + 2x+6}$$

$$= \frac{2x+18}{8x}$$

$$= \frac{2(x+9)}{8x}$$

$$= \frac{x+9}{4x}$$

23. The LCM is $(x+2)(x-5)$.

$$\frac{x-1}{x+2} + \frac{3x-2}{5-x} + \frac{5x^2+15x-11}{x^2-3x-10} = \frac{x-1}{x+2} \cdot \frac{x-5}{x-5} - \frac{3x-2}{x-5} \cdot \frac{x+2}{x+2} + \frac{5x^2+15x-11}{(x+2)(x-5)}$$

$$= \frac{(x-1)(x-5)}{(x+2)(x-5)} - \frac{(3x-2)(x+2)}{(x+2)(x-5)} + \frac{5x^2+15x-11}{(x+2)(x-5)}$$

$$= \frac{x^2-6x+5 - (3x^2+4x-4) + 5x^2+15x-11}{(x+2)(x-5)}$$

$$= \frac{x^2-6x+5 - 3x^2-4x+4 + 5x^2+15x-11}{(x+2)(x-5)}$$

$$= \frac{3x^2+5x-2}{(x+2)(x-5)} = \frac{(3x-1)(x+2)}{(x+2)(x-5)} = \frac{3x-1}{x-5}$$

24. $\dfrac{18x^2+25x-3}{9x^2-28x+3} \div \dfrac{2x^2+9x+9}{x^2-6x+9} = \dfrac{18x^2+25x-3}{9x^2-28x+3} \cdot \dfrac{x^2-6x+9}{2x^2+9x+9}$

$\qquad = \dfrac{(9x-1)(2x+3)}{(9x-1)(x-3)} \cdot \dfrac{(x-3)(x-3)}{(2x+3)(x+3)}$

$\qquad = \dfrac{x-3}{x+3}$

25. $\dfrac{x^2+x-30}{15+2x-x^2} = \dfrac{(x+6)(x-5)}{(5-x)(3+x)} = -\dfrac{x+6}{x+3}$

26. $\dfrac{5}{7} + \dfrac{x}{2} = 2 - \dfrac{x}{7}$

$14\left(\dfrac{5}{7} + \dfrac{x}{2}\right) = 14\left(2 - \dfrac{x}{7}\right)$

$14 \cdot \dfrac{5}{7} + 14 \cdot \dfrac{x}{2} = 14 \cdot 2 - 14 \cdot \dfrac{x}{7}$

$10 + 7x = 28 - 2x$

$9x = 18$

$x = 2$

The solution is 2.

27. $\dfrac{x + \frac{6}{x-5}}{1 + \frac{2}{x-5}} = \dfrac{x + \frac{6}{x-5}}{1 + \frac{2}{x-5}} \cdot \dfrac{x-5}{x-5}$

$\qquad = \dfrac{x(x-5) + \frac{6}{x-5}(x-5)}{1(x-5) + \frac{2}{x-5}(x-5)}$

$\qquad = \dfrac{x^2 - 5x + 6}{x - 5 + 2}$

$\qquad = \dfrac{(x-2)(x-3)}{x-3}$

$\qquad = x - 2$

28. $\dfrac{3x^2+4x-15}{x^2-11x+28} \cdot \dfrac{x^2-5x-14}{3x^2+x-10}$

$\qquad = \dfrac{(3x-5)(x+3)}{(x-7)(x-4)} \cdot \dfrac{(x-7)(x+2)}{(3x-5)(x+2)}$

$\qquad = \dfrac{x+3}{x-4}$

29. $\dfrac{x}{5} = \dfrac{x+12}{9}$

$45\left(\dfrac{x}{5}\right) = 45\left(\dfrac{x+12}{9}\right)$

$9x = 5(x+12)$

$9x = 5x + 60$

$4x = 60$

$x = 15$

The solution is 15.

30. $\dfrac{3}{20} = \dfrac{x}{80}$

$80\left(\dfrac{3}{20}\right) = 80\left(\dfrac{x}{80}\right)$

$12 = x$

The solution is 12.

31. $\dfrac{1 - \frac{1}{x}}{1 - \frac{8x-7}{x^2}} = \dfrac{1 - \frac{1}{x}}{1 - \frac{8x-7}{x^2}} \cdot \dfrac{x^2}{x^2}$

$\qquad = \dfrac{1 \cdot x^2 - \frac{1}{x} \cdot x^2}{1 \cdot x^2 - \frac{8x-7}{x^2} \cdot x^2} = \dfrac{x^2 - x}{x^2 - (8x-7)}$

$\qquad = \dfrac{x^2 - x}{x^2 - 8x + 7} = \dfrac{x(x-1)}{(x-7)(x-1)}$

$\qquad = \dfrac{x}{x-7}$

32. $x - 2y = 15$

$x - 2y + 2y = 2y + 15$

$x = 2y + 15$

33. $\dfrac{6}{a} + \dfrac{9}{b} = \dfrac{6}{a} \cdot \dfrac{b}{b} + \dfrac{9}{b} \cdot \dfrac{a}{a}$

$\qquad = \dfrac{6a}{ab} + \dfrac{9a}{ab}$

$\qquad = \dfrac{6b + 9a}{ab}$

34. $10x^2 - 11x + 3 = (5x-3)(2x-1)$

$20x^2 - 17x + 3 = (5x-3)(4x-1)$

The LCM is $(5x-3)(2x-1)(4x-1)$.

35. $i = \dfrac{100m}{c}$

$c \cdot i = c \cdot \dfrac{100m}{c}$

$ci = 100m$

$\dfrac{ci}{i} = \dfrac{100m}{i}$

$c = \dfrac{100m}{i}$

36. $\dfrac{15}{x} = \dfrac{3}{8}$

$8x\left(\dfrac{15}{x}\right) = 8x\left(\dfrac{3}{8}\right)$

$120 = 3x$

$40 = x$

The solution is 40.

37. $\dfrac{22}{2x+5} = 2$

$(2x+5)\left(\dfrac{22}{2x+5}\right) = (2x+5)(2)$

$22 = 4x + 10$

$12 = 4x$

$3 = x$

The solution is 3.

38. The LCM is $60x$.

$$\frac{x+7}{15x} + \frac{x-2}{20x} = \frac{x+7}{15x} \cdot \frac{4}{4} + \frac{x-2}{20x} \cdot \frac{3}{3}$$
$$= \frac{4x+28}{60x} + \frac{3x-6}{60x}$$
$$= \frac{4x+28+(3x-6)}{60x}$$
$$= \frac{7x+22}{60x}$$

39.
$$\frac{16a^2-9}{16a^2-24a+9} \cdot \frac{8a^2-13a-6}{4a^2-5a-6}$$
$$= \frac{(4a+3)(4a-3)}{(4a-3)(4a-3)} \cdot \frac{(8a+3)(a-2)}{(4a+3)(a-2)}$$
$$= \frac{8a+3}{4a-3}$$

40. $12x^2 + 16x - 3 = (6x-1)(2x+3)$
$6x^2 + 7x - 3 = (3x-1)(2x+3)$
The LCM is $(6x-1)(2x+3)(3x-1)$.

$$\frac{x}{12x^2+16x-3} = \frac{x}{(6x-1)(2x+3)} \cdot \frac{3x-1}{3x-1}$$
$$= \frac{3x^2-x}{(6x-1)(2x+3)(3x-1)}$$

$$\frac{4x^2}{6x^2+7x-3} = \frac{4x^2}{(3x-1)(2x+3)} \cdot \frac{6x-1}{6x-1}$$
$$= \frac{24x^3-4x^2}{(6x-1)(2x+3)(3x-1)}$$

41. $\dfrac{3}{4ab} + \dfrac{5}{4ab} = \dfrac{3+5}{4ab} = \dfrac{8}{4ab} = \dfrac{2 \cdot 2 \cdot 2}{2 \cdot 2 \, ab} = \dfrac{2}{ab}$

42.
$$\frac{20}{x+2} = \frac{5}{16}$$
$$(x+2)(16)\left(\frac{20}{x+2}\right) = (x+2)(16)\left(\frac{5}{16}\right)$$
$$16(20) = (x+2)(5)$$
$$320 = 5x+10$$
$$310 = 5x$$
$$62 = x$$
The solution is 62.

43.
$$\frac{5x}{3} - \frac{2}{5} = \frac{8x}{5}$$
$$15\left(\frac{5x}{3} - \frac{2}{5}\right) = 15\left(\frac{8x}{5}\right)$$
$$25x - 6 = 24x$$
$$-6 = -x$$
$$6 = x$$
The solution is 6.

44.
$$\frac{6a^2b^7}{25x^3y} \div \frac{12a^3b^4}{5x^2y^2} = \frac{6a^2b^7}{25x^3y} \cdot \frac{5x^2y^2}{12a^3b^4}$$
$$= \frac{2 \cdot 3 \, a^2b^7 \cdot 5 \, x^2y^2}{5 \cdot 5x^3y \cdot 2 \cdot 2 \cdot 3 \, a^3b^4}$$
$$= \frac{b^3y}{10ax}$$

45. Strategy
• Time for the apprentice to construct the patio

	Rate	Time	Part
Mason	$\frac{1}{3}$	2	$\frac{2}{3}$
Apprentice	$\frac{1}{t}$	2	$\frac{2}{t}$

• The sum of the parts of the task completed by each mason must equal 1.

Solution
$$\frac{2}{3} + \frac{2}{t} = 1$$
$$3t\left(\frac{2}{3} + \frac{2}{t}\right) = 3t(1)$$
$$2t + 6 = 3t$$
$$6 = t$$

Working alone, it would take the apprentice 6 h to construct the patio.

46. Strategy
To find the distance, write and solve a proportion using x to represent the distance.

Solution
$$\frac{21}{14} = \frac{12}{x}$$
$$\frac{3}{2} = \frac{12}{x}$$
$$2x\left(\frac{3}{2}\right) = 2x\left(\frac{12}{x}\right)$$
$$3x = 24$$
$$x = 8$$
A weight of 12 lb would stretch the spring 8 in.

47. Strategy
• Rate of the wind: r

	Distance	Rate	Time
With wind	2100	$400 + r$	$\dfrac{2100}{400+r}$
Against wind	1900	$400 - r$	$\dfrac{1900}{400-r}$

• The time with the wind equals the time against the wind.

Solution
$$\frac{2100}{400+r} = \frac{1900}{400-r}$$
$$(400+r)(400-r)\left(\frac{2100}{400+r}\right) = (400+r)(400-r)\left(\frac{1900}{400-r}\right)$$
$$2100(400-r) = 1900(400+r)$$
$$840,000 - 2100r = 760,000 + 1900r$$
$$80,000 = 4000r$$
$$20 = r$$
The rate of the wind is 20 mph.

48. Strategy
To find the amount of additional insecticide needed, write and solve a proportion using x to represent the additional number of ounces. The $4 + x$ is the total amount of insecticide.

Solution
$$\frac{4}{2} = \frac{4+x}{10}$$
$$2 = \frac{4+x}{10}$$
$$10(2) = 10\left(\frac{4+x}{10}\right)$$
$$20 = 4 + x$$
$$16 = x$$
16 additional ounces of insecticide are needed.

49. Strategy
• Time to fill the pool using both hoses: t

	Rate	Time	Part
First hose	$\dfrac{1}{15}$	t	$\dfrac{t}{15}$
Second hose	$\dfrac{1}{10}$	t	$\dfrac{t}{10}$

• The sum of the parts of the task completed by each hose must equal 1.

Solution
$$\frac{t}{15} + \frac{t}{10} = 1$$
$$30\left(\frac{t}{15} + \frac{t}{10}\right) = 30 \cdot 1$$
$$2t + 3t = 30$$
$$5t = 30$$
$$t = 6$$
Using both hoses, it would take 6 h to fill the pool.

50. Strategy
• Rate of the bus: r
• Rate of the car: $r + 10$

	Distance	Rate	Time
Car	315	$r + 10$	$\dfrac{315}{r+10}$
Bus	245	r	$\dfrac{245}{r}$

• The time of the bus equals the time of the car.

Solution

$$\frac{245}{r} = \frac{315}{r+10}$$

$$r(r+10)\left(\frac{245}{r}\right) = r(r+10)\left(\frac{315}{r+10}\right)$$

$$245(r+10) = 315r$$

$$245r + 2450 = 315r$$

$$2450 = 70r$$

$$35 = r$$

$$r + 10 = 35 + 10 = 45$$

The rate of the car is 45 mph.

Chapter Test

1.
$$\frac{x^2+3x+2}{x^2+5x+4} \div \frac{x^2-x-6}{x^2+2x-15}$$

$$= \frac{x^2+3x+2}{x^2+5x+4} \cdot \frac{x^2+2x-15}{x^2-x-6}$$

$$= \frac{(x+2)(x+1)}{(x+4)(x+1)} \cdot \frac{(x+5)(x-3)}{(x-3)(x+2)}$$

$$= \frac{x+5}{x+4}$$

2.
$$\frac{2x}{x^2+3x-10} - \frac{4}{x^2+3x-10}$$

$$= \frac{2x-4}{x^2+3x-10}$$

$$= \frac{2(x-2)}{(x+5)(x-2)} = \frac{2}{x+5}$$

3. $6x - 3 = 3(2x - 1)$
$2x^2 + x - 1 = (2x-1)(x+1)$
$\text{LCM} = 3(2x-1)(x+1)$

4.
$$\frac{3}{x+4} = \frac{5}{x+6}$$

$$(x+4)(x+6)\left(\frac{3}{x+4}\right) = (x+4)(x+6)\left(\frac{5}{x+6}\right)$$

$$(x+6)3 = (x+4)5$$

$$3x + 18 = 5x + 20$$

$$-2 = 2x$$

$$-1 = x$$

The solution is -1.

5.
$$\frac{x^3y^4}{x^2-4x+4} \cdot \frac{x^2-x-2}{x^6y^4}$$

$$= \frac{x^3y^4}{(x-2)(x-2)} \cdot \frac{(x-2)(x+1)}{x^6y^4}$$

$$= \frac{x^3y^4(x-2)(x+1)}{(x-2)(x-2)x^6y^4}$$

$$= \frac{x+1}{x^3(x-2)}$$

6.
$$\frac{1+\frac{1}{x}-\frac{12}{x^2}}{1+\frac{2}{x}-\frac{8}{x^2}} = \frac{1+\frac{1}{x}-\frac{12}{x^2}}{1+\frac{2}{x}-\frac{8}{x^2}} \cdot \frac{x^2}{x^2}$$

$$= \frac{1\cdot x^2 + \frac{1}{x}\cdot x^2 - \frac{12}{x^2}\cdot x^2}{1\cdot x^2 + \frac{2}{x}\cdot x^2 - \frac{8}{x^2}\cdot x^2}$$

$$= \frac{x^2+x-12}{x^2+2x-8} = \frac{(x+4)(x-3)}{(x+4)(x-2)} = \frac{x-3}{x-2}$$

7. $x^2 - 2x = x(x-2)$
$x^2 - 4 = (x+2)(x-2)$
The LCM is $x(x+2)(x-2)$.

$$\frac{3}{x^2-2x} = \frac{3}{x(x-2)} \cdot \frac{x+2}{x+2} = \frac{3x+6}{x(x+2)(x-2)}$$

$$\frac{x}{x^2-4} = \frac{x}{(x+2)(x-2)} \cdot \frac{x}{x} = \frac{x^2}{x(x+2)(x-2)}$$

8.
$$3x + 5y + 15 = 0$$
$$3x + 5y - 5y + 15 = 0 - 5y$$
$$3x + 15 = -5y$$
$$3x + 15 - 15 = -5y - 15$$
$$3x = -5y - 15$$
$$\frac{3x}{3} = \frac{-5y-15}{3}$$
$$x = -\frac{5}{3}y - 5$$

9.
$$\frac{6}{x} - 2 = 1$$
$$\frac{6}{x} - 2 + 2 = 1 + 2$$
$$\frac{6}{x} = 3$$
$$x\left(\frac{6}{x}\right) = x(3)$$
$$6 = 3x$$
$$2 = x$$
The solution is 2.

10.
$$\frac{2}{2x-1} - \frac{3}{3x+1} = \frac{2}{2x-1} \cdot \frac{3x+1}{3x+1} - \frac{3}{3x+1} \cdot \frac{2x-1}{2x-1}$$

$$= \frac{6x+2}{(2x-1)(3x+1)} - \frac{6x-3}{(2x-1)(3x+1)}$$

$$= \frac{6x+2-(6x-3)}{(2x-1)(3x+1)}$$

$$= \frac{6x+2-6x+3}{(2x-1)(3x+1)} = \frac{5}{(2x-1)(3x+1)}$$

11. $\dfrac{x^2-x-56}{x^2+8x+7} \div \dfrac{x^2-13x+40}{x^2-4x-5}$

$= \dfrac{x^2-x-56}{x^2+8x+7} \cdot \dfrac{x^2-4x-5}{x^2-13x+40}$

$= \dfrac{(x-8)(x+7)}{(x+1)(x+7)} \cdot \dfrac{(x-5)(x+1)}{(x-5)(x-8)} = 1$

12. $\dfrac{3x}{x^2+5x-24} - \dfrac{9}{x^2+5x-24} = \dfrac{3x-9}{x^2+5x-24}$

$= \dfrac{3x-9}{(x+8)(x-3)} = \dfrac{3(x-3)}{(x+8)(x-3)} = \dfrac{3}{x+8}$

13. $3x^2+6x = 3x(x+2)$

$2x^2+8x+8 = 2(x^2+4x+4)$

$\qquad\qquad = 2(x+2)(x+2)$

$\text{LCM} = (3x)(2)(x+2)(x+2)$

$\qquad\quad = 6x(x+2)^2$

14. $\dfrac{x^2-7x+10}{25-x^2} = \dfrac{(x-5)(x-2)}{(5-x)(5+x)} = -\dfrac{x-2}{x+5}$

15. $\dfrac{3x}{x-3} - 2 = \dfrac{10}{x-3}$

$(x-3)\left(\dfrac{3x}{x-3} - 2\right) = (x-3)\left(\dfrac{10}{x-3}\right)$

$3x - 2(x-3) = 10$

$3x - 2x + 6 = 10$

$x + 6 = 10$

$x = 4$

The solution is 4.

16. $f = v + at$

$f - v = v - v + at$

$f - v = at$

$\dfrac{f-v}{a} = \dfrac{at}{a}$

$\dfrac{f-v}{a} = t$

17. $\dfrac{12x^4y^2}{18xy^7} = \dfrac{\overset{1}{\cancel{2}}\cdot 2 \cdot \overset{1}{\cancel{3}} x^4 y^2}{\underset{1}{\cancel{2}}\cdot \underset{1}{\cancel{3}} 3xy^7} = \dfrac{2x^3}{3y^5}$

18. $\dfrac{2}{2x-1} - \dfrac{1}{x+1}$

$= \dfrac{2}{2x-1} \cdot \dfrac{x+1}{x+1} - \dfrac{1}{x+1} \cdot \dfrac{2x-1}{2x-1}$

$= \dfrac{2x+2}{(x+1)(2x-1)} - \dfrac{(2x-1)}{(x+1)(2x-1)}$

$= \dfrac{2x+2-(2x-1)}{(x+1)(2x-1)}$

$= \dfrac{2x+2-2x+1}{(x+1)(2x-1)}$

$= \dfrac{3}{(x+1)(2x-1)}$

19. $\dfrac{2}{x-2} = \dfrac{12}{x+3}$

$(x-2)(x+3)\left(\dfrac{2}{x-2}\right) = (x-2)(x+3)\left(\dfrac{12}{x+3}\right)$

$(x+3)2 = (x-2)12$

$2x+6 = 12x-24$

$30 = 10x$

$3 = x$

The solution is 3.

20. $\dfrac{x^5y^3}{x^2-x-6} \cdot \dfrac{x^2-9}{x^2y^4} = \dfrac{x^5y^3}{(x+2)(x-3)} \cdot \dfrac{(x+3)(x-3)}{x^2y^4}$

$= \dfrac{x^5y^3(x+3)(x-3)}{(x+2)(x-3)x^2y^4}$

$= \dfrac{x^3(x+3)}{y(x+2)}$

21. $x(1-x) = -x(x-1)$

$(x+1)(x-1) = (x+1)(x-1)$

The LCM is $x(x+1)(x-1)$.

$\dfrac{3y}{x(1-x)} = \dfrac{3y}{-x(x-1)} \cdot \dfrac{x+1}{x+1}$

$= -\dfrac{3xy+3y}{x(x+1)(x-1)}$

$\dfrac{x}{(x+1)(x-1)} = \dfrac{x}{(x+1)(x-1)} \cdot \dfrac{x}{x}$

$= \dfrac{x^2}{x(x+1)(x-1)}$

22. $\dfrac{1-\frac{2}{x}-\frac{15}{x^2}}{1-\frac{25}{x^2}} = \dfrac{1-\frac{2}{x}-\frac{15}{x^2}}{1-\frac{25}{x^2}} \cdot \dfrac{x^2}{x^2}$

$= \dfrac{1\cdot x^2 - \frac{2}{x}\cdot x^2 - \frac{15}{x^2}\cdot x^2}{1\cdot x^2 - \frac{25}{x^2}\cdot x^2}$

$= \dfrac{x^2-2x-15}{x^2-25}$

$= \dfrac{(x-5)(x+3)}{(x-5)(x+5)}$

$= \dfrac{x+3}{x+5}$

23. Strategy

To find the number of pounds, write and solve a proportion using x to represent the additional salt. Then $x + 4$ is the total amount of salt.

Solution

$\dfrac{4}{10} = \dfrac{x+4}{15}$

$30\left(\dfrac{4}{10}\right) = 30\left(\dfrac{x+4}{15}\right)$

$12 = 2(x+4)$

$12 = 2x+8$

$4 = 2x$

$2 = x$

2 additional pounds of salt are needed.

24. Strategy
• Rate of the wind: r

	Distance	Rate	Time
With wind	260	$110 + r$	$\dfrac{260}{110+r}$
Against wind	180	$110 - r$	$\dfrac{180}{110-r}$

• The time with the wind equals the time against the wind.

Solution
$$\frac{260}{110+r} = \frac{180}{110-r}$$
$$(110+r)(110-r)\left(\frac{260}{110+r}\right) = (110+r)(110-r)\left(\frac{180}{110-r}\right)$$
$$260(110-r) = 180(110+r)$$
$$28,600 - 260r = 19,800 + 180r$$
$$8800 = 440r$$
$$20 = r$$
The rate of the wind is 20 mph.

25. Strategy
• Time to fill the tank working together: t

	Rate	Time	Part
First pipe	$\dfrac{1}{9}$	t	$\dfrac{t}{9}$
Second pipe	$\dfrac{1}{18}$	t	$\dfrac{t}{18}$

• The sum of the parts of the task completed by the two pipes is 1.

Solution
$$\frac{t}{9} + \frac{t}{18} = 1$$
$$18\left(\frac{t}{9} + \frac{t}{18}\right) = 18 \cdot 1$$
$$2t + t = 18$$
$$3t = 18$$
$$t = 6$$
It would take 6 min to fill the tank with both pipes working together.

Cumulative Review Exercises

1. $-|-17| = -17$

2. $-\dfrac{3}{4} \cdot (2)^3 = -\dfrac{3}{4} \cdot 8$
$\qquad\qquad = -6$

3. $\left(\dfrac{2}{3}\right)^2 \div \left(\dfrac{3}{2} - \dfrac{2}{3}\right) + \dfrac{1}{2} = \left(\dfrac{2}{3}\right)^2 \div \left(\dfrac{9}{6} - \dfrac{4}{6}\right) + \dfrac{1}{2}$

$\qquad\qquad = \dfrac{4}{9} \div \left(\dfrac{5}{6}\right) + \dfrac{1}{2}$

$\qquad\qquad = \dfrac{4}{9} \cdot \dfrac{6}{5} + \dfrac{1}{2}$

$\qquad\qquad = \dfrac{8}{15} + \dfrac{1}{2}$

$\qquad\qquad = \dfrac{16}{30} + \dfrac{15}{30}$

$\qquad\qquad = \dfrac{31}{30}$

4. $-a^2 + (a-b)^2 - (-2)^2 + (-2-3)^2$
$= -(-2)^2 + (-5)^2$
$= -4 + 25$
$= 21$

5. $-2x - (-3y) + 7x - 5y = -2x + 3y + 7x - 5y$
$= -2x + 7x + 3y - 5y$
$= 5x - 2y$

6. $2[3x - 7(x-3) - 8] = 2[3x - 7x + 21 - 8]$
$= 2[-4x + 13]$
$= -8x + 26$

7. $3 - \dfrac{1}{4}x = 8$
$3 - 3 - \dfrac{1}{4}x = 8 - 3$
$-\dfrac{1}{4}x = 5$
$-4\left(-\dfrac{1}{4}x\right) = -4(5)$
$x = -20$
The solution is -20.

8. $3[x - 2(x-3)] = 2(3 - 2x)$
$3[x - 2x + 6] = 6 - 4x$
$3[-x + 6] = 6 - 4x$
$-3x + 18 = 6 - 4x$
$-3x + 4x + 18 = 6 - 4x + 4x$
$x + 18 = 6$
$x + 18 - 18 = 6 - 18$
$x = -12$
The solution is -12.

9. $16\dfrac{2}{3}\% = \dfrac{1}{6}$
$PB = A$
$\dfrac{1}{6}(60) = A$
$10 = A$
$16\dfrac{2}{3}\%$ of 60 is 10.

10. $\dfrac{5}{9}x < 1$
$\dfrac{9}{5}\left(\dfrac{5}{9}x\right) < \dfrac{9}{5}(1)$
$x < \dfrac{9}{5}$

11. $x - 2 \geq 4x - 47$
$x - 4x - 2 \geq 4x - 4x - 47$
$-3x - 2 \geq -47$
$-3x - 2 + 2 \geq -47 + 2$
$-3x \geq -45$
$\dfrac{-3}{-3} \leq \dfrac{-45}{-3}$
$x \leq 15$

12.

13.

14.

15. $5x + 2y < 6$
$5x - 5x + 2y < -5x + 6$
$2y < -5x + 6$
$y < -\dfrac{5}{2}x + 3$

Sketch a dashed line $y = -\dfrac{5}{2}x + 3$.

Shade below the dashed line.

16. $f(x) = -2x + 11$
$f(-8) = -2(-8) + 11 = 16 + 11 = 27$
$f(-4) = -2(-4) + 11 = 8 + 11 = 19$
$f(0) = -2(0) + 11 = 0 + 11 = 11$
$f(3) = -2(3) + 11 = -6 + 11 = 5$
$f(7) = -2(7) + 11 = -14 + 11 = -3$
The range is $\{-3, 5, 11, 19, 27\}$.

17. $f(x) = 4x - 3$
$f(10) = 4(10) - 3$
$f(10) = 40 - 3$
$f(10) = 37$
The value of $f(10)$ is 37.

18. $6x - y = 1$
$y = 3x + 1$
Substitute for y in equation (1).
$6x - (3x + 1) = 1$
$6x - 3x - 1 = 1$
$3x = 2$
$x = \dfrac{2}{3}$
Solve for y in equation (2).
$y - 3\left(\dfrac{2}{3}\right) + 1$
$y = 2 + 1$
$y = 3$
The solution is $\left(\dfrac{2}{3}, 3\right)$.

19. $2x - 3y = 4$
$4x + y = 1$
Eliminate y.
$2x - 3y = 4$
$3(4x + y) = 3(1)$

$2x - 3y = 4$
$12x + 3y = 3$

$14x = 7$
$x = \dfrac{1}{2}$
Solve for y in equation (2).
$4\left(\dfrac{1}{2}\right) + y = 1$
$2 + y = 1$
$y = -1$
The solution is $\left(\dfrac{1}{2}, -1\right)$.

20. $(3xy^4)(-2x^3 y) = 3(-2)x^{1+3} y^{4+1}$
$= -6x^4 y^5$

21. $(a^4 b^3)^5 = a^{4 \cdot 5} b^{3 \cdot 5}$
$= a^{20} b^{15}$

22. $\dfrac{a^2 b^{-5}}{a^{-1} b^{-3}} = a^{2 - (-1)} b^{-5 - (-3)}$
$= a^3 b^{-2}$
$= \dfrac{a^3}{b^2}$

23. $(a - 3b)(a + 4b)$
$= a^2 + 4ab - 3ab - 12b^2$
$= a^2 + ab - 12b^2$

24. $\dfrac{15b^4 - 5b^2 + 10b}{5b} = \dfrac{15b^4}{5b} - \dfrac{5b^2}{5b} + \dfrac{10b}{5b}$
$= 3b^3 - b + 2$

25.
$$
\begin{array}{r}
x^2 + 2x + 4 \\
x - 2 \overline{\smash{)}\, x^3 + 0x^2 + 0x - 8} \\
\underline{x^3 - 2x^2} \\
2x^2 + 0x \\
\underline{2x^2 - 4x} \\
4x - 8 \\
\underline{4x - 8} \\
0
\end{array}
$$
$(x^3 - 8) \div (x - 2) = x^2 + 2x + 4$

26. $12x^2 - x - 1 = 12x^2 - 4x + 3x - 1$
$= (12x^2 - 4x) + (3x - 1)$
$= 4x(3x - 1) + 1(3x - 1)$
$= (3x - 1)(4x + 1)$

27. $y^2 - 7y + 6 = (y - 6)(y - 1)$

28. $2a^3 + 7a^2 - 15a = a(2a^2 + 7a - 15)$
$= a(2a^2 - 3a + 10a - 15)$
$= a[(2a^2 - 3a) + (10a - 15)]$
$= a[a(2a - 3) + 5(2a - 3)]$
$= a(2a - 3)(a + 5)$

29. $4b^2 - 100 = 4(b^2 - 25) = 4(b + 5)(b - 5)$

30. $(x + 3)(2x - 5) = 0$
$x + 3 = 0 \quad 2x - 5 = 0$
$x = -3 \quad\quad 2x = 5$
$\quad\quad\quad\quad\quad x = \dfrac{5}{2}$
The solutions are -3 and $\dfrac{5}{2}$.

31. $\dfrac{x^2 + 3x - 28}{16 - x^2} = \dfrac{(x + 7)(x - 4)}{(4 + x)(4 - x)} = -\dfrac{x + 7}{x + 4}$

32. $\dfrac{x^2-3x-10}{x^2-4x-12} \div \dfrac{x^2-x-20}{x^2-2x-24} = \dfrac{x^2-3x-10}{x^2-4x-12} \cdot \dfrac{x^2-2x-24}{x^2-x-20}$

$\qquad = \dfrac{(x-5)(x+2)}{(x-6)(x+2)} \cdot \dfrac{(x-6)(x+4)}{(x-5)(x+4)}$

$\qquad = 1$

33. $\dfrac{6}{3x-1} - \dfrac{2}{x+1} = \dfrac{6}{3x-1} \cdot \dfrac{x+1}{x+1} - \dfrac{2}{x+1} \cdot \dfrac{3x-1}{3x-1}$

$\qquad = \dfrac{6x+6}{(3x-1)(x+1)} - \dfrac{6x-2}{(3x-1)(x+1)} = \dfrac{6x+6-(6x-2)}{(3x-1)(x+1)}$

$\qquad = \dfrac{6x+6-6x+2}{(3x-1)(x+1)} = \dfrac{8}{(3x-1)(x+1)}$

34. $\dfrac{4x}{x-3} - 2 = \dfrac{8}{x-3}$

$x-3\left(\dfrac{4x}{x-3} - 2\right) = x-3\left(\dfrac{8}{x-3}\right)$

$4x - 2(x-3) = 8$

$4x - 2x + 6 = 8$

$2x + 6 = 8$

$2x = 2$

$x = 1$

The solution is 1.

35. $f = v + at$

$f - v = v - v + at$

$f - v = at$

$\dfrac{f-v}{t} = \dfrac{at}{t}$

$\dfrac{f-v}{t} = a$

36. The unknown number: x

The difference between five times a number and eighteen	is	the opposite of three

$5x - 18 = -3$

$5x = 15$

$x = 3$

The number is 3.

37. Strategy
• Additional amount: x

	Principal	Rate	Interest
Amount at 7%	5000	0.07	0.07(5000)
Amount at 11%	x	0.11	0.11x
Amount at 9%	5000 + x	0.09	0.09(5000 + x)

• The sum of the interest earned by the two investments equals the interest earned on the total investment.

Solution
$0.07(5000) + 0.11x = 0.09(5000+x)$

$350 + 0.11x = 450 + 0.09x$

$350 + 0.20x = 450$

$0.02x = 100$

$x = 5000$

$5000 more must be invested at 11%.

38. Strategy
• Percent of silver in the alloy: x

	Amount	Percent	Quantity
40% silver	60	0.40	0.40(60)
Silver alloy	120	x	$120x$
Mixture	180	0.60	0.60(180)

• The sum of the quantities before mixing is equal to the quantity after mixing.

Solution
$$0.40(60) + 120x = 0.60(180)$$
$$24 + 120x = 108$$
$$120x = 84$$
$$x = 0.70$$
The silver alloy is 70% silver.

39. Strategy
Base of the triangle: $2x - 2$
Height of the triangle: x
The area of the triangle is 30 in^2.
The equation of for the area of a triangle is
$A = \frac{1}{2}bh$.

Substitute in the equation and solve for x.

Solution
$$A = \frac{1}{2}bh$$
$$30 = \frac{1}{2}(2x - 2)x$$
$$30 = (x - 1)x$$
$$30 = x^2 - x$$
$$0 = x^2 - x - 30$$
$$0 = (x - 6)(x + 5)$$
$$x - 6 = 0 \quad x + 5 = 0$$
$$x = 6 \quad\quad x = -5$$

The solution -5 is not possible.
$2x - 2 = 2(6) - 2 = 12 - 2 = 10$
The height is 6 in.
The base is 10 in.

40. Strategy
• Unknown time for both pipes working together: t

	Rate	Time	Part
First pipe	$\frac{1}{12}$	t	$\frac{t}{12}$
Second pipe	$\frac{1}{24}$	t	$\frac{t}{24}$

• The sum of the parts of the task completed by each pipe must equal 1.

Solution
$$\frac{t}{12} + \frac{t}{24} = 1$$
$$24\left(\frac{t}{12} + \frac{t}{24}\right) = 24(1)$$
$$2t + t = 24$$
$$3t = 24$$
$$t = 8$$

Working together, it would take both pipes 8 min to fill the tank.

Chapter 10: Radical Expressions

Prep Test

1. -14 [1.1.2]

2. $-2x^2y - 4xy^2$ [2.2.2]

3. 14 [3.1.3]

4. $3x - 2 = 5 - 2x$ [3.2.2]
$5x = 7$
$x = \dfrac{7}{5}$

5. x^6 [7.2.1]

6. $(x + y)^2 = (x + y)(x + y)$ [7.3.4]
$= x^2 + xy + xy + y^2$
$= x^2 + 2xy + y^2$

7. $(2x - 3)^2 = (2x - 3)(2x - 3)$ [7.3.4]
$= 4x^2 - 6x - 6x + 9$
$= 4x^2 - 12x + 9$

8. $(a - 5)(a + 5) = a^2 + 5a - 5a - 25$ [7.3.4]
$= a^2 - 25$

9. $(2 - 3v)(2 + 3v) = 4 + 6v - 6v - 9v$ [7.3.4]
$= 4 - 9v^2$

10. $\dfrac{x^2y^2}{9}$ [7.4.1]

Go Figure

Arrange all the guesses on a number line.

$\begin{array}{c}\longleftrightarrow\\ 215 \quad 217 \quad 219\ 220\ 221 \quad 223\end{array}$

Visually, we can see that the difference between 215 and 219 is 4, and the difference between 223 and 219 is also 4. The difference between 217 and 219 is 2, and the difference between 221 and 219 is also 2. The difference between 219 and 220 is 1.
So, the correct number was 219.

Section 10.1

Concept Review 10.1

1. Never true
The square root of a negative number is not a real number.

3. Always true

5. Sometimes true
The square root of a negative number is not a real number, so a and b cannot be negative numbers.

Objective 10.1.1 Exercises

3. $\sqrt{9} = \sqrt{3^2} = 3$
$\sqrt{25} = \sqrt{5^2} = 5$
$\sqrt{81} = \sqrt{9^2} = 9$
$\sqrt{100} = \sqrt{10^2} = 10$
2, 10, 50 are not perfect squares.

7. $\sqrt{16} = \sqrt{2^4} = 2^2 = 4$

9. $\sqrt{49} = \sqrt{7^2} = 7$

11. $\sqrt{32} = \sqrt{2^5} = \sqrt{2^4 \cdot 2} = \sqrt{2^4}\,\sqrt{2} = 2^2\,\sqrt{2} = 4\sqrt{2}$

13. $\sqrt{8} = \sqrt{2^3} = \sqrt{2^2 \cdot 2} = \sqrt{2^2}\,\sqrt{2} = 2\sqrt{2}$

15. $6\sqrt{18} = 6\sqrt{2 \cdot 3^2} = 6\sqrt{3^2}\,\sqrt{2} = 6 \cdot 3\sqrt{2} = 18\sqrt{2}$

17. $5\sqrt{40} = 5\sqrt{2^3 \cdot 5}$
$= 5\sqrt{2^2(2 \cdot 5)}$
$= 5\sqrt{2^2}\,\sqrt{2 \cdot 5}$
$= 5 \cdot 2\sqrt{10} = 10\sqrt{10}$

19. $\sqrt{15} = \sqrt{3 \cdot 5} = \sqrt{15}$

21. $\sqrt{29} = \sqrt{29}$

23. $-9\sqrt{72} = -9\sqrt{2^3 \cdot 3^2}$
$= -9\sqrt{2^2 \cdot 3^2 \cdot 2}$
$= -9\sqrt{2^2 \cdot 3^2}\,\sqrt{2}$
$= -9 \cdot 2 \cdot 3\sqrt{2}$
$= -54\sqrt{2}$

25. $\sqrt{45} = \sqrt{3^2 \cdot 5} = \sqrt{3^2}\,\sqrt{5} = 3\sqrt{5}$

27. $\sqrt{0} = 0$

29. $6\sqrt{128} = 6\sqrt{2^7}$
$= 6\sqrt{2^6 \cdot 2}$
$= 6\sqrt{2^6}\,\sqrt{2}$
$= 6 \cdot 2^3\,\sqrt{2} = 48\sqrt{2}$

31. $\sqrt{105} = \sqrt{3 \cdot 5 \cdot 7} = \sqrt{105}$

33. $\sqrt{900} = \sqrt{2^2 \cdot 3^2 \cdot 5^2} = 2 \cdot 3 \cdot 5 = 30$

35. $5\sqrt{180} = 5\sqrt{2^2 \cdot 3^2 \cdot 5}$
$= 5\sqrt{2^2 \cdot 3^2}\,\sqrt{5}$
$= 5 \cdot 2 \cdot 3\sqrt{5} = 30\sqrt{5}$

37. $\sqrt{250} = \sqrt{2 \cdot 5^3}$
$= \sqrt{5^2(2 \cdot 5)}$
$= \sqrt{5^2} \sqrt{2 \cdot 5}$
$= 5\sqrt{2 \cdot 5} = 5\sqrt{10}$

39. $\sqrt{96} = \sqrt{2^5 \cdot 3}$
$= \sqrt{2^4(2 \cdot 3)}$
$= \sqrt{2^4} \sqrt{2 \cdot 3}$
$= 2^2 \sqrt{2 \cdot 3} = 4\sqrt{6}$

41. $\sqrt{324} = \sqrt{2^2 \cdot 3^4} = 2 \cdot 3^2 = 18$

43. $\sqrt{240} \approx 15.492$

45. $\sqrt{288} \approx 16.971$

47. $\sqrt{245} \approx 15.652$

49. $\sqrt{352} \approx 18.762$

Objective 10.1.2 Exercises

53. $\sqrt{x^6} = x^3$

55. $\sqrt{y^{15}} = \sqrt{y^{14} \cdot y} = \sqrt{y^{14}} \sqrt{y} = y^7 \sqrt{y}$

57. $\sqrt{a^{20}} = a^{10}$

59. $\sqrt{x^4 y^4} = x^2 y^2$

61. $\sqrt{4x^4} = \sqrt{2^2 x^4} = 2x^2$

63. $\sqrt{24x^2} = \sqrt{2^3 \cdot 3 \cdot x^2}$
$= \sqrt{2^2 x^2 \cdot (2 \cdot 3)}$
$= \sqrt{2^2 x^2} \sqrt{2 \cdot 3} = 2x\sqrt{6}$

65. $\sqrt{x^3 y^7} = \sqrt{x^2 y^6 (xy)}$
$= \sqrt{x^2 y^6} \sqrt{xy}$
$= xy^3 \sqrt{xy}$

67. $\sqrt{a^3 b^{11}} = \sqrt{a^2 b^{10}(ab)}$
$= \sqrt{a^2 b^{10}} \sqrt{ab}$
$= ab^5 \sqrt{ab}$

69. $\sqrt{60x^5} = \sqrt{2^2 \cdot 3 \cdot 5 x^5}$
$= \sqrt{2^2 x^4 (3 \cdot 5x)}$
$= \sqrt{2^2 x^4} \sqrt{3 \cdot 5x} = 2x^2 \sqrt{15x}$

71. $\sqrt{49a^4 b^8} = \sqrt{7^2 a^4 b^8} = 7a^2 b^4$

73. $\sqrt{18x^5 y^7} = \sqrt{2 \cdot 3^2 x^5 y^7}$
$= \sqrt{3^2 x^4 y^6 (2xy)}$
$= \sqrt{3^2 x^4 y^6} \sqrt{2xy} = 3x^2 y^3 \sqrt{2xy}$

75. $\sqrt{40x^{11} y^7} = \sqrt{2^3 \cdot 5 x^{11} y^7}$
$= \sqrt{2^2 x^{10} y^6 (2 \cdot 5xy)}$
$= \sqrt{2^2 x^{10} y^6} \sqrt{2 \cdot 5xy}$
$= 2x^5 y^3 \sqrt{10xy}$

77. $\sqrt{80a^9 b^{10}} = \sqrt{2^4 \cdot 5 a^9 b^{10}}$
$= \sqrt{2^4 a^8 b^{10}(5a)}$
$= \sqrt{2^4 a^8 b^{10}} \sqrt{5a}$
$= 2^2 a^4 b^5 \sqrt{5a} = 4a^4 b^5 \sqrt{5a}$

79. $2\sqrt{16a^2 b^3} = 2\sqrt{2^4 a^2 b^3}$
$= 2\sqrt{2^4 a^2 b^2 (b)}$
$= 2\sqrt{2^4 a^2 b^2} \sqrt{b}$
$= 2 \cdot 2^2 ab\sqrt{b} = 8ab\sqrt{b}$

81. $x\sqrt{x^4 y^2} = x \cdot x^2 y = x^3 y$

83. $4\sqrt{20a^4 b^7} = 4\sqrt{2^2 \cdot 5a^4 b^7}$
$= 4\sqrt{2^2 a^4 b^6 (5b)}$
$= 4\sqrt{2^2 a^4 b^6} \sqrt{5b}$
$= 4 \cdot 2a^2 b^3 \sqrt{5b} = 8a^2 b^3 \sqrt{5b}$

85. $3x\sqrt{12x^2 y^7} = 3x\sqrt{2^2 \cdot 3x^2 y^7}$
$= 3x\sqrt{2^2 x^2 y^6 (3y)}$
$= 3x\sqrt{2^2 x^2 y^6} \sqrt{3y}$
$= 3x \cdot 2xy^3 \sqrt{3y} = 6x^2 y^3 \sqrt{3y}$

87. $2x^2 \sqrt{8x^2 y^3} = 2x^2 \sqrt{2^3 x^2 y^3}$
$= 2x^2 \sqrt{2^2 x^2 y^2 (2y)}$
$= 2x^2 \sqrt{2^2 x^2 y^2} \sqrt{2y}$
$= 2x^2 \cdot 2xy\sqrt{2y} = 4x^3 y\sqrt{2y}$

89. $\sqrt{25(a+4)^2} = \sqrt{5^2(a+4)^2} = 5(a+4) = 5a + 20$

91. $\sqrt{4(x+2)^4} = \sqrt{2^2(x+2)^4}$
$= 2(x+2)^2$
$= 2(x^2 + 4x + 4)$
$= 2x^2 + 8x + 8$

93. $\sqrt{x^2 + 4x + 4} = \sqrt{(x+2)^2} = x + 2$

95. $\sqrt{y^2 + 2y + 1} = \sqrt{(y+1)^2} = y + 1$

Applying Concepts 10.1

97. Strategy
Use the given formula for speed, using 1.2 for coefficient of friction, f and 60 ft for length, l.

Solution
$$S = \sqrt{30\,fl}$$
$$= \sqrt{(30)(1.2)(60)}$$
$$= \sqrt{2160}$$
$$= \sqrt{144 \cdot 15}$$
$$= 12\sqrt{15}$$
$$\approx 46.47$$

a. The speed of the car was $12\sqrt{15}$ mph.
b. The approximate speed was 46 mph.

99. $\sqrt{0.0025a^3b^5} = \sqrt{(0.05)^2 a^2 b^4 (ab)}$
$$= 0.05ab^2\sqrt{ab}$$

101. a. $f(x) = \sqrt{2x-1}$
$f(1) = \sqrt{2(1)-1}$
$f(1) = \sqrt{2-1}$
$f(1) = \sqrt{1}$
$f(1) = 1$

b. $f(x) = \sqrt{2x-1}$
$f(5) = \sqrt{2(5)-1}$
$f(5) = \sqrt{10-1}$
$f(5) = \sqrt{9}$
$f(5) = 3$

c. $f(x) = \sqrt{2x-1}$
$f(14) = \sqrt{2(14)-1}$
$f(14) = \sqrt{28-1}$
$f(14) = \sqrt{27}$
$f(14) = \sqrt{3^3}$
$f(14) = \sqrt{3^2 \cdot 3}$
$f(14) = \sqrt{3^2}\sqrt{3}$
$f(14) = 3\sqrt{3}$

103. Strategy
Area of square: 76 cm^2
Use the equation for the area of a square, substituting s for the length of the side of the square.

Solution
$76 = s^2$
$\sqrt{76} = s$
$8.7 \approx s$
The length of the side of the square is approximately 8.7 cm.

105. $4x \geq 0$
$\quad x \geq 0$

107. $x + 5 \geq 0$
$\quad x \geq -5$

109. $5 - 2x \geq 0$
$\quad -2x \geq -5$
$\quad x \leq \dfrac{5}{2}$

111. $x^2 + 1 \geq 0$
All real numbers

Section 10.2

Concept Review 10.2

1. Never true
The radicands are not the same. $5\sqrt{2} + 6\sqrt{3}$ cannot be simplified.

3. Always true

5. Always true

Objective 10.2.1 Exercises

1. $2\sqrt{2} + \sqrt{2} = 3\sqrt{2}$

3. $-3\sqrt{7} + 2\sqrt{7} = -\sqrt{7}$

5. $-3\sqrt{11} - 8\sqrt{11} = -11\sqrt{11}$

7. $2\sqrt{x} + 8\sqrt{x} = 10\sqrt{x}$

9. $8\sqrt{y} - 10\sqrt{y} = -2\sqrt{y}$

11. $-2\sqrt{3b} - 9\sqrt{3b} = -11\sqrt{3b}$

13. $3x\sqrt{2} - x\sqrt{2} = 2x\sqrt{2}$

15. $2a\sqrt{3a} - 5a\sqrt{3a} = -3a\sqrt{3a}$

17. $3\sqrt{xy} - 8\sqrt{xy} = -5\sqrt{xy}$

19. $\sqrt{45} + \sqrt{125} = \sqrt{3^2 \cdot 5} + \sqrt{5^3}$
$$= \sqrt{3^2}\sqrt{5} + \sqrt{5^2}\sqrt{5}$$
$$= 3\sqrt{5} + 5\sqrt{5}$$
$$= 8\sqrt{5}$$

21. $2\sqrt{2} + 3\sqrt{8} = 2\sqrt{2} + 3\sqrt{2^3}$
$$= 2\sqrt{2} + 3\sqrt{2^2}\sqrt{2}$$
$$= 2\sqrt{2} + 3 \cdot 2\sqrt{2}$$
$$= 2\sqrt{2} + 6\sqrt{2} = 8\sqrt{2}$$

23. $5\sqrt{18} - 2\sqrt{75} = 5\sqrt{2 \cdot 3^2} - 2\sqrt{3 \cdot 5^2}$
$$= 5\sqrt{3^2}\sqrt{2} - 2\sqrt{5^2}\sqrt{3}$$
$$= 5 \cdot 3\sqrt{2} - 2 \cdot 5\sqrt{3}$$
$$= 15\sqrt{2} - 10\sqrt{3}$$

25. $5\sqrt{4x} - 3\sqrt{9x} = 5\sqrt{2^2 x} - 3\sqrt{3^2 x}$
$$= 5\sqrt{2^2}\sqrt{x} - 3\sqrt{3^2}\sqrt{x}$$
$$= 5 \cdot 2\sqrt{x} - 3 \cdot 3\sqrt{x}$$
$$= 10\sqrt{x} - 9\sqrt{x} = \sqrt{x}$$

27. $3\sqrt{3x^2} - 5\sqrt{27x^2} = 3\sqrt{3x^2} - 5\sqrt{3^3 x^2}$
$= 3\sqrt{x^2}\sqrt{3} - 5\sqrt{3^2 x^2}\sqrt{3}$
$= 3x\sqrt{3} - 5\cdot 3x\sqrt{3}$
$= 3x\sqrt{3} - 15x\sqrt{3}$
$= -12x\sqrt{3}$

29. $2x\sqrt{xy^2} - 3y\sqrt{x^2 y} = 2x\sqrt{y^2}\sqrt{x} - 3y\sqrt{x^2}\sqrt{y}$
$= 2xy\sqrt{x} - 3xy\sqrt{y}$

31. $3x\sqrt{12x} - 5\sqrt{27x^3} = 3x\sqrt{2^2\cdot 3x} - 5\sqrt{3^3 x^3}$
$= 3x\sqrt{2^2}\sqrt{3x} - 5\sqrt{3^2 x^2}\sqrt{3x}$
$= 3x\cdot 2\sqrt{3x} - 5\cdot 3x\sqrt{3x}$
$= 6x\sqrt{3x} - 15x\sqrt{3x} = -9x\sqrt{3x}$

33. $4y\sqrt{8y^3} - 7\sqrt{18y^5}$
$= 4y\sqrt{2^3 y^3} - 7\sqrt{2\cdot 3^2 y^5}$
$= 4y\sqrt{2^2 y^2}\sqrt{2y} - 7\sqrt{3^2 y^4}\sqrt{2y}$
$= 4y\cdot 2y\sqrt{2y} - 7\cdot 3y^2\sqrt{2y}$
$= 8y^2\sqrt{2y} - 21y^2\sqrt{2y}$
$= -13y^2\sqrt{2y}$

35. $b^2\sqrt{a^5 b} + 3a^2\sqrt{ab^5}$
$= b^2\sqrt{a^4}\sqrt{ab} + 3a^2\sqrt{b^4}\sqrt{ab}$
$= a^2 b^2\sqrt{ab} + 3a^2 b^2\sqrt{ab}$
$= 4a^2 b^2\sqrt{ab}$

37. $4\sqrt{2} - 5\sqrt{2} + 8\sqrt{2} = 7\sqrt{2}$

39. $5\sqrt{x} - 8\sqrt{x} + 9\sqrt{x} = 6\sqrt{x}$

41. $8\sqrt{2} - 3\sqrt{y} - 8\sqrt{2} = -3\sqrt{y}$

43. $8\sqrt{8} - 4\sqrt{32} - 9\sqrt{50}$
$= 8\sqrt{2^3} - 4\sqrt{2^5} - 9\sqrt{2\cdot 5^2}$
$= 8\sqrt{2^2}\sqrt{2} - 4\sqrt{2^4}\sqrt{2} - 9\sqrt{5^2}\sqrt{2}$
$= 8\cdot 2\sqrt{2} - 4\cdot 2^2\sqrt{2} - 9\cdot 5\sqrt{2}$
$= 16\sqrt{2} - 16\sqrt{2} - 45\sqrt{2}$
$= -45\sqrt{2}$

45. $-2\sqrt{3} + 5\sqrt{27} - 4\sqrt{45}$
$= -2\sqrt{3} + 5\sqrt{3^3} - 4\sqrt{3^2\cdot 5}$
$= -2\sqrt{3} + 5\sqrt{3^2}\sqrt{3} - 4\sqrt{3^2}\sqrt{5}$
$= -2\sqrt{3} + 5\cdot 3\sqrt{3} - 4\cdot 3\sqrt{5}$
$= -2\sqrt{3} + 15\sqrt{3} - 12\sqrt{5}$
$= 13\sqrt{3} - 12\sqrt{5}$

47. $4\sqrt{75} + 3\sqrt{48} - \sqrt{99}$
$= 4\sqrt{3\cdot 5^2} + 3\sqrt{2^4\cdot 3} - \sqrt{3^2\cdot 11}$
$= 4\sqrt{5^2}\sqrt{3} + 3\sqrt{2^4}\sqrt{3} - \sqrt{3^2}\sqrt{11}$
$= 4\cdot 5\sqrt{3} + 3\cdot 2^2\sqrt{3} - 3\sqrt{11}$
$= 20\sqrt{3} + 12\sqrt{3} - 3\sqrt{11}$
$= 32\sqrt{3} - 3\sqrt{11}$

49. $\sqrt{25x} - \sqrt{9x} + \sqrt{16x}$
$= \sqrt{5^2 x} - \sqrt{3^2 x} + \sqrt{2^4 x}$
$= \sqrt{5^2}\sqrt{x} - \sqrt{3^2}\sqrt{x} + \sqrt{2^4}\sqrt{x}$
$= 5\sqrt{x} - 3\sqrt{x} + 2^2\sqrt{x}$
$= 5\sqrt{x} - 3\sqrt{x} + 4\sqrt{x}$
$= 6\sqrt{x}$

51. $3\sqrt{3x} + \sqrt{27x} - 8\sqrt{75x}$
$= 3\sqrt{3x} + \sqrt{3^3 x} - 8\sqrt{3\cdot 5^2 x}$
$= 3\sqrt{3x} + \sqrt{3^2}\sqrt{3x} - 8\sqrt{5^2}\sqrt{3x}$
$= 3\sqrt{3x} + 3\sqrt{3x} - 8\cdot 5\sqrt{3x}$
$= 3\sqrt{3x} + 3\sqrt{3x} - 40\sqrt{3x} = -34\sqrt{3x}$

53. $2a\sqrt{75b} - a\sqrt{20b} + 4a\sqrt{45b}$
$= 2a\sqrt{3\cdot 5^2 b} - a\sqrt{2^2\cdot 5b} + 4a\sqrt{3^2\cdot 5b}$
$= 2a\sqrt{5^2}\sqrt{3b} - a\sqrt{2^2}\sqrt{5b} + 4a\sqrt{3^2}\sqrt{5b}$
$= 2a\cdot 5\sqrt{3b} - 2a\sqrt{5b} + 4a\cdot 3\sqrt{5b}$
$= 10a\sqrt{3b} - 2a\sqrt{5b} + 12a\sqrt{5b}$
$= 10a\sqrt{3b} + 10a\sqrt{5b}$

55. $x\sqrt{3y^2} - 2y\sqrt{12x^2} + xy\sqrt{3}$
$= x\sqrt{3y^2} - 2y\sqrt{2^2\cdot 3x^2} + xy\sqrt{3}$
$= x\sqrt{y^2}\sqrt{3} - 2y\sqrt{2^2 x^2}\sqrt{3} + xy\sqrt{3}$
$= xy\sqrt{3} - 2y\cdot 2x\sqrt{3} + xy\sqrt{3}$
$= xy\sqrt{3} - 4xy\sqrt{3} + xy\sqrt{3}$
$= -2xy\sqrt{3}$

57. $3\sqrt{ab^3} + 4a\sqrt{a^2 b} - 5b\sqrt{4ab}$
$= 3\sqrt{ab^3} + 4a\sqrt{a^2 b} - 5b\sqrt{2^2 ab}$
$= 3\sqrt{b^2}\sqrt{ab} + 4a\sqrt{a^2}\sqrt{b} - 5b\sqrt{2^2}\sqrt{ab}$
$= 3b\sqrt{ab} + 4a\cdot a\sqrt{b} - 5b\cdot 2\sqrt{ab}$
$= 3b\sqrt{ab} + 4a^2\sqrt{b} - 10b\sqrt{ab}$
$= -7b\sqrt{ab} + 4a^2\sqrt{b}$

59. $3a\sqrt{2ab^2} - \sqrt{a^2 b^2} + 4b\sqrt{3a^2 b}$
$= 3a\sqrt{b^2}\sqrt{2a} - \sqrt{a^2 b^2} + 4b\sqrt{a^2}\sqrt{3b}$
$= 3ab\sqrt{2a} - ab + 4ab\sqrt{3b}$

Applying Concepts 10.2

61. $5\sqrt{x+2} + 3\sqrt{x+2} = 8\sqrt{x+2}$

63. $\dfrac{1}{2}\sqrt{8x^2y} + \dfrac{1}{3}\sqrt{18x^2y}$

$= \dfrac{1}{2}\sqrt{2^3x^2y} + \dfrac{1}{3}\sqrt{2\cdot3^2x^2y}$

$= \dfrac{1}{2}\sqrt{2^2x^2}\sqrt{2y} + \dfrac{1}{3}\sqrt{3^2x^2}\sqrt{2y}$

$= \dfrac{1}{2}(2x)\sqrt{2y} + \dfrac{1}{3}(3x)\sqrt{2y}$

$= x\sqrt{2y} + x\sqrt{2y}$

$= 2x\sqrt{2y}$

65. $\dfrac{a}{3}\sqrt{54ab^3} + \dfrac{b}{4}\sqrt{96a^3b}$

$= \dfrac{a}{3}\sqrt{2\cdot3^3\,ab^3} + \dfrac{b}{4}\sqrt{2^5\cdot3a^3b}$

$= \dfrac{a}{3}\sqrt{3^2b^2}\sqrt{2\cdot3ab} + \dfrac{b}{4}\sqrt{2^4a^2}\sqrt{2\cdot3ab}$

$= \dfrac{a}{3}(3b)\sqrt{6ab} + \dfrac{b}{4}(4a)\sqrt{6ab}$

$= ab\sqrt{6ab} + ab\sqrt{6ab}$

$= 2ab\sqrt{6ab}$

67. $2\sqrt{8x+4y} - 5\sqrt{18x+9y}$

$= 2\sqrt{4(2x+y)} - 5\sqrt{9(2x+y)}$

$= 2\cdot2\sqrt{2x+y} - 5\cdot3\sqrt{2x+y}$

$= 4\sqrt{2x+y} - 15\sqrt{2x+y}$

$= -11\sqrt{2x+y}$

69. Strategy

Length of first side: $4\sqrt{3}$ cm

Length of second side: $2\sqrt{3}$ cm

Length of third side: $2\sqrt{15}$ cm

Use the equation for the perimeter of a triangle.

Solution

$P = a + b + c$

$P = 4\sqrt{3} + 2\sqrt{3} + 2\sqrt{15}$

$P = 6\sqrt{3} + 2\sqrt{15}$

The perimeter is $(6\sqrt{3} + 2\sqrt{15})$ cm.

71. Strategy

Length: $4\sqrt{5}$ cm

Width: $\sqrt{5}$ cm

Use the equation for the perimeter of a rectangle.

Solution

$P = 2L + 2W$

$P = 2(4\sqrt{5}) + 2(\sqrt{5})$

$P = 8\sqrt{5} + 2\sqrt{5}$

$P = 10\sqrt{5}$

$P \approx 22.4$

The perimeter is about 22.4 cm.

Section 10.3

Concept Review 10.3

1. Always true

3. Always true

5. Sometimes true

A radical expression in simplest form does not have a fraction under the radical sign and does not have a radical in the denominator of a fraction.

Objective 10.3.1 Exercises

1. $\sqrt{5}\cdot\sqrt{5} = \sqrt{5^2} = 5$

3. $\sqrt{3}\cdot\sqrt{12} = \sqrt{36} = \sqrt{2^2\cdot3^2} = 2\cdot3 = 6$

5. $\sqrt{x}\cdot\sqrt{x} = \sqrt{x^2} = x$

7. $\sqrt{xy^3}\cdot\sqrt{x^5y} = \sqrt{x^6y^4} = x^3y^2$

9. $\sqrt{3a^2b^5}\cdot\sqrt{6ab^7} = \sqrt{18a^3b^{12}}$

$\qquad = \sqrt{2\cdot3^2a^3b^{12}}$

$\qquad = \sqrt{3^2a^2b^{12}}\sqrt{2a}$

$\qquad = 3ab^6\sqrt{2a}$

11. $\sqrt{6a^3b^2}\cdot\sqrt{24a^5b} = \sqrt{144a^8b^3}$

$\qquad = \sqrt{2^4\cdot3^2a^8b^3}$

$\qquad = \sqrt{2^4\cdot3^2a^8b^2}\sqrt{b}$

$\qquad = 2^2\cdot3a^4b\sqrt{b}$

$\qquad = 12a^4b\sqrt{b}$

13. $\sqrt{2}\left(\sqrt{2} - \sqrt{3}\right) = \sqrt{2^2} - \sqrt{6} = 2 - \sqrt{6}$

15. $\sqrt{x}\left(\sqrt{x} - \sqrt{y}\right) = \sqrt{x^2} - \sqrt{xy} = x - \sqrt{xy}$

17. $\sqrt{5}\left(\sqrt{10} - \sqrt{x}\right) = \sqrt{50} - \sqrt{5x}$

$\qquad = \sqrt{2\cdot5^2} - \sqrt{5x}$

$\qquad = \sqrt{5^2}\sqrt{2} - \sqrt{5x}$

$\qquad = 5\sqrt{2} - \sqrt{5x}$

19. $\sqrt{8}\left(\sqrt{2} - \sqrt{5}\right) = \sqrt{16} - \sqrt{40}$

$\qquad = \sqrt{2^4} - \sqrt{2^3\cdot5}$

$\qquad = \sqrt{2^4} - \sqrt{2^2}\sqrt{2\cdot5}$

$\qquad = 2^2 - 2\sqrt{10}$

$\qquad = 4 - 2\sqrt{10}$

21. $\left(\sqrt{x} - 3\right)^2 = \sqrt{x^2} - 2\cdot3\sqrt{x} + 3^2 = x - 6\sqrt{x} + 9$

23. $\sqrt{3a}\left(\sqrt{3a} - \sqrt{3b}\right) = \sqrt{3^2a^2} - \sqrt{3^2ab}$

$\qquad = \sqrt{3^2a^2} - \sqrt{3^2}\sqrt{ab}$

$\qquad = 3a - 3\sqrt{ab}$

25. $\sqrt{2ac} \cdot \sqrt{5ab} \cdot \sqrt{10cb} = \sqrt{100a^2 b^2 c^2}$
$= \sqrt{2^2 \cdot 5^2 a^2 b^2 c^2}$
$= 2 \cdot 5abcd$
$= 10abc$

27. $\left(3\sqrt{x} - 2y\right)\left(5\sqrt{x} - 4y\right)$
$= 15\sqrt{x^2} - 12y\sqrt{x} - 10y\sqrt{x} + 8y^2$
$= 15x - 22y\sqrt{x} + 8y^2$

29. $\left(\sqrt{x} - \sqrt{y}\right)\left(\sqrt{x} + \sqrt{y}\right) = \sqrt{x^2} - \sqrt{y^2} = x - y$

31. $\left(2\sqrt{x} + \sqrt{y}\right)\left(5\sqrt{x} + 4\sqrt{y}\right)$
$= 10\sqrt{x^2} + 8\sqrt{xy} + 5\sqrt{xy} + 4\sqrt{y^2}$
$= 10x + 13\sqrt{xy} + 4y$

Objective 10.3.2 Exercises

35. $\dfrac{\sqrt{32}}{\sqrt{2}} = \sqrt{\dfrac{32}{2}} = \sqrt{16} = \sqrt{2^4} = 2^2 = 4$

37. $\dfrac{\sqrt{98}}{\sqrt{2}} = \sqrt{\dfrac{98}{2}} = \sqrt{49} = \sqrt{7^2} = 7$

39. $\dfrac{\sqrt{27a}}{\sqrt{3a}} = \sqrt{\dfrac{27a}{3a}} = \sqrt{9} = \sqrt{3^2} = 3$

41. $\dfrac{\sqrt{15x^3 y}}{\sqrt{3xy}} = \sqrt{\dfrac{15x^3 y}{3xy}} = \sqrt{5x^2} = \sqrt{x^2}\sqrt{5} = x\sqrt{5}$

43. $\dfrac{\sqrt{2a^5 b^4}}{\sqrt{98ab^4}} = \sqrt{\dfrac{2a^5 b^4}{98ab^4}} = \sqrt{\dfrac{a^4}{49}} = \dfrac{\sqrt{a^4}}{\sqrt{7^2}} = \dfrac{a^2}{7}$

45. $\dfrac{1}{\sqrt{3}} = \dfrac{1}{\sqrt{3}} \cdot \dfrac{\sqrt{3}}{\sqrt{3}} = \dfrac{\sqrt{3}}{\sqrt{3^2}} = \dfrac{\sqrt{3}}{3}$

47. $\dfrac{15}{\sqrt{75}} = \dfrac{15}{\sqrt{25 \cdot 3}} = \dfrac{15}{5\sqrt{3}} = \dfrac{15}{5\sqrt{3}} \cdot \dfrac{\sqrt{3}}{\sqrt{3}} = \dfrac{15\sqrt{3}}{15} = \sqrt{3}$

49. $\dfrac{6}{\sqrt{12x}} = \dfrac{6}{\sqrt{4 \cdot 3x}}$
$= \dfrac{6}{2\sqrt{3x}}$
$= \dfrac{6}{2\sqrt{3x}} \cdot \dfrac{\sqrt{3x}}{\sqrt{3x}}$
$= \dfrac{6\sqrt{3x}}{6x} = \dfrac{\sqrt{3x}}{x}$

51. $\dfrac{8}{\sqrt{32x}} = \dfrac{8}{\sqrt{16 \cdot 2x}}$
$= \dfrac{8}{4\sqrt{2x}}$
$= \dfrac{8}{4\sqrt{2x}} \cdot \dfrac{\sqrt{2x}}{\sqrt{2x}}$
$= \dfrac{8\sqrt{2x}}{4 \cdot 2x}$
$= \dfrac{8\sqrt{2x}}{8x} = \dfrac{\sqrt{2x}}{x}$

53. $\dfrac{3}{\sqrt{x}} = \dfrac{3}{\sqrt{x}} \cdot \dfrac{\sqrt{x}}{\sqrt{x}} = \dfrac{3\sqrt{x}}{\sqrt{x^2}} = \dfrac{3\sqrt{x}}{x}$

55. $\dfrac{\sqrt{8x^2 y}}{\sqrt{2x^4 y^2}} = \sqrt{\dfrac{8x^2 y}{2x^4 y^2}}$
$= \sqrt{\dfrac{4}{x^2 y}}$
$= \dfrac{\sqrt{4}}{\sqrt{x^2 y}}$
$= \dfrac{\sqrt{2^2}}{\sqrt{x^2}\sqrt{y}}$
$= \dfrac{2}{x\sqrt{y}}$
$= \dfrac{2}{x\sqrt{y}} \cdot \dfrac{\sqrt{y}}{\sqrt{y}}$
$= \dfrac{2\sqrt{y}}{x\sqrt{y^2}} = \dfrac{2\sqrt{y}}{xy}$

57. $\dfrac{\sqrt{16a}}{\sqrt{49ab}} = \sqrt{\dfrac{16a}{49ab}} = \dfrac{4}{7\sqrt{b}} = \dfrac{4}{7\sqrt{b}} \cdot \dfrac{\sqrt{b}}{\sqrt{b}} = \dfrac{4\sqrt{b}}{7b}$

59. $\dfrac{5\sqrt{18}}{9\sqrt{27}} = \dfrac{5}{9}\sqrt{\dfrac{18}{27}}$
$= \dfrac{5}{9}\sqrt{\dfrac{2}{3}}$
$= \dfrac{5\sqrt{2}}{9\sqrt{3}} \cdot \dfrac{\sqrt{3}}{\sqrt{3}} = \dfrac{5\sqrt{6}}{9 \cdot 3} = \dfrac{5\sqrt{6}}{27}$

61. $\dfrac{\sqrt{3xy}}{\sqrt{27x^3 y^2}} = \sqrt{\dfrac{3xy}{27x^3 y^2}}$
$= \sqrt{\dfrac{1}{9x^2 y}}$
$= \dfrac{1}{3x\sqrt{y}} = \dfrac{1}{3x\sqrt{y}} \cdot \dfrac{\sqrt{y}}{\sqrt{y}} = \dfrac{\sqrt{y}}{3xy}$

63. $\dfrac{\sqrt{4x^2y}}{\sqrt{3xy^3}} = \sqrt{\dfrac{4x^2y}{3xy^3}} = \sqrt{\dfrac{4x}{3y^2}} = \dfrac{\sqrt{4x}}{\sqrt{3y^2}}$

$= \dfrac{\sqrt{2^2x}}{\sqrt{3y^2}} = \dfrac{\sqrt{2^2}\sqrt{x}}{\sqrt{y^2}\sqrt{3}} = \dfrac{2\sqrt{x}}{y\sqrt{3}}$

$= \dfrac{2\sqrt{x}}{y\sqrt{3}} \cdot \dfrac{\sqrt{3}}{\sqrt{3}} = \dfrac{2\sqrt{3x}}{y\sqrt{3^2}} = \dfrac{2\sqrt{3x}}{3y}$

65. $\dfrac{1}{\sqrt{2}-3} = \dfrac{1}{\sqrt{2}-3} \cdot \dfrac{\sqrt{2}+3}{\sqrt{2}+3}$

$= \dfrac{\sqrt{2}+3}{2-9}$

$= \dfrac{\sqrt{2}+3}{-7} = -\dfrac{\sqrt{2}+3}{7}$

67. $\dfrac{3}{5+\sqrt{5}} = \dfrac{3}{5+\sqrt{5}} \cdot \dfrac{5-\sqrt{5}}{5-\sqrt{5}}$

$= \dfrac{3\left(5-\sqrt{5}\right)}{25-5}$

$= \dfrac{15-3\sqrt{5}}{20}$

69. $\dfrac{\sqrt{xy}}{\sqrt{x}-\sqrt{y}} = \dfrac{\sqrt{xy}}{\sqrt{x}-\sqrt{y}} \cdot \dfrac{\sqrt{x}+\sqrt{y}}{\sqrt{x}+\sqrt{y}}$

$= \dfrac{\sqrt{x^2y}+\sqrt{xy^2}}{x-y}$

$= \dfrac{\sqrt{x^2}\sqrt{y}+\sqrt{y^2}\sqrt{x}}{x-y} = \dfrac{x\sqrt{y}+y\sqrt{x}}{x-y}$

71. $\dfrac{5\sqrt{x^2y}}{\sqrt{75xy^2}} = 5\sqrt{\dfrac{x^2y}{75xy^2}}$

$= 5\sqrt{\dfrac{x}{25\cdot 3y}}$

$= \dfrac{5\sqrt{x}}{5\sqrt{3y}} = \dfrac{\sqrt{x}}{\sqrt{3y}} \cdot \dfrac{\sqrt{3y}}{\sqrt{3y}} = \dfrac{\sqrt{3xy}}{3y}$

73. $\dfrac{\sqrt{2}}{\sqrt{2}-\sqrt{3}} = \dfrac{\sqrt{2}}{\sqrt{2}-\sqrt{3}} \cdot \dfrac{\sqrt{2}+\sqrt{3}}{\sqrt{2}+\sqrt{3}}$

$= \dfrac{\sqrt{4}+\sqrt{6}}{2-3}$

$= \dfrac{2+\sqrt{6}}{-1}$

$= -(2+\sqrt{6}) = -2-\sqrt{6}$

75. $\dfrac{\sqrt{5}}{\sqrt{2}-\sqrt{5}} = \dfrac{\sqrt{5}}{\sqrt{2}-\sqrt{5}} \cdot \dfrac{\sqrt{2}+\sqrt{5}}{\sqrt{2}+\sqrt{5}}$

$= \dfrac{\sqrt{10}+\sqrt{25}}{2-5}$

$= \dfrac{\sqrt{10}+5}{-3} = -\dfrac{\sqrt{10}+5}{3}$

77. $\dfrac{\sqrt{x}}{\sqrt{x}+3} = \dfrac{\sqrt{x}}{\sqrt{x}+3} \cdot \dfrac{\sqrt{x}-3}{\sqrt{x}-3}$

$= \dfrac{\sqrt{x^2}-3\sqrt{x}}{\sqrt{x^2}-9}$

$= \dfrac{x-3\sqrt{x}}{x-9}$

79. $\dfrac{5\sqrt{3}-7\sqrt{3}}{4\sqrt{3}} = \dfrac{-2\sqrt{3}}{4\sqrt{3}} = -\dfrac{1}{2}$

81. $\dfrac{5\sqrt{8}-3\sqrt{2}}{\sqrt{2}} = \dfrac{5\sqrt{8}}{\sqrt{2}} - \dfrac{3\sqrt{2}}{\sqrt{2}}$

$= 5\sqrt{4}-3$

$= 5\cdot 2-3 = 10-3 = 7$

83. $\dfrac{3\sqrt{2}-8\sqrt{2}}{\sqrt{2}} = \dfrac{-5\sqrt{2}}{\sqrt{2}} = -5$

85. $\dfrac{2\sqrt{8}+3\sqrt{2}}{\sqrt{32}} = \dfrac{2\sqrt{8}}{\sqrt{32}} + \dfrac{3\sqrt{2}}{\sqrt{32}}$

$= 2\sqrt{\dfrac{8}{32}} + 3\sqrt{\dfrac{2}{32}}$

$= 2\sqrt{\dfrac{1}{4}} + 3\sqrt{\dfrac{1}{16}}$

$= \dfrac{2\sqrt{1}}{\sqrt{4}} + \dfrac{3\sqrt{1}}{\sqrt{16}}$

$= \dfrac{2}{\sqrt{2^2}} + \dfrac{3}{\sqrt{2^4}} = \dfrac{2}{2} + \dfrac{3}{2^2} = 1 + \dfrac{3}{4} = \dfrac{7}{4}$

87. $\dfrac{6-2\sqrt{3}}{4+3\sqrt{3}} = \dfrac{6-2\sqrt{3}}{4+3\sqrt{3}} \cdot \dfrac{4-3\sqrt{3}}{4-3\sqrt{3}}$

$= \dfrac{24-26\sqrt{3}+18}{16-27}$

$= \dfrac{42-26\sqrt{3}}{-11}$

$= -\dfrac{42-26\sqrt{3}}{11}$

89. $\dfrac{2\sqrt{3}-\sqrt{6}}{5\sqrt{3}+2\sqrt{6}} = \dfrac{2\sqrt{3}-\sqrt{6}}{5\sqrt{3}+2\sqrt{6}} \cdot \dfrac{5\sqrt{3}-2\sqrt{6}}{5\sqrt{3}-2\sqrt{6}}$

$= \dfrac{30-9\sqrt{18}+12}{75-24}$

$= \dfrac{42-9\sqrt{18}}{51}$

$= \dfrac{42-27\sqrt{2}}{51}$

$= \dfrac{3(14-9\sqrt{2})}{51}$

$= \dfrac{14-9\sqrt{2}}{17}$

91. $\dfrac{\sqrt{a}-4}{2\sqrt{a}+2} = \dfrac{\sqrt{a}-4}{2\sqrt{a}+2} \cdot \dfrac{2\sqrt{a}-2}{2\sqrt{a}-2}$

$\quad = \dfrac{2a-10\sqrt{a}+8}{4a-4}$

$\quad = \dfrac{2(a-5\sqrt{a}+4)}{2(2a-2)} = \dfrac{a-5\sqrt{a}+4}{2a-2}$

93. $\dfrac{2+\sqrt{y}}{\sqrt{y}-3} = \dfrac{2+\sqrt{y}}{\sqrt{y}-3} \cdot \dfrac{\sqrt{y}+3}{\sqrt{y}+3}$

$\quad = \dfrac{2\sqrt{y}+6+y+3\sqrt{y}}{y-9}$

$\quad = \dfrac{y+5\sqrt{y}+6}{y-9}$

Applying Concepts 10.3

95. $-\sqrt{1.3} \cdot \sqrt{1.3} = -\sqrt{(1.3)^2} = -1.3$

97. $-\sqrt{\dfrac{16}{81}} = -\sqrt{\dfrac{4^2}{9^2}} = -\sqrt{\left(\dfrac{4}{9}\right)^2} = -\dfrac{4}{9}$

99. $\sqrt{2\dfrac{1}{4}} = \sqrt{\dfrac{9}{4}} = \sqrt{\dfrac{3^2}{2^2}} = \sqrt{\left(\dfrac{3}{2}\right)^2} = \dfrac{3}{2}$

101. Strategy

Length: $8+\sqrt{5}$ m

Width: $8-\sqrt{5}$ m

Use the equation for the area of a rectangle.

Solution

$A = LW$

$\quad = (8+\sqrt{5})(8-\sqrt{5})$

$\quad = 64-8\sqrt{5}+8\sqrt{5}-\sqrt{25}$

$\quad = 64-5$

$\quad = 59$

The area is 59 m^2.

103. a. True

b. True

c. False;

$\left(\sqrt{x}+1\right)^2 = \left(\sqrt{x}+1\right)\left(\sqrt{x}+1\right) = x+2\sqrt{x}+1$

d. True

105. $\begin{array}{c|c} \sqrt{x}-\sqrt{x+9} = 1 \\ \hline \sqrt{16}-\sqrt{16+9} & 1 \\ \sqrt{16}-\sqrt{25} & 1 \\ 4-5 & 1 \\ -1 \ne 1 \end{array}$

No, 16 is not a solution of the equation.

Section 10.4

Concept Review 10.4

1. Sometimes true
   A radical equation must have a variable under a radical sign.

3. Always true

5. Sometimes true
   Squaring both sides of an equation may introduce a solution that is not a solution of the original equation.

Objective 10.4.1 Exercises

1. a. $8 = \sqrt{5}+x$ is not a radical equation because there is no variable under the radical sign.

b. $\sqrt{x-7} = 9$ is a radical equation.

c. $\sqrt{x}+4 = 6$ is a radical equation.

d. $12 = \sqrt{3}x$ is not a radical equation because there is no variable under the radical sign.

3. $\sqrt{x} = 5$ $\qquad$ Check: $\begin{array}{c|c} \sqrt{x} = 5 \\ \hline \sqrt{25} & 5 \\ \sqrt{5^2} & 5 \\ & 5 = 5 \end{array}$

$\left(\sqrt{x}\right)^2 = (5)^2$

$x = 25$

The solution is 25.

5. $\sqrt{a} = 12$ $\qquad$ Check: $\begin{array}{c|c} \sqrt{a} = 12 \\ \hline \sqrt{144} & 12 \\ \sqrt{2^4 \cdot 3^2} & 12 \\ 2^2 \cdot 3 & 12 \\ & 12 = 12 \end{array}$

$\left(\sqrt{a}\right)^2 = (12)^2$

$a = 144$

The solution is 144.

7. $\sqrt{5x} = 5$ $\qquad$ Check: $\begin{array}{c|c} \sqrt{5x} = 5 \\ \hline \sqrt{5 \cdot 5} & 5 \\ \sqrt{5^2} & 5 \\ & 5 = 5 \end{array}$

$\left(\sqrt{5x}\right)^2 = (5)^2$

$5x = 25$

$x = 5$

The solution is 5.

9. $\sqrt{4x} = 8$ $\qquad$ Check: $\begin{array}{c|c} \sqrt{4x} = 8 \\ \hline \sqrt{4 \cdot 16} & 8 \\ \sqrt{64} & 8 \\ \sqrt{2^6} & 8 \\ 2^3 & 8 \\ & 8 = 8 \end{array}$

$\left(\sqrt{4x}\right)^2 = (8)^2$

$4x = 64$

$x = 16$

The solution is 16.

11.
$$\sqrt{2x} - 4 = 0$$
$$\sqrt{2x} = 4$$
$$(\sqrt{2x})^2 = (4)^2$$
$$2x = 16$$
$$x = 8$$

Check:
$$\sqrt{2x} - 4 = 0$$

$\sqrt{2 \cdot 8} - 4$	0
$\sqrt{16} - 4$	0
$\sqrt{2^4} - 4$	0
$\sqrt{2^4} - 4$	0
$4 - 4$	0
	$0 = 0$

The solution is 8.

13.
$$\sqrt{4x} + 5 = 2$$
$$\sqrt{4x} = -3$$
$$(\sqrt{4x})^2 = (-3)^2$$
$$4x = 9$$
$$x = \frac{9}{4}$$

Check:
$$\sqrt{4x} + 5 = 2$$

$\sqrt{4 \cdot \dfrac{9}{5}} + 5$	2
$\sqrt{9} + 5$	2
$\sqrt{3^2} + 5$	2
$3 + 5$	2
	$8 \neq 2$

The equation has no solution.

15.
$$\sqrt{3x - 2} = 4$$
$$(\sqrt{3x - 2})^2 = (4)^2$$
$$3x - 2 = 16$$
$$3x = 18$$
$$x = 6$$

Check:
$$\sqrt{3x - 2} = 4$$

$\sqrt{3 \cdot 6 - 2}$	4
$\sqrt{18 - 2}$	4
$\sqrt{16}$	4
$\sqrt{2^4}$	4
2^2	4
	$4 = 4$

The solution is 6.

17.
$$\sqrt{2x + 1} = 7$$
$$(\sqrt{2x + 1})^2 = (7)^2$$
$$2x + 1 = 49$$
$$2x = 48$$
$$x = 24$$

Check:
$$\sqrt{2x + 1} = 7$$

$\sqrt{2 \cdot 24 + 1}$	7
$\sqrt{48 + 1}$	7
$\sqrt{49}$	7
$\sqrt{7^2}$	7
	$7 = 7$

The solution is 24.

19.
$$0 = 2 - \sqrt{3 - x}$$
$$\sqrt{3 - x} = 2$$
$$(\sqrt{3 - x})^2 = (2)^2$$
$$3 - x = 4$$
$$-x = 1$$
$$x = -1$$

Check:
$$0 = 2 - \sqrt{3 - x}$$

0	$2 - \sqrt{3 - (-1)}$
0	$2 - \sqrt{3 + 1}$
0	$2 - \sqrt{4}$
0	$2 - \sqrt{2^2}$
0	$2 - 2$
	$0 = 0$

The solution is −1.

21.
$$\sqrt{5x + 2} = 0$$
$$(\sqrt{5x + 2})^2 = (0)^2$$
$$5x + 2 = 0$$
$$5x = -2$$
$$x = -\frac{2}{5}$$

Check:
$$\sqrt{5x + 2} = 0$$

$\sqrt{5\left(-\dfrac{2}{5}\right) + 2}$	0
$\sqrt{-2 + 2}$	0
$\sqrt{0}$	0
	$0 = 0$

The solution is $-\dfrac{2}{5}$.

23.
$$\sqrt{3x} - 6 = -4$$
$$\sqrt{3x} = 2$$
$$(\sqrt{3x})^2 = (2)^2$$
$$3x = 4$$
$$x = \frac{4}{3}$$

Check:
$$\sqrt{3x} - 6 = -4$$

$\sqrt{3\left(\dfrac{4}{3}\right)} - 6$	−4
$\sqrt{4} - 6$	−4
$\sqrt{2^2} - 6$	−4
$2 - 6$	−4
	$-4 = -4$

The solution is $\dfrac{4}{3}$.

25.
$$0 = \sqrt{3x - 9} - 6$$
$$6 = \sqrt{3x - 9}$$
$$6^2 = (\sqrt{3x - 9})^2$$
$$36 = 3x - 9$$
$$45 = 3x$$
$$15 = x$$

Check:
$$0 = \sqrt{3x - 9} - 6$$

0	$\sqrt{3 \cdot 15 - 9} - 6$
0	$\sqrt{45 - 9} - 6$
0	$\sqrt{36} - 6$
0	$\sqrt{2^2 \cdot 3^2} - 6$
0	$2 \cdot 3 - 6$
0	$6 - 6$
	$0 = 0$

The solution is 15.

27.
$$\sqrt{5x - 1} = \sqrt{3x + 9}$$
$$(\sqrt{5x - 1})^2 = (\sqrt{3x + 9})^2$$
$$5x - 1 = 3x + 9$$
$$2x - 1 = 9$$
$$2x = 10$$
$$x = 5$$

Check:
$$\sqrt{5x - 1} = \sqrt{3x + 9}$$

$\sqrt{5 \cdot 5 - 1}$	$\sqrt{3 \cdot 5 + 9}$
$\sqrt{25 - 1}$	$\sqrt{15 + 9}$
$\sqrt{24}$	$= \sqrt{24}$

The solution is 5.

29.
$$\sqrt{5x - 3} = \sqrt{4x - 2}$$
$$(\sqrt{5x - 3})^2 = (\sqrt{4x - 2})^2$$
$$5x - 3 = 4x - 2$$
$$x - 3 = -2$$
$$x = 1$$

Check:
$$\sqrt{5x - 3} = \sqrt{4x - 2}$$

$\sqrt{5 \cdot 1 - 3}$	$\sqrt{4 \cdot 1 - 2}$
$\sqrt{5 - 3}$	$\sqrt{4 - 2}$
$\sqrt{2}$	$= \sqrt{2}$

The solution is 1.

31.
$$\sqrt{x^2-5x+6}=\sqrt{x^2-8x+9}$$
$$(\sqrt{x^2-5x+6})^2=(\sqrt{x^2-8x+9})^2$$
$$x^2-5x+6=x^2-8x+9$$
$$-5x+6=-8x+9$$
$$3x+6=9$$
$$3x=3$$
$$x=1$$

Check:
$$\sqrt{x^2-5x+6}=\sqrt{x^2-8x+9}$$
$$\sqrt{1^2-5\cdot1+6}\,\bigg|\,\sqrt{1^2-8\cdot1+9}$$
$$\sqrt{1-5+6}\,\bigg|\,\sqrt{1-8+9}$$
$$\sqrt{2}=\sqrt{2}$$

The solution is 1.

33.
$$\sqrt{x}=\sqrt{x+3}-1$$
$$\sqrt{x}+1=\sqrt{x+3}$$
$$(\sqrt{x}+1)^2=(\sqrt{x+3})^2$$
$$x+2\sqrt{x}+1=x+3$$
$$2\sqrt{x}=2$$
$$(2\sqrt{x})^2=2^2$$
$$4x=4$$
$$x=1$$

Check:
$$\sqrt{x}=\sqrt{x+3}-1$$
$$\sqrt{1}\,\bigg|\,\sqrt{1+3}-1$$
$$\sqrt{1}\,\bigg|\,\sqrt{4}-1$$
$$1\,\bigg|\,2-1$$
$$1=1$$

The solution is 1.

35.
$$\sqrt{2x+5}=5-\sqrt{2x}$$
$$(\sqrt{2x+5})^2=(5-\sqrt{2x})^2$$
$$2x+5=25-10\sqrt{2x}+2x$$
$$-20=-10\sqrt{2x}$$
$$2=\sqrt{2x}$$
$$2^2=(\sqrt{2x})^2$$
$$4=2x$$
$$2=x$$

Check:
$$\sqrt{2x+5}=5-\sqrt{2x}$$
$$\sqrt{2(2)+5}\,\bigg|\,5-\sqrt{2\cdot2}$$
$$\sqrt{4+5}\,\bigg|\,5-\sqrt{4}$$
$$\sqrt{9}\,\bigg|\,5-2$$
$$3=3$$

The solution is 2.

37.
$$\sqrt{3x}-\sqrt{3x+7}=1$$
$$-\sqrt{3x+7}=1-\sqrt{3x}$$
$$(-\sqrt{3x+7})^2=(1-\sqrt{3x})^2$$
$$3x+7=1-2\sqrt{3x}+3x$$
$$6=-2\sqrt{3x}$$
$$-3=\sqrt{3x}$$
$$(-3)^2=(\sqrt{3x})^2$$
$$9=3x$$
$$3=x$$

Check:
$$\sqrt{3x}-\sqrt{3x+7}=1$$
$$\sqrt{3(3)}-\sqrt{3(3)+7}\,\bigg|\,1$$
$$\sqrt{9}-\sqrt{9+7}\,\bigg|\,1$$
$$\sqrt{9}-\sqrt{16}\,\bigg|\,1$$
$$3-4\,\bigg|\,1$$
$$-1\neq1$$

The equation has no solution.

Objective 10.4.2 Exercises

39. Strategy

The unknown number: x

The square root of the product of four and the number: $\sqrt{4x}$

Solution
$$\sqrt{4x}+5=7$$
$$\sqrt{4x}=2$$
$$(\sqrt{4x})^2=2^2$$
$$4x=4$$
$$x=1$$

The number is 1.

41. Strategy

The unknown number: x

The square root of the sum of the number and five: $\sqrt{x+5}$

Solution

$$\sqrt{x+5}+2=6$$
$$\sqrt{x+5}=4$$
$$(\sqrt{x+5})^2=4^2$$
$$x+5=16$$
$$x=11$$

The number is 11.

45. Strategy

To find the length of the hypotenuse, solve the Pythagorean Theorem for the hypotenuse.

Solution

$$c=\sqrt{a^2+b^2}$$
$$c=\sqrt{5^2+9^2}$$
$$c=\sqrt{25+81}$$
$$c=\sqrt{106}$$
$$c\approx 10.30$$

The length of the hypotenuse is 10.30 cm.

47. Strategy

To find the length of the other leg, solve the Pythagorean Theorem for the leg.

Solution

$$a=\sqrt{c^2-b^2}$$
$$a=\sqrt{12^2-7^2}$$
$$a=\sqrt{144-49}$$
$$a=\sqrt{95}$$
$$a\approx 9.75$$

The length of the other leg is 9.75 ft.

49. Strategy

To find the length of the diagonal, solve the Pythagorean Theorem for the hypotenuse.

Solution

$$c=\sqrt{a^2+b^2}$$
$$c=\sqrt{5^2+11^2}$$
$$c=\sqrt{25+121}$$
$$c=\sqrt{146}$$
$$c\approx 12.1$$

The length of the diagonal is 12.1 mi.

51. Strategy

Solve the given formula for H, using $C=20$.

Solution

$$C=\sqrt{32H}$$
$$20=\sqrt{32H}$$
$$20^2=(\sqrt{32H})^2$$
$$400=32H$$
$$\frac{400}{32}=\frac{32H}{32}$$
$$12.5=H$$

The depth of the water is 12.5 ft.

53. Strategy

To find the length of the diagonal between home plate and second base, use the equation $c=\sqrt{a^2+b^2}$ and compare $\frac{1}{2}c$ with 46.

Solution

$$c=\sqrt{a^2+b^2}$$
$$c=\sqrt{60^2+60^2}$$
$$c=\sqrt{3600+3600}$$
$$c=\sqrt{7200}$$
$$c\approx 84.85$$
$$\frac{1}{2}c\approx 42.43<46$$

The pitcher's mound is more than halfway from home plate to second base.

55. Strategy

To find the height above the water, replace d in the equation with the given value and solve for h.

Solution

$$\sqrt{1.5h}=d$$
$$\sqrt{1.5h}=4$$
$$(\sqrt{1.5h})^2=4^2$$
$$1.5h=16$$
$$h=\frac{16}{1.5}$$
$$h\approx 10.67$$

The height of the periscope must be 10.67 ft above the water.

57. Strategy

Solve the Pythagorean Theorem for the hypotenuse. Compare this value to 25.

Solution

$$c=\sqrt{a^2+b^2}$$
$$c=\sqrt{6^2+24^2}$$
$$c=\sqrt{36+576}$$
$$c=\sqrt{612}$$
$$c\approx 24.7$$

A 25-foot ladder will be long enough.

59. Strategy
To find the length of the pendulum, replace T in the equation with the given value and solve for L.

Solution

$$T = 2\pi\sqrt{\dfrac{L}{32}}$$

$$3 = 2\pi\sqrt{\dfrac{L}{32}}$$

$$\dfrac{3}{2\pi} = \sqrt{\dfrac{L}{32}}$$

$$\left(\dfrac{3}{2\pi}\right)^2 = \left(\sqrt{\dfrac{L}{32}}\right)^2$$

$$\dfrac{(3)^2}{(2\pi)^2} = \dfrac{L}{32}$$

$$\dfrac{32(9)}{4\pi^2} = L$$

$$7.30 \approx L$$

The length of the pendulum is 7.30 ft.

61. Strategy
To find the distance, use the equation $a = \sqrt{c^2 - b^2}$, where $c = 650$ and $b = 600$.

Solution

$$a = \sqrt{c^2 - b^2}$$

$$a = \sqrt{650^2 - 600^2}$$

$$a = \sqrt{422,500 - 360,000}$$

$$a = \sqrt{62,500}$$

$$a = 250$$

The distance is 250 ft.

63. Strategy
To find the length of the bridge, replace T in the equation with the given value and solve for d.

Solution

$$T = \sqrt{\dfrac{d}{16}}$$

$$2 = \sqrt{\dfrac{d}{16}}$$

$$(2)^2 = \left(\sqrt{\dfrac{d}{16}}\right)^2$$

$$4 = \dfrac{d}{16}$$

$$16(4) = 16\left(\dfrac{d}{16}\right)$$

$$64 = d$$

The height of the bridge is 64 ft.

65. Strategy
To find the height of the screen, use the equation $a = \sqrt{c^2 - b^2}$, where $c = 36$ and $b = 28.8$.

Solution

$$a = \sqrt{c^2 - b^2}$$

$$a = \sqrt{36^2 - (28.8)^2}$$

$$a = \sqrt{1296 - 829.44}$$

$$a = \sqrt{466.56}$$

$$a = 21.6$$

The height of the screen is 21.6 in.

Applying Concepts 10.4

67.

$$\sqrt{\dfrac{5y + 2}{3}} = 3$$

$$\left(\sqrt{\dfrac{5y + 2}{3}}\right)^2 = 3^2$$

$$\dfrac{5y + 2}{3} = 9$$

$$5y + 2 = 27$$

$$5y = 25$$

$$y = 5$$

Check:

$$\sqrt{\dfrac{5y + 2}{3}} = 3$$

$\sqrt{\dfrac{5(5) + 2}{3}}$	3
$\sqrt{\dfrac{25 + 2}{3}}$	3
$\sqrt{\dfrac{27}{3}}$	3
$\sqrt{9}$	3
	$3 = 3$

The solution is 5.

69.

$$\sqrt{9x^2 + 49} + 1 = 3x + 2$$

$$\sqrt{9x^2 + 49} = 3x + 1$$

$$\left(\sqrt{9x^2 + 49}\right)^2 = (3x + 1)^2$$

$$9x^2 + 49 = 9x^2 + 6x + 1$$

$$48 = 6x$$

$$8 = x$$

Check:

$\sqrt{9x^2 + 49} + 1 = 3x + 2$	
$\sqrt{9(8)^2 + 49} + 1$	$3(8) + 2$
$\sqrt{9(64) + 49} + 1$	$24 + 2$
$\sqrt{576 + 49} + 1$	26
$\sqrt{625} + 1$	26
$25 + 1$	26
$26 = 26$	

The solution is 8.

71. Strategy

Hypotenuse: $5\sqrt{2}$ cm

Leg: $4\sqrt{2}$ cm

Solve the Pythagorean Theorem to find the length of the other leg. Then use the equation for the perimeter of a triangle and the equation for the area of a triangle to find the perimeter and area.

Solution

$a = \sqrt{c^2 - b^2}$

$a = \sqrt{(5\sqrt{2})^2 - (4\sqrt{2})^2}$

$a = \sqrt{50 - 32}$

$a = \sqrt{18}$

$a = 3\sqrt{2}$

a. $P = a + b + c$

$P = 3\sqrt{2} + 4\sqrt{2} + 5\sqrt{2}$

$P = 12\sqrt{2}$

The perimeter is $12\sqrt{2}$ cm.

b. $A = \dfrac{1}{2}bh$

$= \dfrac{1}{2}ab$ for the right triangle

$A = \dfrac{1}{2}(3\sqrt{2})(4\sqrt{2})$

$A = 12$

The area is 12 cm^2.

73. $s = \dfrac{1}{2}(a+b+c) = \dfrac{1}{2}(28+31+37) = \dfrac{1}{2}(96) = 48$

$r = \sqrt{\dfrac{(s-a)(s-b)(s-c)}{s}}$

$= \sqrt{\dfrac{(48-28)(48-31)(48-37)}{48}} = \sqrt{\dfrac{(20)(17)(11)}{48}}$

$= \sqrt{\dfrac{3740}{48}} \approx \sqrt{77.917} \approx 8.827$

Area of fountain $\pi r^2 = \pi(8.827)^2 \approx 244.78$

The area of the fountain is approximately 244.78 ft^2.

Chapter Review Exercises

1. $5\sqrt{3} - 16\sqrt{3} = -11\sqrt{3}$

2. $\dfrac{2x}{\sqrt{3} - \sqrt{5}} = \dfrac{2x}{\sqrt{3} - \sqrt{5}} \cdot \dfrac{\sqrt{3} + \sqrt{5}}{\sqrt{3} + \sqrt{5}}$

$= \dfrac{2x\sqrt{3} + 2x\sqrt{5}}{3 - 5}$

$= \dfrac{2x\sqrt{3} + 2x\sqrt{5}}{-2}$

$= -x\sqrt{3} - x\sqrt{5}$

3. $\sqrt{x^2 + 16x + 64} = \sqrt{(x+8)^2} = x + 8$

4. $\begin{aligned} 3 - \sqrt{7x} &= 5 \\ -\sqrt{7x} &= 2 \\ (-\sqrt{7x})^2 &= 2^2 \\ 7x &= 4 \\ x &= \dfrac{4}{7} \end{aligned}$
Check: $\begin{array}{c|c} 3 - \sqrt{7x} = 5 \\ \hline 3 - \sqrt{7 \cdot \dfrac{4}{7}} & 5 \\ 3 - \sqrt{4} & 5 \\ 3 - 2 & 5 \\ 1 \neq 5 \end{array}$

The equation has no solution.

5. $\dfrac{5\sqrt{y} - 2\sqrt{y}}{3\sqrt{y}} = \dfrac{3\sqrt{y}}{3\sqrt{y}} = 1$

6. $6\sqrt{7} + \sqrt{7} = 7\sqrt{7}$

7. $(6\sqrt{a} + 5\sqrt{b})(2\sqrt{a} + 3\sqrt{b})$

$= 12a + 18\sqrt{ab} + 10\sqrt{ab} + 15b$

$= 12a + 28\sqrt{ab} + 15b$

8. $\sqrt{49(x+3)^4} = \sqrt{7^2(x+3)^4}$

$= 7(x+3)^2$

$= 7(x^2 + 6x + 9)$

$= 7x^2 + 42x + 63$

9. $2\sqrt{36} = 2\sqrt{2^2 \cdot 3^2} = 2 \cdot 2 \cdot 3 = 12$

10. $\begin{aligned} \sqrt{b} &= 4 \\ (\sqrt{b})^2 &= 4^2 \\ b &= 16 \end{aligned}$
Check: $\begin{array}{c|c} \sqrt{b} = 4 \\ \hline \sqrt{16} & 4 \\ 4 = 4 \end{array}$

The solution is 16.

11. $9x\sqrt{5} - 5x\sqrt{5} = 4x\sqrt{5}$

12. $(\sqrt{5ab} - \sqrt{7})(\sqrt{5ab} + \sqrt{7}) = (\sqrt{5ab})^2 - (\sqrt{7})^2$

$= 5ab - 7$

13. $\sqrt{2x+9}=\sqrt{8x-9}$

   $(\sqrt{2x+9})^2=(\sqrt{8x-9})^2$

   $2x+9=8x-9$

   $-6x+9=-9$

   $-6x=-18$

   $x=3$

   The solution is 3.

   Check: $\sqrt{2x+9}=\sqrt{8x-9}$

   $\dfrac{\sqrt{2(3)+9}}{\sqrt{6+9}} \Big| \dfrac{\sqrt{8(3)-9}}{\sqrt{24-9}}$

   $\sqrt{15}=\sqrt{15}$

14. $\sqrt{35}=\sqrt{5\cdot7}=\sqrt{35}$

15. $2x\sqrt{60x^3y^3}+3x^2y\sqrt{15xy}$

   $=2x\sqrt{2^2x^2y^2}\sqrt{15xy}+3x^2y\sqrt{15xy}$

   $=2x\cdot2xy\sqrt{15xy}+3x^2y\sqrt{15xy}$

   $=4x^2y\sqrt{15xy}+3x^2y\sqrt{15xy}$

   $=7x^2y\sqrt{15xy}$

16. $\dfrac{\sqrt{3x^3y}}{\sqrt{27xy^5}}=\sqrt{\dfrac{3x^3y}{27xy^5}}=\sqrt{\dfrac{x^2}{9y^4}}$

   $=\sqrt{\dfrac{x^2}{3^2y^4}}$

   $=\dfrac{x}{3y^2}$

17. $(3\sqrt{x}-\sqrt{y})^2=(3\sqrt{x}-\sqrt{y})(3\sqrt{x}-\sqrt{y})$

   $=9x-3\sqrt{xy}-3\sqrt{xy}+y$

   $=9x-6\sqrt{xy}+y$

18. $\sqrt{(a+4)^2}=a+4$

19. $5\sqrt{48}=5\sqrt{2^4\cdot3}=5\sqrt{2^4}\sqrt{3}=5\cdot2^2\sqrt{3}=20\sqrt{3}$

20. $3\sqrt{12x}+5\sqrt{48x}=3\sqrt{2^2\cdot3x}+5\sqrt{2^4\cdot3x}$

   $=3\sqrt{2^2}\sqrt{3x}+5\sqrt{2^4}\sqrt{3x}$

   $=3\cdot2\sqrt{3x}+5\cdot2^2\sqrt{3x}$

   $=6\sqrt{3x}+20\sqrt{3x}$

   $=26\sqrt{3x}$

21. $\dfrac{8}{\sqrt{x}-3}=\dfrac{8}{\sqrt{x}-3}\cdot\dfrac{\sqrt{x}+3}{\sqrt{x}+3}=\dfrac{8\sqrt{x}+24}{x-9}$

22. $\sqrt{6a}(\sqrt{3a}+\sqrt{2a})=\sqrt{18a^2}+\sqrt{12a^2}$

   $=\sqrt{2\cdot3^2a^2}+\sqrt{3\cdot2^2a^2}$

   $=\sqrt{3^2a^2}\sqrt{2}+\sqrt{2^2a^2}\sqrt{3}$

   $=3a\sqrt{2}+2a\sqrt{3}$

23. $3\sqrt{18a^5b}=3\sqrt{2\cdot3^2a^5b}=3\sqrt{3^2a^4\cdot2ab}$

   $=3\sqrt{3^2a^4}\sqrt{2ab}=3\cdot3a^2\sqrt{2ab}$

   $=9a^2\sqrt{2ab}$

24. $-3\sqrt{120}=-3\sqrt{2^3\cdot3\cdot5}=-3\sqrt{2^2}\sqrt{2\cdot3\cdot5}$

   $=-3\cdot2\sqrt{30}=-6\sqrt{30}$

25. $\sqrt{20a^5b^9}-2ab^2\sqrt{45a^3b^5}$

   $=\sqrt{2^2\cdot5a^5b^9}-2ab^2\sqrt{3^2\cdot5a^3b^5}$

   $=\sqrt{2^2a^4b^8(5ab)}-2ab^2\sqrt{3^2a^2b^4(5ab)}$

   $=\sqrt{2^2a^4b^8}\sqrt{5ab}-2ab^2\sqrt{3^2a^2b^4}\sqrt{5ab}$

   $=2a^2b^4\sqrt{5ab}-2ab^2\cdot3ab^2\sqrt{5ab}$

   $=2a^2b^4\sqrt{5ab}-6a^2b^4\sqrt{5ab}$

   $=-4a^2b^4\sqrt{5ab}$

26. $\dfrac{\sqrt{98x^7y^9}}{\sqrt{2x^3y}}=\sqrt{\dfrac{98x^7y^9}{2x^3y}}=\sqrt{49x^4y^8}=\sqrt{7^2x^4y^8}=7x^2y^4$

27. $\sqrt{5x+1}=\sqrt{20x-8}$

   $(\sqrt{5x+1})^2=(\sqrt{20x-8})^2$

   $5x+1=20x-8$

   $-15x+1=-8$

   $-15x=-9$

   $x=\dfrac{-9}{-15}$

   $x=\dfrac{3}{5}$

   The solution is $\dfrac{3}{5}$.

   Check: $\sqrt{5x+1}=\sqrt{20x-8}$

   $\dfrac{\sqrt{5\cdot\frac{3}{5}+1}}{\sqrt{3+1}} \Big| \dfrac{\sqrt{20\cdot\frac{3}{5}-8}}{\sqrt{12-8}}$

   $\sqrt{4} \Big| \sqrt{4}$

   $2=2$

28. $\sqrt{c^{18}} = c^9$

29. $\sqrt{450} = \sqrt{2 \cdot 3^2 \cdot 5^2} = \sqrt{3^2 \cdot 5^2}\sqrt{2} = 3 \cdot 5\sqrt{2} = 15\sqrt{2}$

30. $6a\sqrt{80b} - \sqrt{180a^2 b} + 5a\sqrt{b}$

$= 6a\sqrt{2^4 \cdot 5b} - \sqrt{2^2 \cdot 3^2 \cdot 5a^2 b} + 5a\sqrt{b}$

$= 6a\sqrt{2^4 (5b)} - \sqrt{2^2 \cdot 3^2 a^2 (5b)} + 5a\sqrt{b}$

$= 6a\sqrt{2^4}\sqrt{5b} - \sqrt{2^2 \cdot 3^2 a^2}\sqrt{5b} + 5a\sqrt{b}$

$= 6a \cdot 2^2 \sqrt{5b} - 2 \cdot 3a\sqrt{5b} + 5a\sqrt{b}$

$= 24a\sqrt{5b} - 6a\sqrt{5b} + 5a\sqrt{b}$

$= 18a\sqrt{5b} + 5a\sqrt{b}$

31. $\dfrac{16}{\sqrt{a}} = \dfrac{16}{\sqrt{a}} \cdot \dfrac{\sqrt{a}}{\sqrt{a}} = \dfrac{16\sqrt{a}}{a}$

32.
$$6 - \sqrt{2y} = 2$$
$$-\sqrt{2y} = -4$$
$$\sqrt{2y} = 4$$
$$(\sqrt{2y})^2 = 4^2$$
$$2y = 16$$
$$y = 8$$

Check:
$$\begin{array}{c|c} 6 - \sqrt{2y} = 2 & \\ 6 - \sqrt{2 \cdot 8} & 2 \\ 6 - \sqrt{16} & 2 \\ 6 - 4 & 2 \\ 2 = 2 & \end{array}$$

The solution is 8.

33. $\sqrt{a^3 b^4 c} \cdot \sqrt{a^7 b^2 c^3} = \sqrt{a^{10} b^6 c^4} = a^5 b^3 c^2$

34. $7\sqrt{630} = 7\sqrt{2 \cdot 3^2 \cdot 5 \cdot 7} = 7\sqrt{3^2}\sqrt{2 \cdot 5 \cdot 7}$

$= 7 \cdot 3\sqrt{70} = 21\sqrt{70}$

35. $y\sqrt{24y^6} = y\sqrt{2^3 \cdot 3y^6} = y\sqrt{2^2 y^6 (2 \cdot 3)}$

$= y\sqrt{2^2 y^6}\sqrt{2 \cdot 3} = y \cdot 2y^3 \sqrt{6} = 2y^4\sqrt{6}$

36. $\dfrac{\sqrt{250}}{\sqrt{10}} = \sqrt{\dfrac{250}{10}} = \sqrt{25} = \sqrt{5^2} = 5$

37.
$$\sqrt{x^2 + 5x + 4} = \sqrt{x^2 + 7x - 6}$$
$$(\sqrt{x^2 + 5x + 4})^2 = (\sqrt{x^2 + 7x - 6})^2$$
$$x^2 + 5x + 4 = x^2 + 7x - 6$$
$$5x + 4 = 7x - 6$$
$$-2x + 4 = -6$$
$$-2x = -10$$
$$x = 5$$

Check:
$$\begin{array}{c|c} \sqrt{x^2 + 5x + 4} = & \sqrt{x^2 + 7x - 6} \\ \sqrt{5^2 + 5 \cdot 5 + 4} & \sqrt{5^2 + 7 \cdot 5 - 6} \\ \sqrt{25 + 25 + 4} & \sqrt{25 + 35 - 6} \\ \sqrt{54} = & \sqrt{54} \end{array}$$

The solution is 5.

38. $(4\sqrt{y} - \sqrt{5})(2\sqrt{y} + 3\sqrt{5})$

$= 8y + 12\sqrt{5y} - 2\sqrt{5y} - 3 \cdot 5$

$= 8y + 10\sqrt{5y} - 15$

39. $\sqrt{9900} = \sqrt{2^2 \cdot 3^2 \cdot 5^2 + 11}$

$= 2 \cdot 3 \cdot 5\sqrt{11}$

$= 30\sqrt{11} \approx 99.499$

40. $\sqrt{7} \cdot \sqrt{7} = \sqrt{7^2} = 7$

41. $5x\sqrt{150x^7} = 5x\sqrt{2\cdot3\cdot5^2x^7} = 5x\sqrt{5^2x^6(2\cdot3x)}$

$\qquad = 5x\sqrt{5^2x^6}\sqrt{2\cdot3x} = 5x\cdot5x^3\sqrt{6x}$

$\qquad = 25x^4\sqrt{6x}$

42. $2x^2\sqrt{18x^2y^5} + 6y\sqrt{2x^6y^3} - 9xy^2\sqrt{8x^4y}$

$\qquad = 2x^2\sqrt{2\cdot3^2x^2y^5} + 6y\sqrt{2x^6y^3} - 9xy^2\sqrt{2^3x^4y}$

$\qquad = 2x^2\sqrt{3^2x^2y^4(2y)} + 6y\sqrt{x^6y^2(2y)} - 9xy^2\sqrt{2^2x^4(2y)}$

$\qquad = 2x^2\sqrt{3^2x^2y^4}\sqrt{2y} + 6y\sqrt{x^6y^2}\sqrt{2y} - 9xy^2\sqrt{2^2x^4}\sqrt{2y}$

$\qquad = 2x^2\cdot3xy^2\sqrt{2y} + 6y\cdot x^3y\sqrt{2y} - 9xy^2\cdot2x^2\sqrt{2y}$

$\qquad = 6x^3y^2\sqrt{2y} + 6x^3y^2\sqrt{2y} - 18x^3y^2\sqrt{2y}$

$\qquad = -6x^3y^2\sqrt{2y}$

43. $\dfrac{\sqrt{54a^3}}{\sqrt{6a}} = \sqrt{\dfrac{54a^3}{6a}} = \sqrt{9a^2} = \sqrt{3^2a^2} = 3a$

44. $4\sqrt{250} = 4\sqrt{2\cdot5^3} = 4\sqrt{5^2(2\cdot5)}$

$\qquad = 4\sqrt{5^2}\sqrt{2\cdot5} = 4\cdot5\sqrt{10} = 20\sqrt{10}$

45.
$\sqrt{5x} = 10$ $\qquad$ Check: $\quad\sqrt{5x} = 10$

$(\sqrt{5x})^2 = 10^2$ $\qquad\qquad\qquad \dfrac{\sqrt{5\cdot20}}{} \Big| 10$

$5x = 100$ $\qquad\qquad\qquad\qquad \sqrt{100}\Big| 10$

$x = 20$ $\qquad\qquad\qquad\qquad\qquad 10 = 10$

The solution is 20.

46. $\dfrac{3a\sqrt{3} + 2\sqrt{12a^2}}{\sqrt{27}} = \dfrac{3a\sqrt{3} + 2\sqrt{2^2\cdot3a^2}}{\sqrt{3^3}}$

$\qquad\qquad = \dfrac{3a\sqrt{3} + 2\cdot2a\sqrt{3}}{\sqrt{3^2\cdot3}} = \dfrac{3a\sqrt{3} + 4a\sqrt{3}}{3\sqrt{3}}$

$\qquad\qquad = \dfrac{7a\sqrt{3}}{3\sqrt{3}} = \dfrac{7a}{3}$

47. $\sqrt{2}\cdot\sqrt{50} = \sqrt{100} = \sqrt{10^2} = 10$

48. $\sqrt{36x^4y^5} = \sqrt{2^2\cdot3^2x^4y^5} = \sqrt{2^2\cdot3^2x^4y^4(y)}$

$\qquad = \sqrt{2^2\cdot3^2x^4y^4}\sqrt{y} = 2\cdot3x^2y^2\sqrt{y} = 6x^2y^2\sqrt{y}$

49. $4y\sqrt{243x^{17}y^9} = 4y\sqrt{3^5x^{17}y^9} = 4y\sqrt{3^4x^{16}y^8\cdot3xy}$

$\qquad = 4y\sqrt{3^4x^{16}y^8}\sqrt{3xy} = 4y\cdot3^2x^8y^4 = 36x^8y^5\sqrt{3xy}$

50.
$\sqrt{x+1} - \sqrt{x-2} = 1$ $\qquad$ Check: $\dfrac{\sqrt{x+1} - \sqrt{x-2}}{} = 1$

$\qquad \sqrt{x+1} = 1 + \sqrt{x-2}$ $\qquad\qquad\qquad \dfrac{\sqrt{3+1} - \sqrt{3-2}}{} \Big| 1$

$\qquad (\sqrt{x+1})^2 = (1+\sqrt{x-2})^2$ $\qquad\qquad\qquad \sqrt{4} - \sqrt{1}\Big| 1$

$\qquad x+1 = 1 + 2\sqrt{x-2} + x - 2$ $\qquad\qquad\qquad 2-1\Big| 1$

$\qquad x+1 = -1 + x + 2\sqrt{x-2}$ $\qquad\qquad\qquad\qquad 1 = 1$

$\qquad 2 = 2\sqrt{x-2}$

$\qquad 1 = \sqrt{x-2}$

$\qquad (1)^2 = (\sqrt{x-2})^2$

$\qquad 1 = x - 2$

$\qquad 3 = x$

The solution is 3.

51. $\sqrt{400} = \sqrt{2^4 \cdot 5^2} = 2^2 \cdot 5 = 20$

52. $-4\sqrt{8x} + 7\sqrt{18x} - 3\sqrt{50x} = -4\sqrt{2^3 x} + 7\sqrt{2 \cdot 3^2 x} - 3\sqrt{2 \cdot 5^2 x}$

$$= -4\sqrt{2^2}\sqrt{2x} + 7\sqrt{3^2}\sqrt{2x} - 3\sqrt{5^2}\sqrt{2x}$$
$$= -4 \cdot 2\sqrt{2x} + 7 \cdot 3\sqrt{2x} - 3 \cdot 5\sqrt{2x}$$
$$= -8\sqrt{2x} + 21\sqrt{2x} - 3 \cdot 5\sqrt{2x}$$
$$= -2\sqrt{2x}$$

53.

$0 = \sqrt{10x+4} - 8$ Check: $0 = \sqrt{10x+4} - 8$

$8 = \sqrt{10x+4}$ $0 \,\big|\, \sqrt{10 \cdot 6 + 4} - 8$

$8^2 = (\sqrt{10x+4})^2$ $0 \,\big|\, \sqrt{60+4} - 8$

$64 = 10x + 4$ $0 \,\big|\, \sqrt{64} - 8$

$60 = 10x$ $0 \,\big|\, 8 - 8$

$6 = x$ $0 = 0$

The solution is 6.

54. $\sqrt{3}(\sqrt{12} - \sqrt{3}) = \sqrt{36} - \sqrt{9}$

$$= \sqrt{2^3 \cdot 3^2} - \sqrt{3^2}$$
$$= 2 \cdot 3 - 3$$
$$= 6 - 3$$
$$= 3$$

55. Strategy

First integer: x
Second consecutive odd integer: $x + 2$
The square root of the sum of two consecutive odd integers: $\sqrt{x + x + 2}$

Solution

$\sqrt{x + x + 2} = 10$

$\sqrt{2x + 2} = 10$

$(\sqrt{2x+2})^2 = 10^2$

$2x + 2 = 100$

$2x = 98$

$x = 49$

$x + 2 = 49 + 2 = 51$

The larger integer is 51.

56. Strategy

To find the explorer's weight, substitute 36 for W_a and 8000 d in the given equation and solve for W_0.

Solution

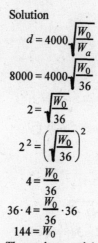

$d = 4000\sqrt{\dfrac{W_0}{W_a}}$

$8000 = 4000\sqrt{\dfrac{W_0}{36}}$

$2 = \sqrt{\dfrac{W_0}{36}}$

$2^2 = \left(\sqrt{\dfrac{W_0}{36}}\right)^2$

$4 = \dfrac{W_0}{36}$

$36 \cdot 4 = \dfrac{W_0}{36} \cdot 36$

$144 = W_0$

The explorer weighs 144 lb on the surface of Earth.

57. Strategy

To find the length of the other leg, solve the Pythagorean Theorem for the unknown leg.

Solution

$a = \sqrt{c^2 - b^2}$

$a = \sqrt{18^2 - 11^2}$

$a = \sqrt{324 - 121}$

$a = \sqrt{203}$

$a \approx 14.25$

The length of the other leg is 14.25 cm.

58. Strategy

To find the radius, substitute 20 for V in the given equation and solve for r.

Solution

$v = 4\sqrt{r}$

$20 = 4\sqrt{r}$

$5 = \sqrt{r}$

$(5)^2 = (\sqrt{r})^2$

$25 = r$

The radius of the corner is 25 ft.

59. Strategy
To find the depth of the water, substitute 30 for V in the given equation and solve for d.

Solution
$$v = 3\sqrt{d}$$
$$30 = 3\sqrt{d}$$
$$10 = \sqrt{d}$$
$$(10)^2 = (\sqrt{d})^2$$
$$100 = d$$
The depth of the water is 100 ft.

60. Strategy
To find the length of the guy wire, solve the Pythagorean Theorem for the unknown of a triangle.

Solution
$$c = \sqrt{a^2 + b^2}$$
$$c = \sqrt{25^2 + 8^2}$$
$$c = \sqrt{625 + 64}$$
$$c = \sqrt{689}$$
$$c \approx 26.25$$
The length of the guy wire is 26.25 ft.

Chapter Test

1. $\sqrt{121x^8 y^2} = \sqrt{11^2 x^8 y^2} = 11x^4 y$

2. $5\sqrt{8} - 3\sqrt{50} = 5\sqrt{2^3} - 3\sqrt{2 \cdot 5^2}$
$$= 5\sqrt{2^2} \cdot \sqrt{2} - 3\sqrt{5^2} \cdot \sqrt{2}$$
$$= 5 \cdot 2\sqrt{2} - 3 \cdot 5\sqrt{2}$$
$$= 10\sqrt{2} - 15\sqrt{2}$$
$$= -5\sqrt{2}$$

3. $\sqrt{3x^2 y}\sqrt{6x^2}\sqrt{2x} = \sqrt{36x^5 y} = \sqrt{2^2 \cdot 3^2 x^5 y}$
$$= \sqrt{2^2 \cdot 3^2 x^4}\sqrt{xy}$$
$$= 2 \cdot 3x^2 \sqrt{xy}$$
$$= 6x^2 \sqrt{xy}$$

4. $\sqrt{45} = \sqrt{3^2 \cdot 5} = \sqrt{3^2} \cdot \sqrt{5} = 3\sqrt{5}$

5. $\sqrt{72x^7 y^2} = \sqrt{2^3 \cdot 3^2 x^7 y^2} = \sqrt{2^2 \cdot 3^2 x^6 y^2} \cdot \sqrt{2x}$
$$= 2 \cdot 3x^3 y\sqrt{2x} = 6x^3 y\sqrt{2x}$$

6. $3\sqrt{8y} - 2\sqrt{72x} + 5\sqrt{18x}$
$$= 3\sqrt{2^3 y} - 2\sqrt{2^3 \cdot 3^2 x} + 5\sqrt{2 \cdot 3^2 x}$$
$$= 3\sqrt{2^2}\sqrt{2y} - 2\sqrt{2^2 \cdot 3^2} \sqrt{2x} + 5\sqrt{3^2} \cdot \sqrt{2x}$$
$$= 3 \cdot 2\sqrt{2y} - 2 \cdot 2 \cdot 3\sqrt{2x} + 5 \cdot 3\sqrt{2x}$$
$$= 6\sqrt{2y} - 12\sqrt{2x} + 15\sqrt{2x}$$
$$= 6\sqrt{2y} + 3\sqrt{2x}$$

7. $(\sqrt{y} + 3)(\sqrt{y} + 5) = \sqrt{y^2} + 5\sqrt{y} + 3\sqrt{y} + 15$
$$= y + 8\sqrt{y} + 15$$

8. $\dfrac{4}{\sqrt{8}} = \dfrac{4}{\sqrt{2^2 \cdot 2}} = \dfrac{4}{2\sqrt{2}} = \dfrac{2}{\sqrt{2}}$
$$= \dfrac{2 \cdot \sqrt{2}}{\sqrt{2} \cdot \sqrt{2}} = \dfrac{2\sqrt{2}}{2} = \sqrt{2}$$

9. $\dfrac{\sqrt{162}}{\sqrt{2}} = \sqrt{\dfrac{162}{2}} = \sqrt{81} = \sqrt{3^4} = 3^2 = 9$

10. $\sqrt{5x - 6} = 7$ Check: $\sqrt{5x - 6} = 7$
$$(\sqrt{5x - 6})^2 = 7^2 \qquad\quad \sqrt{5(11) - 6} \;\big|\; 7$$
$$5x - 6 = 49 \qquad\qquad\quad \sqrt{55 - 6} \;\big|\; 7$$
$$5x = 55 \qquad\qquad\qquad\quad \sqrt{49} \;\big|\; 7$$
$$x = 11 \qquad\qquad\qquad\qquad\quad 7 = 7$$
The solution is 11.

11. $\sqrt{500} \approx 22.361$

12. $\sqrt{32a^5 b^{11}} = \sqrt{2^5 a^5 b^{11}}$
$$= \sqrt{2^4 a^4 b^{10}(2ab)}$$
$$= \sqrt{2^4 a^4 b^{10}}\sqrt{2ab}$$
$$= 2^2 a^2 b^5 \sqrt{2ab}$$
$$= 4a^2 b^5 \sqrt{2ab}$$

13. $\sqrt{a}(\sqrt{a} - \sqrt{b}) = \sqrt{a^2} - \sqrt{ab} = a - \sqrt{ab}$

14. $\sqrt{8x^3 y}\sqrt{10xy^4} = \sqrt{80x^4 y^5} = \sqrt{2^4 \cdot 5x^4 y^5}$
$$= \sqrt{2^4 x^4 y^4}\sqrt{5y} = 2^2 x^2 y^2 \sqrt{5y}$$
$$= 4x^2 y^2 \sqrt{5y}$$

15. $\dfrac{\sqrt{98a^6 b^4}}{\sqrt{2a^3 b^2}} = \sqrt{\dfrac{98a^6 b^4}{2a^3 b^2}} = \sqrt{49a^3 b^2} = \sqrt{7^2 a^3 b^2}$
$$= \sqrt{7^2 a^2 b^2}\sqrt{a} = 7ab\sqrt{a}$$

16. $\sqrt{9x} + 3 = 18$ Check: $\sqrt{9x} + 3 = 18$
$$\sqrt{9x} = 15 \qquad\qquad \sqrt{9(25)} + 3 \;\big|\; 18$$
$$(\sqrt{9x})^2 = 15^2 \qquad\qquad \sqrt{225} + 3 \;\big|\; 18$$
$$9x = 225 \qquad\qquad\qquad 15 + 3 \;\big|\; 18$$
$$x = 25 \qquad\qquad\qquad\qquad 18 = 18$$
The solution is 25.

17. $\sqrt{192x^{13} y^5} = \sqrt{2^6 \cdot 3x^{13} y^5} = \sqrt{2^6 x^{12} y^4 (3xy)}$
$$= \sqrt{2^6 x^{12} y^4}\sqrt{3xy} = 2^3 x^6 y^2 \sqrt{3xy}$$
$$= 8x^6 y^2 \sqrt{3xy}$$

18. $2a\sqrt{2ab^3} + b\sqrt{8a^3 b} - 5ab\sqrt{ab}$
$$= 2a\sqrt{2ab^3} + b\sqrt{2^3 a^3 b} - 5ab\sqrt{ab}$$
$$= 2a\sqrt{b^2}\sqrt{2ab} + b\sqrt{2^2 a^2}\sqrt{2ab} - 5ab\sqrt{ab}$$
$$= 2ab\sqrt{2ab} + 2ab\sqrt{2ab} - 5ab\sqrt{ab}$$
$$= 4ab\sqrt{2ab} - 5ab\sqrt{ab}$$

19. $(\sqrt{a}-2)(\sqrt{a}+2) = \sqrt{a^2}+2\sqrt{a}-2\sqrt{a}-4$
$$= a-4$$

20. $\dfrac{3}{2-\sqrt{5}} = \dfrac{3}{2-\sqrt{5}}\cdot\dfrac{2+\sqrt{5}}{2+\sqrt{5}} = \dfrac{6+3\sqrt{5}}{4-5} = \dfrac{6+3\sqrt{5}}{-1}$$
$$= -6-3\sqrt{5}$$

21.
$$3 = 8-\sqrt{5x}$$
$$-5 = -\sqrt{5x}$$
$$5 = \sqrt{5x}$$
$$5^2 = (\sqrt{5x})^2$$
$$25 = 5x$$
$$5 = x$$
The solution is 5.

Check:
$$3 = 8-\sqrt{5x}$$
$$3 \mid 8-\sqrt{5(5)}$$
$$3 \mid 8-\sqrt{25}$$
$$3 \mid 8-5$$
$$3 = 3$$

22. $\sqrt{3}(\sqrt{6}-\sqrt{x^2}) = \sqrt{18}-\sqrt{3x^2} = \sqrt{2\cdot3^2}-\sqrt{3x^2}$
$$= \sqrt{3^2}\sqrt{2}-\sqrt{x^2}\sqrt{3} = 3\sqrt{2}-x\sqrt{3}$$

23. $3\sqrt{a}-9\sqrt{a} = -6\sqrt{a}$

24. $\sqrt{108} = \sqrt{2^2\cdot3^3}$
$$= \sqrt{2^2\cdot3^2}\sqrt{3}$$
$$= 2\cdot3\sqrt{3} = 6\sqrt{3}$$

25. $\sqrt{63} \approx 7.937$

26. $\dfrac{\sqrt{108a^7b^3}}{\sqrt{3a^4b}} = \sqrt{\dfrac{108a^7b^3}{3a^4b}} = \sqrt{36a^3b^2}$
$$= \sqrt{2^2\cdot3^2a^3b^2} = \sqrt{2^2\cdot3^2a^2b^2}\sqrt{a}$$
$$= 2\cdot3ab\sqrt{a} = 6ab\sqrt{a}$$

27.
$$\sqrt{x}-\sqrt{x+3} = 1$$
$$-\sqrt{x+3} = 1-\sqrt{x}$$
$$(-\sqrt{x+3})^2 = (1-\sqrt{x})^2$$
$$x+3 = 1-2\sqrt{x}+x$$
$$3 = 1-2\sqrt{x}$$
$$2 = -2\sqrt{x}$$
$$-1 = \sqrt{x}$$
$$(-1)^2 = (\sqrt{x})^2$$
$$1 = x$$
The equation has no solution.

Check:
$$\sqrt{x}-\sqrt{x+3} = 1$$
$$\sqrt{1}-\sqrt{1+3} \mid 1$$
$$\sqrt{1}-\sqrt{4} \mid 1$$
$$1-2 \mid 1$$
$$-1 \neq 1$$

28. Strategy
First integer: x
Second consecutive integer: $x+1$
The square root of the sum of two consecutive integers: $\sqrt{x+x+1}$

Solution
$$\sqrt{x+x+1} = 9$$
$$\sqrt{2x+1} = 9$$
$$(\sqrt{2x+1})^2 = 9^2$$
$$2x+1 = 81$$
$$2x = 80$$
$$x = 40$$
The smaller integer is 40

29. **Strategy**
To find the length of the pendulum replace T in the equation with the given value and solve for L.

Solution

$$T = 2\pi\sqrt{\frac{L}{32}}$$

$$2.5 = 2\pi\sqrt{\frac{L}{32}}$$

$$\frac{2.5}{2\pi} = \sqrt{\frac{L}{32}}$$

$$\frac{6.25}{4\pi^2} = \frac{L}{32}$$

$$\frac{32(6.25)}{4\pi^2} = L$$

$$5.07 \approx L$$

The length of the pendulum is 5.07 ft.

30. **Strategy**
To find how high on the building the ladder will reach, use the Pythagorean Theorem. The hypotenuse is the length of the ladder. One leg is the distance from the bottom of the ladder to the building. The other leg is the height above the ground of the point where the ladder touches the building. Solve the Pythagorean Theorem for the leg.

Solution

$$a = \sqrt{c^2 - b^2}$$
$$a = \sqrt{(16)^2 - (5)^2}$$
$$a = \sqrt{256 - 25}$$
$$a = \sqrt{231}$$
$$a \approx 15.2$$

The ladder will reach the building about 15.2 ft above the ground.

Cumulative Review Exercises

1. $\left(\dfrac{2}{3}\right)^2 \cdot \left(\dfrac{3}{4} - \dfrac{3}{2}\right) + \left(\dfrac{1}{2}\right)^2 = \left(\dfrac{2}{3}\right)^2 \cdot \left(\dfrac{3}{4} - \dfrac{6}{4}\right) + \dfrac{1}{4}$

$$= \left(\frac{2}{3}\right)^2 \cdot \left(-\frac{3}{4}\right) + \frac{1}{4}$$

$$= \frac{4}{9} \cdot \left(-\frac{3}{4}\right) + \frac{1}{4}$$

$$= -\frac{1}{3} + \frac{1}{4}$$

$$= -\frac{4}{12} + \frac{3}{12} = -\frac{1}{12}$$

2. $-3[x - 2(3 - 2x) - 5x] + 2x$
$= -3[x - 6 + 4x - 5x] + 2x$
$= -3[-6] + 2x$
$= 2x + 18$

3. $2x - 4[3x - 2(1 - 3x)] = 2(3 - 4x)$
$2x - 4[3x - 2 + 6x] = 6 - 8x$
$2x - 4[9x - 2] = 6 - 8x$
$2x - 36x + 8 = 6 - 8x$
$-34x + 8 = 6 - 8x$
$-34x + 8x + 8 = 6 - 8x + 8x$
$-26x + 8 = 6$
$-26x + 8 - 8 = 6 - 8$
$-26x = -2$
$\dfrac{-26x}{-26} = \dfrac{-2}{-26}$
$x = \dfrac{1}{13}$

The solution is $\dfrac{1}{13}$.

4. $\qquad 3(x - 7) \geq 5x - 12$
$\qquad 3x - 21 \geq 5x - 12$
$3x - 5x - 21 \geq 5x - 5x - 12$
$\qquad -2x - 21 \geq -12$
$-2x - 21 + 21 \geq -12 + 21$
$\qquad -2x \geq 9$
$\qquad \dfrac{-2x}{-2} \leq \dfrac{9}{-2}$
$\qquad x \leq -\dfrac{9}{2}$

5. $(x_1, y_1) = (2, -5)$, $(x_2, y_2) = (-4, 3)$
$m = \dfrac{y_2 - y_1}{x_2 - x_1} = \dfrac{3 - (-5)}{-4 - 2} = \dfrac{8}{-6} = -\dfrac{4}{3}$
The slope of the line is $-\dfrac{4}{3}$.

6. $m = \dfrac{1}{2}$, $(x_1, y_1) = (-2, -3)$
$y - y_1 = m(x - x_1)$
$y - (-3) = \dfrac{1}{2}[x - (-2)]$
$y + 3 = \dfrac{1}{2}(x + 2)$
$y + 3 = \dfrac{1}{2}x + 1$
$y = \dfrac{1}{2}x - 2$

7. $f(x) = \dfrac{5}{2}x - 8$
$f(-4) = \dfrac{5}{2}(-4) - 8$
$f(-4) = -10 - 8$
$f(-4) = -18$
The value of $f(-4) = -18$.

8.

9.
$$2x + y < -2$$
$$2x - 2x + y < -2x - 2$$
$$y < -2x - 2$$

Graph $y = -2x - 2$ as a dotted line. Shade the lower half-plane.

10.

The solution is $(2, -1)$.

11. (1) $\quad 4x - 3y = 1$
(2) $\quad 2x + y = 3$

Solve equation (2) for y.
$$2x + y = 3$$
$$y = 3 - 2x$$

Substitute in equation (1).
$$4x - 3y = 1$$
$$4x - 3(3 - 2x) = 1$$
$$4x - 9 + 6x = 1$$
$$10x - 9 = 1$$
$$10x = 10$$
$$x = 1$$

Substitute x in equation (2).
$$2x + y = 3$$
$$2(1) + y = 3$$
$$2 + y = 3$$
$$y = 1$$

The solution is $(1, 1)$.

12. (1) $\quad 5x + 4y = 7$
(2) $\quad 3x - 2y = 13$

Eliminate y.
$$5x + 4y = 7$$
$$2(3x - 2y) = 2(13)$$

$$5x + 4y = 7$$
$$6x - 4y = 26$$

Add the equations.
$$11x = 33$$
$$x = 3$$

Replace x in equation (1).
$$5x + 4y = 7$$
$$5(3) + 4y = 7$$
$$15 + 4y = 7$$
$$4y = -8$$
$$y = -2$$

The solution is $(3, -2)$.

13. $(-3x^2 y)(-2x^3 y^{-4}) = 6x^5 \cdot y^{-3}$
$$= \frac{6x^5}{y^3}$$

14. $\dfrac{12b^4 - 6b^2 + 2}{-6b^2} = \dfrac{12b^4}{-6b^2} - \dfrac{6b^2}{-6b^2} + \dfrac{2}{-6b^2}$
$$= -2b^2 + 1 - \frac{1}{3b^2}$$

15. $12x^3 y^2 - 9x^2 y^3 = 3x^2 y^2 (4x - 3y)$

16. $9b^2 + 3b - 20 = 9b^2 - 12b + 15b - 20$
$$= (9b^2 - 12b) + (15b - 20)$$
$$= 3b(3b - 4) + 5(3b - 4)$$
$$= (3b - 4)(3b + 5)$$

17. $2a^3 - 16a^2 + 30a = 2a(a^2 - 8a + 15)$
$$= 2a(a - 3)(a - 5)$$

18. $\dfrac{3x^3 - 6x^2}{4x^2 + 4x} \cdot \dfrac{3x - 9}{9x^3 - 45x^2 + 54x}$
$$= \frac{3x^2(x - 2)}{4x(x + 1)} \cdot \frac{3(x - 3)}{9x(x^2 - 5x + 6)}$$
$$= \frac{3x^2(x - 2)}{4x(x + 1)} \cdot \frac{3(x - 3)}{9x(x - 3)(x - 2)}$$
$$= \frac{1}{4(x + 1)}$$

19. $\dfrac{1 - \frac{2}{x} - \frac{15}{x^2}}{1 - \frac{9}{x^2}} = \dfrac{1 - \frac{2}{x} - \frac{15}{x^2}}{1 - \frac{9}{x^2}} \cdot \dfrac{x^2}{x^2}$
$$= \frac{1 \cdot x^2 - \frac{2}{x} \cdot x^2 - \frac{15}{x^2} \cdot x^2}{1 \cdot x^2 - \frac{9}{x^2} \cdot x^2}$$
$$= \frac{x^2 - 2x - 15}{x^2 - 9}$$
$$= \frac{(x + 3)(x - 5)}{(x + 3)(x - 3)} = \frac{x - 5}{x - 3}$$

20. $\dfrac{x + 2}{x - 4} - \dfrac{6}{(x - 4)(x - 3)}$
$$= \frac{x - 3}{x - 3} \cdot \frac{x + 2}{x - 4} - \frac{6}{(x - 4)(x - 3)}$$
$$= \frac{(x - 3)(x + 2) - 6}{(x - 4)(x - 3)}$$
$$= \frac{(x^2 - x - 6) - 6}{(x - 4)(x - 3)}$$
$$= \frac{x^2 - x - 12}{(x - 4)(x - 3)}$$
$$= \frac{(x - 4)(x + 3)}{(x - 4)(x - 3)}$$
$$= \frac{x + 3}{x - 3}$$

21.
$$\frac{x}{2x-5} - 2 = \frac{3x}{2x-5}$$
$$(2x-5)\left(\frac{x}{2x-5} - 2\right) = (2x-5)\left(\frac{3x}{2x-5}\right)$$
$$(2x-5)\cdot\frac{x}{2x-5} - (2x-5)(2) = 3x$$
$$x - (4x-10) = 3x$$
$$x - 4x + 10 = 3x$$
$$-3x + 10 = 3x$$
$$-3x + 3x + 10 = 3x + 3x$$
$$10 = 6x$$
$$\frac{10}{6} = \frac{6x}{6}$$
$$\frac{5}{3} = x$$

The solution is $\frac{5}{3}$.

22.
$$2\sqrt{27a} - 5\sqrt{49a} + 8\sqrt{48a}$$
$$= 2\sqrt{3^3\,a} - 5\sqrt{7^2 a} + 8\sqrt{2^4\cdot 3a}$$
$$= 2\sqrt{3^2}\sqrt{3a} - 5\sqrt{7^2}\sqrt{a} + 8\sqrt{2^4}\sqrt{3a}$$
$$= 2\cdot3\sqrt{3a} - 5\cdot7\sqrt{a} + 8\cdot2^2\sqrt{3a}$$
$$= 6\sqrt{3a} - 35\sqrt{a} + 32\sqrt{3a}$$
$$= 38\sqrt{3a} - 35\sqrt{a}$$

23. $\dfrac{\sqrt{320}}{\sqrt{5}} = \sqrt{\dfrac{320}{5}} = \sqrt{64} = \sqrt{2^6} = 2^3 = 8$

24.
$$\sqrt{2x-3} - 5 = 0$$
$$\sqrt{2x-3} = 5$$
$$(\sqrt{2x-3})^2 = 5^2$$
$$2x - 3 = 25$$
$$2x = 28$$
$$x = 14$$
The solution is 14.

Check:
$$\begin{array}{c|c}
\sqrt{2x-3} - 5 = 0 & \\
\hline
\sqrt{2(14)} - 5 & 0 \\
\sqrt{28-3} - 5 & 0 \\
\sqrt{25} - 5 & 0 \\
5 - 5 & 0 \\
& 0 = 0
\end{array}$$

25. Strategy
To find the integer, write and solve an inequality using N to represent the number.

Solution

Three-eighths of a number	is less than	negative twelve

$$\frac{3}{8}N < -12$$
$$\frac{8}{3}\left(\frac{3}{8}N\right) < \frac{8}{3}(-12)$$
$$N < -32$$

The largest integer that satisfies the inequality is -33.

26. Strategy
Given: $S = 29.40$
$\qquad\quad r = 20\% = 0.20$
Use the equation $S = C + rC$.

Solution
$$S = C + rC$$
$$29.40 = C + 0.20C$$
$$29.40 = 1.20C$$
$$24.50 = C$$
The cost is $24.50.

27. Strategy
Amount of pure water: x

	Amount	Percent	Quantity
Water	x	0.00	$0.00x$
12% solution	40	0.12	$0.12(40)$
5% solution	$x+40$	0.05	$0.05(x+40)$

The sum of the quantities before mixing is equal to the quantity after mixing.

Solution
$$0.00x + 0.12(40) = 0.05(x+40)$$
$$0.12(40) = 0.05(x+40)$$
$$4.8 = 0.05x + 2$$
$$2.8 = 0.05x$$
$$56 = x$$
56 oz of water must be added.

28. Strategy
The smaller number: x
The larger number: $21 - x$

Three more than twice the smaller number	is	The larger number

Solution
$$3 + 2x = 21 - x$$
$$3 + 2x + x = 21 - x + x$$
$$3 + 3x = 21$$
$$-3 + 3 + 3x = 21 - 3$$
$$3x = 18$$
$$x = 6$$
$$21 - x = 21 - 6 = 15$$
The numbers are 6 and 15.

29. Strategy
Time for the large pipe: t
Time for the small pipe: $2t$

	Rate	Time	Part
Small pipe	$2t$	16	$\dfrac{16}{2t}$
Large pipe	t	16	$\dfrac{16}{t}$

The sum of the parts of the task completed by each pipe must equal 1.

Solution
$$\frac{16}{2t}+\frac{16}{t}=1$$
$$2t\left(\frac{16}{2t}+\frac{16}{t}\right)=2t\cdot 1$$
$$16+32=2t$$
$$48=2t$$
$$24=t$$
$$2t=2(24)=48$$

It would take the small pipe, working alone, 48 h.

30. Strategy
First integer: n
Second integer: $n + 1$

Solution

| The square root of the sum of the two consecutive integers | is equal to | 7 |

$$\sqrt{n+n+1}=7$$
$$\sqrt{2n+1}=7$$
$$(\sqrt{2n+1})^{2}=(7)^{2}$$
$$2n+1=49$$
$$2n=48$$
$$n=24$$

The small integer is 24.

Chapter 11: Quadratic Equations

Prep Test

1. $b^2 - 4ac$ [2.1.1]
   $= (-3)^2 - 4(2)(-4)$
   $= 9 + 32$
   $= 41$

2. $5x + 4 = 3$ [3.2.1]
   $5x = -1$
   $x = -\dfrac{1}{5}$

3. $(x + 4)(x - 3)$ [8.2.1]

4. $(2x - 3)^2$ [8.4.1]

5. $x^2 - 10x + 25 = (x - 5)(x - 5)$ [8.4.1]
   $ = (x - 5)^2$
   Yes.

6. $\qquad\dfrac{5}{x - 2} = \dfrac{15}{x}$ [9.5.2]

   $x(x - 2)\left(\dfrac{5}{x - 2}\right) = \left(\dfrac{15}{x}\right)x(x - 2)$
   $5x = 15(x - 2)$
   $5x = 15x - 30$
   $30 = 10x$
   $3 = x$

7. [5.2.2]

8. $\sqrt{28} = \sqrt{4 \cdot 7} = 2\sqrt{7}$ [10.1.1]

9. $|a|$ [10.1.2]

10. Time to walk from beginning to end: t [4.7.1]

	Rate	Time	Distance
First trip	3	$2 - t$	$3(2 - t)$
Second trip	4.5	t	$4.5t$

The distances are the same.

Solution
$3(2 - t) = 4.5t$
$6 - 2t = 4.5t$
$6 = 6.5t$
$0.8 = t$
$4.5t = 3.6 \text{ mi}$
The length of the trail is 3.6 miles..

Go Figure

Multiply each term by its conjugate:

$$\dfrac{1}{\sqrt{1} + \sqrt{2}} \cdot \dfrac{\sqrt{1} - \sqrt{2}}{\sqrt{1} - \sqrt{2}} = \dfrac{\sqrt{1} - \sqrt{2}}{1 - 2} = -\sqrt{1} + \sqrt{2}$$

$$\dfrac{1}{\sqrt{2} + \sqrt{3}} \cdot \dfrac{\sqrt{2} - \sqrt{3}}{\sqrt{2} - \sqrt{3}} = \dfrac{\sqrt{2} - \sqrt{3}}{2 - 3} = -\sqrt{2} + \sqrt{3}$$

$$\vdots$$

$$\dfrac{1}{\sqrt{8} + \sqrt{9}} \cdot \dfrac{\sqrt{8} - \sqrt{9}}{\sqrt{8} - \sqrt{9}} = \dfrac{\sqrt{8} - \sqrt{9}}{8 - 9} = -\sqrt{8} + \sqrt{9}$$

When adding the terms, many are cancelled out:

$-\sqrt{1} + \sqrt{2} - \sqrt{2} + \sqrt{3} - \sqrt{3} + \sqrt{4} \ldots - \sqrt{8} + \sqrt{9}$
$= -\sqrt{1} + \sqrt{9}$
$= -1 + 3$
$= 2$

Section 11.1

Concept Review 11.1

1. Never true
   $2x^2 - 3x + 9 = 0$ is a quadratic equation. A quadratic equation requires an equal sign.

3. Sometimes true
   A quadratic equation may have two distinct real number solutions, a double root, or no real number solutions.

5. Always true

Objective 11.1.1 Exercises

1. $3x^2 - 4x + 1 = 0$
   $a = 3$
   $b = -4$
   $c = 1$

3. $2x^2 - 5 = 0$
   $a = 2$
   $b = 0$
   $c = -5$

5. $6x^2 - 3x = 0$
   $a = 6$
   $b = -3$
   $c = 0$

7. $\quad x^2 - 8 = 3x$
   $x^2 - 3x - 8 = 0$

9. $\quad x^2 = 16$
   $x^2 - 16 = 0$

11. $\qquad 2(x + 3)^2 = 5$
    $2(x^2 + 6x + 9) = 5$
    $2x^2 + 12x + 18 = 5$
    $2x^2 + 12x + 13 = 0$

17. $(x+3)(x-5)=0$

$\quad x+3=0 \quad x-5=0$

$\qquad x=-3 \qquad x=5$

The solutions are –3 and 5.

19. $x(x-7)=0$

$\quad x=0 \quad x-7=0$

$\qquad\qquad x=7$

The solutions are 0 and 7.

21. $(2x+5)(3x-1)=0$

$\quad 2x+5=0 \qquad 3x-1=0$

$\qquad 2x=-5 \qquad 3x=1$

$\qquad x=-\dfrac{5}{2} \qquad x=\dfrac{1}{3}$

The solutions are $-\dfrac{5}{2}$ and $\dfrac{1}{3}$.

23. $x^2+2x-15=0$

$\quad (x-3)(x+5)=0$

$\quad x-3=0 \quad x+5=0$

$\qquad x=3 \qquad x=-5$

The solutions are 3 and –5.

25. $z^2-4z+3=0$

$\quad (z-1)(z-3)=0$

$\quad z-1=0 \quad z-3=0$

$\qquad z=1 \qquad z=3$

The solutions are 1 and 3.

27. $p^2+3p+2=0$

$\quad (p+1)(p+2)=0$

$\quad p+1=0 \quad p+2=0$

$\qquad p=-1 \qquad p=-2$

The solutions are –1 and –2.

29. $x^2-6x+9=0$

$\quad (x-3)(x-3)=0$

$\quad x-3=0 \quad x-3=0$

$\qquad x=3 \qquad x=3$

3 is a double root of the equation. The solution is 3.

31. $6x^2-9x=0$

$\quad 3x(2x-3)=0$

$\quad 3x=0 \quad 2x-3=0$

$\quad x=0 \qquad 2x=3$

$\qquad\qquad x=\dfrac{3}{2}$

The solutions are 0 and $\dfrac{3}{2}$.

33. $r^2-10=3r$

$\quad r^2-3r-10=0$

$\quad (r+2)(r-5)=0$

$\quad r+2=0 \quad r-5=0$

$\qquad r=-2 \qquad r=5$

The solutions are –2 and 5.

35. $3v^2-5v+2=0$

$\quad (3v-2)(v-1)=0$

$\quad 3v-2=0 \quad v-1=0$

$\qquad 3v=2 \qquad v=1$

$\qquad v=\dfrac{2}{3}$

The solutions are $\dfrac{2}{3}$ and 1.

37. $3s^2+8s=3$

$\quad 3s^2+8s-3=0$

$\quad (3s-1)(s+3)=0$

$\quad 3s-1=0 \quad s+3=0$

$\qquad 3s=1 \qquad s=-3$

$\qquad s=\dfrac{1}{3}$

The solutions are $\dfrac{1}{3}$ and –3.

39. $6r^2=12-r$

$\quad 6r^2+r-12=0$

$\quad (2r+3)(3r-4)=0$

$\quad 2r+3=0 \quad 3r-4=0$

$\qquad 2r=-3 \qquad 3r=4$

$\qquad r=-\dfrac{3}{2} \qquad r=\dfrac{4}{3}$

The solutions are $-\dfrac{3}{2}$ and $\dfrac{4}{3}$.

41. $5y^2+11y=12$

$\quad 5y^2+11y-12=0$

$\quad (5y-4)(y+3)=0$

$\quad 5y-4=0 \quad y+3=0$

$\qquad 5y=4 \qquad y=-3$

$\qquad y=\dfrac{4}{5}$

The solutions are $\dfrac{4}{5}$ and –3.

43. $9s^2-6s+1=0$

$\quad (3s-1)(3s-1)=0$

$\quad 3s-1=0 \quad 3s-1=0$

$\qquad 3s=1 \qquad 3s=1$

$\qquad s=\dfrac{1}{3} \qquad s=\dfrac{1}{3}$

$\dfrac{1}{3}$ is a double root of the equation.

The solution is $\dfrac{1}{3}$.

45. $t^2-16=0$

$\quad (t+4)(t-4)=0$

$\quad t+4=0 \quad t-4=0$

$\qquad t=-4 \qquad t=4$

The solutions are –4 and 4.

47.
$$9z^2 - 4 = 0$$
$$(3z - 2)(3z + 2) = 0$$
$$3z - 2 = 0 \quad 3z + 2 = 0$$
$$3z = 2 \quad 3z = -2$$
$$z = \frac{2}{3} \quad z = -\frac{2}{3}$$

The solutions are $\frac{2}{3}$ and $-\frac{2}{3}$.

49.
$$\frac{3x^2}{2} = 4x - 2$$
$$3x^2 = 8x - 4$$
$$3x^2 - 8x + 4 = 0$$
$$(3x - 2)(x - 2) = 0$$
$$3x - 2 = 0 \quad x - 2 = 0$$
$$3x = 2 \quad x = 2$$
$$x = \frac{2}{3}$$

The solutions are $\frac{2}{3}$ and 2.

51.
$$\frac{2x^2}{9} + x = 2$$
$$9\left(\frac{2x^2}{9} + x\right) = 9(2)$$
$$2x^2 + 9x = 18$$
$$2x^2 + 9x - 18 = 0$$
$$(2x - 3)(x + 6) = 0$$
$$2x - 3 = 0 \quad x + 6 = 0$$
$$2x = 3 \quad x = -6$$
$$x = \frac{3}{2}$$

The solutions are $\frac{3}{2}$ and –6.

53.
$$\frac{3}{4}z^2 - z = -\frac{1}{3}$$
$$12\left(\frac{3}{4}z^2 - z\right) = 12\left(-\frac{1}{3}\right)$$
$$9z^2 - 12z = -4$$
$$9z^2 - 12z + 4 = 0$$
$$(3z - 2)(3z - 2) = 0$$
$$3z - 2 = 0 \quad 3z - 2 = 0$$
$$3z = 2 \quad 3z = 2$$
$$z = \frac{2}{3} \quad z = \frac{2}{3}$$

$\frac{2}{3}$ is a double root of the equation.

The solution is $\frac{2}{3}$.

55.
$$p + 18 = p(p - 2)$$
$$p + 18 = p^2 - 2p$$
$$18 = p^2 - 3p$$
$$0 = p^2 - 3p - 18$$
$$0 = (p + 3)(p - 6)$$
$$p + 3 = 0 \quad p - 6 = 0$$
$$p = -3 \quad p = 6$$

The solutions are –3 and 6.

57.
$$s^2 + 5s - 4 = (2s + 1)(s - 4)$$
$$s^2 + 5s - 4 = 2s^2 - 7s - 4$$
$$5s - 4 = s^2 - 7s - 4$$
$$-4 = s^2 - 12s - 4$$
$$0 = s^2 - 12s$$
$$0 = s(s - 12)$$
$$s = 0 \quad s - 12 = 0$$
$$s = 12$$

The solutions are 0 and 12.

Objective 11.1.2 Exercises

59.
$$x^2 = 36$$
$$\sqrt{x^2} = \sqrt{36}$$
$$x = \pm\sqrt{36} = \pm 6$$

The solutions are 6 and –6.

61.
$$v^2 - 1 = 0$$
$$v^2 = 1$$
$$\sqrt{v^2} = \sqrt{1}$$
$$v = \pm\sqrt{1} = \pm 1$$

The solutions are 1 and –1.

63.
$$4x^2 - 49 = 0$$
$$4x^2 = 49$$
$$x^2 = \frac{49}{4}$$
$$\sqrt{x^2} = \sqrt{\frac{49}{4}}$$
$$x = \pm\sqrt{\frac{49}{4}} = \pm\frac{7}{2}$$

The solutions are $\frac{7}{2}$ and $-\frac{7}{2}$.

65.
$$9y^2 = 4$$
$$y^2 = \frac{4}{9}$$
$$\sqrt{y^2} = \sqrt{\frac{4}{9}}$$
$$y = \pm\sqrt{\frac{4}{9}} = \pm\frac{2}{3}$$

The solutions are $\frac{2}{3}$ and $-\frac{2}{3}$.

67. $16v^2 - 9 = 0$

$16v^2 = 9$

$v^2 = \dfrac{9}{16}$

$\sqrt{v^2} = \sqrt{\dfrac{9}{16}}$

$v = \pm\sqrt{\dfrac{9}{16}} = \pm\dfrac{3}{4}$

The solutions are $\dfrac{3}{4}$ and $-\dfrac{3}{4}$.

69. $y^2 + 81 = 0$

$y^2 = -81$

$\sqrt{y^2} = \sqrt{-81}$

$\sqrt{-81}$ is not a real number.

The equation has no real number solution.

71. $w^2 - 24 = 0$

$w^2 = 24$

$\sqrt{w^2} = \sqrt{24}$

$w = \pm\sqrt{24}$

$w = \pm\sqrt{2^3 \cdot 3}$

$w = \pm\sqrt{2^2}\sqrt{2 \cdot 3}$

$w = \pm 2\sqrt{6}$

The solutions are $2\sqrt{6}$ and $-2\sqrt{6}$.

73. $(x-1)^2 = 36$

$\sqrt{(x-1)^2} = \sqrt{36}$

$x - 1 = \pm\sqrt{36} = \pm 6$

$x - 1 = 6 \quad x - 1 = -6$

$x = 7 \qquad x = -5$

The solutions are 7 and -5.

75. $2(x+5)^2 = 8$

$(x+5)^2 = 4$

$\sqrt{(x+5)^2} = \sqrt{4}$

$x + 5 = \pm\sqrt{4} = \pm 2$

$x + 5 = 2 \quad x + 5 = -2$

$x = -3 \qquad x = -7$

The solutions are -3 and -7.

77. $2(x+1)^2 = 50$

$(x+1)^2 = 25$

$\sqrt{(x+1)^2} = \sqrt{25}$

$x + 1 = \pm\sqrt{25} = \pm 5$

$x + 1 = 5 \quad x + 1 = -5$

$x = 4 \qquad x = -6$

The solutions are 4 and -6.

79. $4(x+5)^2 = 64$

$(x+5)^2 = 16$

$\sqrt{(x+5)^2} = \sqrt{16}$

$x + 5 = \pm\sqrt{16} = \pm 4$

$x + 5 = 4 \quad x + 5 = -4$

$x = -1 \qquad x = -9$

The solutions are -1 and -9.

81. $2(x-9)^2 = 98$

$(x-9)^2 = 49$

$\sqrt{(x-9)^2} = \sqrt{49}$

$x - 9 = \pm\sqrt{49} = \pm 7$

$x - 9 = 7 \quad x - 9 = -7$

$x = 16 \qquad x = 2$

The solutions are 16 and 2.

83. $12(x+3)^2 = 27$

$(x+3)^2 = \dfrac{9}{4}$

$\sqrt{(x+3)^2} = \sqrt{\dfrac{9}{4}}$

$x + 3 = \pm\sqrt{\dfrac{9}{4}} = \pm\dfrac{3}{2}$

$x + 3 = \dfrac{3}{2} \qquad x + 3 = -\dfrac{3}{2}$

$x = \dfrac{3}{2} - 3 \qquad x = -\dfrac{3}{2} - 3$

$x = -\dfrac{3}{2} \qquad x = -\dfrac{9}{2}$

The solutions are $-\dfrac{3}{2}$ and $-\dfrac{9}{2}$.

85. $9(x-1)^2 - 16 = 0$

$9(x-1)^2 = 16$

$(x-1)^2 = \dfrac{16}{9}$

$\sqrt{(x-1)^2} = \sqrt{\dfrac{16}{9}}$

$x - 1 = \pm\sqrt{\dfrac{16}{9}}$

$x - 1 = \pm\dfrac{4}{3}$

$x - 1 = \dfrac{4}{3} \qquad x - 1 = -\dfrac{4}{3}$

$x = \dfrac{4}{3} + 1 \qquad x = -\dfrac{4}{3} + 1$

$x = \dfrac{7}{3} \qquad x = -\dfrac{1}{3}$

The solutions are $\dfrac{7}{3}$ and $-\dfrac{1}{3}$.

87. $49(v+1)^2 - 25 = 0$

$49(v+1)^2 = 25$

$(v+1)^2 = \dfrac{25}{49}$

$\sqrt{(v+1)^2} = \sqrt{\dfrac{25}{49}}$

$v+1 = \pm\sqrt{\dfrac{25}{49}}$

$v+1 = \pm\dfrac{5}{7}$

$v+1 = \dfrac{5}{7} \qquad v+1 = -\dfrac{5}{7}$

$v = \dfrac{5}{7} - 1 \qquad v = -\dfrac{5}{7} - 1$

$v = -\dfrac{2}{7} \qquad v = -\dfrac{12}{7}$

The solutions are $-\dfrac{2}{7}$ and $-\dfrac{12}{7}$.

89. $(x-4)^2 - 20 = 0$

$(x-4)^2 = 20$

$\sqrt{(x-4)^2} = \sqrt{20}$

$x-4 = \pm\sqrt{20}$

$x-4 = \pm\sqrt{2^2 \cdot 5}$

$x-4 = \pm\sqrt{2^2}\sqrt{5}$

$x-4 = \pm 2\sqrt{5}$

$x-4 = 2\sqrt{5} \qquad x-4 = -2\sqrt{5}$

$x = 4 + 2\sqrt{5} \qquad x = 4 - 2\sqrt{5}$

The solutions are $4 + 2\sqrt{5}$ and $4 - 2\sqrt{5}$.

91. $(x+1)^2 + 36 = 0$

$(x+1)^2 = -36$

$\sqrt{(x+1)^2} = \sqrt{-36}$

$\sqrt{-36}$ is not a real number.

The equation has no real number solution.

93. $2\left(z - \dfrac{1}{2}\right)^2 = 12$

$\left(z - \dfrac{1}{2}\right)^2 = 6$

$\sqrt{\left(z - \dfrac{1}{2}\right)^2} = \sqrt{6}$

$z - \dfrac{1}{2} = \pm\sqrt{6}$

$z - \dfrac{1}{2} = \sqrt{6} \qquad z - \dfrac{1}{2} = -\sqrt{6}$

$z = \dfrac{1}{2} + \sqrt{6} \qquad z = \dfrac{1}{2} - \sqrt{6}$

The solutions are $\dfrac{1}{2} + \sqrt{6}$ and $\dfrac{1}{2} - \sqrt{6}$.

95. $4\left(x - \dfrac{2}{3}\right)^2 = 16$

$\left(x - \dfrac{2}{3}\right)^2 = 4$

$\sqrt{\left(x - \dfrac{2}{3}\right)^2} = \sqrt{4}$

$x - \dfrac{2}{3} = \pm 2$

$x - \dfrac{2}{3} = 2 \qquad x - \dfrac{2}{3} = -2$

$x = 2 + \dfrac{2}{3} \qquad x = -2 + \dfrac{2}{3}$

$x = \dfrac{8}{3} \qquad x = -\dfrac{4}{3}$

The solutions are $\dfrac{8}{3}$ and $-\dfrac{4}{3}$.

Applying Concepts 11.1

97. $y(y+3) = 28$

$y^2 + 3y = 28$

$y^2 + 3y - 28 = 0$

$(y-4)(y+7) = 0$

$y - 4 = 0 \qquad y + 7 = 0$

$y = 4 \qquad y = -7$

For $y = 4$,

$3y^2 + 5y - 2 = 3(4)^2 + 5(4) - 2$

$= 3(16) + 20 - 2$

$= 48 + 18$

$= 66$

For $y = -7$,

$3y^2 + 5y - 2 = 3(-7)^2 + 5(-7) - 2$

$= 3(49) - 35 - 2$

$= 147 - 37$

$= 110$

The solutions are 66 and 110.

99. $(x^2 + 3)^2 = 25$

$\sqrt{(x^2 + 3)^2} = \sqrt{25}$

$x^2 + 3 = \pm 5$

$x^2 + 3 = 5 \qquad x^2 + 3 = -5$

$x^2 = 5 - 3 \qquad x^2 = -5 - 3$

$x^2 = 2 \qquad x^2 = -8$

$\sqrt{x^2} = \sqrt{2} \qquad \sqrt{x^2} = \sqrt{-8}$

$x = \pm\sqrt{2} \qquad \sqrt{-8}$ is not a real number.

The solutions are $\sqrt{2}$ and $-\sqrt{2}$.

101. $x^2 = x$

$x^2 - x = 0$

$x(x-1) = 0$

$x = 0 \qquad x - 1 = 0$

$x = 1$

The solutions are 0 and 1.

103. $ax^2 - b = 0$

$$ax^2 = b$$

$$x^2 = \frac{b}{a}$$

$$\sqrt{x^2} = \sqrt{\frac{b}{a}}$$

$$x = \pm\sqrt{\frac{b}{a}} = \pm\frac{\sqrt{b}}{\sqrt{a}} \cdot \frac{\sqrt{a}}{\sqrt{a}} = \pm\frac{\sqrt{ab}}{a}$$

The solutions are $\dfrac{\sqrt{ab}}{a}$ and $-\dfrac{\sqrt{ab}}{a}$.

105. $\qquad E = \frac{1}{2}mv^2$

$$250 = \frac{1}{2}(5)v^2$$

$$100 = v^2$$

$$\pm 10 = v$$

The speed is 10 m/s.

Section 11.2

Concept Review 11.2

1. Always true

3. Always true

5. Never true
   For us to complete the square, the leading coefficient must be 1. In this case the leading coefficient is 3.

Objective 11.2.1 Exercises

1. $x^2 + 12x$

$$\left[\frac{1}{2}(12)\right]^2 = 36$$

$$x^2 + 12x + 36$$

$$(x+6)^2$$

3. $x^2 + 10x$

$$\left[\frac{1}{2}(10)\right]^2 = 25$$

$$x^2 + 10x + 25$$

$$(x+5)^2$$

5. $x^2 - x$

$$\left[\frac{1}{2}(-1)\right]^2 = \frac{1}{4}$$

$$x^2 - x + \frac{1}{4}$$

$$\left(x - \frac{1}{2}\right)^2$$

7. $x^2 + 2x - 3 = 0$

$$x^2 + 2x = 3$$

Complete the square.

$$x^2 + 2x + 1 = 3 + 1$$

$$(x+1)^2 = 4$$

$$\sqrt{(x+1)^2} = \sqrt{4}$$

$$x + 1 = \pm 2$$

$$x + 1 = 2 \qquad x + 1 = -2$$

$$x = 1 \qquad\;\; x = -3$$

The solutions are 1 and −3.

9. $v^2 + 4v + 1 = 0$

$$v^2 + 4v = -1$$

Complete the square.

$$v^2 + 4v + 4 = -1 + 4$$

$$(v+2)^2 = 3$$

$$\sqrt{(v+2)^2} = \sqrt{3}$$

$$v + 2 = \pm\sqrt{3}$$

$$v + 2 = \sqrt{3} \qquad v + 2 = -\sqrt{3}$$

$$v = -2 + \sqrt{3} \qquad v = -2 - \sqrt{3}$$

The solutions are $-2 + \sqrt{3}$ and $-2 - \sqrt{3}$.

11. $v^2 - 6v + 13 = 0$

$$v^2 - 6v = -13$$

Complete the square.

$$v^2 - 6v + 9 = -13 + 9$$

$$(v-3)^2 = -4$$

$$\sqrt{(v-3)^2} = \sqrt{-4}$$

$\sqrt{-4}$ is not a real number.
The quadratic equation has no real number solution.

13. $x^2 + 6x = 5$

Complete the square.

$$x^2 + 6x + 9 = 5 + 9$$

$$(x+3)^2 = 14$$

$$\sqrt{(x+3)^2} = \sqrt{14}$$

$$x + 3 = \pm\sqrt{14}$$

$$x + 3 = \sqrt{14} \qquad x + 3 = -\sqrt{14}$$

$$x = -3 + \sqrt{14} \qquad x = -3 - \sqrt{14}$$

The solutions are $-3 + \sqrt{14}$ and $-3 - \sqrt{14}$.

15. $\qquad x^2 = 4x - 4$

$$x^2 - 4x = -4$$

Complete the square.

$$x^2 - 4x + 4 = -4 + 4$$

$$(x-2)^2 = 0$$

$$\sqrt{(x-2)^2} = \sqrt{0}$$

$$x - 2 = 0$$

$$x = 2$$

The solution is 2.

17. $z^2 = 2z + 1$

$z^2 - 2z = 1$

Complete the square.

$z^2 - 2z + 1 = 1 + 1$

$(z-1)^2 = 2$

$\sqrt{(z-1)^2} = \sqrt{2}$

$z - 1 = \pm\sqrt{2}$

$z - 1 = \sqrt{2}$ $z - 1 = -\sqrt{2}$

$z = 1 + \sqrt{2}$ $z = 1 - \sqrt{2}$

The solutions are $1 + \sqrt{2}$ and $1 - \sqrt{2}$.

19. $p^2 + 3p = 1$

Complete the square.

$p^2 + 3p + \dfrac{9}{4} = 1 + \dfrac{9}{4}$

$\left(p + \dfrac{3}{2}\right)^2 = \dfrac{13}{4}$

$\sqrt{\left(p + \dfrac{3}{2}\right)^2} = \sqrt{\dfrac{13}{4}}$

$p + \dfrac{3}{2} = \pm\dfrac{\sqrt{13}}{2}$

$p + \dfrac{3}{2} = \dfrac{\sqrt{13}}{2}$ $p + \dfrac{3}{2} = -\dfrac{\sqrt{13}}{2}$

$p = -\dfrac{3}{2} + \dfrac{\sqrt{13}}{2}$ $p = -\dfrac{3}{2} - \dfrac{\sqrt{13}}{2}$

$p = \dfrac{-3 + \sqrt{13}}{2}$ $p = \dfrac{-3 - \sqrt{13}}{2}$

The solutions are $\dfrac{-3 + \sqrt{13}}{2}$ and $\dfrac{-3 - \sqrt{13}}{2}$.

21. $w^2 + 7w = 8$

Complete the square.

$w^2 + 7w + \dfrac{49}{4} = 8 + \dfrac{49}{4}$

$\left(w + \dfrac{7}{2}\right)^2 = \dfrac{81}{4}$

$\sqrt{\left(w + \dfrac{7}{2}\right)^2} = \sqrt{\dfrac{81}{4}}$

$w + \dfrac{7}{2} = \pm\dfrac{9}{2}$

$w + \dfrac{7}{2} = \dfrac{9}{2}$ $w + \dfrac{7}{2} = -\dfrac{9}{2}$

$w = \dfrac{9}{2} - \dfrac{7}{2}$ $w = -\dfrac{9}{2} - \dfrac{7}{2}$

$w = 1$ $w = -8$

The solutions are 1 and –8.

23. $x^2 + 6x + 4 = 0$

$x^2 + 6x = -4$

Complete the square.

$x^2 + 6x + 9 = -4 + 9$

$(x+3)^2 = 5$

$\sqrt{(x+3)^2} = \sqrt{5}$

$x + 3 = \pm 5$

$x + 3 = \sqrt{5}$ $x + 3 = -\sqrt{5}$

$x = -3 + \sqrt{5}$ $x = -3 - \sqrt{5}$

The solutions are $-3 + \sqrt{5}$ and $-3 - \sqrt{5}$.

25. $r^2 - 8r = -2$

Complete the square.

$r^2 - 8r + 16 = -2 + 16$

$(r-4)^2 = 14$

$\sqrt{(r-4)^2} = \sqrt{14}$

$r - 4 = \pm\sqrt{14}$

$r - 4 = \sqrt{14}$ $r - 4 = -\sqrt{14}$

$r = 4 + \sqrt{14}$ $r = 4 - \sqrt{14}$

The solutions are $4 + \sqrt{14}$ and $4 - \sqrt{14}$.

27. $t^2 - 3t = -2$

Complete the square.

$t^2 - 3t + \dfrac{9}{4} = -2 + \dfrac{9}{4}$

$\left(t - \dfrac{3}{2}\right)^2 = \dfrac{1}{4}$

$\sqrt{\left(t - \dfrac{3}{2}\right)^2} = \sqrt{\dfrac{1}{4}}$

$t - \dfrac{3}{2} = \pm\dfrac{1}{2}$

$t - \dfrac{3}{2} = \dfrac{1}{2}$ $t - \dfrac{3}{2} = -\dfrac{1}{2}$

$t = 2$ $t = 1$

The solutions are 2 and 1.

29. $w^2 = 3w + 5$

$w^2 - 3w = 5$

Complete the square.

$w^2 - 3w + \dfrac{9}{4} = 5 + \dfrac{9}{4}$

$\left(w - \dfrac{3}{2}\right)^2 = \dfrac{29}{4}$

$\sqrt{\left(w - \dfrac{3}{2}\right)^2} = \sqrt{\dfrac{29}{4}}$

$w - \dfrac{3}{2} = \pm\dfrac{\sqrt{29}}{2}$

$w - \dfrac{3}{2} = \dfrac{\sqrt{29}}{2}$ $w - \dfrac{3}{2} = -\dfrac{\sqrt{29}}{2}$

$w = \dfrac{3 + \sqrt{29}}{2}$ $w = \dfrac{3 - \sqrt{29}}{2}$

The solutions are $\dfrac{3 + \sqrt{29}}{2}$ and $\dfrac{3 - \sqrt{29}}{2}$.

31. $x^2 - x - 1 = 0$

$x^2 - x = 1$

Complete the square.

$x^2 - x + \dfrac{1}{4} = 1 + \dfrac{1}{4}$

$\left(x - \dfrac{1}{2}\right)^2 = \dfrac{5}{4}$

$\sqrt{\left(x - \dfrac{1}{2}\right)^2} = \sqrt{\dfrac{5}{4}}$

$x - \dfrac{1}{2} = \pm\dfrac{\sqrt{5}}{2}$

$x - \dfrac{1}{2} = \dfrac{\sqrt{5}}{2}$ $x - \dfrac{1}{2} = -\dfrac{\sqrt{5}}{2}$

$x = \dfrac{1 + \sqrt{5}}{2}$ $x = \dfrac{1 - \sqrt{5}}{2}$

The solutions are $\dfrac{1 + \sqrt{5}}{2}$ and $\dfrac{1 - \sqrt{5}}{2}$.

33. $y^2 - 5y + 3 = 0$

$y^2 - 5y = -3$

Complete the square.

$y^2 - 5y + \dfrac{25}{4} = -3 + \dfrac{25}{4}$

$\left(y - \dfrac{5}{2}\right)^2 = \dfrac{13}{4}$

$\sqrt{\left(y - \dfrac{5}{2}\right)^2} = \sqrt{\dfrac{13}{4}}$

$y - \dfrac{5}{2} = \pm\dfrac{\sqrt{13}}{2}$

$y - \dfrac{5}{2} = \dfrac{\sqrt{13}}{2}$ $y - \dfrac{5}{2} = -\dfrac{\sqrt{13}}{2}$

$y = \dfrac{5 + \sqrt{13}}{2}$ $y = \dfrac{5 - \sqrt{13}}{2}$

The solutions are $\dfrac{5 + \sqrt{13}}{2}$ and $\dfrac{5 - \sqrt{13}}{2}$.

35. $v^2 + v - 3 = 0$

$v^2 + v = 3$

Complete the square.

$v^2 + v + \dfrac{1}{4} = 3 + \dfrac{1}{4}$

$\left(v + \dfrac{1}{2}\right)^2 = \dfrac{13}{4}$

$\sqrt{\left(v + \dfrac{1}{2}\right)^2} = \sqrt{\dfrac{13}{4}}$

$v + \dfrac{1}{2} = \pm\dfrac{\sqrt{13}}{2}$

$v + \dfrac{1}{2} = \dfrac{\sqrt{13}}{2}$ $v + \dfrac{1}{2} = -\dfrac{\sqrt{13}}{2}$

$v = \dfrac{-1 + \sqrt{13}}{2}$ $v = \dfrac{-1 - \sqrt{13}}{2}$

The solutions are $\dfrac{-1 + \sqrt{13}}{2}$ and $\dfrac{-1 - \sqrt{13}}{2}$.

37. $y^2 = 7 - 10y$

$y^2 + 10y = 7$

Complete the square.

$y^2 + 10y + 25 = 7 + 25$

$(y + 5)^2 = 32$

$\sqrt{(y + 5)^2} = \sqrt{32}$

$y + 5 = \pm\sqrt{32}$

$y + 5 = \pm 4\sqrt{2}$

$y + 5 = 4\sqrt{2}$ $y + 5 = -4\sqrt{2}$

$y = -5 + 4\sqrt{2}$ $y = -5 - 4\sqrt{2}$

The solutions are $-5 + 4\sqrt{2}$ and $-5 - 4\sqrt{2}$.

39. $s^2 + 3s = -1$

Complete the square.

$s^2 + 3s + \dfrac{9}{4} = -1 + \dfrac{9}{4}$

$\left(s + \dfrac{3}{2}\right)^2 = \dfrac{5}{4}$

$\sqrt{\left(s + \dfrac{3}{2}\right)^2} = \sqrt{\dfrac{5}{4}}$

$s + \dfrac{3}{2} = \pm\dfrac{\sqrt{5}}{2}$

$s + \dfrac{3}{2} = \dfrac{\sqrt{5}}{2}$ $s + \dfrac{3}{2} = -\dfrac{\sqrt{5}}{2}$

$s = -\dfrac{3}{2} + \dfrac{\sqrt{5}}{2}$ $s = -\dfrac{3}{2} - \dfrac{\sqrt{5}}{2}$

$s = \dfrac{-3 + \sqrt{5}}{2}$ $s = \dfrac{-3 - \sqrt{5}}{2}$

The solutions are $\dfrac{-3 + \sqrt{5}}{2}$ and $\dfrac{-3 - \sqrt{5}}{2}$.

41. $t^2 - t = 4$

Complete the square.

$t^2 - t + \dfrac{1}{4} = 4 + \dfrac{1}{4}$

$\left(t - \dfrac{1}{2}\right)^2 = \dfrac{17}{4}$

$\sqrt{\left(t - \dfrac{1}{2}\right)^2} = \sqrt{\dfrac{17}{4}}$

$t - \dfrac{1}{2} = \pm\dfrac{\sqrt{17}}{2}$

$t - \dfrac{1}{2} = \dfrac{\sqrt{17}}{2}$ $t - \dfrac{1}{2} = -\dfrac{\sqrt{17}}{2}$

$t = \dfrac{1}{2} + \dfrac{\sqrt{17}}{2}$ $t = \dfrac{1}{2} - \dfrac{\sqrt{17}}{2}$

$t = \dfrac{1 + \sqrt{17}}{2}$ $t = \dfrac{1 - \sqrt{17}}{2}$

The solutions are $\dfrac{1 + \sqrt{17}}{2}$ and $\dfrac{1 - \sqrt{17}}{2}$.

43. $x^2 - 3x + 5 = 0$

$x^2 - 3x = -5$

Complete the square.

$x^2 - 3x + \dfrac{9}{4} = -5 + \dfrac{9}{4}$

$\left(x - \dfrac{3}{2}\right)^2 = -\dfrac{11}{4}$

$\sqrt{\left(x - \dfrac{3}{2}\right)^2} = \sqrt{-\dfrac{11}{4}}$

$\sqrt{-\dfrac{11}{4}}$ is not a real number.

The quadratic equation has no real number solution.

45. $2t^2 - 3t + 1 = 0$

$2t^2 - 3t = -1$

$\dfrac{1}{2}(2t^2 - 3t) = \dfrac{1}{2}(-1)$

$t^2 - \dfrac{3}{2}t = -\dfrac{1}{2}$

Complete the square.

$t^2 - \dfrac{3}{2}t + \dfrac{9}{16} = -\dfrac{1}{2} + \dfrac{9}{16}$

$\left(t - \dfrac{3}{4}\right)^2 = \dfrac{1}{16}$

$\sqrt{\left(t - \dfrac{3}{4}\right)^2} = \sqrt{\dfrac{1}{16}}$

$t - \dfrac{3}{4} = \pm\dfrac{1}{4}$

$t - \dfrac{3}{4} = \dfrac{1}{4}$ $t - \dfrac{3}{4} = -\dfrac{1}{4}$

$t = \dfrac{1}{4} + \dfrac{3}{4}$ $t = -\dfrac{1}{4} + \dfrac{3}{4}$

$t = 1$ $t = \dfrac{1}{2}$

The solutions are 1 and $\dfrac{1}{2}$.

47. $2r^2 + 5r = 3$

$\dfrac{1}{2}(2r^2 + 5r) = \dfrac{1}{2}(3)$

$r^2 + \dfrac{5}{2}r = \dfrac{3}{2}$

Complete the square.

$r^2 + \dfrac{5}{2}r + \dfrac{25}{16} = \dfrac{3}{2} + \dfrac{25}{16}$

$\left(r + \dfrac{5}{4}\right)^2 = \dfrac{49}{16}$

$\sqrt{\left(r + \dfrac{5}{4}\right)^2} = \sqrt{\dfrac{49}{16}}$

$r + \dfrac{5}{4} = \pm\dfrac{7}{4}$

$r + \dfrac{5}{4} = \dfrac{7}{4}$ $r + \dfrac{5}{4} = -\dfrac{7}{4}$

$r = \dfrac{7}{4} - \dfrac{5}{4}$ $r = -\dfrac{7}{4} - \dfrac{5}{4}$

$r = \dfrac{1}{2}$ $r = -3$

The solutions are $\dfrac{1}{2}$ and -3.

49. $4v^2 - 4v - 1 = 0$

$4v^2 - 4v = 1$

$\dfrac{1}{4}(4v^2 - 4v) = \dfrac{1}{4}(1)$

$v^2 - v = \dfrac{1}{4}$

Complete the square.

$v^2 - v + \dfrac{1}{4} = \dfrac{1}{4} + \dfrac{1}{4}$

$\left(v - \dfrac{1}{2}\right)^2 = \dfrac{1}{2}$

$\sqrt{\left(v - \dfrac{1}{2}\right)^2} = \sqrt{\dfrac{1}{2}}$

$v - \dfrac{1}{2} = \pm\dfrac{\sqrt{2}}{2}$

$v - \dfrac{1}{2} = \dfrac{\sqrt{2}}{2}$ $v - \dfrac{1}{2} = -\dfrac{\sqrt{2}}{2}$

$v = \dfrac{1}{2} + \dfrac{\sqrt{2}}{2}$ $v = \dfrac{1}{2} - \dfrac{\sqrt{2}}{2}$

$v = \dfrac{1 + \sqrt{2}}{2}$ $v = \dfrac{1 - \sqrt{2}}{2}$

The solutions are $\dfrac{1 + \sqrt{2}}{2}$ and $\dfrac{1 - \sqrt{2}}{2}$.

51.
$$4z^2 - 8z = 1$$
$$\frac{1}{4}(4z^2 - 8z) = \frac{1}{4}(1)$$
$$z^2 - 2z = \frac{1}{4}$$
Complete the square.
$$z^2 - 2z + 1 = \frac{1}{4} + 1$$
$$(z-1)^2 = \frac{5}{4}$$
$$\sqrt{(z-1)^2} = \sqrt{\frac{5}{4}}$$
$$z - 1 = \pm\frac{\sqrt{5}}{2}$$
$$z - 1 = \frac{\sqrt{5}}{2} \qquad z - 1 = -\frac{\sqrt{5}}{2}$$
$$z = 1 + \frac{\sqrt{5}}{2} \qquad z = 1 - \frac{\sqrt{5}}{2}$$
$$z = \frac{2+\sqrt{5}}{2} \qquad z = \frac{2-\sqrt{5}}{2}$$
The solutions are $\dfrac{2+\sqrt{5}}{2}$ and $\dfrac{2-\sqrt{5}}{2}$.

53.
$$3y - 5 = (y-1)(y-2)$$
$$3y - 5 = y^2 - 3y + 2$$
$$-5 = y^2 - 6y + 2$$
$$-7 = y^2 - 6y$$
$$y^2 - 6y = -7$$
Complete the square
$$y^2 - 6y + 9 = -7 + 9$$
$$(y-3)^2 = 2$$
$$\sqrt{(y-3)^2} = \sqrt{2}$$
$$y - 3 = \pm\sqrt{2}$$
$$y - 3 = \sqrt{2} \qquad y - 3 = -\sqrt{2}$$
$$y = 3 + \sqrt{2} \qquad y = 3 - \sqrt{2}$$
The solutions are $3 + \sqrt{2}$ and $3 - \sqrt{2}$.

55.
$$\frac{x^2}{4} - \frac{x}{2} = 3$$
$$4\left(\frac{x^2}{4} - \frac{x}{2}\right) = 4(3)$$
$$x^2 - 2x = 12$$
Complete the square.
$$x^2 - 2x + 1 = 12 + 1$$
$$(x-1)^2 = 13$$
$$\sqrt{(x-1)^2} = \sqrt{13}$$
$$x - 1 = \pm\sqrt{13}$$
$$x - 1 = \sqrt{13} \qquad x - 1 = -\sqrt{13}$$
$$x = 1 + \sqrt{13} \qquad x = 1 - \sqrt{13}$$
The solutions are $1 + \sqrt{13}$ and $1 - \sqrt{13}$.

57.
$$\frac{2x^2}{3} = 2x + 3$$
$$\frac{2x^2}{3} - 2x = 3$$
$$3\left(\frac{2x^2}{3} - 2x\right) = 3(3)$$
$$2x^2 - 6x = 9$$
$$\frac{1}{2}(2x^2 - 6x) = \frac{1}{2}(9)$$
$$x^2 - 3x = \frac{9}{2}$$
Complete the square
$$x - 3x + \frac{9}{4} = \frac{9}{2} + \frac{9}{4}$$
$$\left(x - \frac{3}{2}\right)^2 = \frac{27}{4}$$
$$\sqrt{\left(x - \frac{3}{2}\right)^2} = \sqrt{\frac{27}{4}}$$
$$x - \frac{3}{2} = \pm\frac{3\sqrt{3}}{2}$$
$$x - \frac{3}{2} = \frac{3\sqrt{3}}{2} \qquad x - \frac{3}{2} = -\frac{3\sqrt{3}}{2}$$
$$x = \frac{3}{2} + \frac{3\sqrt{3}}{2} \qquad x = \frac{3}{2} - \frac{3\sqrt{3}}{2}$$
$$x = \frac{3+3\sqrt{3}}{2} \qquad x = \frac{3-3\sqrt{3}}{2}$$
The solutions are $\dfrac{3+3\sqrt{3}}{2}$ and $\dfrac{3-3\sqrt{3}}{2}$.

59.
$$\frac{x}{3} + \frac{3}{x} = \frac{8}{3}$$
$$3x\left(\frac{x}{3} + \frac{3}{x}\right) = 3x\left(\frac{8}{3}\right)$$
$$x^2 + 9 = 8x$$
$$x^2 - 8x = -9$$
Complete the square
$$x^2 - 8x + 16 = -9 + 16$$
$$(x-4)^2 = 7$$
$$\sqrt{(x-4)^2} = \sqrt{7}$$
$$x - 4 = \pm\sqrt{7}$$
$$x - 4 = \sqrt{7} \qquad x - 4 = -\sqrt{7}$$
$$x = 4 + \sqrt{7} \qquad x = 4 - \sqrt{7}$$
The solutions are $4 + \sqrt{7}$ and $4 - \sqrt{7}$.

292   **Chapter 11:** *Quadratic Equations*

61. $y^2 + 3y = 5$

Complete the square.

$$y^2 + 3y + \frac{9}{4} = 5 + \frac{9}{4}$$

$$\left(y + \frac{3}{2}\right)^2 = \frac{29}{4}$$

$$\sqrt{\left(y + \frac{3}{2}\right)^2} = \sqrt{\frac{29}{4}}$$

$$y + \frac{3}{2} = \pm\frac{\sqrt{29}}{2}$$

$$y + \frac{3}{2} = \frac{\sqrt{29}}{2} \qquad y + \frac{3}{2} = -\frac{\sqrt{29}}{2}$$

$$y = -\frac{3}{2} + \frac{\sqrt{29}}{2} \qquad y = -\frac{3}{2} - \frac{\sqrt{29}}{2}$$

$$y = \frac{-3 + \sqrt{29}}{2} \qquad y = \frac{-3 - \sqrt{29}}{2}$$

$$y \approx \frac{-3 + 5.3851}{2} \qquad y \approx \frac{-3 - 5.3851}{2}$$

$$y \approx \frac{2.3851}{2} \qquad y \approx \frac{-8.3851}{2}$$

$$y \approx 1.193 \qquad y \approx -4.193$$

The solutions are approximately 1.193 and −4.193.

63. $2z^2 - 3z = 7$

$$\frac{1}{2}(2z^2 - 3z) = \frac{1}{2}(7)$$

$$z^2 - \frac{3}{2}z = \frac{7}{2}$$

Complete the square.

$$z^2 - \frac{3}{2}z + \frac{9}{16} = \frac{7}{2} + \frac{9}{16}$$

$$\left(z - \frac{3}{4}\right)^2 = \frac{65}{16}$$

$$\sqrt{\left(z - \frac{3}{4}\right)^2} = \sqrt{\frac{65}{16}}$$

$$z - \frac{3}{4} = \pm\frac{\sqrt{65}}{4}$$

$$z - \frac{3}{4} = \frac{\sqrt{65}}{4} \qquad z - \frac{3}{4} = -\frac{\sqrt{65}}{4}$$

$$z = \frac{3}{4} + \frac{\sqrt{65}}{4} \qquad z = \frac{3}{4} - \frac{\sqrt{65}}{4}$$

$$z = \frac{3 + \sqrt{65}}{4} \qquad z = \frac{3 - \sqrt{65}}{4}$$

$$z \approx \frac{3 + 8.0622}{4} \qquad z \approx \frac{3 - 8.0622}{4}$$

$$z \approx \frac{11.0622}{4} \qquad z \approx \frac{-5.0622}{4}$$

$$z \approx 2.766 \qquad z \approx -1.266$$

The solutions are approximately 2.766 and −1.266.

65. $4x^2 + 6x - 1 = 0$

$$4x^2 + 6x = 1$$

$$\frac{1}{4}(4x^2 + 6x) = \frac{1}{4}(1)$$

$$x^2 + \frac{3}{2}x = \frac{1}{4}$$

Complete the square.

$$x^2 + \frac{3}{2}x + \frac{9}{16} = \frac{1}{4} + \frac{9}{16}$$

$$\left(x + \frac{3}{4}\right)^2 = \frac{13}{16}$$

$$\sqrt{\left(x + \frac{3}{4}\right)^2} = \sqrt{\frac{13}{16}}$$

$$x + \frac{3}{4} = \pm\frac{\sqrt{13}}{4}$$

$$x + \frac{3}{4} = \frac{\sqrt{13}}{4} \qquad x + \frac{3}{4} = -\frac{\sqrt{13}}{4}$$

$$x = -\frac{3}{4} + \frac{\sqrt{13}}{4} \qquad x = -\frac{3}{4} - \frac{\sqrt{13}}{4}$$

$$x = \frac{-3 + \sqrt{13}}{4} \qquad x = \frac{-3 - \sqrt{13}}{4}$$

$$x \approx \frac{-3 + 3.6056}{4} \qquad x \approx \frac{-3 - 3.6056}{4}$$

$$x \approx \frac{0.6056}{4} \qquad x \approx \frac{-6.6056}{4}$$

$$x \approx 0.151 \qquad x \approx -1.651$$

The solutions are approximately 0.151 and −1.651.

Applying Concepts 11.2

67. $b^2 - 6b + 7 = 0$

$$b^2 - 6b = -7$$

Complete the square.

$$b^2 - 6b + 9 = -7 + 9$$

$$(b - 3)^2 = 2$$

$$\sqrt{(b - 3)^2} = \sqrt{2}$$

$$b - 3 = \pm\sqrt{2}$$

$$b = 3 \pm \sqrt{2}$$

Evaluate $2b^2$.

$$2b^2 = 2\left(3 + \sqrt{2}\right)^2 \qquad 2b^2 = 2\left(3 - \sqrt{2}\right)^2$$

$$= 2\left(9 + 6\sqrt{2} + 2\right) \qquad = 2\left(9 - 6\sqrt{2} + 2\right)$$

$$= 2\left(11 + 6\sqrt{2}\right) \qquad = 2\left(11 - 6\sqrt{2}\right)$$

$$= 22 + 12\sqrt{2} \qquad = 22 - 12\sqrt{2}$$

The solutions are $22 + 12\sqrt{2}$ and $22 - 12\sqrt{2}$.

69. $\sqrt{2x+7} - 4 = x$

$\sqrt{2x+7} = x+4$

$(\sqrt{2x+7})^2 = (x+4)^2$

$2x+7 = x^2 + 8x + 16$

$0 = x^2 + 6x + 9$

$0 = (x+3)(x+3)$

$x+3 = 0$

$x = -3$

-3 is a double root.

The solution is -3.

Check:

$\sqrt{2x+7} - 4 = x$	
$\sqrt{2(-3)+7} - 4$	-3
$\sqrt{-6+7} - 4$	-3
$\sqrt{1} - 4$	-3
$1 - 4$	-3
$-3 = -3$	

-3 checks as a solution.

71. $\dfrac{x-2}{3} + \dfrac{2}{x+2} = 4$

$3(x+2)\left[\dfrac{x-2}{3} + \dfrac{2}{x+2}\right] = 3(x+2)(4)$

$(x-2)(x+2) + 6 = 12(x+2)$

$x^2 - 4 + 6 = 12x + 24$

$x^2 + 2 = 12x + 24$

$x^2 - 12x = 22$

Complete the square.

$x^2 - 12x + 36 = 22 + 36$

$(x-6)^2 = 58$

$\sqrt{(x-6)^2} = \sqrt{58}$

$x - 6 = \pm\sqrt{58}$

$x = 6 \pm \sqrt{58}$

The solutions are $6 + \sqrt{58}$ and $6 - \sqrt{58}$.

Section 11.3

Concept Review 11.3

1. Always true

3. Never true
The equation must have a zero on one side of the equation. $4x^2 - 3x - 9 = 0$ is in standard form.

5. Never true
$\dfrac{6 \pm 2\sqrt{3}}{2} = \dfrac{2(3 \pm \sqrt{3})}{2} = 3 \pm \sqrt{3}$

Objective 11.3.1 Exercises

3. $z^2 + 6z - 7 = 0$

$a = 1,\ b = 6,\ c = -7$

$z = \dfrac{-b \pm \sqrt{b^2 - 4ac}}{2a}$

$z = \dfrac{-6 \pm \sqrt{(6)^2 - 4(1)(-7)}}{2 \cdot 1}$

$= \dfrac{-6 \pm \sqrt{36+28}}{2} = \dfrac{-6 \pm \sqrt{64}}{2} = \dfrac{-6 \pm 8}{2}$

$z = \dfrac{-6+8}{2} \qquad z = \dfrac{-6-8}{2}$

$= \dfrac{2}{2} = 1 \qquad = \dfrac{-14}{2} = -7$

The solutions are 1 and -7.

5. $w^2 = 3w + 18$

$w^2 - 3w - 18 = 0$

$a = 1,\ b = -3,\ c = -18$

$w = \dfrac{-b \pm \sqrt{b^2 - 4ac}}{2a}$

$w = \dfrac{-(-3) \pm \sqrt{(-3)^2 - 4(1)(-18)}}{2 \cdot 1}$

$= \dfrac{3 \pm \sqrt{9+72}}{2} = \dfrac{3 \pm \sqrt{81}}{2} = \dfrac{3 \pm 9}{2}$

$w = \dfrac{3+9}{2} \qquad w = \dfrac{3-9}{2}$

$= \dfrac{12}{2} = 6 \qquad = \dfrac{-6}{2} = -3$

The solutions are 6 and -3.

7. $t^2 - 2t = 5$

$t^2 - 2t - 5 = 0$

$a = 1,\ b = -2,\ c = -5$

$t = \dfrac{-b \pm \sqrt{b^2 - 4ac}}{2a}$

$t = \dfrac{-(-2) \pm \sqrt{(-2)^2 - 4(1)(-5)}}{2 \cdot 1}$

$= \dfrac{2 \pm \sqrt{4+20}}{2} = \dfrac{2 \pm \sqrt{24}}{2} = \dfrac{2 \pm 2\sqrt{6}}{2} = 1 \pm \sqrt{6}$

The solutions are $1 + \sqrt{6}$ and $1 - \sqrt{6}$.

9. $t^2 + 6t - 1 = 0$

$a = 1,\ b = 6,\ c = -1$

$t = \dfrac{-b \pm \sqrt{b^2 - 4ac}}{2a}$

$t = \dfrac{-6 \pm \sqrt{(6)^2 - 4(1)(-1)}}{2 \cdot 1}$

$= \dfrac{-6 \pm \sqrt{36+4}}{2} = \dfrac{-6 \pm \sqrt{40}}{2}$

$= \dfrac{-6 \pm 2\sqrt{10}}{2} = -3 \pm \sqrt{10}$

The solutions are $-3 + \sqrt{10}$ and $-3 - \sqrt{10}$.

11. $w^2 + 3w + 5 = 0$
$a = 1, b = 3, c = 5$

$$w = \frac{-b \pm \sqrt{b^2 - 4ac}}{2a}$$

$$w = \frac{-3 \pm \sqrt{(3)^2 - 4(1)(5)}}{2 \cdot 1}$$

$$= \frac{-3 \pm \sqrt{9 - 20}}{2} = \frac{-3 \pm \sqrt{-11}}{2}$$

$\sqrt{-11}$ is not a real number.
The quadratic equation has no real number solution.

13. $\qquad w^2 = 4w + 9$

$w^2 - 4w - 9 = 0$
$a = 1, b = -4, c = -9$

$$w = \frac{-b \pm \sqrt{b^2 - 4ac}}{2a}$$

$$w = \frac{-(-4) \pm \sqrt{(-4)^2 - 4(1)(-9)}}{2 \cdot 1}$$

$$= \frac{4 \pm \sqrt{16 + 36}}{2} = \frac{4 \pm \sqrt{52}}{2}$$

$$= \frac{4 \pm 2\sqrt{13}}{2} = 2 \pm \sqrt{13}$$

The solutions are $2 + \sqrt{13}$ and $2 - \sqrt{13}$.

15. $p^2 - p = 0$
$a = 1, b = -1, c = 0$

$$p = \frac{-b \pm \sqrt{b^2 - 4ac}}{2a}$$

$$p = \frac{-(-1) \pm \sqrt{(-1)^2 - 4(1)(0)}}{2 \cdot 1}$$

$$= \frac{1 \pm \sqrt{1}}{2} = \frac{1 \pm 1}{2}$$

$$p = \frac{1 + 1}{2} = \frac{2}{2} = 1 \quad p = \frac{1 - 1}{2} = \frac{0}{2} = 0$$

The solutions are 1 and 0.

17. $4t^2 - 4t - 1 = 0$
$a = 4, b = -4, c = -1$

$$t = \frac{-b \pm \sqrt{b^2 - 4ac}}{2a}$$

$$t = \frac{-(-4) \pm \sqrt{(-4)^2 - 4(4)(-1)}}{2 \cdot 4}$$

$$= \frac{4 + \sqrt{16 + 16}}{8} = \frac{4 \pm \sqrt{32}}{8}$$

$$= \frac{4 \pm 4\sqrt{2}}{8} = \frac{1 \pm \sqrt{2}}{2}$$

The solutions are $\frac{1 + \sqrt{2}}{2}$ and $\frac{1 - \sqrt{2}}{2}$.

19. $4t^2 - 9 = 0$
$a = 4, b = 0, c = -9$

$$t = \frac{-b \pm \sqrt{b^2 - 4ac}}{2a}$$

$$t = \frac{-0 \pm \sqrt{(0)^2 - 4(4)(-9)}}{2 \cdot 4}$$

$$= \frac{\pm\sqrt{144}}{8} = \frac{\pm 12}{8} = \pm\frac{3}{2}$$

$$t = \frac{12}{8} = \frac{3}{2} \quad t = \frac{-12}{8} = -\frac{3}{2}$$

The solutions are $\frac{3}{2}$ and $-\frac{3}{2}$.

21. $3x^2 - 6x + 2 = 0$
$a = 3, b = -6, c = 2$

$$x = \frac{-b \pm \sqrt{b^2 - 4ac}}{2a}$$

$$x = \frac{-(-6) \pm \sqrt{(-6)^2 - 4(3)(2)}}{2 \cdot 3}$$

$$= \frac{6 \pm \sqrt{36 - 24}}{6} = \frac{6 \pm \sqrt{12}}{6}$$

$$= \frac{6 \pm 2\sqrt{3}}{6} = \frac{3 \pm \sqrt{3}}{3}$$

The solutions are $\frac{3 + \sqrt{3}}{3}$ and $\frac{3 - \sqrt{3}}{3}$.

23. $\qquad 3t^2 = 2t + 3$

$3t^2 - 2t - 3 = 0$
$a = 3, b = -2, c = -3$

$$t = \frac{-b \pm \sqrt{b^2 - 4ac}}{2a}$$

$$t = \frac{-(-2) \pm \sqrt{(-2)^2 - 4(3)(-3)}}{2 \cdot 3}$$

$$= \frac{2 \pm \sqrt{4 + 36}}{6} = \frac{2 \pm \sqrt{40}}{6}$$

$$= \frac{2 \pm 2\sqrt{10}}{6} = \frac{1 \pm \sqrt{10}}{3}$$

The solutions are $\frac{1 + \sqrt{10}}{3}$ and $\frac{1 - \sqrt{10}}{3}$.

25. $2x^2 + x + 1 = 0$
$a = 2, b = 1, c = 1$

$$x = \frac{-b \pm \sqrt{b^2 - 4ac}}{2a}$$

$$x = \frac{-1 \pm \sqrt{(1)^2 - 4(2)(1)}}{2 \cdot 2}$$

$$= \frac{-1 \pm \sqrt{1 - 8}}{4} = \frac{-1 \pm \sqrt{-7}}{4}$$

$\sqrt{-7}$ is not a real number.
The quadratic equation has no real number solution.

27.
$$2y^2 + 3 = 8y$$
$$2y^2 - 8y + 3 = 0$$
$$a = 2,\ b = -8,\ c = 3$$
$$y = \frac{-b \pm \sqrt{b^2 - 4ac}}{2a}$$
$$y = \frac{-(-8) \pm \sqrt{(-8)^2 - 4(2)(3)}}{2 \cdot 2}$$
$$= \frac{8 \pm \sqrt{64 - 24}}{4} = \frac{8 \pm \sqrt{40}}{4}$$
$$= \frac{8 \pm 2\sqrt{10}}{4} = \frac{4 \pm \sqrt{10}}{2}$$

The solutions are $\dfrac{4 + \sqrt{10}}{2}$ and $\dfrac{4 - \sqrt{10}}{2}$.

29.
$$3t^2 = 7t + 6$$
$$3t^2 - 7t - 6 = 0$$
$$a = 3,\ b = -7,\ c = -6$$
$$t = \frac{-b \pm \sqrt{b^2 - 4ac}}{2a}$$
$$t = \frac{-(-7) \pm \sqrt{(-7)^2 - 4(3)(-6)}}{2 \cdot 3}$$
$$= \frac{7 \pm \sqrt{49 + 72}}{6} = \frac{7 \pm \sqrt{121}}{6} = \frac{7 \pm 11}{6}$$
$$t = \frac{7 + 11}{6} \qquad t = \frac{7 - 11}{6}$$
$$= \frac{18}{6} = 3 \qquad = \frac{-4}{6} = -\frac{2}{3}$$

The solutions are 3 and $-\dfrac{2}{3}$.

31.
$$3y^2 - 4 = 5y$$
$$3y^2 - 5y - 4 = 0$$
$$a = 3,\ b = -5,\ c = -4$$
$$y = \frac{-b \pm \sqrt{b^2 - 4ac}}{2a}$$
$$y = \frac{-(-5) \pm \sqrt{(-5)^2 - 4(3)(-4)}}{2 \cdot 3}$$
$$= \frac{5 \pm \sqrt{25 + 48}}{6} = \frac{5 \pm \sqrt{73}}{6}$$

The solutions are $\dfrac{5 + \sqrt{73}}{6}$ and $\dfrac{5 - \sqrt{73}}{6}$.

33.
$$3x^2 = x + 3$$
$$3x^2 - x - 3 = 0$$
$$a = 3,\ b = -1,\ c = -3$$
$$x = \frac{-b \pm \sqrt{b^2 - 4ac}}{2a}$$
$$x = \frac{-(-1) \pm \sqrt{(-1)^2 - 4(3)(-3)}}{2 \cdot 3}$$
$$= \frac{1 \pm \sqrt{1 + 36}}{6} = \frac{1 \pm \sqrt{37}}{6}$$

The solutions are $\dfrac{1 + \sqrt{37}}{6}$ and $\dfrac{1 - \sqrt{37}}{6}$.

35.
$$5d^2 - 2d - 8 = 0$$
$$a = 5,\ b = -2,\ c = -8$$
$$d = \frac{-b \pm \sqrt{b^2 - 4ac}}{2a}$$
$$d = \frac{-(-2) \pm \sqrt{(-2)^2 - 4(5)(-8)}}{2 \cdot 5}$$
$$= \frac{2 \pm \sqrt{4 + 160}}{10} = \frac{2 \pm \sqrt{164}}{10}$$
$$= \frac{2 \pm 2\sqrt{41}}{10} = \frac{1 \pm \sqrt{41}}{5}$$

The solutions are $\dfrac{1 + \sqrt{41}}{5}$ and $\dfrac{1 - \sqrt{41}}{5}$.

37.
$$5z^2 + 11z = 12$$
$$5z^2 + 11z - 12 = 0$$
$$a = 5,\ b = 11,\ c = -12$$
$$z = \frac{-b \pm \sqrt{b^2 - 4ac}}{2a}$$
$$z = \frac{-11 \pm \sqrt{(11)^2 - 4(5)(-12)}}{2 \cdot 5}$$
$$= \frac{-11 \pm \sqrt{121 + 240}}{10} = \frac{-11 \pm \sqrt{361}}{10} = \frac{-11 \pm 19}{10}$$
$$z = \frac{-11 + 19}{10} \qquad z = \frac{-11 - 19}{10}$$
$$= \frac{8}{10} = \frac{4}{5} \qquad = \frac{-30}{10} = -3$$

The solutions are $\dfrac{4}{5}$ and -3.

39.
$$v^2 + 6v + 1 = 0$$
$$a = 1,\ b = 6,\ c = 1$$
$$v = \frac{-b \pm \sqrt{b^2 - 4ac}}{2a}$$
$$v = \frac{-6 \pm \sqrt{(6)^2 - 4(1)(1)}}{2 \cdot 1}$$
$$= \frac{-6 \pm \sqrt{36 - 4}}{2} = \frac{-6 \pm \sqrt{32}}{2}$$
$$= \frac{-6 \pm 4\sqrt{2}}{2} = -3 \pm 2\sqrt{2}$$

The solutions are $-3 + 2\sqrt{2}$ and $-3 - 2\sqrt{2}$.

41.
$$4t^2 - 12t - 15 = 0$$
$$a = 4,\ b = -12,\ c = -15$$
$$t = \frac{-b \pm \sqrt{b^2 - 4ac}}{2a}$$
$$t = \frac{-(-12) \pm \sqrt{(-12)^2 - 4(4)(-15)}}{2 \cdot 4}$$
$$= \frac{12 \pm \sqrt{144 + 240}}{8} = \frac{12 \pm \sqrt{384}}{8}$$
$$= \frac{12 \pm 8\sqrt{6}}{8} = \frac{3 \pm 2\sqrt{6}}{2}$$

The solutions are $\dfrac{3 + 2\sqrt{6}}{2}$ and $\dfrac{3 - 2\sqrt{6}}{2}$.

43.
$$2x^2 = 4x - 5$$
$$2x^2 - 4x + 5 = 0$$
$$a = 2, b = -4, c = 5$$
$$x = \frac{-b \pm \sqrt{b^2 - 4ac}}{2a}$$
$$x = \frac{-(-4) \pm \sqrt{(-4)^2 - 4(2)(5)}}{2 \cdot 2}$$
$$= \frac{4 \pm \sqrt{16 - 40}}{4} = \frac{4 \pm \sqrt{-24}}{4}$$

$\sqrt{-24}$ is not a real number.
The quadratic equation has no real number solution.

45.
$$9y^2 + 6y - 1 = 0$$
$$a = 9, b = 6, c = -1$$
$$y = \frac{-b \pm \sqrt{b^2 - 4ac}}{2a}$$
$$y = \frac{-6 \pm \sqrt{(6)^2 - 4(9)(-1)}}{2 \cdot 9}$$
$$= \frac{-6 \pm \sqrt{36 + 36}}{18} = \frac{-6 \pm \sqrt{72}}{18}$$
$$= \frac{-6 \pm 6\sqrt{2}}{18} = \frac{-1 \pm \sqrt{2}}{3}$$

The solutions are $\dfrac{-1 + \sqrt{2}}{3}$ and $\dfrac{-1 - \sqrt{2}}{3}$.

47.
$$6s^2 - s - 2 = 0$$
$$a = 6, b = -1, c = -2$$
$$s = \frac{-b \pm \sqrt{b^2 - 4ac}}{2a}$$
$$s = \frac{-(-1) \pm \sqrt{(-1)^2 - 4(6)(-2)}}{2 \cdot 6}$$
$$= \frac{1 \pm \sqrt{1 + 48}}{12} = \frac{1 \pm \sqrt{49}}{12} = \frac{1 \pm 7}{12}$$
$$s = \frac{1 + 7}{12} \qquad s = \frac{1 - 7}{12}$$
$$= \frac{8}{12} = \frac{2}{3} \qquad = \frac{-6}{12} = -\frac{1}{2}$$

The solutions are $\dfrac{2}{3}$ and $-\dfrac{1}{2}$.

49.
$$4p^2 + 16p = -11$$
$$4p^2 + 16p + 11 = 0$$
$$a = 4, b = 16, c = 11$$
$$p = \frac{-b \pm \sqrt{b^2 - 4ac}}{2a}$$
$$p = \frac{-16 \pm \sqrt{(16)^2 - 4(4)(11)}}{2 \cdot 4}$$
$$= \frac{-16 \pm \sqrt{256 - 176}}{8} = \frac{-16 \pm \sqrt{80}}{8}$$
$$= \frac{-16 \pm 4\sqrt{5}}{8} = \frac{-4 \pm \sqrt{5}}{2}$$

The solutions are $\dfrac{-4 + \sqrt{5}}{2}$ and $\dfrac{-4 - \sqrt{5}}{2}$.

51.
$$4x^2 = 4x + 11$$
$$4x^2 - 4x - 11 = 0$$
$$a = 4, b = -4, c = -11$$
$$x = \frac{-b \pm \sqrt{b^2 - 4ac}}{2a}$$
$$x = \frac{-(-4) \pm \sqrt{(-4)^2 - 4(4)(-11)}}{2 \cdot 4}$$
$$= \frac{4 \pm \sqrt{16 + 176}}{8} = \frac{4 \pm \sqrt{192}}{8}$$
$$= \frac{4 \pm 8\sqrt{3}}{8} = \frac{1 \pm 2\sqrt{3}}{2}$$

The solutions are $\dfrac{1 + 2\sqrt{3}}{2}$ and $\dfrac{1 - 2\sqrt{3}}{2}$.

53.
$$4p^2 = -12p - 9$$
$$4p^2 + 12p + 9 = 0$$
$$a = 4, b = 12, c = 9$$
$$p = \frac{-b \pm \sqrt{b^2 - 4ac}}{2a}$$
$$p = \frac{-12 \pm \sqrt{(12)^2 - 4(4)(9)}}{2 \cdot 4}$$
$$= \frac{-12 \pm \sqrt{144 - 144}}{8} = \frac{-12}{8} = -\frac{3}{2}$$

The solution is $-\dfrac{3}{2}$.

55.
$$9v^2 = -30v - 23$$
$$9v^2 + 30v + 23 = 0$$
$$a = 9, b = 30, c = 23$$
$$v = \frac{-b \pm \sqrt{b^2 - 4ac}}{2a}$$
$$v = \frac{-30 \pm \sqrt{(30)^2 - 4(9)(23)}}{2 \cdot 9}$$
$$= \frac{-30 \pm \sqrt{900 - 828}}{18} = \frac{-30 \pm \sqrt{72}}{18}$$
$$= \frac{-30 \pm 6\sqrt{2}}{18} = \frac{-5 \pm \sqrt{2}}{3}$$

The solutions are $\dfrac{-5 + \sqrt{2}}{3}$ and $\dfrac{-5 - \sqrt{2}}{3}$.

57.
$$\frac{x^2}{2} - \frac{x}{3} = 1$$
$$6\left(\frac{x^2}{2} - \frac{x}{3}\right) = 6(1)$$
$$3x^2 - 2x = 6$$
$$3x^2 - 2x - 6 = 0$$
$$a = 3, b = -2, c = -6$$
$$x = \frac{-b \pm \sqrt{b^2 - 4ac}}{2a}$$
$$x = \frac{-(-2) \pm \sqrt{(-2)^2 - 4(3)(-6)}}{2 \cdot 3}$$
$$= \frac{2 \pm \sqrt{4 + 72}}{6} = \frac{2 \pm \sqrt{76}}{6}$$
$$= \frac{2 \pm 2\sqrt{19}}{6} = \frac{1 \pm \sqrt{19}}{3}$$

The solutions are $\dfrac{1 + \sqrt{19}}{3}$ and $\dfrac{1 - \sqrt{19}}{3}$.

59.
$$\frac{2x^2}{5} = x + 1$$
$$5\left(\frac{2x^2}{5}\right) = 5(x + 1)$$
$$2x^2 = 5x + 5$$
$$2x^2 - 5x - 5 = 0$$
$$a = 2, b = -5, c = -5$$
$$x = \frac{-b \pm \sqrt{b^2 - 4ac}}{2a}$$
$$x = \frac{-(-5) \pm \sqrt{(-5)^2 - 4(2)(-5)}}{2 \cdot 2}$$
$$= \frac{5 \pm \sqrt{25 + 40}}{4} = \frac{5 \pm \sqrt{65}}{4}$$

The solutions are $\dfrac{5 + \sqrt{65}}{4}$ and $\dfrac{5 - \sqrt{65}}{4}$.

61.
$$\frac{x}{5} + \frac{5}{x} = \frac{12}{5}$$
$$5x\left(\frac{x}{5} + \frac{5}{x}\right) = 5x\left(\frac{12}{5}\right)$$
$$x^2 + 25 = 12x$$
$$x^2 - 12x + 25 = 0$$
$$a = 1, b = -12, c = 25$$
$$x = \frac{-b \pm \sqrt{b^2 - 4ac}}{2a}$$
$$x = \frac{-(-12) \pm \sqrt{(-12)^2 - 4(1)(25)}}{2 \cdot 1}$$
$$= \frac{12 \pm \sqrt{144 - 100}}{2} = \frac{12 \pm \sqrt{44}}{2}$$
$$= \frac{12 \pm 2\sqrt{11}}{2} = 6 \pm \sqrt{11}$$

The solutions are $6 + \sqrt{11}$ and $6 - \sqrt{11}$.

63.
$$x^2 - 2x - 21 = 0$$
$$a = 1, b = -2, c = -21$$
$$x = \frac{-b \pm \sqrt{b^2 - 4ac}}{2a}$$
$$x = \frac{-(-2) \pm \sqrt{(-2)^2 - 4(1)(-21)}}{2 \cdot 1}$$
$$= \frac{2 \pm \sqrt{4 + 84}}{2} = \frac{2 \pm \sqrt{88}}{2}$$
$$= \frac{2 \pm 2\sqrt{22}}{2} = 1 \pm \sqrt{22}$$
$$x = 1 + \sqrt{22} \qquad x = 1 - \sqrt{22}$$
$$\approx 1 + 4.690 \qquad \approx 1 - 4.690$$
$$\approx 5.690 \qquad \approx -3.690$$

The solutions are approximately 5.690 and −3.690.

65.
$$s^2 - 6s - 13 = 0$$
$$a = 1, b = -6, c = -13$$
$$s = \frac{-b \pm \sqrt{b^2 - 4ac}}{2a}$$
$$s = \frac{-(-6) \pm \sqrt{(-6)^2 - 4(1)(-13)}}{2 \cdot 1}$$
$$= \frac{6 \pm \sqrt{36 + 52}}{2} = \frac{6 \pm \sqrt{88}}{2}$$
$$= \frac{6 \pm 2\sqrt{22}}{2} = 3 \pm \sqrt{22}$$
$$s = 3 + \sqrt{22} \qquad s = 3 - \sqrt{22}$$
$$\approx 3 + 4.690 \qquad \approx 3 - 4.690$$
$$\approx 7.690 \qquad \approx -1.690$$

The solutions are approximately 7.690 and −1.690.

67. $2p^2 - 7p - 10 = 0$

$a = 2, b = -7, c = -10$

$p = \dfrac{-b \pm \sqrt{b^2 - 4ac}}{2a}$

$p = \dfrac{-(-7) \pm \sqrt{(-7)^2 - 4(2)(-10)}}{2 \cdot 2}$

$= \dfrac{7 \pm \sqrt{49 + 80}}{4} = \dfrac{7 \pm \sqrt{129}}{4}$

$p = \dfrac{7 + \sqrt{129}}{4} \qquad p = \dfrac{7 - \sqrt{129}}{4}$

$\approx \dfrac{7 + 11.3578}{4} \qquad \approx \dfrac{7 - 11.3578}{4}$

$\approx 4.589 \qquad\quad \approx -1.089$

The solutions are approximately 4.589 and -1.089.

69. $4z^2 + 8z - 1 = 0$

$a = 4, b = 8, c = -1$

$z = \dfrac{-b \pm \sqrt{b^2 - 4ac}}{2a}$

$z = \dfrac{-8 \pm \sqrt{(8)^2 - 4(4)(-1)}}{2 \cdot 4}$

$= \dfrac{-8 \pm \sqrt{64 + 16}}{8} = \dfrac{-8 \pm \sqrt{80}}{8}$

$= \dfrac{-8 \pm 4\sqrt{5}}{8} = \dfrac{-2 \pm \sqrt{5}}{2}$

$z = \dfrac{-2 + \sqrt{5}}{2} \qquad z = \dfrac{-2 - \sqrt{5}}{2}$

$\approx \dfrac{-2 + 2.236}{2} \qquad \approx \dfrac{-2 - 2.236}{2}$

$\approx 0.118 \qquad\quad \approx -2.118$

The solutions are approximately 0.118 and -2.118.

71. $5v^2 - v - 5 = 0$

$a = 5, b = -1, c = -5$

$v = \dfrac{-b \pm \sqrt{b^2 - 4ac}}{2a}$

$v = \dfrac{-(-1) \pm \sqrt{(-1)^2 - 4(5)(-5)}}{2 \cdot 5}$

$= \dfrac{1 \pm \sqrt{1 + 100}}{10} = \dfrac{1 \pm \sqrt{101}}{10}$

$v = \dfrac{1 + \sqrt{101}}{10} \qquad v = \dfrac{1 - \sqrt{101}}{10}$

$\approx \dfrac{1 + 10.0499}{10} \qquad \approx \dfrac{1 - 10.0499}{10}$

$\approx 1.105 \qquad\quad \approx -0.905$

The solutions are approximately 1.105 and -0.905.

Applying Concepts 11.3

73. $\dfrac{x+2}{3} - \dfrac{4}{x-2} = 2$

$3(x-2)\left(\dfrac{x+2}{3} - \dfrac{4}{x-2}\right) = 3(x-2)(2)$

$(x-2)(x+2) - 3(4) = 6(x-2)$

$x^2 - 4 - 12 = 6x - 12$

$x^2 - 6x - 4 = 0$

$a = 1, b = -6, c = -4$

$x = \dfrac{-b \pm \sqrt{b^2 - 4ac}}{2a}$

$x = \dfrac{-(-6) \pm \sqrt{(-6)^2 - 4(1)(-4)}}{2(1)}$

$= \dfrac{6 \pm \sqrt{36 + 16}}{2} = \dfrac{6 \pm \sqrt{52}}{2}$

$= \dfrac{6 \pm 2\sqrt{13}}{2} = 3 \pm \sqrt{13}$

The solutions are $3 + \sqrt{13}$ and $3 - \sqrt{13}$.

75. a. False

$x = \sqrt{12 - x}$

$x^2 = \left(\sqrt{12 - x}\right)^2$

$x^2 = 12 - x$

$x^2 + x - 12 = 0$

$(x+4)(x-3) = 0$

$x + 4 = 0 \qquad x - 3 = 0$

$x = -4 \qquad x = 3$

-4 and 3 are solutions to $x^2 = 12 - x$ and are possible solutions to $x = \sqrt{12 - x}$.

Check:

$x = -4 \qquad\qquad x = 3$

$x = \sqrt{12 - x} \qquad\qquad x = \sqrt{12 - x}$

$\begin{array}{c|c} -4 & \sqrt{12 - (-4)} \\ & \sqrt{12 + 4} \\ & \sqrt{16} \\ & 4 \end{array} \qquad \begin{array}{c|c} 3 & \sqrt{12 - 3} \\ & \sqrt{9} \\ & 3 \\ & 3 = 3 \end{array}$

$-4 \neq 4$

3 is the solution to $x = \sqrt{12 - x}$.

b. False

$\sqrt{16} + \sqrt{9} = 4 + 3$

$\qquad\qquad = 7$

$16 + 9 = 25$

$7^2 \neq 25$

c. False

$\sqrt{9} = 3$

d. True

Section 11.4

Concept Review 11.4

1. Always true

3. Never true
This is a second-degree equation. The graph of a second degree equation is a parabola.

5. Never true
The coefficient of x^2 is negative. The parabola opens down.

Objective 11.4.1 Exercises

5. $y = x^2 - 3$
$a = 1$
a is positive.
The graph opens up.

7. $y = \frac{1}{2}x^2 - 2$
$a = \frac{1}{2}$
a is positive.
The graph opens up.

9. $y = x^2 - 2x + 3$
$a = 1$
a is positive.
The graph opens up.

11. $f(x) = x^2 - 2x + 1$
$f(3) = (3)^2 - 2(3) + 1$
$= 9 - 6 + 1$
$= 4$

13. $f(x) = 4 - x^2$
$f(-3) = 4 - (-3)^2$
$= 4 - 9$
$= -5$

15. $f(x) = -x^2 + 5x - 6$
$f(-4) = -(-4)^2 + 5(-4) - 6$
$= -16 - 20 - 6$
$= -42$

17.

19.

21.

23.

25.

27.

29.

31.

33.

35. Find the x-intercepts,
$$y = x^2 - 5x + 6$$
$$0 = x^2 - 5x + 6$$
$$x^2 - 5x + 6 = 0$$
$$(x-2)(x-3) = 0$$
$$x - 2 = 0 \quad x - 3 = 0$$
$$x = 2 \quad\quad x = 3$$
Find the y-intercept,
$$y = x^2 - 5x + 6$$
$$y = (0)^2 - 5(0) + 6$$
$$y = 6$$
The x-intercepts are (2, 0) and (3, 0), the y-intercept is (0, 6).

37. Find the *x*-intercepts,
$$y = 9 - x^2$$
$$0 = -x^2 + 9$$
$$x^2 - 9 = 0$$
$$(x + 3)(x - 3) = 0$$
$$x + 3 = 0 \quad x - 3 = 0$$
$$x = -3 \quad x = 3$$
Find the *y*-intercept,
$$y = 9 - x^2$$
$$y = 9 - (0)^2$$
$$y = 9$$
The *x*-intercepts are (–3, 0) and (3, 0), the *y*-intercept is (0, 9).

39. Find the *x*-intercepts,
$$y = x^2 + 2x - 6$$
$$0 = x^2 + 2x - 6$$
$$x^2 + 5x - 6 = 0$$
$$x = \frac{-2 \pm \sqrt{(2)^2 - 4(1)(-6)}}{2(1)}$$
$$= \frac{-2 \pm \sqrt{4 + 24}}{2}$$
$$= \frac{-2 \pm \sqrt{28}}{2}$$
$$= \frac{-2 \pm 2\sqrt{7}}{2}$$
$$= -1 \pm \sqrt{7}$$
Find the *y*-intercept,
$$y = x^2 + 2x - 6$$
$$y = (0)^2 + 2(0) - 6$$
$$y = -6$$

The *x*-intercepts are $\left(-1 + \sqrt{7}, 0\right)$ and $\left(-1 - \sqrt{7}, 0\right)$, the *y*-intercept is (0, –6).

41. Find the *x*-intercepts,
$$f(x) = 2x^2 - x - 3$$
$$0 = 2x^2 - x - 3$$
$$2x^2 - x - 3 = 0$$
$$(2x - 3)(x + 1) = 0$$
$$2x - 3 = 0 \quad x + 1 = 0$$
$$2x = 3 \quad x = -1$$
$$x = \frac{3}{2}$$
Find the *y*-intercept,
$$f(x) = 2x^2 - x - 3$$
$$f(0) = 2(0)^2 - (0) - 3$$
$$f(0) = -3$$

The *x*-intercepts are (–1, 0) and $\left(\frac{3}{2}, 0\right)$, the *y*-intercept is (0, –3).

43. Find the *x*-intercepts,
$$y = 4 - x - x^2$$
$$0 = -x^2 - x + 4$$
$$-x^2 - x + 4 = 0$$
$$x = \frac{1 \pm \sqrt{(-1)^2 - 4(-1)(4)}}{2(-1)}$$
$$= \frac{1 \pm \sqrt{1 + 16}}{-2}$$
$$= \frac{1 \pm \sqrt{17}}{-2}$$
Find the *y*-intercept,
$$y = 4 - x - x^2$$
$$y = 4 - (0) - (0)^2$$
$$y = 4$$

The *x*-intercepts are $\left(\frac{1 + \sqrt{17}}{-2}, 0\right)$ and $\left(\frac{1 - \sqrt{17}}{-2}, 0\right)$, the *y*-intercept is (0, 4).

45.

47.

49.

51.

53.

55.

Applying Concepts 11.4

57. $y+1=(x-4)^2$
$y+1=x^2-8x+16$
$y=x^2-8x+15$

59. $y-4=2(x-3)^2$
$y-4=2(x^2-6x+9)$
$y-4=2x^2-12x+18$
$y=2x^2-12x+22$

61. $y=x^3-x^2-6x$
$0=x^3-x^2-6x$
$x^3-x^2-6x=0$
$x(x^2-x-6)=0$
$x(x-3)(x+2)=0$
$x=0 \quad x-3=0 \quad x+2=0$
$\quad\quad\quad x=3 \quad\quad x=-2$
The x-intercepts are $(-2, 0)$, $(0, 0)$, and $(3, 0)$.

63. $y=x^3+x^2-4x-4$
$0=x^3+x^2-4x-4$
$x^3+x^2-4x-4=0$
$x^2(x+1)-4(x+1)=0$
$(x^2-4)(x+1)=0$
$(x+2)(x-2)(x+1)=0$
$x+2=0 \quad x-2=0 \quad x+1=0$
$\quad x=-2 \quad\quad x=2 \quad\quad x=-1$
The x-intercepts are $(-2, 0)$, $(-1, 0)$, and $(2, 0)$.

65. linear

67. neither

69. quadratic

71. $y=0.0087x^2-0.6125x+9.9096$
For the year 2020, $x=120$.
$y=0.0087(120)^2-0.6125(120)+9.9096$
$y=125.28-73.5+9.9096$
$y\approx 62$
The model predicts that a first-class stamp will cost 62¢ in the year 2020.

73. $y=2x^2-3x-2$
$12=2x^2-3x-2$
$2x^2-3x-14=0$
$(2x-7)(x+2)=0$
$2x-7=0 \quad x+2=0$
$2x=7 \quad\quad x=-2$
$x=\dfrac{7}{2}$
Since the point (x_1, y_1) is in quadrant II,
$x_1=-2$.

Section 11.5

Concept Review 11.5

1. Never true
Three more than twice the width is written as $2W+3$.

3. Always true

5. Always true

Objective 11.5.1 Exercises

1. Strategy
• This is a consecutive integer problem.
• First consecutive odd integer: n.
Second consecutive odd integer: $n+2$
• The sum of the squares of the two consecutive positive odd integers is 130.

Solution
$n^2+(n+2)^2=130$
$n^2+n^2+4n+4=130$
$2n^2+4n-126=0$
$2(n^2+2n-63)=0$
$2(n+9)(n-7)=0$
$(n+9)(n-7)=0$
$n+9=0 \quad n-7=0$
$n=-9 \quad\quad n=7$
The solution $n=-9$ is not possible because -9 is not a positive integer.
$n+2=7+2=9$
The integers are 7 and 9.

3. Strategy
• This is an integer problem.
• First integer: x
Second integer: $x+4$
• The product of the two integers is 60.

Solution
$x(x+4)=60$
$x^2+4x=60$
$x^2+4x-60=0$
$(x+10)(x-6)=0$
$x+10=0 \quad x-6=0$
$x=-10 \quad\quad x=6$
$x+4=-10+4=-6$
$x+4=6+4=10$
The integers are -10 and -6 or 6 and 10.

5. Strategy
   - This is a geometry problem.
   - Width of the rectangle: x
     Length of the rectangle: $x + 2$
   The area is 24 ft^2.
   - Use the equation $A = LW$.

   Solution
   $$A = LW$$
   $$24 = (x+2)x$$
   $$24 = x^2 + 2x$$
   $$0 = x^2 + 2x - 24$$
   $$0 = (x+6)(x-4)$$
   $$x+6 = 0 \qquad x - 4 = 0$$
   $$x = -6 \qquad x = 4$$
   The solution $x = -6$ is not possible.
   $$x + 2 = 4 + 2 = 6$$
   The width is 4 ft.
   The length is 6 ft.

7. Strategy
   - This is a geometry problem.
   - Width of the rectangle: x
     Length of the rectangle: $2x$.
   The area is 5000 ft^2.
   - Use the equation $A = LW$.

   Solution
   $$A = LW$$
   $$5000 = x(2x)$$
   $$5000 = 2x^2$$
   $$2500 = x^2$$
   $$\sqrt{2500} = \sqrt{x^2}$$
   $$\pm 50 = x$$
   The solution $x = -50$ is not possible.
   $$2x = 2(50) = 100$$
   The width is 50 ft.
   The length is 100 ft.

9. Strategy
   - This is a work problem.
   - Time for the second computer to solve the equation: t
   Time for the first computer to solve the equation: $t + 21$

	Rate	Time	Part
Second computer	$\dfrac{1}{t}$	10	$\dfrac{10}{t}$
First computer	$\dfrac{1}{t+21}$	10	$\dfrac{10}{t+21}$

   - The sum of the parts of the task completed by each computer must equal 1.

   Solution
   $$\frac{10}{t} + \frac{10}{t+21} = 1$$
   $$t(t+21)\left(\frac{10}{t} + \frac{10}{t+21}\right) = t(t+21)(1)$$
   $$10(t+21) + 10t = t(t+21)$$
   $$10t + 210 + 10t = t^2 + 21t$$
   $$0 = t^2 + t - 210$$
   $$0 = (t+15)(t-14)$$
   $$t + 15 = 0 \qquad t - 14 = 0$$
   $$t = -15 \qquad t = 14$$
   The solution -15 is not possible.
   $$t + 21 = 14 + 21 = 35$$
   The first computer can solve the equation in 35 min.
   The second computer can solve the equation in 14 min.

11. Strategy
   • This is a work problem.
   • Time for using the second engine: t
   Time for using the first engine: $t + 6$

	Rate	Time	Part
Second engine	$\frac{1}{t}$	4	$\frac{4}{t}$
First engine	$\frac{1}{t+6}$	4	$\frac{4}{t+6}$

   • The sum of the parts of the task completed by each engine must equal 1.

Solution

$$\frac{4}{t} + \frac{4}{t+6} = 1$$

$$t(t+6)\left(\frac{4}{t} + \frac{4}{t+6}\right) = t(t+6)(1)$$

$$4(t+6) + 4t = t(t+6)$$

$$4t + 24 + 4t = t^2 + 6t$$

$$0 = t^2 - 2t - 24$$

$$0 = (t-6)(t+4)$$

$$t - 6 = 0 \quad t + 4 = 0$$
$$t = 6 \qquad t = -4$$

The solution -4 is not possible.
$t + 6 = 6 + 6 = 12$
Using the first engine, it would take 12 h.
Using the second engine, it would take 6 h.

13. Strategy
   • This is a distance-rate problem.
   • Rate of the plane in calm air: r

	Distance	Rate	Time
With wind	375	$r + 25$	$\frac{375}{r+25}$
Against wind	375	$r - 25$	$\frac{375}{r-25}$

   • The time spent traveling against the wind is 2 h more than the time spent traveling with the wind.

Solution

$$\frac{375}{r-25} = 2 + \frac{375}{r+25}$$

$$(r-25)(r+25)\left(\frac{375}{r-25}\right) = (r-25)(r+25)\left(2 + \frac{375}{r+25}\right)$$

$$375(r+25) = 2(r-25)(r+25) + 375(r-25)$$

$$375r + 9375 = 2(r^2 - 625) + 375r - 9375$$

$$375r + 9375 = 2r^2 - 1250 + 375r - 9375$$

$$0 = 2r^2 - 20,000$$

$$0 = 2(r^2 - 10,000)$$

$$0 = r^2 - 10,000$$

$$0 = (r+100)(r-100)$$

$$r + 100 = 0 \qquad r - 100 = 0$$
$$r = -100 \qquad r = 100$$

The solution -100 is not possible.
The rate of the plane in calm air is 100 mph.

15. Strategy
- This is a distance-rate problem.
- Rate for the first 150 mi: r
- Rate for the next 35 mi: $r - 15$

	Distance	Rate	Time
First 150 mi	150	r	$\dfrac{150}{r}$
Next 35 mi	35	$r - 15$	$\dfrac{35}{r-15}$

- The total time for the trip was 4 h.

Solution
$$\frac{150}{r} + \frac{35}{r-15} = 4$$
$$r(r-15)\left(\frac{150}{r} + \frac{35}{r-15}\right) = r(r-15)(4)$$
$$150(r-15) + 35r = 4r(r-1)$$
$$150r - 2250 + 35r = 4r^2 - 60r$$
$$0 = 4r^2 - 245r + 2250$$
$$0 = (4r - 45)(r - 50)$$

$4r - 45 = 0 \quad r - 50 = 0$
$4r = 45 \qquad \quad r = 50$
$r = \dfrac{45}{4}$

The solution $\dfrac{45}{4}$ is not possible because the rate for the next 35 mi would be a negative number. The rate of the motorist for the first 150 mi was 50 mph.

17. Strategy
To find the time, substitute 32 for h in the equation $h = 48t - 16t^2$ and solve for t.

Solution
$$h = 48t - 16t^2$$
$$32 = 48t - 16t^2$$
$$16t^2 - 48t + 32 = 0$$
$$16(t^2 - 3t + 2) = 0$$
$$t^2 - 3t + 2 = 0$$
$$(t-1)(t-2) = 0$$

$t - 1 = 0 \quad t - 2 = 0$
$t = 1 \qquad t = 2$

The arrow will be 32 ft above the ground at 1 s and 2 s.

19. Strategy
To find the maximum speed, substitute 150 for d in the equation $d = 0.0056v^2 + 0.14v$ and solve for v.

Solution
$$d = 0.0056v^2 + 0.14v$$
$$150 = 0.0056v^2 + 0.14v$$
$$0 = 0.0056v^2 + 0.14v - 150$$
$$a = 0.0056, \ b = 0.14, \ c = -150$$
$$v = \frac{-b \pm \sqrt{b^2 - 4ac}}{2a}$$
$$v = \frac{-0.14 \pm \sqrt{(0.14)^2 - 4(0.0056)(-150)}}{2(0.0056)}$$
$$v = \frac{-0.14 \pm \sqrt{0.0196 + 3.36}}{0.0112}$$
$$v = \frac{-0.14 \pm \sqrt{3.3796}}{0.0112}$$
$$v \approx 151.6 \text{ or } -176.6$$

Since the speed cannot be negative, -176.6 cannot be a solution.
The maximum speed a driver can be traveling is 151.6 km/h.

21. Strategy
To find the distance, substitute 5 for y in the equation $y = -0.005x^2 + 1.2x + 10$ and solve for x.

Solution
$$y = -0.005x^2 + 1.2x + 10$$
$$5 = -0.005x^2 + 1.2x + 10$$
$$0 = -0.005x^2 + 1.2x + 5$$
$$0 = -0.005(x^2 - 240x - 1000)$$
$$0 = x^2 - 240x - 1000$$
$$a = 1, \ b = -240, \ c = -1000$$
$$x = \frac{-b \pm \sqrt{b^2 - 4ac}}{2a}$$
$$x = \frac{-(-240) \pm \sqrt{(-240)^2 - 4(1)(-1000)}}{2 \cdot 1}$$
$$x = \frac{240 \pm \sqrt{57,600 + 4000}}{2}$$
$$x = \frac{240 \pm \sqrt{61,600}}{2}$$
$$x \approx 244.1 \text{ or } -4.1$$

Since the distance cannot be negative, -4.1 cannot be a solution.
The hose is 244.1 ft from the tugboat.

23. Strategy

To find the time, substitute 10 for h in the equation $h = -16t^2 + 32t + 6.5$ and solve for t.

Solution

$$h = -16t^2 + 32t + 6.5$$
$$10 = -16t^2 + 32t + 6.5$$
$$0 = -16t^2 + 32t - 3.5$$
$$a = -16, \ b = 32, \ c = -3.5$$
$$t = \frac{-b \pm \sqrt{b^2 - 4ac}}{2a}$$
$$t = \frac{-32 \pm \sqrt{(32)^2 - 4(-16)(-3.5)}}{2(-16)}$$
$$t = \frac{-32 \pm \sqrt{1024 - 224}}{-32}$$
$$t = \frac{-32 \pm \sqrt{800}}{-32}$$
$$t \approx 0.12 \text{ or } 1.88$$

The ball will be 10 ft above the ground twice, on the way up and on the way down.
The ball hits the basket 1.88 s after it is released.

Applying Concepts 11.5

25. Strategy

To find the radius, solve the equation $V = \pi r^2 h$ for r, using $V = 1500$ and $h = 15$.

Solution

$$V = \pi r^2 h$$
$$1500 = \pi r^2 \cdot 15$$
$$\frac{1500}{15\pi} = \frac{15\pi r^2}{15\pi}$$
$$\frac{100}{\pi} = r^2$$
$$\sqrt{\frac{100}{\pi}} = \sqrt{r^2}$$
$$\pm 5.64 \approx r$$

Since the radius cannot be negative, -5.64 cannot be a solution. The radius of the can is 5.64 cm.

27. Strategy

Let x represent the first integer. Then the other integers are $x + 1$, $x + 2$, and $x + 3$.

Solution

$$x^2 + (x+1)^2 + (x+2)^2 + (x+3)^2 = 86$$
$$x^2 + x^2 + 2x + 1 + x^2 + 4x + 4 + x^2 + 6x + 9 = 86$$
$$4x^2 + 12x + 14 = 86$$
$$4x^2 + 12x - 72 = 0$$
$$4(x^2 + 3x - 18) = 0$$
$$x^2 + 3x - 18 = 0$$
$$(x + 6)(x - 3) = 0$$
$$x + 6 = 0 \qquad x - 3 = 0$$
$$x = -6 \qquad x = 3$$

The integers are either -6, -5, -4, and -3 or 3, 4, 5, and 6.

29. Strategy

- Radius of the small pizza: r
Radius of the large pizza: $2r - 1$
- Area of the small pizza: πr^2
Area of the large pizza: $\pi(2r - 1)^2$
- The difference in the areas is 33π in^2.

Solution

$$\pi(2r - 1)^2 - \pi r^2 = 33\pi$$
$$\frac{\pi[(2r-1)^2 - r^2]}{\pi} = \frac{33\pi}{\pi}$$
$$(2r - 1)^2 - r^2 = 33$$
$$4r^2 - 4r + 1 - r^2 = 33$$
$$3r^2 - 4r - 32 = 0$$
$$a = 3, \ b = -4, \ c = -32$$
$$r = \frac{-b \pm \sqrt{b^2 - 4ac}}{2a}$$
$$r = \frac{-(-4) \pm \sqrt{(-4)^2 - 4(3)(-32)}}{2 \cdot 3}$$
$$r = \frac{4 \pm \sqrt{16 + 384}}{6}$$
$$r = \frac{4 \pm \sqrt{400}}{6}$$
$$r = \frac{4 \pm 20}{6}$$
$$r = 4 \text{ or } -2.67$$

Since the radius cannot be negative, -2.67 cannot be a solution.
The radius of the small pizza is 4 in., so the radius of the large pizza is $2(4) - 1 = 7$ in.

31. Strategy

To find the dimensions of the cardboard, find an expression to represent the volume of the box and solve for x, using 512 for the volume.
$$\text{Volume} = (x - 2 \cdot 2)(x - 2 \cdot 2)(2)$$
$$\text{Volume} = 2(x - 4)(x - 4)$$

Solution

$$512 = 2(x - 4)(x - 4)$$
$$512 = 2(x^2 - 8x + 16)$$
$$512 = 2x^2 - 16x + 32$$
$$0 = 2x^2 - 16x - 480$$
$$0 = 2(x^2 - 8x - 240)$$
$$0 = x^2 - 8x - 240$$
$$0 = (x - 20)(x + 12)$$
$$x - 20 = 0 \qquad x + 12 = 0$$
$$x = 20 \qquad x = -12$$

Since the dimensions of the box cannot be negative, -12 is not a solution.
The dimensions of the cardboard are 20 in. by 20 in.

Chapter Review Exercises

1. $b^2 - 16 = 0$
$$b^2 = 16$$
$$\sqrt{b^2} = \sqrt{16}$$
$$b = \pm 4$$
The solutions are 4 and –4.

2. $x^2 - x - 3 = 0$
$a = 1, \ b = -1, \ c = -3$
$$x = \frac{-b \pm \sqrt{b^2 - 4ac}}{2a}$$
$$x = \frac{-(-1) \pm \sqrt{(-1)^2 - 4(1)(-3)}}{2 \cdot 1}$$
$$= \frac{1 \pm \sqrt{1 + 12}}{2}$$
$$= \frac{1 \pm \sqrt{13}}{2}$$
The solutions are $\dfrac{1 + \sqrt{13}}{2}$ and $\dfrac{1 - \sqrt{13}}{2}$.

3. $x^2 - 3x - 5 = 0$
$a = 1, \ b = -3, \ c = -5$
$$x = \frac{-b \pm \sqrt{b^2 - 4ac}}{2a}$$
$$x = \frac{-(-3) \pm \sqrt{(-3)^2 - 4(1)(-5)}}{2 \cdot 1}$$
$$= \frac{3 \pm \sqrt{9 + 20}}{2}$$
$$= \frac{3 \pm \sqrt{29}}{2}$$
The solutions are $\dfrac{3 + \sqrt{29}}{2}$ and $\dfrac{3 - \sqrt{29}}{2}$.

4. $49x^2 = 25$
$$x^2 = \frac{25}{49}$$
$$\sqrt{x^2} = \sqrt{\frac{25}{49}}$$
$$x = \pm \frac{5}{7}$$
The solutions are $\dfrac{5}{7}$ and $-\dfrac{5}{7}$.

5.

6.

7. $x^2 = 10x - 2$
$x^2 - 10x + 2 = 0$
$a = 1, \ b = -10, \ c = 2$
$$x = \frac{-b \pm \sqrt{b^2 - 4ac}}{2a}$$
$$x = \frac{-(-10) \pm \sqrt{(-10)^2 - 4(1)(2)}}{2 \cdot 1}$$
$$= \frac{10 \pm \sqrt{100 - 8}}{2}$$
$$= \frac{10 \pm \sqrt{92}}{2}$$
$$= \frac{10 \pm 2\sqrt{23}}{2}$$
$$= 5 \pm \sqrt{23}$$
The solutions are $5 + \sqrt{23}$ and $5 - \sqrt{23}$.

8. $6x(x + 1) = x - 1$
$6x^2 + 6x = x - 1$
$6x^2 + 5x + 1 = 0$
$(2x + 1)(3x + 1) = 0$
$2x + 1 = 0 \qquad 3x + 1 = 0$
$2x = -1 \qquad 3x = -1$
$x = -\dfrac{1}{2} \qquad x = -\dfrac{1}{3}$
The solutions are $-\dfrac{1}{2}$ and $-\dfrac{1}{3}$.

9. $4y^2 + 9 = 0$
$$4y^2 = -9$$
$$y^2 = -\frac{9}{4}$$
$$\sqrt{y^2} = \sqrt{-\frac{9}{4}}$$
$$y^2 = \pm\sqrt{-\frac{9}{4}}$$
$\sqrt{-\dfrac{9}{4}}$ is not a real number.
The equation has no real number solution.

10. $5x^2 + 20x + 12 = 0$

$a = 5,\ b = 20,\ c = 12$

$$x = \frac{-b \pm \sqrt{b^2 - 4ac}}{2a}$$

$$x = \frac{-20 \pm \sqrt{(20)^2 - 4(5)(12)}}{2 \cdot 5}$$

$$= \frac{-20 \pm \sqrt{400 - 240}}{10}$$

$$= \frac{-20 \pm \sqrt{160}}{10}$$

$$= \frac{-20 \pm 4\sqrt{10}}{10}$$

$$= \frac{-10 \pm 2\sqrt{10}}{5}$$

The solutions are $\dfrac{-10 + 2\sqrt{10}}{5}$ and $\dfrac{-10 - 2\sqrt{10}}{5}$.

11. $x^2 - 4x + 1 = 0$

$a = 1,\ b = -4,\ c = 1$

$$x = \frac{-b \pm \sqrt{b^2 - 4ac}}{2a}$$

$$x = \frac{-(-4) \pm \sqrt{(-4)^2 - 4(1)(1)}}{2 \cdot 1}$$

$$= \frac{4 \pm \sqrt{16 - 4}}{2}$$

$$= \frac{4 \pm \sqrt{12}}{2}$$

$$= \frac{4 \pm 2\sqrt{3}}{2}$$

$$= 2 \pm \sqrt{3}$$

The solutions are $2 + \sqrt{3}$ and $2 - \sqrt{3}$.

12. $x^2 - x = 30$

$x^2 - x - 30 = 0$

$(x + 5)(x - 6) = 0$

$x + 5 = 0 \quad x - 6 = 0$

$\quad x = -5 \qquad x = 6$

The solutions are -5 and 6.

13. $6x^2 + 13x - 28 = 0$

$(3x - 4)(2x + 7) = 0$

$3x - 4 = 0 \quad 2x + 7 = 0$

$3x = 4 \qquad 2x = -7$

$x = \dfrac{4}{3} \qquad x = -\dfrac{7}{2}$

The solutions are $\dfrac{4}{3}$ and $-\dfrac{7}{2}$.

14. $x^2 = 40$

$\sqrt{x^2} = \sqrt{40}$

$x = \pm\sqrt{40}$

$x = \pm 2\sqrt{10}$

The solutions are $2\sqrt{10}$ and $-2\sqrt{10}$.

15. $3x^2 - 4x = 1$

$3x^2 - 4x - 1 = 0$

$a = 3,\ b = -4,\ c = -1$

$$x = \frac{-b \pm \sqrt{b^2 - 4ac}}{2a}$$

$$x = \frac{-(-4) \pm \sqrt{(-4)^2 - 4(3)(-1)}}{2 \cdot 3}$$

$$= \frac{4 \pm \sqrt{16 + 12}}{6}$$

$$= \frac{4 \pm \sqrt{28}}{6}$$

$$= \frac{4 \pm 2\sqrt{7}}{6}$$

$$= \frac{2 \pm \sqrt{7}}{3}$$

The solutions are $\dfrac{2 + \sqrt{7}}{3}$ and $\dfrac{2 - \sqrt{7}}{3}$.

16. $x^2 - 2x - 10 = 0$

$a = 1,\ b = -2,\ c = -10$

$$x = \frac{-b \pm \sqrt{b^2 - 4ac}}{2a}$$

$$x = \frac{-(-2) \pm \sqrt{(-2)^2 - 4(1)(-10)}}{2 \cdot 1}$$

$$= \frac{2 \pm \sqrt{4 + 40}}{2}$$

$$= \frac{2 \pm \sqrt{44}}{2}$$

$$= \frac{2 \pm 2\sqrt{11}}{2}$$

$$= 1 \pm \sqrt{11}$$

The solutions are $1 + \sqrt{11}$ and $1 - \sqrt{11}$.

17. $x^2 - 12x + 27 = 0$

$(x - 3)(x - 9) = 0$

$x - 3 = 0 \quad x - 9 = 0$

$x = 3 \qquad x = 9$

The solutions are 3 and 9.

18. $(x - 7)^2 = 81$

$\sqrt{(x - 7)^2} = \sqrt{81}$

$x - 7 = \pm\sqrt{81}$

$x - 7 = \pm 9$

$x - 7 = 9 \quad x - 7 = -9$

$\quad x = 16 \qquad x = -2$

The solutions are 16 and -2.

19.

20.

21. $(y+4)^2 - 25 = 0$

$$(y+4)^2 = 25$$

$$\sqrt{(y+4)^2} = \sqrt{25}$$

$$y+4 = \pm\sqrt{25}$$

$$y+4 = \pm 5$$

$$y+4 = 5 \qquad y+4 = -5$$

$$y = 1 \qquad\quad y = -9$$

The solutions are 1 and −9.

22. $\quad 4x^2 + 16x = 7$

$$4x^2 + 16x - 7 = 0$$

$$a = 4,\ b = 16,\ c = -7$$

$$x = \frac{-b \pm \sqrt{b^2 - 4ac}}{2a}$$

$$x = \frac{-16 \pm \sqrt{(16)^2 - 4(4)(-7)}}{2 \cdot 4}$$

$$= \frac{-16 \pm \sqrt{256 + 112}}{8}$$

$$= \frac{-16 \pm \sqrt{368}}{8}$$

$$= \frac{-16 \pm 4\sqrt{23}}{8}$$

$$= \frac{-4 \pm \sqrt{23}}{2}$$

The solutions are $\dfrac{-4 + \sqrt{23}}{2}$ and $\dfrac{-4 - \sqrt{23}}{2}$.

23. $24x^2 + 34x + 5 = 0$

$$(6x + 1)(4x + 5) = 0$$

$$6x + 1 = 0 \qquad 4x + 5 = 0$$

$$6x = -1 \qquad\ 4x = -5$$

$$x = -\frac{1}{6} \qquad x = -\frac{5}{4}$$

The solutions are $-\dfrac{1}{6}$ and $-\dfrac{5}{4}$.

24. $\qquad x^2 = 4x - 8$

$$x^2 - 4x + 8 = 0$$

$$a = 1,\ b = -4,\ c = 8$$

$$x = \frac{-b \pm \sqrt{b^2 - 4ac}}{2a}$$

$$x = \frac{-(-4) \pm \sqrt{(-4)^2 - 4(1)(8)}}{2 \cdot 1}$$

$$= \frac{4 \pm \sqrt{16 - 32}}{2}$$

$$= \frac{4 \pm \sqrt{-16}}{2}$$

$\sqrt{-16}$ is not a real number.
The equation has no real number solution.

25. $\qquad x^2 - 5 = 8x$

$$x^2 - 8x - 5 = 0$$

$$a = 1,\ b = -8,\ c = -5$$

$$x = \frac{-b \pm \sqrt{b^2 - 4ac}}{2a}$$

$$x = \frac{-(-8) \pm \sqrt{(-8)^2 - 4(1)(-5)}}{2 \cdot 1}$$

$$= \frac{8 \pm \sqrt{64 + 20}}{2}$$

$$= \frac{8 \pm \sqrt{84}}{2}$$

$$= \frac{8 \pm 2\sqrt{21}}{2}$$

$$= 4 \pm \sqrt{21}$$

The solutions are $4 + \sqrt{21}$ and $4 - \sqrt{21}$.

26. $25(2x^2 - 2x + 1) = (x + 3)^2$

$$50x^2 - 50x + 25 = x^2 + 6x + 9$$

$$49x^2 - 56x + 16 = 0$$

$$(7x - 4)(7x - 4) = 0$$

$$7x - 4 = 0 \qquad 7x - 4 = 0$$

$$7x = 4 \qquad\quad 7x = 4$$

$$x = \frac{4}{7} \qquad\quad x = \frac{4}{7}$$

The solution is $\dfrac{4}{7}$.

27. $\left(x - \dfrac{1}{2}\right)^2 = \dfrac{9}{4}$

$$\sqrt{\left(x - \frac{1}{2}\right)^2} = \sqrt{\frac{9}{4}}$$

$$x - \frac{1}{2} = \pm\sqrt{\frac{9}{4}}$$

$$x - \frac{1}{2} = \pm\frac{3}{2}$$

$$x - \frac{1}{2} = \frac{3}{2} \qquad x - \frac{1}{2} = -\frac{3}{2}$$

$$x = \frac{3}{2} + \frac{1}{2} \qquad x = -\frac{3}{2} + \frac{1}{2}$$

$$x = \frac{4}{2} \qquad\qquad x = -\frac{2}{2}$$

$$x = 2 \qquad\qquad x = -1$$

The solutions are 2 and −1.

28. $\qquad 16x^2 = 30x - 9$

$$16x^2 - 30x + 9 = 0$$

$$(8x - 3)(2x - 3) = 0$$

$$8x - 3 = 0 \qquad 2x - 3 = 0$$

$$8x = 3 \qquad\quad 2x = 3$$

$$x = \frac{3}{8} \qquad\quad x = \frac{3}{2}$$

The solutions are $\dfrac{3}{8}$ and $\dfrac{3}{2}$.

29.
$$x^2 + 7x = 3$$
$$x^2 + 7x - 3 = 0$$
$$a = 1,\ b = 7,\ c = -3$$
$$x = \frac{-b \pm \sqrt{b^2 - 4ac}}{2a}$$
$$x = \frac{-7 \pm \sqrt{(7)^2 - 4(1)(-3)}}{2 \cdot 1}$$
$$= \frac{-7 \pm \sqrt{49 + 12}}{2}$$
$$= \frac{-7 \pm \sqrt{61}}{2}$$

The solutions are $\dfrac{-7 + \sqrt{61}}{2}$ and $\dfrac{-7 - \sqrt{61}}{2}$.

30.
$$12x^2 + 10 = 29x$$
$$12x^2 - 29x + 10 = 0$$
$$(12x - 5)(x - 2) = 0$$
$$12x - 5 = 0 \qquad x - 2 = 0$$
$$12x = 5 \qquad\quad x = 2$$
$$x = \frac{5}{12}$$

The solutions are $\dfrac{5}{12}$ and 2.

31.
$$4(x - 3)^2 = 20$$
$$(x - 3)^2 = 5$$
$$\sqrt{(x - 3)^2} = \sqrt{5}$$
$$x - 3 = \pm\sqrt{5}$$
$$x - 3 = \sqrt{5} \qquad x - 3 = -\sqrt{5}$$
$$x = 3 + \sqrt{5} \qquad x = 3 - \sqrt{5}$$

The solutions are $3 + \sqrt{5}$ and $3 - \sqrt{5}$.

32.
$$x^2 + 8x - 3 = 0$$
$$a = 1,\ b = 8,\ c = -3$$
$$x = \frac{-b \pm \sqrt{b^2 - 4ac}}{2a}$$
$$x = \frac{-8 \pm \sqrt{8^2 - 4(1)(-3)}}{2 \cdot 1}$$
$$= \frac{-8 \pm \sqrt{64 + 12}}{2}$$
$$= \frac{-8 \pm \sqrt{76}}{2}$$
$$= \frac{-8 \pm 2\sqrt{19}}{2}$$
$$= -4 \pm \sqrt{19}$$

The solutions are $-4 + \sqrt{19}$ and $-4 - \sqrt{19}$.

33.

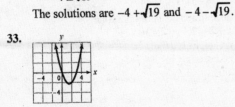

34.

35.
$$(x + 9)^2 = x + 11$$
$$x^2 + 18x + 81 = x + 11$$
$$x^2 + 17x + 70 = 0$$
$$(x + 7)(x + 10) = 0$$
$$x + 7 = 0 \qquad x + 10 = 0$$
$$x = -7 \qquad\quad x = -10$$

The solutions are -7 and -10.

36.
$$(x - 2)^2 - 24 = 0$$
$$(x - 2)^2 = 24$$
$$\sqrt{(x - 2)^2} = \sqrt{24}$$
$$x - 2 = \pm\sqrt{24}$$
$$x - 2 = \pm 2\sqrt{6}$$
$$x - 2 = 2\sqrt{6} \qquad x - 2 = -2\sqrt{6}$$
$$x = 2 + 2\sqrt{6} \qquad x = 2 - 2\sqrt{6}$$

The solutions are $2 + 2\sqrt{6}$ and $2 - 2\sqrt{6}$.

37.
$$x^2 + 6x + 12 = 0$$
$$a = 1,\ b = 6,\ c = 12$$
$$x = \frac{-b \pm \sqrt{b^2 - 4ac}}{2a}$$
$$x = \frac{-6 \pm \sqrt{(6)^2 - 4(1)(12)}}{2 \cdot 1}$$
$$= \frac{-6 \pm \sqrt{36 - 48}}{2}$$
$$= \frac{-6 \pm \sqrt{-12}}{2}$$

$\sqrt{-12}$ is not a real number.
The equation has no real number solution.

38.
$$x^2 + 6x - 2 = 0$$
$$a = 1,\ b = 6,\ c = -2$$
$$x = \frac{-b \pm \sqrt{b^2 - 4ac}}{2a}$$
$$x = \frac{-6 \pm \sqrt{(6)^2 - 4(1)(-2)}}{2 \cdot 1}$$
$$= \frac{-6 \pm \sqrt{36 + 8}}{2}$$
$$= \frac{-6 \pm \sqrt{44}}{2}$$
$$= \frac{-6 \pm 2\sqrt{11}}{2}$$
$$= -3 \pm \sqrt{11}$$

The solutions are $-3 + \sqrt{11}$ and $-3 - \sqrt{11}$.

39.
$$18x^2 - 52x = 6$$
$$18x^2 - 52x - 6 = 0$$
$$2(9x^2 - 26x - 3) = 0$$
$$9x^2 - 26x - 3 = 0$$
$$(9x + 1)(x - 3) = 0$$
$$9x + 1 = 0 \qquad x - 3 = 0$$
$$9x = -1 \qquad x = 3$$
$$x = -\frac{1}{9}$$

The solutions are $-\dfrac{1}{9}$ and 3.

40.
$$2x^2 + 5x = 1$$
$$2x^2 + 5x - 1 = 0$$
$$a = 2,\ b = 5,\ c = -1$$
$$x = \frac{-b \pm \sqrt{b^2 - 4ac}}{2a}$$
$$x = \frac{-5 \pm \sqrt{(5)^2 - 4(2)(-1)}}{2 \cdot 2}$$
$$= \frac{-5 \pm \sqrt{25 + 8}}{4}$$
$$= \frac{-5 \pm \sqrt{33}}{4}$$

The solutions are $\dfrac{-5 + \sqrt{33}}{4}$ and $\dfrac{-5 - \sqrt{33}}{4}$.

41.

42.
$$x^2 + 3 = 9x$$
$$x^2 - 9x + 3 = 0$$
$$a = 1,\ b = -9,\ c = 3$$
$$x = \frac{-b \pm \sqrt{b^2 - 4ac}}{2a}$$
$$x = \frac{-(-9) \pm \sqrt{(-9)^2 - 4(1)(3)}}{2 \cdot 1}$$
$$= \frac{9 \pm \sqrt{81 - 12}}{2}$$
$$= \frac{9 \pm \sqrt{69}}{2}$$

The solutions are $\dfrac{9 + \sqrt{69}}{2}$ and $\dfrac{9 - \sqrt{69}}{2}$.

43.
$$2x^2 + 5 = 7x$$
$$2x^2 - 7x + 5 = 0$$
$$(2x - 5)(x - 1) = 0$$
$$2x - 5 = 0 \qquad x - 1 = 0$$
$$2x = 5 \qquad x = 1$$
$$x = \frac{5}{2}$$

The solutions are $\dfrac{5}{2}$ and 1.

44.

45. Strategy
• This is a distance-rate problem.
• Rate of the air balloon in calm air: r

	Distance	Rate	Time
Against the wind	60	$r - 5$	$\dfrac{60}{r - 5}$
With wind	60	$r + 5$	$\dfrac{60}{r + 5}$

• The time traveling against the wind is an hour more than the time traveling with the wind.

Solution
$$\frac{60}{r - 5} = 1 + \frac{60}{r + 5}$$
$$(r - 5)(r + 5)\left(\frac{60}{r - 5}\right) = (r - 5)(r + 5)\left(1 + \frac{60}{r + 5}\right)$$
$$60(r + 5) = (r - 5)(r + 5) + 60(r - 5)$$
$$60r + 300 = r^2 - 25 + 60r - 300$$
$$0 = r^2 - 625$$
$$0 = (r - 25)(r + 25)$$
$$r - 25 = 0 \qquad r + 25 = 0$$
$$r = 25 \qquad r = -25$$

The solution -25 is not possible.
The rate of the air balloon in calm air is 25 mph.

46. Strategy
• This is a geometry problem.
• Length of the base: x
Height: $2x + 2$
The area is 20 m^2.
• Use the equation $A = \dfrac{1}{2}bh$.

Solution
$$A = \frac{1}{2}bh$$
$$20 = \frac{1}{2}x(2x + 2)$$
$$20 = x^2 + x$$
$$0 = x^2 + x - 20$$
$$0 = (x + 5)(x - 4)$$
$$x + 5 = 0 \qquad x - 4 = 0$$
$$x = -5 \qquad x = 4$$

The solution -5 is not possible.
$2x + 2 = 2(4) + 2 = 10$
The height is 10 m.
The base is 4 m.

47. Strategy
• This is a consecutive integer problem.
• First odd positive integer: n
Second odd positive integer: $n + 2$
• The sum of the squares of the two consecutive positive integers is 34.

Solution
$$n^2 + (n+2)^2 = 34$$
$$n^2 + n^2 + 4n + 4 = 34$$
$$2n^2 + 4n + 4 = 34$$
$$2n^2 + 4n - 30 = 0$$
$$2(n^2 + 2n - 15) = 0$$
$$2(n+5)(n-3) = 0$$
$$n + 5 = 0 \quad n - 3 = 0$$
$$n = -5 \qquad n = 3$$

The solution -5 is not possible, since it is not a positive integer.
$$n + 2 = 3 + 2 = 5$$
The integers are 3 and 5.

48. Strategy
• This is a distance-rate problem.
• The rate at which the campers rowed: r

	Distance	Rate	Time
Downstream	12	$r+1$	$\dfrac{12}{r+1}$
Upstream	12	$r-1$	$\dfrac{12}{r-1}$

• The total time for the trip was 5 h.

Solution
$$\frac{12}{r+1} + \frac{12}{r-1} = 5$$
$$(r+1)(r-1)\left(\frac{12}{r+1} + \frac{12}{r-1}\right) = (r+1)(r-1)(5)$$
$$12(r-1) + 12(r+1) = (r^2 - 1)(5)$$
$$12r - 12 + 12r + 12 = 5r^2 - 5$$
$$0 = 5r^2 - 24r - 5$$
$$0 = (5r+1)(r-5)$$
$$5r + 1 = 0 \quad r - 5 = 0$$
$$5r = -1 \qquad r = 5$$
$$r = -\frac{1}{5}$$

The solution $-\dfrac{1}{5}$ is not possible.

The rate at which the campers rowed was 5 mph.

49. Strategy
To find the time, substitute 12 for h in the equation $h = 32t - 16t^2$ and solve for t.

Solution
$$h = 32t - 16t^2$$
$$12 = 32t - 16t^2$$
$$16t^2 - 32t + 12 = 0$$
$$4(4t^2 - 8t + 3) = 0$$
$$4t^2 - 8t + 3 = 0$$
$$(2t - 1)(2t - 3) = 0$$
$$2t - 1 = 0 \quad 2t - 3 = 0$$
$$2t = 1 \qquad 2t = 3$$
$$t = \frac{1}{2} \qquad t = \frac{3}{2}$$

The objective will be 12 ft above the ground at $\dfrac{1}{2}$ s and $\dfrac{3}{2}$ s.

50. Strategy
• This is a work problem.
• Time for larger drain to empty the tank: t
Time for smaller drain to empty the tank: $t + 8$

	Rate	Time	Part
Larger drain	$\dfrac{1}{t}$	3	$\dfrac{3}{t}$
Smaller drain	$\dfrac{1}{t+8}$	3	$\dfrac{3}{t+8}$

• The sum of the parts of the task completed must equal 1.

Solution
$$\frac{3}{t} + \frac{3}{t+8} = 1$$
$$t(t+8)\left(\frac{3}{t} + \frac{3}{t+8}\right) = t(t+8)(1)$$
$$3(t+8) + 3t = t(t+8)$$
$$3t + 24 + 3t = t^2 + 8t$$
$$0 = t^2 + 2t - 24$$
$$0 = (t - 4)(t + 6)$$
$$t - 4 = 0 \quad t + 6 = 0$$
$$t = 4 \qquad t = -6$$

The solution -6 is not possible.
$$t + 8 = 4 + 8 = 12$$
It would take the larger drain 4 h to empty the tank. It would take the smaller drain 12 h to empty the tank.

Chapter Test

1. $3(x+4)^2 - 60 = 0$

$$3(x+4)^2 = 60$$
$$(x+4)^2 = 20$$
$$\sqrt{(x+4)^2} = \sqrt{20}$$
$$x+4 = \pm 2\sqrt{5}$$

$x+4 = 2\sqrt{5} \qquad x+4 = -2\sqrt{5}$
$\quad x = -4 + 2\sqrt{5} \qquad x = -4 - 2\sqrt{5}$

The solutions are $-4 + 2\sqrt{5}$ and $-4 - 2\sqrt{5}$.

2. $2x^2 + 8x = 3$

$2x^2 + 8x - 3 = 0$
$a = 2,\ b = 8,\ c = -3$

$$x = \frac{-b \pm \sqrt{b^2 - 4ac}}{2a}$$

$$x = \frac{-8 \pm \sqrt{(8)^2 - 4(2)(-3)}}{2 \cdot 2}$$

$$= \frac{-8 \pm \sqrt{64 + 24}}{4}$$

$$= \frac{-8 \pm \sqrt{88}}{4}$$

$$= \frac{-8 \pm 2\sqrt{22}}{4}$$

$$= \frac{-4 \pm \sqrt{22}}{2}$$

The solutions are $\dfrac{-4 + \sqrt{22}}{2}$ and $\dfrac{-4 - \sqrt{22}}{2}$.

3. $3x^2 + 7x = 20$

$3x^2 + 7x - 20 = 0$
$(3x - 5)(x + 4) = 0$
$3x - 5 = 0 \qquad x + 4 = 0$
$\quad 3x = 5 \qquad\quad x = -4$
$\quad\ x = \dfrac{5}{3}$

The solutions are $\dfrac{5}{3}$ and -4.

4. $x^2 - 3x = 6$

$x^2 - 3x - 6 = 0$
$a = 1,\ b = -3,\ c = -6$

$$x = \frac{-b \pm \sqrt{b^2 - 4ac}}{2a}$$

$$x = \frac{-(-3) \pm \sqrt{(-3)^2 - 4(1)(-6)}}{2 \cdot 1}$$

$$= \frac{3 \pm \sqrt{9 + 24}}{2} = \frac{3 \pm \sqrt{33}}{2}$$

The solutions are $\dfrac{3 + \sqrt{33}}{2}$ and $\dfrac{3 - \sqrt{33}}{2}$.

5. $x^2 + 4x - 16 = 0$
$a = 1,\ b = 4,\ c = -16$

$$x = \frac{-b \pm \sqrt{b^2 - 4ac}}{2a}$$

$$x = \frac{-4 \pm \sqrt{(4)^2 - 4(1)(-16)}}{2 \cdot 1}$$

$$= \frac{-4 \pm \sqrt{16 + 64}}{2}$$

$$= \frac{-4 \pm \sqrt{80}}{2}$$

$$= \frac{-4 \pm 4\sqrt{5}}{2}$$

$$= -2 \pm 2\sqrt{5}$$

The solutions are $-2 + 2\sqrt{5}$ and $-2 - 2\sqrt{5}$.

6.

7. $x^2 + 4x + 2 = 0$
$a = 1,\ b = 4,\ c = 2$

$$x = \frac{-b \pm \sqrt{b^2 - 4ac}}{2a}$$

$$x = \frac{-4 \pm \sqrt{(4)^2 - 4(1)(2)}}{2 \cdot 1}$$

$$= \frac{-4 \pm \sqrt{16 - 8}}{2} = \frac{-4 \pm \sqrt{8}}{2}$$

$$= \frac{-4 \pm 2\sqrt{2}}{2} = -2 \pm \sqrt{2}$$

The solutions are $-2 + \sqrt{2}$ and $-2 - \sqrt{2}$.

8. $x^2 + 3x = 8$

$$x^2 + 3x + \frac{9}{4} = 8 + \frac{9}{4}$$

$$\left(x + \frac{3}{2}\right)^2 = \frac{41}{4}$$

$$\sqrt{\left(x + \frac{3}{2}\right)^2} = \sqrt{\frac{41}{4}}$$

$$x + \frac{3}{2} = \pm \frac{\sqrt{41}}{2}$$

$x + \dfrac{3}{2} = \dfrac{\sqrt{41}}{2} \qquad x + \dfrac{3}{2} = -\dfrac{\sqrt{41}}{2}$

$\quad x = -\dfrac{3}{2} + \dfrac{\sqrt{41}}{2} \qquad x = -\dfrac{3}{2} - \dfrac{\sqrt{41}}{2}$

$\quad x = \dfrac{-3 + \sqrt{41}}{2} \qquad x = \dfrac{-3 - \sqrt{41}}{2}$

The solutions are $\dfrac{-3 + \sqrt{41}}{2}$ and $\dfrac{-3 - \sqrt{41}}{2}$.

9. $2x^2 - 5x - 3 = 0$
$(2x + 1)(x - 3) = 0$
$2x + 1 = 0 \qquad x - 3 = 0$
$\qquad 2x = -1 \qquad x = 3$
$\qquad x = -\frac{1}{2}$

The solutions are $-\frac{1}{2}$ and 3.

10. $2x^2 - 6x + 1 = 0$
$a = 2, b = -6, c = 1$
$x = \dfrac{-b \pm \sqrt{b^2 - 4ac}}{2a}$

$x = \dfrac{-(-6) \pm \sqrt{(-6)^2 - 4(2)(1)}}{2 \cdot 2}$

$= \dfrac{6 \pm \sqrt{36 - 8}}{4}$

$= \dfrac{6 \pm \sqrt{28}}{4} = \dfrac{6 \pm 2\sqrt{7}}{4}$

$= \dfrac{3 \pm \sqrt{7}}{2}$

The solutions are $\dfrac{3 + \sqrt{7}}{2}$ and $\dfrac{3 - \sqrt{7}}{2}$.

11. $3x^2 - x = 1$
$3x^2 - x - 1 = 0$
$a = 3, b = -1, c = -1$
$x = \dfrac{-b \pm \sqrt{b^2 - 4ac}}{2a}$

$x = \dfrac{-(-1) \pm \sqrt{(-1)^2 - 4(3)(-1)}}{2 \cdot 3}$

$= \dfrac{1 \pm \sqrt{1 + 12}}{6} = \dfrac{1 \pm \sqrt{13}}{6}$

The solutions are $\dfrac{1 + \sqrt{13}}{6}$ and $\dfrac{1 - \sqrt{13}}{6}$.

12. $2(x - 5)^2 = 36$
$(x - 5)^2 = 18$
$\sqrt{(x - 5)^2} = \sqrt{18}$
$x - 5 = \pm 3\sqrt{2}$
$x = 5 \pm 3\sqrt{2}$

The solutions are $5 + 3\sqrt{2}$ and $5 - 3\sqrt{2}$.

13. $x^2 - 6x - 5 = 0$
$a = 1, b = -6, c = -5$
$x = \dfrac{-b \pm \sqrt{b^2 - 4ac}}{2a}$

$x = \dfrac{-(-6) + \sqrt{(-6)^2 - 4(1)(-5)}}{2 \cdot 1}$

$x = \dfrac{6 \pm \sqrt{36 + 20}}{2} = \dfrac{6 \pm \sqrt{56}}{2}$

$= \dfrac{6 \pm 2\sqrt{14}}{2} = 3 \pm \sqrt{14}$

The solutions are $3 + \sqrt{14}$ and $3 - \sqrt{14}$.

14. $x^2 - 5x = 1$
$x^2 - 5x - 1 = 0$
$a = 1, b = -5, c = -1$
$x = \dfrac{-b \pm \sqrt{b^2 - 4ac}}{2a}$

$x = \dfrac{-(-5) \pm \sqrt{(-5)^2 - 4(1)(-1)}}{2 \cdot 1}$

$= \dfrac{5 \pm \sqrt{25 + 4}}{2} = \dfrac{5 \pm \sqrt{29}}{2}$

The solutions are $\dfrac{5 + \sqrt{29}}{2}$ and $\dfrac{5 - \sqrt{29}}{2}$.

15. $x^2 - 5x = 2$
$x^2 - 5x - 2 = 0$
$a = 1, b = -5, c = -2$
$x = \dfrac{-b \pm \sqrt{b^2 - 4ac}}{2a}$

$x = \dfrac{-(-5) \pm \sqrt{(-5)^2 - 4(1)(-2)}}{2 \cdot 1}$

$x = \dfrac{5 \pm \sqrt{25 + 8}}{2} = \dfrac{5 \pm \sqrt{33}}{2}$

The solutions are $\dfrac{5 + \sqrt{33}}{2}$ and $\dfrac{5 - \sqrt{33}}{2}$.

16. $6x^2 - 17x = -5$
$6x^2 - 17x + 5 = 0$
$(2x - 5)(3x - 1) = 0$
$2x - 5 = 0 \qquad 3x - 1 = 0$
$\qquad 2x = 5 \qquad\quad 3x = 1$
$\qquad x = \frac{5}{2} \qquad\quad x = \frac{1}{3}$

The solutions are $\frac{5}{2}$ and $\frac{1}{3}$.

17. $x^2 + 3x - 7 = 0$
$a = 1, b = 3, c = -7$
$x = \dfrac{-b \pm \sqrt{b^2 - 4ac}}{2a}$

$x = \dfrac{-3 \pm \sqrt{(3)^2 - 4(1)(-7)}}{2 \cdot 1}$

$= \dfrac{-3 \pm \sqrt{9 + 28}}{2} = \dfrac{-3 \pm \sqrt{37}}{2}$

The solutions are $\dfrac{-3 + \sqrt{37}}{2}$ and $\dfrac{-3 - \sqrt{37}}{2}$.

18. $2x^2 - 4x - 5 = 0$
$a = 2,\ b = -4,\ c = -5$

$$x = \frac{-b \pm \sqrt{b^2 - 4ac}}{2a}$$

$$x = \frac{-(-4) \pm \sqrt{(-4)^2 - 4(2)(-5)}}{2 \cdot 2}$$

$$= \frac{4 \pm \sqrt{16 + 40}}{4}$$

$$= \frac{4 \pm \sqrt{56}}{4} = \frac{4 \pm 2\sqrt{14}}{4}$$

$$= \frac{2 \pm \sqrt{14}}{2}$$

The solutions are $\dfrac{2 + \sqrt{14}}{2}$ and $\dfrac{2 - \sqrt{14}}{2}$.

19. $2x^2 - 3x - 2 = 0$
$(2x + 1)(x - 2) = 0$
$2x + 1 = 0 \qquad x - 2 = 0$
$\quad\ \ 2x = -1 \qquad\ \ x = 2$
$\quad\ \ \ x = -\dfrac{1}{2}$

The solutions are $-\dfrac{1}{2}$ and 2.

20.

21. $3x^2 - 2x = 3$
$3x^2 - 2x - 3 = 0$
$a = 3,\ b = -2,\ c = -3$

$$x = \frac{-b \pm \sqrt{b^2 - 4ac}}{2a}$$

$$x = \frac{-(-2) \pm \sqrt{(-2)^2 - 4(3)(-3)}}{2 \cdot 3}$$

$$= \frac{2 \pm \sqrt{4 + 36}}{6} = \frac{2 \pm \sqrt{40}}{6}$$

$$= \frac{2 \pm 2\sqrt{10}}{6} = \frac{1 \pm \sqrt{10}}{3}$$

The solutions are $\dfrac{1 + \sqrt{10}}{3}$ and $\dfrac{1 - \sqrt{10}}{3}$.

22. Strategy
• This is a geometry problem.
• Width of the rectangle: x
Length of the rectangle: $2x - 2$
The area is $40\ \text{ft}^2$.
• Use the equation $A = LW$.

Solution
$A = LW$
$40 = (2x - 2)(x)$
$40 = 2x^2 - 2x$
$0 = 2x^2 - 2x - 40$
$0 = 2(x^2 - x - 20)$
$0 = x^2 - x - 20$
$0 = (x - 5)(x + 4)$
$x - 5 = 0 \quad x + 4 = 0$
$\quad x = 5 \qquad\ \ x = -4$

The solution -4 is not possible.
$2x - 2 = 2(5) - 2 = 8$
The width of the rectangle is 5 ft.
The length of the rectangle is 8 ft.

23. Strategy
• This is a distance-rate problem.
• Rate of the boat in calm water: r

	Distance	Rate	Time
With current	60	$r + 1$	$\dfrac{60}{r + 1}$
Against current	60	$r - 1$	$\dfrac{60}{r - 1}$

• The time to travel against the current is 1 h more than the time to travel with the current.

Solution

$$\frac{60}{r + 1} + 1 = \frac{60}{r - 1}$$

$$(r + 1)(r - 1)\left(\frac{60}{r + 1} + 1\right) = (r + 1)(r - 1)\left(\frac{60}{r - 1}\right)$$

$$60(r - 1) + (r + 1)(r - 1) = 60(r + 1)$$

$$60r - 60 + r^2 - 1 = 60r + 60$$

$$r^2 + 60r - 61 = 60r + 60$$

$$r^2 - 121 = 0$$

$$(r + 11)(r - 11) = 0$$

$r + 11 = 0 \qquad r - 11 = 0$
$\quad\ r = -11 \qquad\ \ r = 11$

The solution -11 is not possible.
The rate of the boat in calm water is 11 mph.

24. Strategy
- This is a consecutive integer problem.
- First odd integer: n
  Second odd integer: $n + 2$
  Third odd integer: $n + 4$
- The sum of the squares of the three consecutive odd integers is 83.

Solution
$$n^2 + (n+2)^2 + (n+4)^2 = 83$$
$$n^2 + (n^2 + 4n + 4) + (n^2 + 8n + 16) = 83$$
$$3n^2 + 12n - 63 = 0$$
$$3(n^2 + 4n - 21) = 0$$
$$3(n + 7)(n - 3) = 0$$
$$n + 7 = 0 \qquad n - 3 = 0$$
$$n = -7 \qquad n = 3$$
$$n + 2 = -7 + 2 = -5 \quad n + 2 = 3 + 2 = 5$$
The middle odd integer is -5 or 5.

25. Strategy
- This is a distance-rate problem.
- Rate of the jogger for the first 7 mi: r

	Distance	Rate	Time
First part of run	7	r	$\dfrac{7}{r}$
Second part of run	8	$r - 3$	$\dfrac{8}{r-3}$

- The total time for the 15 mi run was 3 h.

Solution
$$\frac{7}{r} + \frac{8}{r-3} = 3$$
$$r(r-3) \cdot \left(\frac{7}{r} + \frac{8}{r-3}\right) = r(r-3) \cdot 3$$
$$7(r-3) + 8r = 3r(r-3)$$
$$7r - 21 + 8r = 3r^2 - 9r$$
$$15r - 21 = 3r^2 - 9r$$
$$0 = 3r^2 - 24r + 21$$
$$0 = 3(r^2 - 8r + 7)$$
$$0 = 3(r - 7)(r - 1)$$
$$r - 7 = 0 \qquad r - 1 = 0$$
$$r = 7 \qquad r = 1$$
$$r - 3 = 7 - 3 = 4 \quad r - 3 = 1 - 3 = -2$$
The solution -2 is not possible.
The rate of the jogger for the last 8 mi was 4 mph.

Cumulative Review Exercises

1. $2x - 3[2x - 4(3 - 2x) + 2] - 3$
$= 2x - 3[2x - 12 + 8x + 2] - 3$
$= 2x - 3[10x - 10] - 3$
$= 2x - 30x + 30 - 3$
$= -28x + 27$

2.
$$-\frac{3}{5}x = -\frac{9}{10}$$
$$-\frac{5}{3}\left(-\frac{3}{5}x\right) = -\frac{5}{3}\left(-\frac{9}{10}\right)$$
$$x = \frac{3}{2}$$
The solution is $\frac{3}{2}$.

3. $2x - 3(4x - 5) = -3x - 6$
$2x - 12x + 15 = -3x - 6$
$-10x + 15 = -3x - 6$
$-10x + 3x + 15 = -3x + 3x - 6$
$-7x + 15 = -6$
$-7x + 15 - 15 = -6 - 15$
$-7x = -21$
$\dfrac{-7x}{-7} = \dfrac{-21}{-7}$
$x = 3$
The solution is 3.

4. $2x - 3(2 - 3x) > 2x - 5$
$2x - 6 + 9x > 2x - 5$
$11x - 6 > 2x - 5$
$-2x + 11x - 6 > -2x + 2x - 5$
$9x - 6 > -5$
$9x - 6 + 6 > -5 + 6$
$9x > 1$
$\dfrac{9x}{9} > \dfrac{1}{9}$
$x > \dfrac{1}{9}$

5.
x-intercept	y-intercept
$4x - 3(0) = 12$	$4(0) - 3y = 12$
$4x = 12$	$-3y = 12$
$x = 3$	$y = -4$
$(3, 0)$	$(0, -4)$

6. $m = -\dfrac{4}{3} \qquad (x_1, y_1) = (-3, 2)$
$$y - y_1 = m(x - x_1)$$
$$y - 2 = -\frac{4}{3}[x - (-3)]$$
$$y - 2 = -\frac{4}{3}(x + 3)$$
$$y - 2 = -\frac{4}{3}x - 4$$
$$y = -\frac{4}{3}x - 2$$

7. The domain is $\{-2, -1, 0, 1, 2, 3\}$.
The range is $\{-8, -1, 0, 1, 8\}$.
No ordered pairs have the same first coordinate.
The relation is a function.

8. $f(x) = -3x + 10$
$f(-9) = -3(-9) + 10$
$f(-9) = 27 + 10$
$f(-9) = 37$
The value of $f(-9)$ is 37.

9.

10. $2x - 3y > 6$

$-3y > -2x + 6$

$y < \dfrac{2}{3}x - 2$

Sketch a dashed line

$y = \dfrac{2}{3}x - 2$

Shaded below the dashed line.

11. (1) $\quad 3x - y = 5$

(2) $\qquad y = 2x - 3$

Substitute in equation (1).

$3x - y = 5$

$3x - (2x - 3) = 5$

$3x - 2x + 3 = 5$

$x + 3 = 5$

$x = 2$

Substitute 2 for x in equation (2).

$y = 2x - 3$

$y = 2(2) - 3$

$y = 4 - 3$

$y = 1$

The solution is (2, 1).

12. (1) $\quad 3x + 2y = 2$

(2) $\quad 5x - 2y = 14$

Add the equations.

$8x = 16$

$x = 2$

Substitute 2 for x in equation (1).

$3x + 2y = 2$

$3(2) + 2y = 2$

$6 + 2y = 2$

$2y = -4$

$y = -2$

The solution is (2, –2).

13. $\dfrac{(2a^{-2}b)^2}{-3a^{-5}b^4} = \dfrac{2^2 a^{-4}b^2}{-3a^{-5}b^4}$

$= \dfrac{4a^{-4}b^2}{-3a^{-5}b^4} = -\dfrac{4a}{3b^2}$

14.

$$\begin{array}{r} x+2 \\ x-2 \overline{\big)\,x^2 + 0x - 8} \\ \underline{x^2 - 2x} \\ 2x - 8 \\ \underline{2x - 4} \\ -4 \end{array}$$

$(x^2 - 8) \div (x - 2) = x + 2 - \dfrac{4}{x - 2}$

15. $4y(x - 4) - 3(x - 4) = (x - 4)(4y - 3)$

16. $3x^3 + 2x^2 - 8x = x(3x^2 + 2x - 8)$

$\qquad = x(3x^2 - 4x + 6x - 8)$

$\qquad = x[(3x^2 - 4x) + (6x - 8)]$

$\qquad = x[x(3x - 4) + 2(3x - 4)]$

$\qquad = x(3x - 4)(x + 2)$

17. $\dfrac{3x^2 - 6x}{4x - 6} \div \dfrac{2x^2 + x - 6}{6x^2 - 24x}$

$= \dfrac{3x^2 - 6x}{4x - 6} \cdot \dfrac{6x^2 - 24x}{2x^2 + x - 6}$

$= \dfrac{3x(x - 2)}{2(2x - 3)} \cdot \dfrac{6x(x - 4)}{(2x - 3)(x + 2)}$

$= \dfrac{9x^2(x - 2)(x - 4)}{(2x - 3)^2(x + 2)}$

18. $\dfrac{x}{2(x - 1)} - \dfrac{1}{(x - 1)(x + 1)}$

$= \dfrac{x}{2(x - 1)} \cdot \dfrac{x + 1}{x + 1} - \dfrac{1}{(x - 1)(x + 1)} \cdot \dfrac{2}{2}$

$= \dfrac{x^2 + x}{2(x - 1)(x + 1)} - \dfrac{2}{(x - 1)(x + 1)}$

$= \dfrac{x^2 + x - 2}{2(x - 1)(x + 1)} = \dfrac{(x - 1)(x + 2)}{2(x - 1)(x + 1)}$

$= \dfrac{x + 2}{2(x + 1)}$

19. $\dfrac{1 - \frac{7}{x} + \frac{12}{x^2}}{2 - \frac{1}{x} - \frac{15}{y^2}} = \dfrac{1 - \frac{7}{x} + \frac{12}{x^2}}{2 - \frac{1}{x} - \frac{15}{x^2}} \cdot \dfrac{x^2}{x^2}$

$= \dfrac{1 \cdot x^2 - \frac{7}{x} \cdot x^2 + \frac{12}{x^2} \cdot x^2}{2 \cdot x^2 - \frac{1}{x} \cdot x^2 - \frac{15}{x^2} \cdot x^2}$

$= \dfrac{x^2 - 7x + 12}{2x^2 - x - 15}$

$= \dfrac{(x - 3)(x - 4)}{(2x + 5)(x - 3)}$

$= \dfrac{x - 4}{2x + 5}$

20.
$$\frac{x}{x+6}=\frac{3}{x}$$
$$x(x+6)\left(\frac{x}{x+6}\right)=x(x+6)\left(\frac{3}{x}\right)$$
$$x\cdot x=(x+6)\cdot 3$$
$$x^2=3x+18$$
$$x^2-3x-18=0$$
$$(x+3)(x-6)=0$$
$$x+3=0 \quad x-6=0$$
$$x=-3 \quad x=6$$
The solutions are -3 and 6.

21. $(\sqrt{a}-\sqrt{2})(\sqrt{a}+\sqrt{2})=\sqrt{a}^2-\sqrt{2}^2$
$$=a-2$$

22.
$$3=8-\sqrt{5x}$$
$$-5=-\sqrt{5x}$$
$$5=\sqrt{5x}$$
$$5^2=(\sqrt{5x})^2$$
$$25=5x$$
$$5=x$$

Check: $3=8-\sqrt{5x}$

3	$8-\sqrt{5(5)}$
3	$8-\sqrt{25}$
3	$8-5$
	$3=3$

The solution is 5.

23.
$$2x^2-7x=-3$$
$$2x^2-7x+3=0$$
$$(2x-1)(x-3)=0$$
$$2x-1=0 \quad x-3=0$$
$$2x=1 \quad x=3$$
$$x=\frac{1}{2}$$
The solutions are $\frac{1}{2}$ and 3.

24.
$$3(x-2)^2=36$$
$$(x-2)^2=12$$
$$\sqrt{(x-2)^2}=\sqrt{12}$$
$$x-2=\pm 2\sqrt{3}$$
$$x=2\pm 2\sqrt{3}$$
The solutions are $2+2\sqrt{3}$ and $2-2\sqrt{3}$.

25. $3x^2-4x-5=0$
$a=3, b=-4, c=-5$
$$x=\frac{-b\pm\sqrt{b^2-4ac}}{2a}$$
$$x=\frac{-(-4)\pm\sqrt{(-4)^2-4(3)(-5)}}{2\cdot 3}$$
$$=\frac{4\pm\sqrt{16+60}}{6}$$
$$=\frac{4\pm\sqrt{76}}{6}$$
$$=\frac{4\pm 2\sqrt{19}}{6}$$
$$=\frac{2\pm\sqrt{19}}{3}$$

The solutions are $\frac{2+\sqrt{19}}{3}$ and $\frac{2-\sqrt{19}}{3}$.

26.

27. Strategy
• To find the cost, solve the basic percent equation using $A=\$5.22$ and $P=7\frac{1}{4}\%=0.0725$. The base is unknown.

Solution
$$PB=A$$
$$0.0725B=5.22$$
$$\frac{0.0725B}{0.0725}=\frac{5.22}{0.0725}$$
$$B=72$$
The cost of the textbook is $72.

28. Strategy
• Cost per pound of mixture: x

	Amount	Cost	Value
Cashews	20	7	20(7)
Peanuts	50	3.50	50(3.50)
Mixture	70	x	$70x$

• The sum of the values before mixing equals the value after mixing.

Solution
$$20(7)+50(3.50)=70x$$
$$140+175=70x$$
$$315=70x$$
$$4.50=x$$
The cost of the mixture is $4.50 per pound.

29. Strategy

To find the number of additional shares, write and solve a proportion using x to represent the number of additional shares. Then $100 + x$ is the total number of shares.

Solution

$$\frac{215}{100} = \frac{752.50}{100 + x}$$

$$100(100 + x)\left(\frac{215}{100}\right) = 100(100 + x)\left(\frac{752.50}{100 + x}\right)$$

$$(100 + x)215 = 100(752.50)$$

$$21,500 + 215x = 75,250$$

$$215x = 53,750$$

$$x = 250$$

250 additional shares are required.

30. Strategy

• Rate of the plane in still air: r
• Rate of the wind: w

	Rate	Time	Distance
With wind	$r + w$	3	$3(r + w)$
Against wind	$r - w$	4.5	$4.5(r - w)$

• The distance traveled with the wind is 720 mi. The distance traveled against the wind is 720 mi.

Solution

$$3(r + w) = 720$$

$$4.5(r - w) = 720$$

$$\frac{1}{3} \cdot 3(r + w) = \frac{1}{3} \cdot 720$$

$$\frac{1}{4.5} \cdot 4.5(r - w) = \frac{1}{4.5} \cdot 720$$

$$r + w = 240$$

$$r - w = 160$$

$$2r = 400$$

$$r = 200$$

$$r + w = 240$$

$$200 + w = 240$$

$$w = 40$$

The rate of the plane in still air is 200 mph. The rate of the wind is 40 mph.

31. Strategy

Score on fifth test: x

Write and solve an inequality using x to represent the student's score on the fifth test.

Solution

The sum of the five test scores is greater than or equal to 400 points

$$70 + 91 + 85 + 77 + x \geq 400$$

$$323 + x \geq 400$$

$$x \geq 77$$

The student must receive a score of 77 or better.

32. Strategy

To find the length of the guy wire, use the Pythagorean Theorem. One leg is the distance from the bottom of the wire to the base of the telephone pole. The other leg is the distance from the top of the wire to the base of the telephone pole. The guy wire is the hypotenuse. Solve the Pythagorean Theorem for the hypotenuse.

Solution

$$c = \sqrt{a^2 + b^2}$$

$$c = \sqrt{(10)^2 + (30)^2}$$

$$c = \sqrt{100 + 900}$$

$$c = \sqrt{1000}$$

$$c \approx 31.62$$

The length of the guy wire is approximately 31.62 m.

33. Strategy

• First odd integer: x
  Second odd integer: $x + 2$
  Third odd integer: $x + 4$
• The sum of the squares of the three consecutive integers is 155.

Solution

$$x^2 + (x + 2)^2 + (x + 4)^2 = 155$$

$$x^2 + x^2 + 4x + 4 + x^2 + 8x + 16 = 155$$

$$3x^2 + 12x + 20 = 155$$

$$3x^2 + 12x - 135 = 0$$

$$3(x^2 + 4x - 45) = 0$$

$$3(x + 9)(x - 5) = 0$$

$$x + 9 = 0 \quad x - 5 = 0$$

$$x = -9 \quad x = 5$$

$$x + 2 = -9 + 2 = -7 \quad x + 2 = 5 + 2 = 7$$

The middle integer is 7 or –7.

Final Exam

1. $-|-3| = -3$

2. $-15 - (-12) - 3 = -15 + 12 - 3$
   $$= -3 - 3$$
   $$= -6$$

3. $\frac{1}{8} = \frac{1}{8}(100\%) = \frac{100}{8}\% = 12.5\%$

4. $-2^4(-2)^4 = -16(16)$
   $$= -256$$

5. $-7 - \frac{12 - 15}{2 - (-1)} \cdot (-4) = -7 - \frac{-3}{3} \cdot (-4)$
   $$= -7 - (-1) \cdot (-4)$$
   $$= -7 - (4)$$
   $$= -11$$

6. $\dfrac{a^2 - 3b}{2a - 2b^2} = \dfrac{(3)^2 - 3(-2)}{2(3) - 2(-2)^2}$

$= \dfrac{9 - 3(-2)}{2(3) - 2(4)}$

$= \dfrac{9 + 6}{6 - 8}$

$= \dfrac{15}{-2} = -\dfrac{15}{2}$

7. $6x - (-4y) - (-3x) + 2y = 6x + 4y + 3x + 2y$
$= 9x + 6y$

8. $(-15z)\left(-\dfrac{2}{5}\right) = 6z$

9. $-2[5 - 3(2x - 7) - 2x] = -2[5 - 6x + 21 - 2x]$
$= -2[-8x + 26]$
$= 16x - 52$

10. $20 = -\dfrac{2}{5}x$

$-\dfrac{5}{2}(20) = -\dfrac{5}{2}\left(-\dfrac{2}{5}x\right)$

$-50 = x$

The solution is -50.

11. $4 - 2(3x + 1) = 3(2 - x) + 5$
$4 - 6x - 2 = 6 - 3x + 5$
$-6x + 2 = -3x + 11$
$-6x + 3x + 2 = -3x + 3x + 11$
$-3x + 2 = 11$
$-3x + 2 - 2 = 11 - 2$
$-3x = 9$
$\dfrac{-3x}{-3} = \dfrac{9}{-3}$
$x = -3$

The solution is -3.

12. $PB = A$
$0.19(80) = A$
$15.2 = A$
19% of 80 is 15.2.

13. $4 - x \geq 7$
$4 - 4 - x \geq 7 - 4$
$-x \geq 3$
$-1(-x) \leq -1(3)$
$x \leq -3$

14. $2 - 2(y - 1) \leq 2y - 6$
$2 - 2y + 2 \leq 2y - 6$
$-2y + 4 \leq 2y - 6$
$-2y - 2y + 4 \leq 2y - 2y - 6$
$-4y + 4 \leq -6$
$-4y + 4 - 4 \leq -6 - 4$
$-4y \leq -10$
$\dfrac{-4y}{-4} \geq \dfrac{-10}{-4}$
$y \geq \dfrac{5}{2}$

15. $(x_1, y_1) = (-1, -3), (x_2, y_2) = (2, -1)$
$m = \dfrac{y_2 - y_1}{x_2 - x_1} = \dfrac{-1 - (-3)}{2 - (-1)} = \dfrac{2}{3}$
The slope of the line is $\dfrac{2}{3}$.

16. $m = -\dfrac{2}{3}, (x_1, y_1) = (3, -4)$
$y - y_1 = m(x - x_1)$
$y - (-4) = -\dfrac{2}{3}(x - 3)$
$y + 4 = -\dfrac{2}{3}x + 2$
$y = -\dfrac{2}{3}x - 2$

17.

18.

19. $f(x) = -x + 5$
$f(-6) = -(-6) + 5 = 6 + 5 = 11$
$f(-3) = -(-3) + 5 = 3 + 5 = 8$
$f(0) = -(0) + 5 = 0 + 5 = 5$
$f(3) = -3 + 5 = 2$
$f(6) = -6 + 5 = -1$
The domain is $\{-1, 2, 5, 8, 11\}$.

20.

21. (1) $y = 4x - 7$
(2) $y = 2x + 5$
Substitute into equation (2).
$4x - 7 = 2x + 5$
$2x - 7 = 5$
$2x = 12$
$x = 6$
Substitute x in equation (1).
$y = 4x - 7$
$y = 4(6) - 7$
$y = 24 - 7$
$y = 17$
The solution is $(6, 17)$.

22. (1) $4x - 3y = 11$

(2) $2x + 5y = -1$

Eliminate x and add the equations.

$$4x - 3y = 11 \qquad 4x - 3y = 11$$
$$\underline{-2(2x + 5y) = -1(-2)} \quad \underline{-4x - 10y = 2}$$
$$-13y = 13$$
$$y = -1$$

Replace y in equation (2).

$$2x + 5y = -1$$
$$2x + 5(-1) = -1$$
$$2x - 5 = -1$$
$$2x = 4$$
$$x = 2$$

The solution is $(2, -1)$.

23. $(2x^2 - 5x + 1) - (5x^2 - 2x - 7)$

$= (2x^2 - 5x + 1) + (-5x^2 + 2x + 7)$

$= -3x^2 - 3x + 8$

24. $(-3xy^3)^4 = (-3)^4 x^4 y^{12} = 81x^4 y^{12}$

25.
$$\begin{array}{r} 3x^2 - x - 2 \\ 2x + 3 \\ \hline 9x^2 - 3x - 6 \\ 6x^3 - 2x^2 - 4x \\ \hline 6x^3 + 7x^2 - 7x - 6 \end{array}$$

26. $\dfrac{(-2x^2 y^3)^3}{(-4x^{-1}y^4)^2} = \dfrac{(-2)^3 x^6 y^9}{(-4)^2 x^{-2} y^8}$

$= \dfrac{-8x^6 y^9}{16x^{-2} y^8}$

$= -\dfrac{x^8 y}{2}$

27. $(4x^{-2}y)^3 (2xy^{-2})^{-2} = (4^3 x^{-6} y^3)(2^{-2} x^{-2} y^4)$

$= \left(\dfrac{4^3 y^3}{x^6}\right)\left(\dfrac{y^4}{2^2 x^2}\right)$

$= \dfrac{64y^7}{4x^8} = \dfrac{16y^7}{x^8}$

28. $\dfrac{12x^3 y^2 - 16x^2 y^2 - 20y^2}{4xy^2}$

$= \dfrac{12x^3 y^2}{4xy^2} - \dfrac{16x^2 y^2}{4xy^2} - \dfrac{20y^2}{4xy^2}$

$= 3x^2 - 4x - \dfrac{5}{x}$

29.
$$\begin{array}{r} 5x - 12 \\ x + 2 \enclose{longdiv}{5x^2 - 2x - 1} \\ \underline{5x^2 + 10x} \\ -12x - 1 \\ \underline{-12x - 24} \\ 23 \end{array}$$

$(5x^2 - 2x - 1) \div (x + 2) = 5x - 12 + \dfrac{23}{x + 2}$

30. $0.000000039 = 3.9 \cdot 10^{-8}$

31. $2a(4 - x) - 6(x - 4) = 2a(4 - x) + 6(4 - x)$

$= (4 - x)(2a + 6)$

$= 2(4 - x)(a + 3)$

32. $x^2 - 5x - 6 = (x - 6)(x + 1)$

33. $2x^2 - x - 3 = 2x^2 - 3x + 2x - 3$

$= (2x^2 - 3x) + (2x - 3)$

$= x(2x - 3) + 1(2x - 3)$

$= (2x - 3)(x + 1)$

34. $6x^2 - 5x - 6 = 6x^2 + 4x - 9x - 6$

$= (6x^2 + 4x) - (9x + 6)$

$= 2x(3x + 2) - 3(3x + 2)$

$= (3x + 2)(2x - 3)$

35. $8x^3 - 28x^2 + 12x = 4x(2x^2 - 7x + 3)$

$= 4x(2x^2 - x - 6x + 3)$

$= 4x[(2x^2 - x) - (6x - 3)]$

$= 4x[x(2x - 1) - 3(2x - 1)]$

$= 4x(2x - 1)(x - 3)$

36. $25x^2 - 16 = (5x)^2 - (4)^2$

$= (5x + 4)(5x - 4)$

37. $75y - 12x^2 y = 3y(25 - 4x^2)$

$= 3y[(5)^2 - (2x)^2]$

$= 3y(5 + 2x)(5 - 2x)$

38.
$$2x^2 = 7x - 3$$
$$2x^2 - 7x + 3 = 0$$
$$(2x - 1)(x - 3) = 0$$
$$2x - 1 = 0 \qquad x - 3 = 0$$
$$2x = 1 \qquad\quad x = 3$$
$$x = \frac{1}{2}$$

The solutions are $\dfrac{1}{2}$ and 3.

39. $\dfrac{2x^2 - 3x + 1}{4x^2 - 2x} \cdot \dfrac{4x^2 + 4x}{x^2 - 2x + 1}$

$= \dfrac{\cancel{(2x - 1)}\,\cancel{(x - 1)}}{2x\,\cancel{(2x - 1)}} \cdot \dfrac{4x(x + 1)}{\cancel{(x - 1)}\,(x - 1)}$

$= \dfrac{2(x + 1)}{x - 1}$

40. The LCM is $(x+3)(2x-5)$.

$$\frac{5}{x+3} - \frac{3x}{2x-5} = \frac{5}{x+3} \cdot \frac{2x-5}{2x-5} - \frac{3x}{2x-5} \cdot \frac{x+3}{x+3}$$

$$= \frac{10x-25}{(x+3)(2x-5)} - \frac{3x^2+9x}{(2x-5)(x+3)}$$

$$= \frac{10x-25-(3x^2+9x)}{(2x-5)(x+3)}$$

$$= \frac{10x-25-3x^2-9x}{(2x-5)(x+3)}$$

$$= \frac{-3x^2+x-25}{(2x-5)(x+3)}$$

41.

$$\frac{x-\frac{3}{2x-1}}{1-\frac{2}{2x-1}} = \frac{x-\frac{3}{2x-1}}{1-\frac{2}{2x-1}} \cdot \frac{2x-1}{2x-1}$$

$$= \frac{x(2x-1) - \frac{3}{2x-1}(2x-1)}{1(2x-1) - \frac{2}{2x-1}(2x-1)}$$

$$= \frac{2x^2-x-3}{2x-1-2}$$

$$= \frac{2x^2-x-3}{2x-3}$$

$$= \frac{\overset{1}{\cancel{(2x-3)}}(x+1)}{\underset{1}{\cancel{2x-3}}}$$

$$= x+1$$

42.

$$\frac{5x}{3x-5} - 3 = \frac{7}{3x-5}$$

$$(3x-5)\left(\frac{5x}{3x-5} - 3\right) = (3x-5)\left(\frac{7}{3x-5}\right)$$

$$5x - 3(3x-5) = 7$$

$$5x - 9x + 15 = 7$$

$$-4x + 15 = 7$$

$$-4x = -8$$

$$x = 2$$

43.

$$a = 3a - 2b$$

$$a - 3a = 3a - 3a - 2b$$

$$a - 3a = -2b$$

$$-2a = -2b$$

$$\frac{-2a}{-2} = \frac{-2b}{-2}$$

$$a = b$$

44. $\sqrt{49x^6} = \sqrt{7^2 x^6} = 7x^3$

45.

$$2\sqrt{27a} + 8\sqrt{48a} = 2\sqrt{3^3 a} + 8\sqrt{2^4 \cdot 3a}$$

$$= 2\sqrt{3^2 \cdot 3a} + 8\sqrt{2^4 \cdot 3a}$$

$$= 2\sqrt{3^2}\sqrt{3a} + 8\sqrt{2^4}\sqrt{3a}$$

$$= 2 \cdot 3\sqrt{3a} + 8 \cdot 2^2 \sqrt{3a}$$

$$= 6\sqrt{3a} + 32\sqrt{3a}$$

$$= 38\sqrt{3a}$$

46.

$$\frac{\sqrt{3}}{\sqrt{5}-2} = \frac{\sqrt{3}}{\sqrt{5}-2} \cdot \frac{\sqrt{5}+2}{\sqrt{5}+2}$$

$$= \frac{\sqrt{3}(\sqrt{5}+2)}{5-4} = \frac{\sqrt{15}+2\sqrt{3}}{1}$$

$$= \sqrt{15} + 2\sqrt{3}$$

47.

$$\sqrt{x+4} - \sqrt{x-1} = 1$$

$$\sqrt{x+4} = 1 + \sqrt{x-1}$$

$$(\sqrt{x+4})^2 = (1+\sqrt{x-1})^2$$

$$x+4 = 1 + 2\sqrt{x-1} + x - 1$$

$$x+4 = 2\sqrt{x-1} + x$$

$$4 = 2\sqrt{x-1}$$

$$2 = \sqrt{x-1}$$

$$(2)^2 = (\sqrt{x-1})^2$$

$$4 = x - 1$$

$$5 = x$$

The solution is 5.

48.

$$(x-3)^2 = 7$$

$$\sqrt{(x-3)^2} = \pm\sqrt{7}$$

$$x - 3 = \pm\sqrt{7}$$

$$x = 3 \pm \sqrt{7}$$

The solutions are $3 + \sqrt{7}$ and $3 - \sqrt{7}$.

49.

$$4x^2 - 2x - 1 = 0$$

$$a = 4,\ b = -2,\ c = -1$$

$$x = \frac{-(-2) \pm \sqrt{(-2)^2 - 4(4)(-1)}}{2 \cdot 4}$$

$$= \frac{2 \pm \sqrt{4+16}}{8}$$

$$= \frac{2 \pm \sqrt{20}}{8}$$

$$= \frac{2 \pm 2\sqrt{5}}{8}$$

$$= \frac{1 \pm \sqrt{5}}{4}$$

50.

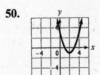

51. Strategy

The unknown number: x

Twice the number: $2x$

The difference between the number and two: $x-2$

Three times the difference between the number and two: $3(x-2)$

Solution

$$2x + 3(x-2) = 2x + 3x - 6$$

$$= 5x - 6$$

52. Strategy
To find the original value, solve the basic percent equation for the base. The percent is 80% = 0.80. The amount is 2400.

Solution
$$PB = A$$
$$0.80B = 2400$$
$$\frac{0.80B}{0.80} = \frac{2400}{0.80}$$
$$B = 3000$$
The original value of the machine is $3000.

53. Strategy
• Number of quarters: x
Number of dimes: $3x$

	Number	Value	Total Value
Quarters	x	25	$25x$
Dimes	$3x$	10	$10(3x)$

• The total value of all the coins is 1100 cents.

Solution
$$25x + 10(3x) = 1100$$
$$25x + 30x = 1100$$
$$55x = 1100$$
$$x = 20$$
$$3x = 3(20) = 60$$
There are 60 dimes in the bank.

54. Strategy
• First angle: $x + 10$
Second angle: x
Third angle: $x + 20$
• The sum of the angles of a triangle is 180°.

Solution
$$x + x + 10 + x + 20 = 180$$
$$3x + 30 = 180$$
$$3x = 150$$
$$x = 50$$
$$x + 10 = 50 + 10 = 60$$
$$x + 20 = 50 + 20 = 70$$
The angles are 50°, 60°, and 70°.

55. Strategy
Given: $C = \$900$
$S = \$1485$
Unknown markup rate: r
Use the equation $S = C + rC$.

Solution
$$S = C + rC$$
$$1485 = 900 + 900r$$
$$1485 - 900 = 900 - 900 + 900r$$
$$585 = 900r$$
$$\frac{585}{900} = \frac{900r}{900}$$
$$0.65 = r$$
The markup rate is 65%.

56. Strategy
• Amount of money to be invested at 11%: x

	Principal	Rate	Interest
Amount at 8%	3000	0.08	0.08(3000)
Amount at 11%	x	0.11	$0.11x$
Amount at 10%	$x + 3000$	0.10	$0.10(x + 3000)$

• The total interest earned is 10% of the total investment.

Solution
$$0.08(3000) + 0.11x = 0.10(x + 3000)$$
$$240 + 0.11x = 0.10x + 300$$
$$240 + 0.01x = 300$$
$$0.01x = 60$$
$$x = 6000$$
$6000 must be invested at 11%.

57. Strategy
• Cost per pound for the mixture: x

	Amount	Cost	Value
Peanuts	4	2	2(4)
Walnuts	2	5	5(2)
Mixture	6	x	$6x$

• The sum of the values before mixing equals the value after mixing.

Solution
$$2(4) + 5(2) = 6x$$
$$8 + 10 = 6x$$
$$18 = 6x$$
$$3 = x$$
The cost per pound for the mixture is $3.

58. Strategy
• Percent concentration of acid in the mixture: x

	Amount	Percent	Quantity
60% acid	20	0.60	0.60(20)
20% acid	30	0.20	0.20(30)
Mixture	50	x	$50x$

• The sum of the quantities before mixing is equal to the quantity after mixing.

Solution
$$0.60(20) + 0.20(30) = 50x$$
$$12 + 6 = 50x$$
$$18 = 50x$$
$$0.36 = x$$
The percent concentration of acid in the mixture is 36%.

59. Strategy
- Rate for first part of trip: x
  Rate for second part of trip: $2x$

	Rate	Time	Distance
First part	x	1	$1x$
Second part	$2x$	1.5	$1.5(2x)$

- The total distance is 860 km.

Solution
$$1x + 1.5x(2x) = 860$$
$$1x + 3x = 860$$
$$4x = 860$$
$$x = 215$$
The rate for the first part of the trip is 215 km/h.
$$d = rt$$
$$d = 215(1)$$
$$d = 215$$
The distance for the first part of the trip is 215 km.

60. Strategy
- Rate of the boat in calm water: r
  Rate of the current: c

	Rate	Time	Distance
With current	$r + c$	25	$2.5(r + c)$
Against current	$r - c$	5	$5(r - c)$

- The distance traveled with the current is 50 mi. The distance traveled against the current is 50 mi.

Solution

$$2.5(r+c) = 50 \quad \frac{1}{2.5} \cdot 2.5(r+c) = 50 \cdot \frac{1}{2.5}$$
$$5(r-c) = 50 \quad \frac{1}{5} \cdot 5(r-c) = 50 \cdot \frac{1}{5}$$
$$r + c = 20$$
$$r - c = 10$$
$$2r = 30$$
$$r = 15$$

$$r + c = 20$$
$$15 + c = 20$$
$$c = 5$$

The rate of the boat in calm water is 15 mph. The rate of the current is 5 mph.

61. Strategy
- Width of the rectangle: x
  Length of the rectangle: $x + 5$
- Use the equation for the area of a rectangle.

Solution
$$LW = A$$
$$x(x + 5) = 50$$
$$x^2 + 5x = 50$$
$$x^2 + 5x - 50 = 0$$
$$(x + 10)(x - 5) = 0$$
$$x + 10 = 0 \qquad x - 5 = 0$$
$$x = -10 \qquad x = 5$$
The solution -10 is not possible.
$$x + 5 = 5 + 5 = 10$$
The width is 5 m. The length is 10 m.

62. Strategy
To find the number of ounces, write and solve a proportion using x to represent the unknown number of ounces.

Solution
$$\frac{15}{2} = \frac{120}{x}$$
$$2x\left(\frac{15}{2}\right) = 2x\left(\frac{120}{x}\right)$$
$$15x = 240$$
$$x = 16$$
The required amount of dye is 16 oz.

63. Strategy
- Time to complete the task working together: t

	Rate	Time	Part
Chef	1	t	$\dfrac{t}{1}$
Apprentice	1.5	t	$\dfrac{t}{1.5}$

- The sum of the parts of the task completed by each must equal 1.

Solution
$$\frac{t}{1} + \frac{t}{1.5} = 1$$
$$1.5\left(\frac{t}{1} + \frac{t}{1.5}\right) = 1.5(1)$$
$$1.5t + t = 1.5$$
$$2.5t = 1.5$$
$$t = 0.6$$
The time to complete the task working together is 0.6 h (36 min).

64. Strategy

To find the leg of the triangle, substitute $c = 14$ and $b = 8$ in the formula $a = \sqrt{c^2 - b^2}$ and solve for a.

Solution

$$a = \sqrt{c^2 - b^2}$$
$$a = \sqrt{14^2 - 8^2}$$
$$a = \sqrt{196 - 64}$$
$$a = \sqrt{132}$$
$$a \approx 11.5$$

The leg of the triangle is approximately 11.5 cm.

65. Strategy

• Rate of the wind: w

	Distance	Rate	Time
Against wind	500	$225 - w$	$\dfrac{500}{225 - w}$
With wind	500	$225 + w$	$\dfrac{500}{225 + w}$

• The time spent traveling against the wind is $\dfrac{1}{2}$ h more than the time spent traveling with the wind.

Solution

$$\frac{500}{225 - w} = \frac{1}{2} + \frac{500}{225 + w}$$

$$2(225 - w)(225 + w)\left(\frac{500}{225 - w}\right) = 2(225 - w)(225 + w)\left(\frac{1}{2} + \frac{500}{225 + w}\right)$$

$$2(500)(225 + w) = (225 + w)(225 - w) + 2(500)(225 - w)$$

$$1000(225 + w) = 50{,}625 - w^2 + 1000(225 - w)$$

$$225{,}000 + 1000w = 50{,}625 - w^2 + 225{,}000 - 1000w$$

$$w^2 + 2000w - 50{,}625 = 0$$

$$(w + 2025)(w - 25) = 0$$

$$w + 2025 = 0 \qquad w - 25 = 0$$

$$w = -2025 \qquad w = 25$$

The solution −2025 is not possible.

The rate of the wind is 25 mph.